94.95
7SE
#183

Aviation Weather Surveillance Systems

Progress in Astronautics and Aeronautics

Editor-in-Chief
Paul Zarchan
Charles Stark Draper Laboratory, Inc.

Editorial Board

John D. Binder
The MathWorks, Inc.

Michael D. Griffin
Orbital Sciences Corporation

Steven A. Brandt
U.S. Air Force Academy

Philip D. Hattis
Charles Stark Draper Laboratory, Inc.

Luigi De Luca
Politecnico di Milano

Ahmed K. Noor
NASA Langley Research Center

Leroy S. Fletcher
Texas A&M University

Albert C. Piccirillo
ANSER, Inc.

Allen E. Fuhs
Carmel, California

Anthony M. Springer
NASA Marshall Space Flight Center

Vigor Yang
Pennsylvania State University

Aviation Weather Surveillance Systems
Advanced Radar and Surface Sensors for Flight Safety and Air Traffic Management

Pravas R. Mahapatra

Indian Institute of Science
Bangalore, India

with
Richard J. Doviak
Vladislav Mazur
Dusan Zrnič
National Severe Storms Laboratory, Norman, Oklahoma

Volume 183
PROGRESS IN ASTRONAUTICS AND AERONAUTICS

Paul Zarchan, Editor-in-Chief
Charles Stark Draper Laboratory, Inc.
Cambridge, Massachusetts

Copublished by the
American Institute of Aeronautics and Astronautics, Inc.
1801 Alexander Bell Drive, Reston, Virginia 20191-4344
and the Institution of Electrical Engineers, Michael Faraday House,
Six Hills Way, Stevenage, Herts, SG1 2AY UK

Copublished by:

The American Institute of Aeronautics and Astronautics
1801 Alexander Bell Drive
Suite 500
Reston, VA 20191-4344
USA

and

The Institution of Electrical Engineers,
Michael Faraday House,
Six Hills Way, Stevenage,
Herts. SG1 2AY, United Kingdom

© 1999: The Institution of Electrical Engineers

This publication is copyright under the Berne Convention and the Universal Copyright Convention. All rights reserved. Apart from any fair dealing for the purposes of research or private study, or criticism or review, as permitted under the Copyright, Designs and Patents Act, 1988, this publication may be reproduced, stored or transmitted, in any forms or by any means, only with the prior permission in writing of the Institution of Electrical Engineers (IEE) or in the case of reprographic reproduction in accordance with the terms of licences issued by the Copyright Licensing Agency. Inquiries concerning reproduction outside those terms should be sent to the IEE at the address above.

While the author and the publishers believe that the information and guidance given in this work are correct, all parties must rely upon their own skill and judgment when making use of them. Neither the author nor the publishers assume any liability to anyone for any loss or damage caused by any error or omission in the work, whether such error or omission is the result of negligence or any other cause. Any and all such liability is disclaimed.

The moral right of the author to be identified as author of this work has been asserted by him/her in accordance with the Copyright, Designs and Patents Act 1988.

ISBN 1-56347-340-2

Printed in England by Short Run Press Ltd., Exeter

to my parents

Contents

Preface		xiii
Acknowledgments		xv
Abbreviations		xvii
Symbols		xxi
1	**Introduction**	1
	1.1 Aviation and electronics: a symbiotic relationship	1
	1.2 Phases in evolution of aircraft navigation	2
	1.3 Modern aviation weather surveillance	4
	1.4 Scope and organisation of the book	5
	1.5 References	9
2	**Basic background of aviation**	11
	2.1 Goal of aviation systems	11
	2.2 Phases of aircraft flight	11
	2.2.1 Terminal area operations	12
	2.2.2 *En route* operations	14
	2.3 Mechanics of aircraft flight	14
	2.4 Aircraft navigation systems	21
	2.4.1 *En route* navigation: dead-reckoning systems	23
	2.4.2 *En route* navigation: position fixing systems	24
	2.4.3 Aircraft landing guidance systems	25
	2.5 Air traffic control and air traffic services	26
	2.6 Radars in aircraft navigation and air traffic control	29
	2.7 Aeronautical communication systems	32
	2.8 Summary	34
	2.9 References	35
3	**Atmospheric effects on aviation**	37
	3.1 Weather as a factor in aviation	37

viii *Aviation weather surveillance systems*

3.2 Overall effects of weather on aviation	37
3.2.1 Safety	37
3.2.2 Comfort	38
3.2.3 Schedule-keeping	39
3.2.4 Efficiency	40
3.2.5 Economy	40
3.2.6 Combination of factors	41
3.3 Atmospheric phenomena involving air motion	42
3.3.1 Wind shear	43
3.3.2 Turbulence	50
3.4 Hydrometeorological phenomena	57
3.4.1 Rain	59
3.4.2 Snow	60
3.4.3 Hail	61
3.5 Aircraft icing	61
3.6 Low visibility	67
3.7 Atmospheric electrical phenomena	69
3.8 Need for improved aviation weather information	72
3.9 Summary	74
3.10 References	75
4 Origins of harmful atmospheric effects on aircraft	**79**
4.1 General	79
4.2 Structure of atmosphere	79
4.3 Thunderstorms: nature, initiation and evolution	81
4.4 Thunderstorm parameters	85
4.5 Phenomena associated with thunderstorms	86
4.5.1 Divergence and convergence	87
4.5.2 Turbulence	89
4.5.3 Downburst	91
4.5.4 Cyclonic motion and tornadoes	91
4.5.5 Rain	93
4.5.6 Hail	93
4.5.7 Lightning, electric fields and atmospherics	95
4.5.8 Icing	97
4.5.9 Poor visibility	97
4.5.10 Overall thunderstorm scenario	98
4.6 Gust fronts and related phenomena	98
4.6.1 Characteristics	99
4.6.2 Outflow-induced waves and bores	101
4.7 Macrobursts and microbursts	105
4.7.1 Microburst types	106
4.7.2 Characteristics	108
4.7.3 Asymmetry	111
4.8 Other sources of atmospheric hazard	111

	4.9 Summary	115
	4.10 References	115
5	**Requirements of systems for aviation weather surveillance**	**121**
	5.1 General	121
	5.2 Types of weather surveillance systems for aviation	122
	5.2.1 *In situ* and remote sensing	122
	5.2.2 Ground-based, airborne and spaceborne sensors	124
	5.3 Spatial coverage	125
	5.4 Data update rates	129
	5.5 Spatial resolution	130
	5.6 Data processing and display systems	132
	5.6.1 Stages in data processing	132
	5.6.2 Display of aviation weather data	134
	5.6.3 Requirements of data processing and display systems	135
	5.7 Automated operation	137
	5.8 Selection of primary sensors	138
	5.8.1 Atmospheric parameters monitored for aviation	138
	5.8.2 Primary sensors for modern aviation weather surveillance	139
	5.9 Summary	140
	5.10 References	141
6	**Doppler weather radar as a primary aviation weather sensor**	**143**
	6.1 General	143
	6.2 Basic aspects	144
	6.2.1 Weather radar resolution	145
	6.2.2 Mapping of weather fields	150
	6.2.3 Scattering by raindrops and radar reflectivity of weather	154
	6.2.4 Radar echoes from clear air	157
	6.2.5 Weather attenuation of radar signals	158
	6.2.6 Operating frequencies of weather radars	162
	6.3 Conventional weather radar	164
	6.3.1 Reflectivity measurement: radar range equation	165
	6.3.2 Estimation of rain rates	172
	6.3.3 WSR-57 radar	178
	6.4 Motivation for developing modern weather sensors	180
	6.5 Doppler weather radar: basics	181
	6.5.1 Basic principle and limitation	182
	6.5.2 Atmospheric wind tracers	183
	6.6 Doppler weather radar: primary data products	185
	6.6.1 Spectral moments of weather echo signals	186
	6.6.2 Doppler weather radar system features and architecture	190
	6.6.3 Computation of basic data products	194
	6.6.3.1 Reflectivity	195
	6.6.3.2 Mean radial velocity	197

	6.6.3.3 Doppler velocity spectrum width	199
	6.6.3.4 Some general aspects of Doppler moment estimation	200
	6.6.4 Display of basic products	201
	6.6.5 Derivation of vector wind fields	204
6.7	Summary	208
6.8	References	212

Colour plates   **215**

7 Modern Doppler weather radars for aviation   245
   7.1 General   245
   7.2 WSR-88D system   246
      7.2.1 Architecture   246
      7.2.2 Parameters   247
      7.2.3 System features   247
      7.2.4 Data products   250
      7.2.5 Performance   251
   7.3 Range and velocity ambiguities   252
      7.3.1 Nature of problem   252
      7.3.2 Minimisation of range overlays   257
         7.3.2.1 Low elevation angles   257
         7.3.2.2 Middle elevation angles   257
         7.3.2.3 High elevation angles   258
      7.3.3 Velocity dealiasing   258
      7.3.4 Advanced ambiguity resolution methods   258
      7.3.5 Potential and futuristic methods   260
         7.3.5.1 Spectral decomposition   260
         7.3.5.2 Triple-PRF radar observation   261
         7.3.5.3 Staggered PRT scheme   262
         7.3.5.4 Random phase transmission   263
         7.3.5.5 Systematic discrete phase coding   263
         7.3.5.6 Single-pulse Doppler estimation   264
   7.4 Other special considerations   264
      7.4.1 Coverage   265
      7.4.2 Siting for terminal area surveillance   266
         7.4.2.1 Resolution   266
         7.4.2.2 Range coverage   266
         7.4.2.3 Low-altitude coverage   267
         7.4.2.4 Zone of blindness   267
         7.4.2.5 Range ambiguity and overlaid echoes   268
         7.4.2.6 Airport configuration   268
         7.4.2.7 Comparison of siting alternatives   269
      7.4.3 Scanning strategies and modes   270
      7.4.4 Data lag   274
      7.4.5 Comparison with air route surveillance radar   274

	7.5	Terminal Doppler weather radar (TDWR)	275
	7.6	Airport surveillance radar with weather channel	281
	7.7	Summary	284
	7.8	References	285
8	**Other sensors and systems for aviation weather**		**289**
	8.1	General	289
	8.2	Wind profilers	290
		8.2.1 Conventional wind profiling	290
		8.2.2 Radar wind profilers	291
	8.3	Radio-acoustic sounding systems (RASS)	302
		8.3.1 Basic system	302
		8.3.2 RASS augmentation for sensing aircraft icing conditions	304
	8.4	Low-level wind shear alert system (LLWAS)	305
		8.4.1 Concept and basic configuration	306
		8.4.2 Enhanced system	307
	8.5	Airborne wind shear detection	309
		8.5.1 *In situ* sensing	310
		8.5.2 Forward-looking remote sensing	311
	8.6	Airborne turbulence measurement	316
	8.7	Automated weather observing systems	317
	8.8	Radiometric satellite observation	320
	8.9	Airport visibility measurement	325
	8.10	Summary	330
	8.11	References	332
9	**Integrated system approaches**		**337**
	9.1	General	337
	9.2	Integrated terminal weather system	338
		9.2.1 Data integration	339
		9.2.2 Automated operation and fully processed output	340
		9.2.3 Performance enhancement, versatility and adaptability	341
		9.2.4 Predictive capability	341
	9.3	Aviation gridded forecast system	342
	9.4	Aviation weather products generator	343
	9.5	Summary	348
	9.6	References	349
10	**Automatic detection and tracking of hazardous weather features**		**351**
	10.1	General	351
	10.2	Basis of automated weather feature detection	352
	10.3	Thunderstorm cells	353
	10.4	Mesocyclones	356
	10.5	Gust fronts	363
	10.6	Storm outflows and microbursts	369
	10.7	Summary	372
	10.8	References	373

xii *Aviation weather surveillance systems*

SPECIAL TOPICS IN AVIATION WEATHER SURVEILLANCE

11 Atmospheric turbulence and its detection by radar — 375
   11.1 General — 375
   11.2 Wind shear and turbulence in meteorological events — 376
      11.2.1 Thunderstorms — 377
      11.2.2 Thermal plumes — 378
      11.2.3 K-H waves — 379
   11.3 Detection of turbulence with Doppler radar — 380
   11.4 Statistical theory of turbulence — 384
      11.4.1 Correlation and spectral functions in the inertial subrange — 388
      11.4.2 Filtering by the radar's weighting function — 391
      11.4.3 Variance of point and average velocities — 395
   11.5 Doppler spectrum width and eddy dissipation rate — 396
   11.6 Eddy dissipation rates in thunderstorms — 398
   11.7 Avoiding turbulence — 400
   11.8 Summary — 403
   11.9 References — 404

12 Lightning and aviation — 407
   12.1 General — 407
   12.2 Lightning, electric fields and atmospherics — 407
   12.3 Lightning–aircraft interaction — 411
   12.4 Weather conditions and lightning strikes to aircraft — 418
   12.5 Detection and surveillance of lightning phenomena — 423
   12.6 Lightning threats to aircraft: what else do we need to know? — 425
   12.7 Summary — 425
   12.8 References — 426

13 Polarisation diversity radars — 429
   13.1 General — 429
   13.2 Description — 429
   13.3 Basic definitions — 431
   13.4 Propagation effects — 432
   13.5 Rainfall measurement — 434
   13.6 Hail detection — 438
   13.7 Automatic classification and quantification of precipitation — 439
   13.8 Status and prospects for aviation use — 441
   13.9 Summary — 441
   13.10 References — 442

Index — 445

Preface

While writing this book I had the feeling of chasing and trying to catch the pieces of an exploding bombshell. So rapid and diversified has been the growth of the subject of this book in the recent past, that there was a distinct fear of the book becoming obsolete even before it was completed. I have therefore taken the approach of focusing more on the fundamental aspects of the aviation weather problem and generic solutions to them. Specific equipment and systems are referred to essentially to illustrate the capabilities and potential of modern aviation weather surveillance systems, as well as the problems encountered in performing the surveillance function. The specific systems also serve to provide a realistic flavour to the description. One casualty of such an essentially generic approach has been the relative lack of reference to specific software and algorithms that perform many of the intelligent tasks described in this book. But given the fluidity of the software scene which undergoes rapid and continual upgrading, substantial coverage of specific software would be impractical for a book of this nature.

The book is written with scientists, engineers, airline technology managers, civil aviation planners and other interested meteorological and aviation personnel in mind. Most of the material presented here should be of value in the training programmes of aviation operators including pilots and air traffic controllers.

The literature in the area of aviation weather surveillance is vast but scattered among a wide variety of sources. This highly interdisciplinary area of activity draws personnel and information from diverse scientific and technological fields which are fundamentally different from one another, each with its own distinct methodology, focus and even jargon. This book is the first attempt to make a synthesis of such scattered information and to present it to personnel with diverse backgrounds in a coherent manner for independent and self-contained reading. The information contained here is graduated in such a way that the serious technically minded reader can apprise himself or herself of many details of modern aviation weather surveillance, while the lay reader can still get a fair appreciation of the intricacies of the interplay of various apparently unrelated factors in the common task of aviation quality improvement.

Pravas R. Mahapatra

Acknowledgments

I am indebted to many in bringing this book to the present form. First and foremost, I wish to express my deepest gratitude to Drs Dusan Zrnic' and Dick Doviak of the US National Severe Storms Laboratory (NSSL), who have long been my friends, philosophers and guides in this area of inquiry, and who, along with their colleague and my friend Dr Vlad Mazur, have contributed the last three 'guest chapters' of the book, and also other graphical material that has greatly enriched this work. I have also been fortunate in receiving inspiration, both philosophical and practical, from my special friend and former director of NSSL, Dr Ed Kessler. A large number of professional colleagues and manufacturers' representatives from around the world have generously contributed and permitted their material to be included in this book, for which I am indebted to them. Input from my colleague Prof. S. P. Govindaraju on aeroplane flight parameters appearing in Chapter 2 is gratefully acknowledged. Also greatly appreciated is the help on multiple occasions rendered by my long-time friend and colleague Dr M. Sachidananda in obtaining hard-to-get reference material. Special thanks are due to the editorial team at the IEE, especially John St Aubyn, Jonathan Simpson and Fiona MacDonald who have provided superlative support on all aspects connected with the processing of the manuscript. I am grateful to the reviewers of the manuscript who have read the book with meticulous care and made corrections and useful suggestions which have improved the quality of the work. Finally, but importantly, I must express words of sentimental gratitude to my wife Purnima and children Satya and Pooja who have not only cheerfully borne the deprivation of my attention during the thick of this project, but even cheered me up during my long nocturnal writing sessions.

Abbreviations

ACARS	ARINC communication and retrieval system
ACAS	airborne collision avoidance system
A/D	analogue-to-digital (converter)
ADAS	AWOS data acquisition system
ADF	automatic direction finder
AGC	automatic gain control
AGFS	aviation gridded forecast system
AIV	aviation impact variable
ARINC	Aeronautical Radio, Inc.
ARSR	air route surveillance radar
ARTCC	air route traffic control centre
ARTS	automated radar terminal system
ASD	aircraft situation display
ASDE	airport surface detection equipment
ASOS	automated surface observing system
ASR	airport surveillance radar
ATC	air traffic control
ATCRBS	air traffic control radar beacon system
ATM	air traffic management
ATMS	advanced traffic management system
AV	airport visibility
AWOS	automated weather observation system
AWPG	aviation weather products generator
CAA	Civil Aviation Authority (UK)
DME	distance measuring equipment
EFAS	en route flight advisory service
FAA	Federal Aviation Administration (USA)
FAR	Federal Aviation Regulation (USA)
FAST	fore/aft scanning technique
FDP	flight data processing (or processor)
FFT	fast-Fourier transform
GOES	geostationary operational environmental satellite
GPS	global positioning system
HF	high frequency
ICAO	International Civil Aviation Organisation
IFR	instrument flight rules

ILS	instrument landing system
INMARSAT	international maritime satellite organisation
ITWR	interim terminal doppler radar
ITWS	integrated terminal weather system
JAWS	joint airport weather studies
K–H	Kelvin–Helmholtz
LAPS	local analysis and prediction system
LDR	linear depolarisation ratio
LLWAS	low level windshear alert system
LLWAS-NE	low level windshear alert system with network expansion
LORAN	long-range navigation system
LST	local standard time
MAP	mesoscale analysis and prediction (system)
MKS	metre-kilogram-second
MLS	microwave landing system
MOPA	master oscillator power amplifier
MST	mesospheric-stratospheric-tropospheric (radar)
MVD	median volume diameter (of a population of droplets)
NACA	National Advisory Committee on Aeronautics (USA)
NASA	National Aeronautics and Space Administration (USA)
NAWPG	national aviation weather products generator
NCAR	National Center for Atmospheric Research
NDB	nondirectional beacon
NEXRAD	next-generation radar (forerunner of WSR-88D)
NIMROD	Northern Illinois Meteorological Research on Downbursts
PAR	precision approach radar
Pirep	pilot report
POSH	probability of severe hail
PPI	plan-position indicator
PRF	pulse repetition frequency
PRI	pulse repetition interval (same as PRT)
PRT	pulse repetition time (same as PRI)
PUP	principal user processor
RAMS	regional atmospheric modelling system
RASS	radio-acoustic sounding system
RAWPG	regional aviation weather products generator
RDA	radar data acquisition (unit or subsystem)
RDASC	radar data acquisition status control
RDP	radar data processor (or processing)
RHI	range-height indicator
RPG	radar product generator
RVR	runway visual range
RWP	real-time weather processor
SAV	state-of-the-atmosphere variable
SSR	secondary surveillance radar
ST	stratospheric-tropospheric (radar/profiler)
STC	sensitivity-time control
STOL	short takeoff and landing
TACAN	tactical air navigation (system)

TASS	terminal area surveillance system
TCAS	traffic alert collision avoidance system
TDWR	terminal doppler weather radar
TOA	time of arrival
TRACON	terminal radar control
TRSB	time reference scanning beam
TVAD	tangential velocity azimuth display
TVS	tornado vortex signature
UHF	ultra-high frequency
USAF	United States Air Force
V/STOL	vertical/short takeoff and landing
VAD	velocity azimuth display
VFR	visual flight rules
VHF	very high frequency
VLF	very low frequency
VOR	very-high-frequency omnirange
VORTAC	colocated VOR and TACAN systems
WISP	winter icing and storms project
WSR–88D	weather surveillance radar – 1988 Doppler

Symbols

a, b	constants used to specify Z–R relationship
c	speed of light
d	diameter (of hailstones); distance from sensor; thickness of sheared layer
d_{max}	maximum distance of radar for observing weather at a minimum height h_{min}
e	base of natural logarithm
e_{max}	highest elevation angle of the radar antenna during the scan cycle
f_c	carrier frequency
f_d	Doppler frequency
f_{dm}	mean Doppler frequency (of an ensemble of scatterers)
f_r	pulse repetition frequency
f_N	Nyquist frequency
$f^4(\theta, \phi)$	normalised two-way power pattern of antenna
g	acceleration due to gravity
h	height (clearance) of radar beam above ground (for straight-line propagation)
h'	height (clearance) of radar beam above ground (considering atmospheric refraction)
h_{min}	minimum height of observation of weather phenomena
j	$\sqrt{-1}$
k	R'/R (~4/3)
k_m	sample number (in frequency domain) corresponding to the mean Doppler frequency
l_c	reference length (length of wing chord at designated cross-section)
m	refractive index of water
n	refractive index of air
n_r	total number of resolution volumes or pixels in a full circle of scan
r	range (distance) to a given resolution volume
$\mathbf{r}$	position vector
r_{aa}	radius of airport area
r_b	maximum radius of blind zone
r_m	maximum range (e.g. of radar)

Symbols

$r_{max(aa)}$	maximum permitted distance of radar from centre of airport area (while providing resolution ρ_{aa} over the entire airport area)
$r_{max(ta)}$	maximum permitted distance of radar from centre of terminal area (while providing resolution ρ_{ta} over the entire terminal area)
$r_{max(\rho)}$	maximum permitted distance of radar from the common centre of airport and terminal areas while meeting the resolution requirement everywhere within terminal area
r_{ta}	radius of terminal area
r_u	(maximum) unambiguous range
s_{hh}	complex scattering coefficient of a hydrometeor considering the horizontally polarised component of scattered radiation caused by a horizontally polarised incident radiation
s_{hv}	complex scattering coefficient of a hydrometeor considering the horizontally polarised component of scattered radiation caused by a vertially polarised incident radiation
s_{vv}	complex scattering coefficient of a hydrometeor considering the vertically polarised component of scattered radiation caused by a vertically polarised incident radiation
t	two-way propagation delay corresponding to a given radar target or resolution volume
t_m	electromagnetic propagation time delay (two-way) corresponding to maximum range r_m
t_s	sampling instant
u, v, w	orthogonal wind components (w being vertical)
$\mathbf{v}$	wind vector
v	horizontal wind speed
$\mathbf{v}_h$	horizontal wind velocity vector
v_l	longitudinal velocity component
v_t	transverse velocity component
v_r	radial component of scatterer velocity relative to radar
v_{rm}	mean radial velocity (of an ensemble of scatterers) (used as v in Chap. 11)
v_u	(maximum) unambiguous velocity
w	weight of aircraft
w_t	terminal velocity (of raindrops)
x_i	samples of (complex) receiver voltage output
B	receiver bandwidth
C	radar constant
C_D	drag coefficient
C_L	lift coefficient
$C_{L\,max}$	maximum lift coefficient
C_{Mp}	pitching moment coefficient
C_n^2	turbulent structure parameter of refractive index
D	drag force; diameter of raindrop

Symbols   xxiii

F	F-factor (related to wind shear)
$\bar{F}$	equivalent average F-factor
$F(\mathbf{K})$	Fourier transform of weighting function I_n
G	antenna gain
H_{max}	maximum altitude of weather surveillance
H_{DR}	hail detection signal
I	in-phase component of radar signal
I_n	a normalised weighting function
K	attenuation coefficient in dB/km; magnitude of $\mathbf{K}$
$\mathbf{K}$	wavenumber
K_g	gust alleviation factor
K_s	attenuation coefficient due to snow
K_w	constant related to refractive index of water
K_{DP}	specific differential phase shift
L	lift force; logarithm of ratio of echo power and single-lag autocorrelation estimate; length (distance) interval
L_a	amplitude loss factor (one-way)
L_f	receiver filtering loss (or finite bandwidth loss) factor
L_s	system loss factor
M	number of radar pulses used for signal processing (Doppler moment estimation)
M_p	pitching moment
N	number of raindrops per unit spatial volume per unit diameter interval; noise power
N_d	number of hailstones per cubic metre of spatial volume per millimetre size interval
P	instantaneous power level of microwave radiation through air
P_0	initial power level of microwave radiation through air
P_a	average transmitted power
P_i	power associated with the ith signal sample
P_r	received power (by radar)
P_{ro}	receiver output power
P_t	transmitted power (peak)
$\hat{P}$	estimated radar echo power
Q	quadrature component of radar signal
R	radius of the earth; rainfall rate; autocorrelation of the signal sample sequence
R'	fictitious radius of the earth to account for atmospheric refraction of radar beam
R_g	Richardson number
R_s	rate of snowfall
S	reference area (wing platform area)
S_k	kth sample of the periodogram
S_m	receiver sensitivity (minimum detectable signal)
S_{xx} etc.	wind shear components
T	aircraft engine thrust
T_r	pulse repetition time (or pulse repetition interval)
T_l	temperature of the layer in which a wave evolves

U_{de}	derived gust velocity
V, V_6	radar resolution volume (the subscropt 6 explicitly denotes 6-dB thresholding)
V_a	airspeed of aircraft
$\mathbf{V}_a$	airspeed vector of aircraft
V_∞	speed of aircraft relative to undisturbed air
W	magnitude of the range weighting function (of processing filter)
$\mathbf{W}$	absolute (inertial) wind vector
W_w	wind speed perturbation due to weather factors
W_x	horizontal wind component (usually along aeroplane ground track)
W_y	horizontal wind component perpendicular to W_x
W_z	vertical component of wind (positive downwards)
Z	reflectivity factor
Z_{DP}	reflectivity difference (between two different polarisations)
Z_{DR}	differential reflectivity
Z_e	equivalent reflectivity factor
Z_h	reflectivity factor at horizontal polarisation
Z_v	reflectivity factor at vertical polarisation
α	angle of attack; constant in K–R relationship
α_{stall}	stall angle of attack
β	profiler beam tilt angle; constant in K–R relationship
δ_{ij}	Kronecker delta function ($\delta_{ij}=0$, $i \neq j$; $\delta_{ij}=1$, $i=j$)
ε	extinction coefficient (or specific attenuation); normalised turbulent energy dissipation rate
η	reflectivity
θ_b	beam width
$\dot{\theta}$	antenna scan rate
θ	potential temperature
λ	wavelength (of radar signal)
ρ	vector separation between two points (in the atmosphere)
ρ_∞	density of undisturbed air
ρ_a	density of ambient air
ρ_{aa}	linear resolution requirement over airport area
ρ_d	duty ratio
$\rho_{hv}(0)$	correlation coefficient of polarised signal at zero lag
ρ_{sn}	signal-to-noise ratio
ρ_{ta}	linear resolution requirement over the terminal area (outside the airport area)
σ_b	backscattering cross section of rain drop
σ_f	spectrum width component due to fall speed differences among hydrometeors
σ_o	spectrum width component due to oscillation(s) of hydrometeors
σ_r	spectrum width component due to rotation (scanning) of radar beam
σ_s	spectrum width component due to wind shear
σ_t	spectrum width component due to turbulence

σ_v	spectrum width of radial (Doppler) velocity (for an ensemble of scatterers)
σ_θ^2	second central moment of two-way antenna power pattern
σ_r^2	second central moment of two-way range weighting function
σ_p^2	variance of the velocity at a point
τ	radar pulse width
ϕ_{hh}	two-way phase shift with horizontally polarised transmission and reception
ϕ_{vv}	two-way phase shift with vertically polarised transmission and reception
ϕ_{DP}	differential phase (between two different polarisations)
Δh	change in aircraft altitude
Δn	turbulence-induced incremental vertical acceleration of aircraft; perturbation in refractive index of air due to turbulence
Δr	range resolution
ΔN	number of raindrops per unit spatial volume having diameter between D and $D + \Delta D$
ΔV	difference in wind speed (e.g. across gust fronts and microbursts)
Λ	Fourier wavelength (i.e. scale of the velocity perturbation)
$\boldsymbol{\Phi}$	spectral density (tensor)

Chapter 1
Introduction

1.1 Aviation and electronics: a symbiotic relationship

Aviation must rank among the most important technological developments of the twentieth century. Like electronics, which is another pervasive discipline of science and technology that has had its birth and explosive growth during this century, aviation has had a radical effect on our lives, the profundity of which is of civilisational dimensions. Indeed, aviation has done to transportation what electronics has done to communication: shrink distances and time, and reduce the wide world to a 'global village'.

Aviation and electronics have not been mutually exclusive developments. Except for a few initial years during which aeroplane makers struggled to understand the mechanical principles of flight and make flying machines using these principles, the two disciplines have grown hand in hand, with electronics aiding aviation in achieving higher levels of complexity and quality, and in turn benefiting from the impetus provided by aviation applications. The use of electronic devices and systems in aeroplanes and their associated operations is obvious to anyone having anything to do with aviation, including the common passenger, though it can safely be said that the actual role of electronics in modern aviation is far more pervasive and profound than meets the layman's eye. The reverse path of this symbiotic relationship, i.e. the beneficial effects of the developments in aviation on electronics is, however, generally not so well known.

The main reason that aviation and its allied activities have spurred developments in their related and supportive areas is the highly investment-intensive nature of aerospace operations. Designing and developing new aircraft, both civil and military, as well as spacecraft, entails high costs, and the unit costs of these vehicles are also very high. Consequently, elaborate systems support is used to prevent loss or damage to aerospace vehicles and to enhance the efficiency of their usage. This fact provides the justification for relatively large investments in developing these supporting systems, especially because the magnitude of such investments is usually a fraction of the overall aerospace system costs. It is no surprise that a large part of the systems designed in support of modern aviation relies heavily on electronics.

In the aviation domain, the most demanding performance requirements are usually associated with military aircraft, and large investments are made by many countries in this sector in the interest of national security and superiority. The diversity and performance demands placed on military aircraft have caused the development of numerous new sciences and technologies which have subsequently been passed on to benefit the civilian aviation sector, and even the nonaviation industrial sector. A simple example is the technology of electroluminescence, which was first devised to illuminate the instrument dials of military aircraft for night flying, but is now used in wrist-watches to provide uniform and clear visibility in the dark with low power consumption. Similarly, in the domain of theoretical and system sciences, Kalman filtering, which is now a very well founded discipline with diverse applications, received its first impetus from the requirements of aerospace navigation.

The performance requirements of civilian aircraft are generally far less stringent than those of many types of military aircraft such as fighter planes, but their design and operation pose challenges of a different class. Civilian aircraft, especially commercial airliners, usually fly over long distances and must ensure a very high degree of passenger safety and comfort. Further, the global nature of civil aviation operations necessitates the use of support systems that are standardised and open (in terms of known technologies and system characteristics). A large number of electronic systems have been built to satisfy these requirements. Among these are a number of modern instruments designed to sense weather and other atmospheric phenomena of importance to aviation. These instruments and their associated computing, display and communication hardware and software make up a range of systems with a strong potential to facilitate every segment of aviation activity. The design and development of multiple sophisticated and dedicated aviation weather surveillance systems form a recent field of activity, and the equipment resulting from such activity is gaining widespread acceptance. This book attempts to cover this emerging and interesting area of scientific and technological activity from a systems perspective, with appropriate reference to its scientific, technological and operational aspects.

1.2 Phases in evolution of aircraft navigation

The Wright brothers etched their names in history on 17 December 1903 when the aeroplane they designed took off from the ground and stayed in the air for a few seconds under its own power. That aeroplane and its flight were, of course, primitive by today's standards and did little more than prove (the very important fact) that the flight of self-propelled heavier-than-air machines is possible. Aviation has not looked back since, and the size, speed, sophistication and diversity of aircraft have steadily and rapidly increased during the relatively short history of aviation. This has been accompanied by an equally steady growth in the reach, complexity and versatility of the aviation operations themselves. In support of such growth, aircraft navigation equipment and procedures have also undergone progressive development and sophistication during this period.

In the very early years after the first flight by the Wright brothers, the main attention of aircraft designers was devoted to ensuring the integrity and stability of the aircraft as a structural and aerodynamic entity. This resulted in the design and construction of aircraft which were sufficiently reliable to perform economically important tasks including the transport of passengers and cargo. Simultaneously the aircraft were also adapted to the role of machines of war.

Growing economic and military use of aircraft led to increasing emphasis on the operational and mission-related aspects of flight. As long as flights were relatively few and far between, the main concern in ensuring the success of aircraft missions was to navigate individual aircraft so as to make them reach their destinations reliably. Through relentless research and development, design improvements and progress in instrumentation and the procedural framework, the range of ambient conditions under which safe flight was possible was progressively widened. These included night flying, transoceanic flying and flying under increasingly adverse conditions of weather and wind patterns.

The next major phase of aircraft navigation system evolution was in response to the growing number and diversity of aircraft populating the skies. As the air traffic density increased globally, and reached particularly high levels within certain aviation corridors, the most important focus of navigation shifted to the avoidance of collisions and conflicts between aircraft sharing a common air space. With ever-growing aviation activity, ensuring adequate separation between aircraft during all phases of flight remains the prime focus of air traffic control. A complex procedural framework, supported by appropriate instrumentation and communication networks, has been built up and co-ordinated globally to ensure the smooth flow of air traffic (e.g. Litchford, 1975; Perry and Wallich, 1986; US Department of Transportation, 1989). The effectiveness of modern air traffic control systems is proven by the fact that commercial aviation is now among the safest forms of transportation in a statistical sense (Fischetti and Perry, 1986).

Yet the spectre of aviation disasters has not gone away. Vivid visuals of horrifying airliner crashes flash across the world's television screens at periodic intervals, inspiring a feeling of awe and jolting our sense of complacency. The reasons that air disasters make bold headlines are many, ranging from economic to social and psychological, and even political. First, the size, cost and capacity of individual airliners have increased dramatically, with the largest aircraft costing well over a hundred million dollars apiece, and carrying up to 400 passengers. Individual airliner crashes therefore cost heavily in terms of men, material and money. Then, in many societies it is only the upper crust of the population that gets the 'privilege' of air travel, and air disasters often result in the death of the rich and the powerful from all sections of society, including government and business. Many heads of states and governments have lost their lives in aeroplane accidents, with profound effects on the polity and economy of their countries. Thus the consequential effects of air crashes may be many times more than the direct losses resulting from them. Finally, psychologically speaking, humans have a sense of in-built insecurity about leaving the *terra firma*, and each air crash reinforces a popular perception of the unsafe nature of air travel.

Then there are other segments of aviation covered within the navigation framework. Although accidents involving civilian airliners attract the most media attention, airliners numerically constitute only a small fraction of the total air traffic. Military and general aviation aircraft constitute the other important segments of aviation. Accidents in these sectors of aviation are less reported, but are far more numerous than airliner accidents. Some of these aeroplanes, especially advanced military aircraft, can be as or even more expensive than commercial airliners. From the point of view of navigation and air traffic control, each aircraft, irrespective of its size or type, is a navigational entity and contributes to the load on the air traffic control system.

There is thus a clearly felt need and mandate to minimise damage to aviation from all sources, and not merely collisions between aircraft. This enhanced and broadened role is the main characteristic of the latest phase in air navigation, and has led to the term 'air traffic management' (ATM) being increasingly used in preference to the original term, 'air traffic control' (ATC). Fortunately, parallel developments in science and technology, notably in the fields of electronics and informatics, have made available vastly more potent tools to the aviation sector to cope with its enhanced role of comprehensive air traffic management.

This book focuses on one pervasive factor that influences aviation of all types in all its phases, and impinges on air traffic management as a weighty factor. It discusses the effects of atmospheric phenomena and processes on aviation, and presents a comprehensive study of modern electronic systems designed in the service of aviation from the weather point of view.

1.3 Modern aviation weather surveillance

Weather information for aviation within given air spaces has traditionally been provided by meteorological agencies of the governments in charge of the respective air spaces. Meteorology is now a highly developed science that utilises a variety of instruments for data acquisition, and sophisticated weather models and supercomputers for prediction of meteorological conditions. The instruments include *in situ* sensors on or near the surface of the earth, and remote observing devices such as weather radars and satellites. Vividly colourful and detailed radar-derived weather fields routinely shown by television channels, high-resolution satellite imagery of global and regional scale cloud patterns, animated visuals showing the evolution and propagation of weather features, and increasingly accurate weather forecasts have enhanced popular awareness and confidence with regard to the science of meteorology. In quantitative terms meteorological instrumentation has been able to provide accurate data for use in many fields of activity such as hydrology, dam design and construction, irrigation, agriculture, outdoor sports and adventure, and military operations.

While the growth and reach of general meteorological facilities and information have been impressive, general meteorological data products cannot be used directly to satisfy all the weather needs of aviation. This is because the requirements of weather information from the aviation point of view differ in many significant ways from those for other areas of human activities. The differences pertain to the nature of the parameters of interest, the spatial and temporal scales of observation, and the mode of dissemination and utilisation of information. These aspects will be covered in detail in the later chapters. Such special requirements have made aviation weather a distinct discipline of study.

Weather has always been of important concern for aviation, but the scientific and engineering disciplines relating to aviation weather studies and instrumentation have received a major boost in the past two or three decades. During this period there has been rapid progress in the finer understanding of the nature of different types of atmospheric processes that are significant from the aviation point of view. As a result of focused and directed research, fairly clear insight has been gained into the interaction of atmospheric processes with aircraft in flight, the process of detection of severe weather phenomena, and the estimation of their hazard potential. This insight has not only helped reinforce the motivation, but also provided the knowledge input necessary for the design and development of advanced surveillance systems for the mitigation of the deleterious effects of weather on aviation.

Aviation weather surveillance is a multidisciplinary field of activity, drawing from at least three major disciplines in science and technology. First, it relates to the aeronautical sciences for us to be able to understand and quantify the hazard potential of different types of meteorological phenomena with regard to different types and phases of flight. Second, it naturally relates to the broader science of meteorology. Finally, because of the stringent demands imposed by aviation applications, aviation weather surveillance depends heavily on electronics for sensing, processing, communication and display of weather information. The purpose of this book is to give a flavour of this interdisciplinary nature of the subject while emphasising and detailing the system aspects of such surveillance.

1.4 Scope and organisation of the book

This book is designed to provide a comprehensive introduction to the science, sensors and systems that go into making the modern aviation weather surveillance systems. It is expected to provide the reader with an insight into the fundamentals of the various disciplines involved, and their complex interplay. The difficult task of presenting an interdisciplinary subject has been attempted by arranging the material in a logical and stepwise manner.

Aviation weather surveillance and information systems must fit into, and be compatible with, the broader system structure intended to facilitate aircraft navigation. Keeping this in mind, the basic background of aviation systems and operations is presented in Chapter 2, with particular emphasis on air traffic control, air traffic services, and such important system elements as radars and communication subsystems.

The susceptibility of aircraft to different types of adverse atmospheric effects is presented next. It is emphasised in Chapter 3 that weather impacts nearly all aspects of aviation operation: safety, passenger comfort, schedule-keeping and operating efficiency and economy. The types of atmospheric hazard that hamper flight operations are listed, and their specific effects on aviation are discussed at a level of depth and detail necessary for proper appreciation of the developments of the subsequent chapters.

The atmospheric processes that adversely impact aviation owe their origins to certain weather phenomena, and these are discussed in Chapter 4 with a view to familiarising the reader with the details of the weather environments encountered in aviation operations and their relative hazard potential for flight. This knowledge also helps in appreciating the detection aspects of aviation-hazardous weather phenomena. In particular, the thunderstorm environment, which is the origin of diverse phenomena affecting aviation adversely, is analysed in detail. Insight is provided into the nature of the precipitation and wind fields associated with various types of hazardous features, which would help in understanding the radar signatures of these phenomena discussed in later chapters.

Following the discussion of the basic aspects concerning aviation and weather, the generic requirements of systems intended for aviation weather surveillance are highlighted in Chapter 5. The relative merits of *in situ* and remote sensing are presented, and the need for the latter mode of observation as the primary data source for modern aviation weather surveillance systems is established. The important attributes of such systems are their spatial coverage and resolution, data processing speeds and refreshment rates, information content and user-friendliness of displays, and facility of automated operation. The chapter discusses the possible primary sensors for modern aviation weather surveillance systems with regard to these attributes.

The cornerstone of modern aviation weather surveillance systems is the Doppler weather radar, and the characteristics and capabilities of this device are discussed in Chapter 6. Radars meant for atmospheric observation differ significantly from those used to observe 'point targets' such as aircraft, and the special features of weather radars are discussed in detail in this chapter. In particular, the echo processes from air laden with precipitation particles and air without them (the so-called 'clear air') receive considerable attention. The relationship between the various significant parameters of the radar and its ability to detect atmospheric phenomena is brought out to help the reader in making a proper choice of the type of radar appropriate for a given aviation weather surveillance application. The distinguishing feature of the modern generation of weather radars compared to the earlier generation, which are still in service in large parts of the world, is their ability to sense wind fields in addition to precipitation intensity. These radars are also characterised by the high accuracy of weather parameter estimation. The principles involved in Doppler radar measurement of precipitation, wind speeds and turbulence are introduced in the chapter.

Chapter 7 deals with some of the actual Doppler radar systems developed in recent years for weather observation, especially for aviation support. Three representative radars are discussed in detail: the WSR-88D, the Terminal Doppler Weather Radar, and the ASR-9 with a dedicated weather channel. Each of these represents a particular type of application: the first is ideal for long-range weather surveillance as necessary for *en route* air navigation, the second for focused weather surveillance of terminal areas, and the third offers a combination of air traffic control and weather functions. Specific issues such as ambiguity problems, coverage and siting issues, and scanning strategies are discussed relative to these radars.

A number of sensors other than radars are also very useful in the case of modern aviation weather surveillance. Such sensors can provide useful aviation weather information by themselves, and very usefully supplement and augment radar data. Chapter 8 covers a few such sensors and sensor systems. Among them are wind profilers which yield the vertical distribution of horizontal wind at a given location, radio acoustic sounding systems which are capable of providing the temperature profiles of the atmosphere, the low-level wind shear alert system that warns of the existence of gust fronts and microbursts in airport areas, automated weather observing systems, and satellites as observers of local weather phenomena.

The modern aviation weather sensors are potent by themselves, but a much higher level of performance and utility can be derived by combining data from these sensors in a synergistic way. Fortunately, modern data handling hardware and software provide such capabilities, and these are being used extensively in the service of aviation. Chapter 9 discusses a few of the integrated aviation weather systems being developed that combine data from multiple sources, and add further value by way of forecasting and systematic data organisation. The Integrated Terminal Weather System is one such system that performs high-level data integration from multiple sources and provides finished weather products to be used directly by aviation operators without the need for interpretation by specialist meteorologists. The Aviation Gridded Forecast System generates and organises forecast data in a four-dimensional space-time grid which is fine enough for aviation use. The Aviation Weather Products Generator provides high-level weather products to each user in an individualised form best suited for his or her use.

Chapter 10 discusses an important aspect of modern aviation weather surveillance systems: the automatic detection of weather phenomena hazardous to aviation. Performance of intelligent functions such as detection, recognition and hazard estimation of weather features eliminates the human element with its attendant fallibilities, and speeds up the weather data interpretation process to a level consistent with the needs of modern aviation. The automatic detection process is based on the recognition of patterns or signatures of various types of weather phenomena in specific data fields, especially those generated by modern Doppler radars. In particular, the detection of storm cells, rotational phenomena such as mesocyclones and tornadoes, velocity discontinuities such as those caused by gust fronts, and divergent wind fields of the type arising from microbursts, receive specific attention in the chapter.

The next three chapters cover selected special topics of importance to modern and futuristic aviation weather surveillance systems, and each has been contributed by a leading expert in the field. Although the topics of these three chapters have been mentioned briefly earlier in the book, the treatment here is made in greater depth.

Chapter 11 deals with the important topic of atmospheric turbulence, covering its associated definitions, characterisation, detection by modern Doppler weather radars, and avoidance by aircraft. The turbulence phenomenon is explained in statistical terms, and the effects of the radar parameters on its detection are modelled. The concept of eddy dissipation rate is explained, and the rate is used as a classifier of the turbulence level from the aviation point of view. In particular, attention is devoted to eddy dissipation rates in the thunderstorm environment, which is the most hazardous for aviation.

Lightning, which is another important atmospheric phenomenon affecting aviation, receives closer attention in Chapter 12. A physical understanding of the complex electrical process of lightning and its interaction with aircraft in flight is conveyed. Further, observational and statistical data on lightning strikes on aircraft, obtained from extensive and careful experimentation in recent times, are presented to generate an appreciation of the conditions under which such strikes are more probable.

The last of the special topics covers a new generation of atmospheric sensors: coherent radars with polarimetric capability. By radiating and/or receiving signals with diverse polarisation characteristics, such radars can monitor more details of atmospheric processes than ordinary (single-polarisation) Doppler radars. Even when measuring the more 'classical' parameters like reflectivity, polarimetric radars have the advantage of being relatively insensitive to many corrupting effects such as rain attenuation and the variability of drop size statistics which affect the accuracy of past and present weather radars. A systematic use of the additional parameters sensed by polarimetric radars can enable not only the identification of the types of hydrometeors present in a given precipitation field, but their individual quantification. A concise discussion of the working of polarimetric radars, their capabilities and their potential advantages for aviation weather surveillance are presented in Chapter 13 of the book.

1.5 References

FISCHETTI, M.A., and PERRY, T.S. (1986): 'Our burdened skies', *IEEE Spectrum*, **23**, (11), pp. 36–37

LITCHFORD, G.B. (1975): 'Avoiding mid-air collisions', *IEEE Spectrum*, **12**, (9), pp. 41–48

PERRY, T.S., and WALLICH, P. (1986): 'A matter of margins', *IEEE Spectrum*, **23**, (11), pp. 38-49

US DEPARTMENT OF TRANSPORTATION (1989): 'Air traffic control'. Federal Aviation Administration Report 7110.65 F, 21 September 1989

Chapter 2
Basic background of aviation

2.1 Goal of aviation systems

The basic purpose of aviation systems is airborne transportation of people and material. The primary component of aviation systems consists of the aircraft themselves, which actually perform this task. By its very nature, aviation is a highly organised and co-ordinated activity. The prime focus of such organisation and co-ordination is to ensure the twin objectives of flight safety and efficient use of airspace and other resources deployed in support of aviation. The objectives of major aviation systems are typified by the US Federal Aviation Act of 1958, which defined the mission of the Federal Aviation Agency as (Fischetti, 1986) '... to provide the regulation and promotion of civil aviation in such a manner as to best foster its development and safety, and to provide for the safe and efficient use of airspace'.

Aviation safety and efficiency are affected by a number of factors, among which weather is a major one. The purpose of this book is to discuss the various effects of weather on aviation, and the modern electronic systems devised to combat the harmful effects. It is obvious that such systems will not work in isolation, but in harmonious conjunction with other electronic systems used in service of aviation, sharing resources (such as communication channels, displays, information processing power, electromagnetic spectrum and operating personnel) with them, and operating within the overall framework of the policies and procedures that govern aviation. It is thus necessary to understand some basic features of the aviation process, aircraft flight, and the general electronic environment in which aircraft operate, in order to appreciate the nuances of weather effects on aviation and the features of the systems used for their alleviation. The sections of this chapter that follow are intended to provide different aspects of such background. The purpose here is, however, not to cover all aspects of aviation, but essentially those which may have a bearing on the focus of this book.

2.2 Phases of aircraft flight

The sequence of stages through which the flight of an aircraft passes, starting from parking at the origin, consists of taxiing, takeoff, climb, cruise, descent, landing and taxiing, before again parking at the destination. The flight environment and parameters are different in each of these stages, and the nature of weather phenomena varies with altitude. Consequently, the nature and severity of weather hazards are different for the various stages of flight. However, broadly speaking, the totality of flight operations can be divided into two distinct phases from the point of view of weather susceptibility. These are discussed in the following subsections.

2.2.1 Terminal area operations

Terminal area operations are the operations carried out within or in the vicinity of designated areas around airports called terminal areas. These operations normally include takeoff, initial part of climbing, final part of descent, touchdown, taxiing and parking. In addition, such nonroutine operations as aborted landings (consisting of approach, 'flare' and ascent), in-flight holding of aircraft before landing, etc. are also included among terminal area operations.

Air navigation in terminal areas is characterised by high-density and dynamic operations. Further, flight operations in this area are conducted at low altitudes and speeds, leaving limited margin in these parameters to effect recovery from weather-induced disturbances. Another important feature of terminal area operations is the significant use of visual inputs by pilots, in addition to support from a variety of navigational aids. For weather surveillance purposes, the terminal radar control (TRACON) phase of ATC operations may also be considered as part of the terminal area operations.

The terminal area corresponding to an airport has a radius of 56 km around the centre of its runway complex. The parameters of flight within this area differ for different types of aircraft. For most fixed-wing aircraft, the preferred angle of approach for landing (glide-slope angle) is about 3° and the takeoff angle is of the order of 6°. Helicopters and other special types of aircraft (e.g. STOL or V/STOL) may land and take off at much steeper angles, which may even be close to the vertical. Landing speeds for different aircraft are specified so as to provide a certain safety margin over their respective stall speeds. The margin is normally in the range of 25–30% in the case of airliners, transport planes and most other civilian aircraft.

A combination of factors such as operations close to ground, limited margin over stall speeds, high traffic density, and a highly dynamic operational scenario ensures that terminal area operations constitute the most accident-prone phase of flight. A broad analysis of aviation accidents (Fischetti and Perry, 1986) shows that a vast majority of them (~93%) occur during operations associated with the terminal area (climb and descent are included among them, though part of these two operations may be outside the terminal area). The same study also showed that the stages of flight which are most affected by inclement weather are the takeoff-and-climb and the descent-and-approach phases. Many other studies lend further credence to such inferences (e.g. Rudich, 1986). Table 2.1 lists the percentage of air carrier accidents in the USA occurring in different phases of flight during the

Table 2.1   US air carrier accidents in 1990, initiated during different phases of flight (National Transportation Safety Board, 1993)

Type of air carrier operation	Phase of flight operation	Percentage of accidents
14 CFR 121, 125, 127 (no. of aircraft involved: 25)	Takeoff ground run	20
	Descent–normal	16
	Taxi	8
	Landing taxi	8
	Takeoff climb	4
	Climb to cruise	4
	Approach	4
	Landing	4
	Landing roll	4
14 CFR 135 (scheduled) (no. of aircraft involved: 16)	Landing	31.3
	Approach	12.5
	Taxi	12.3
	Takeoff	6.3
	Descent	6.3
14 CFR 135 (nonscheduled) (no. of aircraft involved: 108)	Takeoff	25.9
	Landing	17.6
	Approach	11.1
	Descent	5.6
	Manoeuvre	4.6
	Climb	4.6
	Taxi	3.7

Explanatory note on types of aviation operation:
Title 14 CFR Part 121: Air carriers such as major airlines and cargo haulers that fly large transport aircraft
Title 14 CFR Part 125: Large, privately owned aircraft not held out for hire
Title 14 CFR Part 127: Helicopters used as scheduled air carriers
Title 14 CFR Part 135: Commercial air carriers commonly referred to as commuter airlines and air taxis

year 1990. The table clearly establishes that an overwhelming majority of aircraft accidents occur during takeoff, landing and other low-altitude operations.

Weather vulnerability during takeoff, landing and other low-altitude operations is not confined to aircraft alone. Many other types of aerospace vehicles, including space boosters, reusable space vehicles such as the Space Shuttle, and rocket-powered strategic and tactical missiles as well as cruise missiles are subject to low-altitude weather effects. In fact, adverse weather is

14 Aviation weather surveillance systems

Table 2.2 Ranges of flight parameters of different types of aircraft

Type of aircraft	Cruise/operating altitude, ft		Cruise/operating speed, km/h	
	Minimum	Maximum	Minimum	Maximum
Propeller-driven	A few thousands	~15 000	~100	~500
Subsonic airliners and transport aircraft with turbojet and turbofan engines	~20 000	~40 000	~800	~1000
Supersonic transport	~60 000	~70 000	~2000	~3000
High-performance fighters	~10 000	~70 000	~700	~2500
Long-range bombers	~20 000	~70 000	~800	~1000
Helicopters	Near ground	~20 000	~0	~250

among the most frequent disrupting factors for launch and landing or recovery of space missions.

2.2.2 En route operations

Flight operations *en route* are characterised by relatively high altitudes and speeds of flight under nearly steady operating conditions, low traffic densities, and greater separation between aircraft compared to the terminal phase of flight. Aircraft usually dwell for relatively long periods of time in the *en route* phase with minimal pilot action. A general feature of *en route* flight is the availability of high margins of speed and altitude for aircraft to recover from disturbances.

En route flight parameters vary widely for different classes of aircraft. The most important of these parameters are the flight altitudes and speeds, the ranges for which are indicated in Table 2.2 for different classes of aircraft. While studying the susceptibility of particular types of aircraft to atmospheric effects, those phenomena that occur at the normal operating heights of such aircraft are of utmost importance. Aircraft speeds and dynamic parameters are of importance in investigating the effects of atmospheric phenomena of different temporal and spatial scales on flight.

2.3 Mechanics of aircraft flight

To study the effect of the atmosphere on the flight of aircraft, it is necessary to understand the various forces and moments that act on an aircraft during flight, and the effect of each of these on the aircraft. A complete treatment of these aspects, which constitute an important branch of aeronautical engineering, is outside the scope of this book. There are many good books on flight mechanics and related topics (e.g. Etkin, 1982; Nelson, 1989) to which the interested reader may refer for a complete understanding of the subject. An excellent introductory coverage of the subject is made by Anderson (1989). A brief discussion of the forces and moments affecting flight is provided here for the sake of completeness, and to serve as elementary information for the nonaeronautical scientist.

Unlike surface vehicles which are supported by the normal reaction force of the ground (or the buoyancy of water, in the case of ships) in both static and mobile states, heavier-than-air machines must derive the forces which sustain and control flight through dynamic interaction with the surrounding air mass. Again, unlike surface vehicles which usually have only two principal degrees of freedom (forward/backward motion and turning about a vertical axis), aircraft have six degrees of freedom, of which three are translational and three rotational. The flight of aircraft is therefore a fairly complex phenomenon, dependent on a delicate and dynamic balance of forces and moments acting on the aircraft.

The most important, and usually the single largest, component of force acting on an aircraft in flight is the *lift*, which helps the aircraft maintain its altitude against the pull of gravity and determines its rate of ascent or descent. Aircraft generate their lift force by the motion of an *aerofoil* through the air at a certain *angle of attack* (see Fig. 2.1). The motion of the aerofoil through air causes a pressure differential to develop between its upper and lower surfaces. The integral of this differential pressure over the lifting surface(s) acts as the overall lift force on the aircraft.

In fixed-wing aircraft, the main lifting surface is attached rigidly to the body in the form of wings, and hence the entire aircraft has to move forward at a certain minimum speed to generate the lift necessary to take off and to sustain flight. For cruising flight along a straight line at constant speed and altitude, the lift force must be just sufficient to balance the weight of the aircraft. For performing manoeuvres such as turns, takeoff and climb, the aircraft would need to generate additional lift.

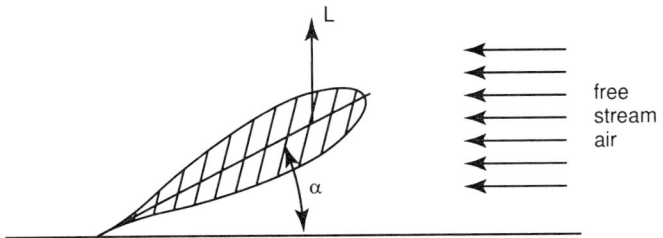

Figure 2.1 Airflow impinging on an aerofoil at an angle of attack α generates lift force L

The lift force L acting on an aircraft in flight is expressed in the form of the following simple equation:

$$L = \frac{1}{2} \rho_\infty V_\infty^2 S C_L \qquad (2.1)$$

where

ρ_∞ = density of the undisturbed[1] surrounding air
V_∞ = speed of the aircraft relative to the undisturbed surrounding air
S = a reference area, usually the plan form area of the wings
C_L = a dimensionless quantity called the *lift coefficient*.

The simple eqn. 2.1 indicates that the lift on an aircraft at any given altitude (where the air density ρ_∞ may be assumed constant) has a square-law dependence on the aircraft speed relative to the surrounding air. The actual dependence may be somewhat different, especially over large speed changes, since C_L is not exactly a constant but varies with speed, the variation being more pronounced as the flight speed approaches the local speed of sound.

The lift coefficient C_L has a linear dependence on the angle of attack α up to a certain value, after which the curve rapidly flattens to reach a maximum value C_{Lmax} and then droops precipitously (Fig. 2.2). This droop, signifying a rapid loss of lift with increasing angle of attack, is caused by the streamlined airflow separating itself from the upper wing surface, and creating a region of erratic or turbulent flow, as shown in Fig. 2.3. An aircraft in this condition is

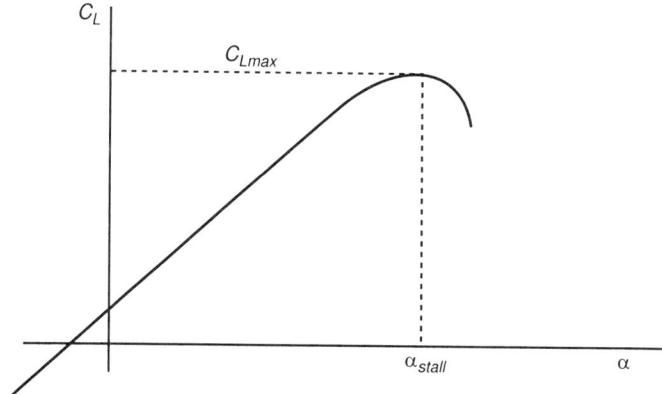

Figure 2.2 Typical dependence of lift coefficient of an aerofoil on the angle of attack

---

[1] The term 'undisturbed' is used here in the context of the disturbance to the air caused by the flight of the aircraft itself, and does not pertain to the disturbances due to weather and other factors. An aircraft in flight would create a flow field around itself and thus affect the state of the air in its own vicinity relative to the undisturbed air far away. Instruments mounted on the aircraft to measure air speed and density (e.g. a Pitot-static tube) produce readings corrupted by the flight-induced flow disturbance, and must be corrected to yield the true values

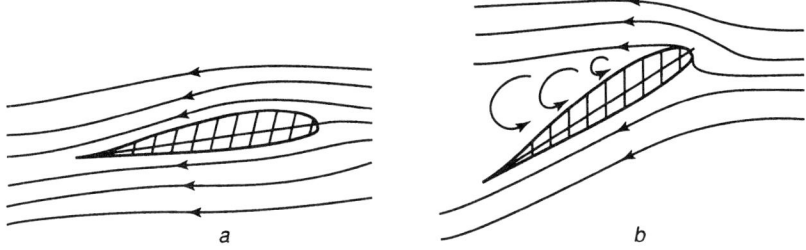

Figure 2.3 *The smooth, streamlined flow of air past an aerofoil at low angle of attack (a), and separated flow accompanied by turbulence at high angles of attack (b)*

said to be *stalling*. The angle of attack corresponding to the beginning of stall is called the *stall angle of attack*, α_{stall}.

If the speed of an aircraft decreases in flight, as during landing, its lift coefficient must increase as the square of the speed loss ratio to maintain the lift, as per eqn. 2.1. As is apparent from the C_L-α curve of Fig. 2.2, an increase in the lift coefficient is possible by increasing the aircraft's angle of attack, which in turn can be achieved by tilting the aircraft body about its lateral axis to make the wings more and more oblique with respect to the direction of the incident air flow. A higher lift coefficient may also be obtained by increasing the curvature of the wing aerofoil by deploying a moving surface called *flap*, as shown schematically in Fig. 2.4. However, the process of increasing the lift coefficient cannot be continued beyond the stall limit, on crossing which the aircraft would suddenly lose height. This explains why aircraft are most vulnerable to weather hazards during takeoff and landing. During these flight phases the aircraft fly at relatively low speeds, close to their stall speeds (*stall speed* is the minimum speed of an aircraft at which sustained flight is possible without stalling). Under such flight conditions, sudden loss of airspeed and/or changes in angle of attack due to strong, and especially unexpected,

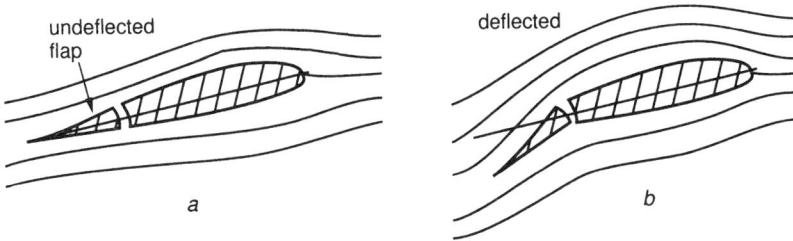

Figure 2.4 *Schematic diagram of flap operation: in raised position (a) the flap acts as a part of the main aerofoil, and when lowered (b), it increases the effective curvature of the aerofoil, causes greater deflection of the air stream, and thereby enhances lift*

atmospheric air currents or changes therein can induce stall with little time or altitude margin to recover.

The square-law dependence of lift on speed amplifies the effect of wind perturbations due to weather and other factors. If an aircraft flying at a speed V_a relative to still air encounters a weather-induced wind perturbation of speed W_w in an opposite direction, then its speed relative to the local air would be $(V_a + W_w)$, and the ratio of the wind-perturbed lift to the still-air lift would be $(V_a + W_w)^2/(V_a)^2$ or $[1 + (W_w/V_a)]^2$. Thus the fractional perturbation in lift would be about twice the wind perturbation (as a fraction of the aircraft speed) if the latter is small, and more than twice if the wind perturbation is a significant fraction of the aircraft speed. This latter condition occurs frequently in the case of small aircraft, which usually fly at low speeds, and during low-speed operations such as takeoff and landing in the case of larger aircraft.

An important parameter related to lift is the *wing loading*, defined as the weight of the aircraft per unit plan area of its wings, or w/S. The wing loading is a function of the performance level of the aircraft. High-speed and high-performance flight vehicles such as fighter aircraft are normally designed for higher levels of wing loading, while the loading is small for low-speed general aviation aircraft.

Another major force acting on an aircraft in flight is the *drag*, which is caused by the resistance of air to its motion. All parts of the aircraft exposed to air flow contribute to the total drag acting on it. The main contribution to drag acting on an aircraft at low subsonic speeds comes from *skin friction*. If there is significant separation of the airflow from the wing surface, then a *pressure drag* component must be considered. At high subsonic, sonic and supersonic speeds, the main component of drag is the *wave drag* caused by the formation of shock waves. In addition, at all speeds, there is a component of drag called *induced drag* or *lift-dependent drag* which increases with increase in the lift generated by the aircraft wings.

The drag force has an expression similar to lift, and is given as

$$D = \frac{1}{2} \rho_\infty V_\infty^2 S C_D \qquad (2.2)$$

where C_D is a dimensionless *drag coefficient*. Thus, like lift, drag has a square-law dependence on speed, and hence has a similar sensitivity to wind speed changes due to weather. At low subsonic speeds the drag coefficient remains nearly constant, but increases rapidly near the sonic speed.

The propulsive force or *thrust* provided by the aircraft's power plant (engine) is the third major force involved in flight, which essentially counteracts the drag and makes sustained forward motion possible. In addition, the thrust also provides the additional force for forward acceleration necessary during certain stages of the flight. Small general aviation aircraft operating at low speeds are often powered by reciprocating engines driving a propeller to generate the thrust. The thrust of such power plants is the highest at zero forward velocity (i.e. when the aircraft is at rest, with engine and propeller running). The thrust decreases as the aircraft speed increases, the rate of fall being more rapid as the flight speed approaches the speed of sound. In contrast, turbojet and turbofan engines, which are used to power aircraft of larger sizes and speeds, deliver a relatively constant thrust with velocity.

While the forces acting on the aircraft help maintain it in equilibrium as a point mass, its body attitudes are the result of a balance of moments. The attitudes are measured as angles about three orthogonal axes related to aircraft body. The fore-and-aft axis of the aircraft is called the *roll axis*, the lateral axis parallel to the line joining the wing tips is the *pitch axis*, and the third orthogonal axis (which is vertical when the aircraft remains in its natural horizontal position) is the *yaw axis* (see Fig. 2.5). The most important of the moments are those in the longitudinal plane of symmetry of the aircraft, which determine its pitch attitude. The pitching moment is given by the expression

$$M_p = \frac{1}{2} \rho_\infty V_\infty^2 S l_c C_{Mp} \tag{2.3}$$

where C_{Mp} is the *pitching moment coefficient* and l_c is a reference length, usually

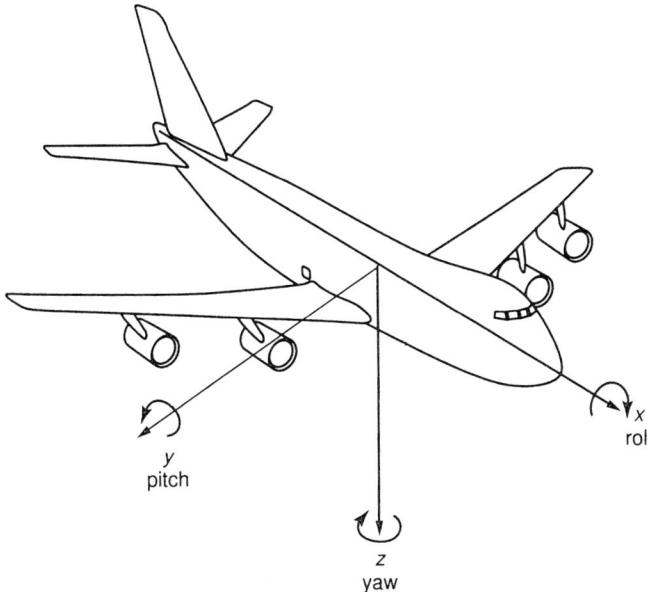

Figure 2.5 Aircraft body axes

the root-to-tip length (or 'chord length') of the aerofoil forming a designated cross-section of the aircraft wing.

In addition to these main forces and moments, an aircraft in flight is subject to a number of other forces and moments. Some of them are deliberately generated through pilot or autopilot action. These are called *control forces* and *control moments*. A balance of all the moments acting on the aircraft determines its spatial orientation or *attitude* in flight. Further, together with the inertial parameters of the aircraft (mass, moments of inertia about the three principal axes), the forces and moments acting on the aircraft determine its dynamic behaviour in flight.

Most aircraft are designed to remain *statically stable* under normal flying conditions. This means that the aircraft will maintain its attitude in flight without the continuous or frequent application of control moments. The main contribution to attitudinal stability in the longitudinal vertical or *pitch* plane usually comes from a pair of *horizontal tail planes*. The stability in the horizontal or *yaw* plane is provided by the *vertical tail plane*, also called *vertical fin*. The horizontal tail planes and the vertical fin (occasionally there are two of them) are lifting surfaces, rather like small wings with aerofoil sections, fixed to the aircraft near its tail end. In rare aircraft configurations, especially at supersonic speeds, separate horizontal tail planes are not provided. In such cases their function is performed by the wings, which usually cover a large part of the length of the aircraft's fuselage at their roots. A familiar example of this configuration among civilian aircraft is found in the Anglo-French Concorde.

Besides having static stability in pitch and yaw, each aircraft must also be stable in attitude about its fore-and-aft axis, called the *roll* axis. Inherent roll stability is provided by proper choice of the wing *dihedral* (angle that the wings make with the horizontal when the aircraft is standing level), wing *sweep* (angle by which each wing is tilted backwards with respect to a plane normal to the aircraft's fore-and-aft axis), vertical position of the wings on the fuselage, and the size and shape of the vertical tail fin.

Certain flying qualities of the aircraft, notably its controllability under wide variations of internal load distributions, may be improved if it is not constrained to be statically stable inherently. Such aircraft are provided with mechanisms for *active stabilisation*, which sense any deviation of the aircraft from its desired attitude, and generate the control forces and moments necessary to counteract the deviation. To ensure the necessary speed of response and to minimise pilot fatigue, the active stabilisation mechanism is incorporated within a comprehensive computer-based control system, popularly known as the fly-by-wire flight control system. Such control systems are being used in most modern high-performance military aircraft, as well as airliners such as the recent models designed and produced by Airbus Industries and Boeing Airplane Company.

The forces and moments acting on a flying aircraft depend on its attitude, velocity vector, and their rates. In turn, these forces and moments also determine the very same quantities, i.e. the angular and linear rates of motion of the aircraft. This feedback relationship between forces and moments, on the one hand, and displacements on the other, results in oscillations of various types when an aircraft is subjected to sudden disturbances from its stable flying conditions. An important type of oscillatory mode of flight, called the *phugoid mode*, occurs in the longitudinal plane of symmetry of the aircraft and involves periodic variation of the aircraft's altitude, pitch attitude and forward velocity. The phugoid mode may be excited by sudden variation in air speed and/or density, besides control inputs. Other dynamic modes of aircraft motion include those involving *weathercock oscillations* about the yaw axis, and *roll oscillations* about the longitudinal (fore-and-aft) axis. A combination of lateral (normally accompanied by roll) and directional oscillations gives rise to *dutch roll*.

Further contributions to oscillatory behaviour come from the flexibility of the different structural, aerodynamic and control elements of the aircraft, including the wings and the fuselage. A necessary condition for normal flight of an aircraft is its *dynamic stability* which ensures that oscillations, once induced, die down on their own, though the damping process may be further aided by programmed control action on the part of the pilot or autopilot.

Two velocity parameters are of great importance in describing the motion of an aircraft. The *ground speed* of the aircraft refers to its speed (or, more generally, its velocity) relative to the ground underneath. This speed parameter is important for determining the aircraft's motion between fixed points on the earth. The ground distance travelled by an aircraft during a given interval equals the time-integral of its ground speed. In contrast, the *air speed* of an aircraft is its speed (or, more generally, its velocity) relative to the ambient air. Air speed is the speed parameter that determines the aerodynamic forces (such as lift and drag) and moments acting on the aircraft.

From the above definitions it should be easy to see that the ground speed of an aircraft would be equal to the sum of its air speed and the velocity of the ambient air in an earth-fixed frame, when all the quantities are considered vectorially. Thus, when an aircraft encounters headwind (wind in a direction opposite to that of the flight) during flight, its air speed would be more than its ground speed, and when flying under tailwind (wind along the flight direction) condition, its air speed would be less than its ground speed. Aircraft pilots routinely use this fact for enhancing the fuel economy of their operations by choosing those altitudes for cruise flight (within permissible limits) where they encounter maximum tailwind. Under such flight conditions, the aircraft can cover more ground distance for a given air speed (and hence drag penalty or fuel expenditure). If there is no tailwind within the permissible altitude band, then the best choice is that altitude where the headwind is the least.

Beneficial use of ambient wind conditions are also made in deciding certain airport operations. The most important among them is the decision regarding the direction of takeoff and landing on particular runways. Of the two possible directions of takeoff and landing on a given runway, the preferred one would be the direction in which the aircraft encounters headwind. In such a case, the aircraft can be airborne or touch down (each of which requires a given air speed on the part of the aircraft) at a relatively low ground speed (compared to the tailwind condition), thus requiring less acceleration for takeoff and less braking effort for stopping as the case may be. As will be shown later (for example, in Sections 3.2.5 and 7.5), detecting and predicting wind shifts along runways is of great economic benefit to aviation, and is an important requirement of modern aviation weather surveillance systems.

As mentioned above, the difference between the air speed and ground speed vectors of an aircraft is the earth-referenced velocity of the air in the vicinity of the aircraft. To a first approximation, an earth-fixed frame may be considered as an inertial frame in a local sense. Thus, the ground speed of an aircraft would be close to its inertial speed, which it would tend to maintain as per Newton's first law of motion. Thus, when an aircraft encounters sudden winds or wind changes during flight, it is the air speed of the aircraft that changes first, causing corresponding changes in the aerodynamic forces and moments (as already shown, the percentage change in forces/moments would be over twice that of velocity perturbations). These sudden changes in external forces and moments act as disturbances or transient inputs which have a jolting or rumbling effect on the aircraft and usually also trigger oscillations. Changes in the ground speed of aircraft following wind disturbances occur more slowly, in response to the altered aerodynamic forces and/or control inputs (either by the human pilot or autopilot).

Since the earth-referenced local velocity of the air is of such crucial importance to flight, direct sensing of this quantity is of great use to aviation. As the later chapters show, accurate remote sensing of this parameter by ground-based sensors, and the use of this information in the detection of aviation-hazardous atmospheric phenomena, are major strengths of modern aviation weather surveillance systems.

2.4 Aircraft navigation systems

While the forces and moments discussed in the preceding section make sustained, stable and controlled flight possible, it is necessary for an aircraft to navigate properly to fulfil its mission and reach its destination, both by itself and in the presence of other aircraft. Aircraft are guided along their path by using position information from a variety of instruments and aids. Some of them are entirely carried on board the aircraft, working independently, and others work in co-operation with facilities located on the ground or in space. Devices helping aircraft to determine their own position during flight function according to two generically different principles: *dead-reckoning* and *position fixing.*

2.4.1 En route navigation: dead-reckoning systems

In dead-reckoning, the current position of the aircraft is found by starting from a position with known co-ordinates, and integrating the incremental displacements along the way up to the present instant. The oldest aircraft dead-reckoning system is the *air-data system* which senses the forward velocity using a Pitot tube and the direction by magnetic compass, and integrates with respect to time the components of velocity along orthogonal directions (typically north and east) on a horizontal plane. The altitude of the aircraft is sensed through barometric altimeters. Numerous corrections are employed to estimate velocities and positions as accurately as possible. However, the air data system basically measures the air speed of the aircraft, and a knowledge of the ambient wind vector is necessary to derive its ground speed. This constitutes a drawback of the system, since current and accurate wind information is often not available all along the flight path. For this and other reasons, air data systems are not considered to be very accurate dead-reckoning devices. They are used as primary dead-reckoning devices by small general-aviation aircraft, and as a supplementary navigation information source in more sophisticated aircraft. A device used for directly sensing the ground speed of aircraft for dead-reckoning purposes is the *Doppler navigator,* which projects multiple directed microwave beams towards the ground and estimates the three components of velocity by processing the Doppler frequency shift of the ground echo signal in each beam due to the motion of the aircraft.

The most commonly used and accurate dead-reckoning device used in aircraft today is the *inertial navigator,* which measures the acceleration components of the aircraft along three orthogonal axes using three separate accelerometers. The orientation of these axes with respect to the reference directions on the earth (e.g. north, east, vertical) is measured (and may be controlled) using gyroscopes. Double integration of the appropriately resolved acceleration components, using the initial conditions at the starting point or known way points, provides continuous updates of the aircraft's position on the earth.

Inertial navigation technology has improved steadily over the years. Modern inertial systems using ring laser gyroscopes and fast digital computers provide position accuracies of the order of one nautical mile (1.8 km) per hour of flight. Inertial navigation systems are in use in all classes of aircraft in civilian and military domains, except small and low-cost general aviation aircraft.

Dead-reckoning devices are generally completely self-contained within the aircraft, and work autonomously. A generic drawback of pure dead-reckoning systems is that their position errors grow with time, and hence navigation becomes less accurate for longer flights. To obviate this difficulty, inertial navigation units of aircraft are usually augmented by continuous or intermittent data from one or more position-fixing devices.

2.4.2 En route navigation: position fixing systems

In position fixing, measurement of the parameters of the signal from one or more ground stations helps establish *lines of position*, and the intersection of these lines provides position estimates of the aircraft. A pair of navigational aids (abbreviated as *navaids*) commonly used for this purpose are the *very-high-frequency omnirange* (VOR) and the *distance measuring equipment* (DME), which respectively measure the direction and distance of the aircraft relative to selected ground stations. The VOR, operating within the frequency band of 108–117.95 MHz, works by measuring on board the aircraft the phase difference between a reference signal, which has a constant phase at all points within the range of the ground transmitter, and a variable signal whose phase depends on the direction of the aircraft with respect to the ground station. The airborne DME receiver estimates distance from a compatible ground station by measuring the time difference between a transmitted signal and the response signal returned by a transponder on the ground. The ground components of VOR and DME systems are often collocated to give a common ground reference for aircraft positioning. The *Tactical Air Navigation System* (TACAN) is a navaid of military origin, operating in the vicinity of 1000 MHz in frequency, and performing the same function as a VOR/DME pair. A station containing the VOR and TACAN systems, called *VORTAC*, can cater to both civilian and military aircraft.

Nondirectional beacons (NDBs) are simple ground-based transmitters with horizontally omnidirectional radiation that provide signals for guidance of aircraft. *Automatic direction finders* (ADFs) receive signals from NDBs via loop antennas and compute the direction of these NDBs relative to the aircraft by using the behaviour of the receiving pattern of the antennas.

Another class of position fixing navigational aids is based on the principle of hyperbolic navigation. When two ground stations radiate signals in a synchronised manner, the path of an airborne receiver which receives the two signals at a constant time difference is a hyperbola. Different values of time difference define a family of hyperbolas. The intersection of two or more such hyperbolas generated by pairs of ground stations can be used to define the aircraft position. Important navigational aids using the hyperbolic principle are the *LORAN-C* and the *OMEGA* systems, the latter providing a global coverage.

A more modern position fixing system providing global coverage is the space-based *global positioning system* (GPS). The system utilises 24 satellites (21 operational and three active spares) placed on six equispaced 12 h circular earth orbits, inclined 55° with respect to the equatorial plane, in such a way that at least four satellites (usually five or more) are 'visible' above the horizon at any time over most parts of the world. The position of each satellite is known precisely and is transmitted as a part of the coded signals from the satellite. A receiver on earth obtains the 'pseudorange' to each of four chosen satellites by noting the time of arrival of each signal on a local clock. A three-dimensional triangulation yields the position of the receiver, and also an estimate of the bias error of the local clock relative to the GPS system clock. The GPS system provides horizontal positioning accuracy of the order of 100 m to general civilian users worldwide, making it an excellent aid for global *en route* navigation. Much better accuracies can be obtained by a differential mode of operation of the GPS, permitting its use in precise relative navigation applications such as landing.

2.4.3 Aircraft landing guidance systems

The radionavigational aids named above provide guidance to aircraft in their *en route* phases of flight. However, a different type of guidance is required for aircraft during their landing phases. Landing guidance systems are characterised by high accuracy requirements and the ability to handle the large number of closely spaced aircraft approaching the landing threshold. Two main types of aircraft landing systems are in use: the conventional *instrument landing system* (ILS) and the more recent *microwave landing system* (MLS).

The ILS incorporates two main subsystems. The *localiser*, the antenna of which is located on the extended centreline of the runway, operates in the very-high-frequency (VHF) band and provides left–right guidance to the aircraft, i.e. information regarding the lateral departure of the aircraft from a vertical plane through the runway centreline. The error signal is obtained as the difference between two beams whose equal-power contours intersect along the runway centreline. The *glideslope* subsystem operates on a similar principle, but employs ultra-high frequencies (UHF) radiated from an antenna assembly located offset from the runway. The glideslope equipment provides vertical or up–down guidance to the aircraft, the error being null along a designated glideslope plane. The glideslope angle is typically set at about 3° above the horizontal at most airports, but may be much higher at certain locations, e.g. the London City Airport, where it is as high as 7.5° (Middleton, 1989, p. 158).

The ILS by itself does not provide information to the aircraft regarding its distance from the landing threshold. This third dimension of spatial information is obtained by the aircraft at two or three discrete points along the glidepath as it passes directly over 75 MHz marker beacons which have fixed, vertically pointing fan beams perpendicular to the runway direction. Additionally, accurate height information over the local terrain may be obtained on board from radio altimeters, which sense height by the return time for a signal bounced from the earth, during both the instrument-guided part of the descent and the visually guided part that usually follows.

The ILS has two major drawbacks. Because it defines only one fixed glidepath (as the intersection of the null planes of the glideslope and localiser equipment), all aircraft, irrespective of their speed, size and descent characteristics, must queue up for landing along this path. This leads to path congestion and reduced traffic handling capacity of runways. Further, the ILS performance is susceptible to the presence of hangars, buildings and hills in the vicinity. Such structures give rise to unwanted reflections that interfere with the antenna beams, producing 'course bends' which are kinks and undulations in the electrically defined glidepaths.

The MLS was conceived to obviate these difficulties. The MLS, using frequencies of the order of 5000 MHz, operates on the time reference scanning beam (TRSB) principle. A vertical fan beam, sweeping back and forth ('nodding') horizontally, generates signal 'pips' in the airborne receiver, the time difference between which is a function of the angular displacement of the aircraft from the runway centreline. A horizontal fan beam, nodding vertically, provides the angular elevation information to the aircraft. The MLS has lower susceptibility to multipath effects due to buildings and hills, permits multi-aircraft approaches, a wide range of descent slope angles, even approaches with curved noise-abatement flight profiles, and provides guidance during overshoots and aborted landings. The MLS is operated in conjunction with a precision DME which indicates the distance of the aircraft from the landing threshold. This combination provides full three-dimensional landing guidance to aircraft.

In spite of its technical advantages, however, the MLS has not proved itself to be very popular. Perhaps its advantages over the ILS do not justify the high cost of replacing the existing ILSs. Another probable reason is the expectation of GPS-based landing techniques providing more flexible, autonomous and cost-effective landing solutions in the foreseeable future.

Landing guidance systems are characterised by their ability to guide aircraft down to certain heights and distances from the runway without the pilot having to see the runway. The different categories of landings defined in this connection are shown in Table 2.3 (Kayton and Fried, 1997, Chapter 13). These categories have significance in relation to atmospheric phenomena, specifically those that affect visibility. The poorer the general condition of visibility at an airport, the more would be the demand on the landing guidance system (i.e. the lower and closer the system would be called upon to guide the aircraft automatically), requiring the landing capability of a higher category.

2.5 Air traffic control and air traffic services

While navigation systems essentially serve to guide individual aircraft along their paths, *air traffic control* (ATC) procedures and facilities are necessary to regulate the movement of aircraft in such a way as to promote safe and

Table 2.3 Categories of aircraft landing

Landing type	Decision height, ft	Visibility, m	Runway visual range,[a] m
Category I	≮200	≮800	≮550
Category II	≮100		≮350
Category IIIA ('see to land')	<100		≮200
Category IIIB ('see to taxi')	<50		≮50
Category IIIC ('zero visibility')[b]			

[a] For a discussion of *runway visual range*, see Section 8.9
[b] Not approved anywhere in the world as of 1996

Notes:
1. Different units are used for height and distances in keeping with normal practice
2. Blank entries mean that the parameter is not specified
3. Besides the parameters specified in this Table, there are additional stipulations, including the type of airport and airborne equipment, for specific landing categories

efficient use of air space. The primary emphasis of the ATC service provided by the appropriate aviation authority is on the avoidance of collision and conflict between aircraft in the air, and collision of aircraft with other aircraft and obstructions in designated areas on the ground. However, the broad goals of ATC include the minimisation of hazards and other impediments to orderly and expeditious air traffic, encompassing those due to adverse weather factors. The generic term *air traffic service* includes the ATC service as well as flight information and alerting services (among others) which, in turn, cover information regarding weather.

ATC rules, procedures, conventions and equipment have been standardised to a large extent through the International Civil Aviation Organisation (ICAO), but significant national, regional and local variations and augmentations do exist. In the meteorological context, an example of international standards is provided by International Civil Aviation Organisation (1995).

In its early days aviation depended entirely on visual clues for navigation to destination, avoidance of collisions and avoidance of bad weather. In modern times visual navigation is resorted to typically by pilots of small low-cost and low-performance aircraft. More sophisticated aircraft use visual navigation as a back-up in case of failure of instruments on board, and nearly all aircraft depend on visual clues during the final phase of landing. Rules that govern the procedures for flight under visual conditions are called *visual flight rules* (VFR). VFR flight operations are permitted only under conditions of good visibility and separation from cloud masses. In contrast, flights aided by on-board instruments that indicate position, altitude, attitude, heading and other navigational parameters can be conducted under a much wider range of environmental conditions. Instrument-based flights are governed by *instrument flight rules* (IFR). For VFR flights the responsibility for avoidance of collision rests with the pilot, whereas the air traffic controllers play a central role in the separation of aircraft under IFR.

For ATC purposes the air space is divided into *controlled* and *uncontrolled* regions depending on the availability (or nonavailability) of air traffic control services. The controlled air space is further divided into *control zones, control areas, terminal areas, airways, upper air space, special rule air space*, etc. which may be separated from one another geographically or by altitude. All flights are not subject to ATC, but IFR flights within controlled air space, and all flights within certain designated air spaces, are required to participate in the ATC system. Numerically speaking, a majority of aircraft may not participate in the ATC process, as is the case in countries such as the US and those in Western Europe where general aviation activity is widespread. However, those that do usually consist of the more important segment of aviation, comprising scheduled flights (airline, commuter, cargo), corporate and military aircraft, and general aviation aircraft operating under IFR. Even when VFR pilots are not controlled by a certain ATC installation, but have the responsibility of visually avoiding conflict with other aircraft, they may choose to utilise some of the services offered by the regulating authority.

There are three broad divisions of ATC. An *aerodrome control unit* provides services mainly to aircraft in the vicinity of the aerodrome traffic zone and on the airport ground manoeuvring area. Among other important functions, this unit provides essential aerodrome information which includes airport weather conditions, the extent and state of snow and ice on runways, taxiways and manoeuvring areas (together with any remedial measures adopted such as sweeping and sanding, and estimate of braking action), standing water on runways, etc. *Approach control* concerns itself with approaching aircraft prior to their transfer to aerodrome control, and departing aircraft moving out of aerodrome control. *Area control* provides air traffic services outside the air spaces coming under the purview of approach control and aerodrome control. *Air route traffic control centres* (ARTCCs) providing this service within delineated stretches of controlled air space are responsible for the control of airways route networks and the larger terminal areas. The basic function of the ARTCC is to ensure separation of aircraft travelling between airports. In this phase of flight, *en route flight advisory service* (EFAS) provides timely weather information through specifically designated service stations.

Although the primary means and responsibility for the avoidance of collision among aircraft rest with the ATC authorities, in recent times aircraft are being fitted with autonomous *Traffic Alert and Collision Avoidance Systems* (TCAS) to further minimise the probability of collisions and conflicts (the European equivalent of the system is called the *Airborne Collision Avoidance System*, or ACAS). The system has two variants. TCAS I is a basic version for small aircraft, and provides a warning to the pilot of the proximity of another aircraft equipped with a TCAS or SSR transponder (see Section 2.6). TCAS II is a more sophisticated equipment intended primarily for airliners, with the built-in ability for determining relative three-dimensional separation from nearby TCAS-equipped aircraft, evaluation of threat, and indication of optimum manoeuvre for its avoidance.

For an aircraft to be provided with ATC services during a particular flight, a *flight plan* is filed with the ATC authorities, typically 30–60 min before the estimated departure time. Indeed, a flight plan must be submitted in the case of all flights within certain designated areas and across international borders, and all IFR flights within controlled air space. In the case of organised operators such as airliners and military aviation authorities, the organisation generally files the flight plan, whereas pilots of personal and corporate aircraft contact a *flight service station* to file the plan and receive weather briefings. In the case of repetitive and predictable flights such as scheduled airline flights, a single repetitive plan may be filed.

The flight plan contains information regarding the aircraft (type, identification, callsign, radar response code), type of flight (scheduled, general aviation, military), type of flight rule (IFR or VFR), flight details (origin, destination, route, heights, estimated time at boundaries of flight information regions, alternate airports), and other information such as the aircraft endurance, number of passengers, type of survival equipment carried, etc. The flight plan enables the provision of air traffic services to the aircraft, and is of crucial importance for the success of *search and rescue services* in the event of an accident.

2.6 Radars in aircraft navigation and air traffic control

Radar plays a central role in aircraft navigation and ATC operations. As we shall see later, radar is also the key sensing device for advanced aviation weather surveillance. Although different types of radars are used for these two functions, because of their generic similarity they offer scope for sharing of functions and hardware, and also hold the potential for conflict and interference. Thus it is necessary to have an appreciation of the overall role of radar in aviation, which this section aims to provide.

Radars are inherently capable of providing precise data regarding the plan position of the aircraft relative to the radar location. With some additional hardware, they can also sense the height of individual aircraft within their zone of operation. Further, the high degree of resolution (ability to 'see' closely spaced aircraft as different entities) provided by the radar makes it an ideal sensor for enforcing aircraft separation in a controlled air space. Finally, a major advantage of the radar as an ATC tool is that radar observations and data are available directly at the air traffic controller's console. In the absence of radar, aircraft position and height data would have to be transmitted by the pilot to the controller over the radio link. Such data are relatively inaccurate and increase controller workload. Because of the inaccuracy of the pilot-reported position data, *procedural air traffic control systems* based on such data have to enforce larger aircraft separation to ensure safety against collision. In contrast, *radar separation* permits aircraft to come closer to one another, thus enhancing the traffic capacities of given volumes of air space. This is an important advantage in crowded terminal areas and busy air corridors.

Radars are used in nearly all phases of aircraft navigation and ATC. The *precision approach radar* (PAR) was designed in the 1940s as an aid to landing in all weather conditions. The landing guidance function was subsequently taken over largely by the ILS (and more recently by the MLS to a limited extent), but because of their mobility PARs are in limited military use in some countries.

Airport surveillance radars (ASRs) are very important elements of the ATC systems and are to be found at most major airports around the world, both civilian and military. The ASR, operating in the S-band (~3 GHz frequency) and with a range of ~180 km, has a predominant role in aircraft position determination and separation during the approach control phase. ASRs have undergone continuous evolution during the decades of their operation. The most sophisticated in the series of radar types is the ASR-9 of the United States, which is a significant improvement over the ASR-7 and ASR-8 radars currently in use in that country and elsewhere. A special feature of the ASR-9 is its dedicated weather channel which enables the air traffic controller to directly obtain accurate quantitative weather data regarding the controlled air space in addition to the air traffic data, and to use them in the ATC process effectively. This aspect will be covered in much more detail in a later chapter (see Section 7.6).

Separation among *en route* aircraft is maintained with the aid of *air route surveillance radars* (ARSRs). Relative to the ASRs these radars have high transmitted power, low pulse repetition frequency, and large and slowly scanning antennas, providing a longer operating range of the order of 450 km. ARSR-2 and ARSR-3 are currently used radars in this category, and ARSR-4 is the most modern system in the series. The operating frequency of ARSRs lies in the L band (~1.3 GHz) because of the requirement of long range detection of aircraft and penetration of extended rain fields.

Airport surface detection equipment (ASDE) are essentially short range radar systems designed to detect and display the locations of aircraft and moving vehicles on the surface of the aerodrome. Because of the high definition required in such a ground mapping application, the preferred frequencies of operation of these devices lie in the J (10–20 GHz), and particularly in the Ku (12–18 GHz), bands. The ASDE is extremely useful in monitoring surface vehicular and aircraft traffic at busy airports, and is essential for safety during weather conditions such as heavy snow and fog which impair visual observation of the runways, taxiways and apron.

Radars used for ATC are of two generic types. *Primary radars* are those that rely on the radiated microwave energy reflected from the skin of aircraft for their detection. In contrast, *secondary radars* work in conjunction with transponders located on aircraft. The transponders receive the radar's transmitted signal and radiate coded signals for reception by the radar receiver. As the power radiated by the transponder is far more than the feeble echo from the aircraft skin, secondary radars can 'see' out to longer ranges with less transmitted power than primary radars. Further, the coded response by the transponders conveys many types of useful information such as aircraft identification and altitude.

Indeed, the development of the *secondary surveillance radar* (SSR), also called the *air traffic control radar beacon system* (ATCRBS), was a major milestone in the evolution of ATC technology. The system uses two ground-based antennas, one fixed to the top of the rotating primary radar antenna, and the other placed on the ground next to the rotating antenna. The rotating SSR antenna has a horizontal orientation and a vertical fan beam. The ground-fixed antenna is vertical, with an omnidirectional pattern in a horizontal plane, and is used for suppression of sidelobe effects, which are a major nuisance in ATC radars.

The rotating antenna transmits pairs of pulses, the spacing between which sequentially takes one of six values between 3 and 25 µs. Each of these six 'modes' corresponds to a different application. The transmission from the SSR ground antenna is at a carrier frequency of 1030 MHz. Transponders on board aircraft are set to reply to one of the transmitted modes at a carrier frequency of 1090 MHz. The response pulse train contains an identification code selected by the pilot, and may transmit a code signifying emergency if needed. In response to an interrogation in Mode C (pulse pair spacing 21 µs) an aircraft fitted with an altitude-encoding transponder transmits an additional frame of pulses containing information regarding the aircraft's altitude. Because of a common transmission frequency among all SSRs the beacons often get triggered by more than one ground transmitter, and all the resulting responses are received by each SSR ground receiver. This phenomenon gives rise to random dots or 'fruits' on the air traffic controller's *plan position indicator* (PPI) display, causing interference in his ability to identify and separate aircraft. A *defruiter* circuit effectively removes the fruit from the display.

A more advanced form of the SSR is the *monopulse SSR*. The ground antenna of this system has sum, difference and omnidirectional patterns, and transponder replies are received in all these three patterns. The monopulse SSR provides its antenna boresight position at every instant like other radars; in addition, it measures the offset of individual aircraft from the boresight based on the ratio of the difference-to-sum signal amplitudes received. These two quantities together provide a much more accurate absolute measurement of the angular position of the aircraft (with respect to true north) than the nonmonopulse SSRs. The omnidirectional pattern is used for sidelobe suppression. Another advantage of the monopulse SSR is that it requires a much smaller number of interrogations to produce a reply. This greatly reduces the probability of interference with transponders of other aircraft in the vicinity, and makes it possible to selectively address individual aircraft (see discussion on Mode S system in the following section).

The main advantages of the SSR are the enhanced detection range, reliable echoes and the availability of additional information in the coded response signal. The last-mentioned advantage results in highly informative and organised displays which show alpha-numeric codes of each aircraft's identity and parameters beside its PPI representation. In modern and futuristic ATC systems SSR response data can be processed directly by computers to provide more advanced displays and automated ATC services.

Considerable progress has been made towards incorporating automation into ATC services based on radar data. The *radar data processing* (RDP) system is used in the ARTCCs. It provides for data input from multiple radar sites, mosaic capability, computer validation of data and selection of the best data for display, automatic aircraft tracking, display of flight information, and automatic radar hand-off capabilities. The system works in conjunction with an automated *flight data processing* (FDP) system, which processes the flight plan data. In the terminal areas, the system used is the *automated radar terminal system* (ARTS) which uses data from the standard ATC radar systems such as those of the ASR-*x* series, and automatically performs radar data acquisition (both primary and secondary), processing, display and recording functions. The ARTS output is overlaid on the controller's PPI such that in the event of malfunction of the ARTS computer, the basic primary and secondary radar position data are still available on the display.

There is a distinct drive towards further automation of air traffic services and, as we shall see later, the weather surveillance systems evolving in the context of aviation are designed for a high level of automated operation and data generation.

Besides the ground-based radars mentioned above, aircraft navigation and safety are also aided by a number of airborne radars. The role of the airborne Doppler navigation radar has been cited before in Section 2.4.1. Another radar of great help in aviation is a forward-looking radar mounted in the nose of most large airliners and civil and military transport aircraft which detects and maps any adverse weather ahead of the aircraft. More will be said about this radar in a later section. Other high-performance military aircraft rely on on-board radars in a variety of ways for navigation and mission fulfilment, but this aspect is beyond the scope of our discussion here.

2.7 Aeronautical communication systems

Another important segment of aircraft electronic systems consists of aeronautical communication systems. Communication systems carry information between ground stations and aircraft, and among aircraft themselves. Thus they constitute a very important element in the navigation, ATC and safety aspects of aviation. Among other types of information, aeronautical communication systems carry weather information and data.

The most pervasive mode of transmission and reception of ATC information involves voice communication over radio links. ATC systems normally employ *simplex communication* in which a single frequency is used for both transmission and reception. The frequency may be in the high-frequency (HF), very-high-frequency (VHF) or ultra-high-frequency (UHF) band. Signals in the HF band are not limited to the line of sight, and hence permit long-range communication. These are useful for ARTCCs controlling transoceanic flights. VHF is the most commonly used frequency band for ATC communication with civilian aircraft. Military aircraft use both VHF and UHF bands. Each air traffic controller is normally provided with one or more channels of radio communication, and also wire and/or radio telephone links with other controllers in the vicinity. The management of these channels to minimise interference and confusion is achieved through a *voice switching system*. An internationally agreed standard for pronunciation of letters and numbers, as well as formats for reporting various types of information and data, has been evolved to maximise communication efficiency between speakers with different accents and dialects in the presence of communication and background noise, interference and disturbances.

Although voice communication is indispensable for ATC services, modern aviation operations, with growing emphasis on automation, are increasingly relying on nonvoice modes of communication for transfer of data, information and warnings. Such communication is normally carried out via transponders. Transponders were briefly mentioned in the last section in connection with the SSR. The most advanced of the SSR transponders is the *Mode S* transponder. Mode S refers to a *selective mode* of communication between the aircraft and the ground installation. The selective mode is contrasted from the open *broadcast mode* of communication in the sense that aircraft equipped with the Mode S transponder can be addressed individually, whereas in a broadcast mode all aircraft with compatible equipment receive and respond to common transmitted messages and/or queries. Selective addressing helps the interrogator to confine its interrogation and communication only to those aircraft for which it has ATC responsibility. This minimises the possibility of ground system saturation and provides control over the temporal spacing of aircraft responses to prevent overlap.

The Mode S system, which may be described as a co-operative surveillance and communication system for air traffic control, consists of a sensor and a transponder. The sensor is a monopulse SSR which provides accurate range and azimuth location of aircraft, and a higher level of resolution among aircraft. Each aircraft equipped with a Mode S transponder is assigned a unique discrete coded address. The transponder accepts an appropriately coded surveillance interrogation and provides a reply protocol which has built-in capability for data transmission between aircraft and ground. At a 4 Mbit/s interrogation data rate and 1 Mbit/s response data rate, the data transmission capacity of the Mode S link is quite high. A description of the functioning and signal format of the Mode S system is given by Orlando and Druilhet (1986).

The Mode S transponder has many operating modes, and is compatible with other ATCRBS interrogation modes. The Mode S transponder may be linked to an on-board computer to transmit both raw and processed data covering a large number of parameters from the aircraft to the ground computer directly. Such a facility, coupled with high data rates, makes the Mode S system a far more potent data transfer link than the earlier transponders. Because of these advantages the Mode S link has a distinct role to play in the exchange of modern weather surveillance system data between the aircraft and the ATC installations.

In recent years satellite communications has been playing an increasing role in maintaining contact with aircraft in flight. Satellites can cater to the voice communication needs of aircraft passengers as well as cockpit communication needs. Satellites are particularly advantageous in the aircraft communication context because of their fairly uniform global coverage, which is of special value for navigation over oceanic routes. Satellites such as those operated by the London-based International Maritime Satellite Organization (INMARSAT) provide specialised aeronautical communication services.

The trend towards greater reliance on digital data communication is prompted in good measure by the desire to reduce the workload of pilots and air traffic controllers. Voice communication is estimated to account for 31 to 43% of the controller's overall workload (Sokkappa, 1987). As more activities such as detailed weather information dissemination are added to the controller function, this workload is bound to increase further. This problem may be alleviated to a significant extent by establishing an integrated air–ground communication system for exchange of digital data. In such a system, the data exchange over land masses can occur through the Mode S link, while oceanic operations can utilise satellite communication. Extensive satellite data link support over developed land masses such as those of Europe or the continental USA is more difficult because of the high traffic density. The use of dedicated satellite systems for the purpose offers one solution to this problem.

2.8 Summary

Some fundamental aspects of aviation that have a bearing on the understanding of weather effects on aircraft in flight have been presented in this chapter. The chapter also provides an overview of the broader system framework in which modern aviation weather surveillance systems must operate. Aircraft flight has two distinct phases, one involving terminal area operations and the other involving *en route* operations, each with its own types of flight environment and weather vulnerabilities. In general, flight operations over and within terminal areas involve a higher level of weather susceptibility than *en route* operations. Stable flight of aircraft in the atmosphere is achieved through a rather delicate balance of forces and moments acting on the aircraft, most of which are aerodynamically generated. Rapid changes in atmospheric parameters along the aircraft flight path caused by weather phenomena disturb the stable flight conditions, resulting in oscillatory motion which may be catastrophic.

Modern aircraft are guided along their course and complete their missions successfully with the help of elaborate navigation systems. Some navigational aids are contained within the aircraft and operate autonomously, while others operate in conjunction with ground-based facilities (more recently, also space-based facilities), and may even require operator intervention routinely. Modern aircraft navigation systems have the capability of guiding aircraft over the most part of their paths under all weather conditions and ambient light levels.

Elaborate air traffic control systems and air traffic services have been set up to ensure orderly and efficient operation of aviation systems. International co-ordination among the various agencies providing such services ensures a high degree of procedural uniformity and hardware commonality among the services globally. The main emphasis of air traffic control is to ensure adequate separation between aircraft to prevent collisions and conflicts. The provision of appropriate weather information and warning to pilots and controllers is an important part of air traffic services.

Radars form an important class of sensors in aircraft navigation, air traffic control and airport operations. As will be seen later, specially designed radars are also the primary tools in modern aviation weather surveillance. Data from such sensors would have to be communicated between ground facilities and aircraft in near-real-time, and nonverbal forms of communication, especially data communication, are essential for the efficient utilisation of modern weather sensor data. The chapter includes a discussion of the aeronautical communication systems which may be involved in the transmission of weather information.

2.9 References

ANDERSON, J.D. (1989): 'Introduction to flight' (McGraw-Hill International Editions, Singapore, 3rd edn.)
ETKIN, B. (1982): 'Dynamics of flight: stability and control' (Wiley, New York)
FISCHETTI, M.A. (1986): 'New times at the FAA', *IEEE Spectrum*, **23**, (11), pp. 67–69
FISCHETTI, M.A., and PERRY, T.S. (1986): 'Our burdened skies', *IEEE Spectrum*, **23**, (11), pp. 36–37
INTERNATIONAL CIVIL AVIATION ORGANISATION (1995): 'International standards and recommended practices, Meteorological services for international air navigation, Annexure 3 to the Convention on International Civil Aviation', 12th edn., July 1995
KAYTON, M., and FRIED, W.R. (1997): 'Avionics navigation systems' (Wiley, New York, 2nd edn.)
MIDDLETON, D.H. (Ed.) (1989): 'Avionic systems' (Longman Scientific and Technical, London)

NATIONAL TRANSPORTATION SAFETY BOARD (1993): 'Annual review of aircraft accident data, US carrier operations, Calendar year 1990'. Report no. NTSB/ARC-93/02

NELSON, R.C. (1989): 'Flight stability and automatic control' (McGraw-Hill, New York)

ORLANDO, V.A., and DROUILHET, P.R. (1986): 'Mode S beacon system: functional description', Lincoln Laboratory Report

RUDICH, R.D. (1986): 'Weather-involved U.S. air carrier accidents 1962–1984: A compendium and brief summary'. Proceedings of AIAA 24th Aerospace Sciences Meeting, Reno, NV, Paper AIAA-86-0327

SOKKAPPA, B.G. (1987): 'An automated en route ATC towards increased user benefits'. Air Traffic Controller Association Proceedings, 1987, pp. 203–209

Chapter 3
Atmospheric effects on aviation

3.1 Weather as a factor in aviation

Aircraft fly within the atmosphere and are wholly dependent on it for the generation of the aerodymamic forces that sustain and regulate flight. Aeroplanes are also propelled by air-breathing engines that ingest air from the atmosphere to support combustion and generate thrust. Further, all navigational and communication signals, including visual, must penetrate a layer of the atmosphere before reaching the aircraft. The same is true for signals from aircraft to ground-based facilities and controllers. The aviation process is therefore strongly affected by the state of the atmosphere.

Atmospheric processes are very diverse in terms of their origin, physical nature, spatial and temporal scales, and intensity. However, from the point of view of effects on aviation, they may be classified into five different groups: (i) phenomena involving physical motion of air; (ii) hydrometeorological phenomena; (iii) phenomena inducing and facilitating ice formation on aircraft surfaces; (iv) phenomena causing low visibility; and (v) phenomena involving atmospheric electricity. The following sections discuss different facets of the complex interaction between weather and aviation, the characteristics of different types of atmospheric phenomena, and their effects on aviation.

3.2 Overall effects of weather on aviation

The atmospheric phenomena of different types listed in the preceding section affect aircraft in fundamentally different ways, but their overall effect on aviation is a degradation in safety, efficiency, and economy of operation. These effects are discussed in the following subsections, before we move on to throw light on the individual types of aviation-significant atmospheric phenomena.

3.2.1 Safety

In considering the effect of weather phenomena on aviation, the safety aspect comes to mind first. This is because the first priority of any civilian aviation system is to ensure passenger safety. Another reason, as mentioned before, is the shocking effect of air disasters on the popular psyche. However, the fear of hazardous weather in the aviation safety context is not merely psychological. Numerous systematic studies have shown weather to be the main cause or at least a strong factor in a large number of aircraft accidents.

In earlier years, when flight was dependent to a greater extent on visual clues, and accident investigation methods and tools were less developed, many fatal accidents due to a variety of factors including weather were often attributed to pilot error. With more scientific methods of accident investigation, the true extent of the effect of weather on aviation safety began to be realised. In recent times, weather has been established as a cause or factor in a large fraction of accidents across the entire spectrum of aviation activity.

The Civil Aviation Authority (CAA) of the UK determined weather as a causal factor in 30 out of a total of 206 fixed-wing aircraft accidents in 1977 (Civil Aviation Authority, 1978). Included in this total were 16 accidents involving public transport aircraft, of which five (i.e. >30%) were caused by adverse weather factors. In addition, CAA listed 27 more accidents as having been caused by pilot errors under unfavourable weather and wind conditions.

Accident data from the USA also show weather as a predominant cause of aviation accidents. A survey by Rudich (1986) listed as many as 11 fatal weather-related accidents involving US air carriers alone between 1962 and 1984. The accident data for the period 1985–1990, discussed in some detail here, provides a good idea of the role of weather in aviation. For example, according to National Transportation Safety Board (1993a) data, during the year 1990, in the 14 CFR 135 category (commercial air carriers commonly referred to as commuter airlines and air taxis) under 'unscheduled' operation, a total of 108 aircraft were involved in accidents, of which 28 were fatal. Adverse weather elements were determined as a cause or factor in 30 of the 108 accidents, amounting to 27.8%. The percentage of weather-induced accidents was much higher among the fatal accidents, being 16 out of 28, or 57.1%. These statistics and those for other segments of aviation activity during the period from 1985 to 1990 are summarised in Table 3.1. The Table shows that in certain years and certain categories of aviation, as much as 75% or even 100% of the accidents may be related to adverse weather.

3.2.2 Comfort

Although fatal air crashes, occurring periodically, are the most visible manifestation of the effect of adverse weather on aviation, weather factors influence aviation in many more subtle but profound ways, and on a more continuous basis. An important aspect of weather–aviation interaction relates to passenger comfort.

Weather-induced rough flights, capable of causing disorders in the cabin, serious passenger discomfort and even injuries to passengers, are a matter of common experience. Rough flights are usually caused by atmospheric turbulence, either in association with clouds or rain activity, or under clear

Table 3.1 *Percentage of accidents in which weather is a broad cause or factor in US aviation accidents during 1985–1990*

Period ⇒ Type of aviation operation[a] ⇓	1990		1985–1989 (cumulative)	
	All accidents	Fatal accidents	All accidents	Fatal accidents
14 CFR 121, 125, 127[b]	28	50	29.8	14.3
14 CFR 135 (scheduled)[b]	43.8	75	29	46.2[d]
14 CFR 135 (nonscheduled)[b]	27.8	57.1	33.8	45.3
General aviation[c]	21.7	25.9	24.9	34.2

[a] For explanation on the type of operation, see Table 2.1
[b] Source: National Transportation Safety Board (1993a)
[c] Source: National Transportation Safety Board (1993b)
[d] Among these, weather was a cause or factor in 100% of the fatal accidents during the year 1989 (National Transportation Safety Board, 1993c)

weather conditions. Passenger discomfort may also result from sudden loss of height by airplanes encountering 'air pockets' or masses of air moving with speeds significantly different from that of the ambient air. When unpredictable wind variations occur near the ground while the airplane is landing, a hard landing may result, if not an accident, with attendant passenger discomfort.

3.2.3 Schedule-keeping

One nonfatal but highly important area in which weather factors adversely affect aviation, especially the operation of scheduled airliners, is the punctuality of flights. Inclement weather adversely affects the operational readiness of runways and other airport facilities, as well as that of aircraft themselves, resulting in flight delays. Further, since flight schedules are serially arranged using common aircraft and routes, delays usually propagate even to areas unaffected by weather until they are absorbed by the slack in the system. Since the emphasis in modern aviation systems is on efficiency, which necessarily means low slack or idle capacity, the cascading effect due to delays can be quite severe.

An early and thorough study utilising data collected by the US Federal Aviation Administration (FAA) from a large number of flights showed a strong effect of weather in disrupting flight schedules (Bromley, 1977). For example, during the year 1975, of a total of 31 672 flight delays ≥ 30 min, as many as 27 047, amounting to >85%, were weather-related. Similar statistics were obtained for the five previous years, with weather-caused delays varying between 65% and 90%, and the absolute numbers of these delays exceeding 30 000 in each of the years. In the less developed parts of the world, where surveillance, communication and other airport facilities are often inadequate, weather is known to play havoc with airline flight schedules.

3.2.4 Efficiency

Aviation is a highly capital-intensive enterprise. Hence there is utmost emphasis on the efficient utilisation of resources. In the aviation context, resources are deployed in two main segments: the aircraft themselves and the ground facilities that support their operation. Weather impacts adversely on the efficient utilisation of resources in both these segments.

The main indicators of operating efficiency of aviation systems are the traffic handling capabilities of airports and air corridors, and the average fraction of time that individual aircraft remain in operation. Weather phenomena reduce the efficiency of aviation operations by adversely affecting the traffic handling capacities of airports. Factors such as heavy rain, fog, strong crosswinds or wind gradients, heavy snow and runway icing render entire airports or individual runways nonoperational for significant lengths of time, reducing the total number of takeoffs and landings. In addition, adverse weather conditions may necessitate a higher degree of separation between aircraft in flight, reducing the traffic rate along takeoff and landing corridors. Further, adverse weather may keep aircraft grounded for a larger fraction of time, resulting in lower utilisation factors of both airports and aircraft.

3.2.5 Economy

All the deleterious effects of weather ultimately affect the operating economy of aviation-based enterprises, especially that of commercial air carriers. The most visible item of financial loss in an air disaster is due to the loss of aircraft and the subsequent cost of replacement which may run into hundreds of millions of dollars in single instances in the case of modern airliners (as also transport aircraft and high-end military aircraft). Other important items of cost associated with airline accidents arise from possible damage to ground facilities and other public and private property, increased cost of insurance, accident investigation, litigation, and compensation to crew and passengers. In addition to these direct and/or closely connected costs, accidents also involve consequential or social costs which may be many times higher. Loss of life in accidents results in social and commercial dislocation, and loss of important cargo can have severe consequential implications. This effect is even more severe in relatively poor economies where the privilege of air travel and transportation is reserved essentially for the most important people and cargo.

Passenger discomfort and poor schedule-keeping also result in high levels of direct and consequential costs. Bumpy flights lead to the erosion of passenger confidence and preference for specific airlines or air routes, or for air travel in general. In extreme cases, bumpy flights can cause injury to passengers and damage to their belongings even if the aircraft as a whole may not be damaged.

The economic implications of poor schedule-keeping are even more severe for airlines. Significant delays in takeoffs and landings lead to revenue loss due to the reduced number of operations. They also result in low aircraft utilisation factors, necessitating larger capital and recurring expenditures for maintaining a given level of airline operation. Further, unscheduled delays on the part of passengers lead to individual schedule disruption and loss of productivity. In the modern hub-and-spoke model of airline operations, where given pairs of points are often connected through multiple flights, delays in one leg usually get compounded by more missed flights, which further amplifies consequent losses. One estimate (Fischetti and Perry, 1986) put the cost to the US business community due to loss of productivity caused by delayed flights at 1.5 billion dollars in 1985 alone. As mentioned before, a majority of flight delays are usually caused by weather.

The costs incurred due to loss of aviation system efficiency are equally serious. The most significant ways in which weather affects aviation efficiency are related to aircraft route modification to avoid hazardous weather zones and partial or total shutdown of airports for landing and takeoff operations. Diversion of flight paths to skirt hazardous weather entails consumption of additional fuel and loss of time. Airport or runway closure often necessitates in-flight holding of aircraft for considerable periods, again leading to wastage of fuel and time. The loss is much higher when the closure of airports or runways forces flights to be diverted to different airports altogether.

A 1979 US study in this connection is highly instructive (PROFS, 1979). It showed that the economic value of being able to forecast weather over a 12 h period at just four airports, those at New York, Los Angeles, Chicago and Denver alone, would amount to 25 million dollars per year. Another estimate in 1984 by the Transportation Systems Research Center in Boston indicated that the availability of real-time wind-shift information in the control tower at Stapleton Airport in Denver resulted in fuel savings of approximately US$375 000 during a 45-day period (Serafin, 1986).

Indeed, the most significant bottleneck in improving the traffic handling capacity of air traffic control systems is due to the reduction in capacity of airports occurring under instrument flight rule (IFR) conditions. It has been estimated that benefits of >3.5 billion dollars can be obtained through a modest increase of 16% in the IFR terminal capacity in the US (Evans, 1991). A large part of this gain is through savings in petroleum products, which is a nonrenewable resource. A major contribution to these gains can come from proper weather information made available in a timely manner.

3.2.6 Combination of factors

Ideally, the goal of systems designed for the alleviation of hazardous weather effects should be to minimise the influence of weather on all the aspects of aviation mentioned in the foregoing subsections. However, the simultaneous achievement of the goals of safety, comfort, schedule-keeping, efficiency and economy is complicated by the existence of conflicts between some of these goals.

An example will serve to illustrate such conflicts. It is well known that weather-related air disasters could be greatly minimised by the simple rule of avoiding flight operations through regions of adverse weather (Kessler, 1988; 1990). Indeed this is a classical safety instruction given to most pilots, the prudent use of which has promoted safety against weather hazards in the past, and has relevance even today. However, in a dense, dynamic and interrelated aviation scenario, routine avoidance of weather zones with hazard potential, allowing for the necessary safety margins, would result in the disruption of a large number of flights, leading to poor schedule-keeping, efficiency and economy.

As another example, overall airport downtimes may be reduced by providing greater numbers of widely spaced runways in airports, so that while some runways are inoperative due to bad weather, others with adequate traffic handling capacity may remain operational. However, such an approach would greatly increase both capital and recurring costs associated with airports, which will ultimately reflect in the operating cost of airlines and other users of these facilities.

It is clear that these conflicting goals in aviation cannot be satisfied optimally from the weather point of view with the classical surveillance instrumentation, information channels and procedures. This realisation has provided the motive force for the development of a whole array of sensors and systems to minimise the adverse impact of weather on aviation.

The general philosophy and guiding principle of these modern aviation weather surveillance systems are predicated on the assumption that the best approach to meeting the multiple goals of aviation betterment is through a significant improvement in the quality, quantity, timeliness and reliability of the weather observations available to the flight management system. Further, forecasts based on such observations, valid over timescales consistent with the decision times in aviation operations (e.g. flight diversions or detours, runway closures, changing the direction of takeoff/landing, etc.) would also greatly streamline aviation operations in general, leading to better efficiency and economy. Such an 'observation-intensive approach to local weather forecasting, with timely use of current data, in which remote sensing plays a dominant role' is termed *nowcasting* (Browning, 1982). The modern developments in aviation weather technology are guided essentially by a nowcasting approach, dependent on sophisticated electronic systems for remote sensing, processing, display and communication of weather data.

3.3 Atmospheric phenomena involving air motion

It is clear from the discussions of the preceding chapter (see Section 2.3) that the most important determinant of the forces and moments acting on an aircraft in flight is the vector velocity of the surrounding air mass relative to the aircraft. This means that both the magnitude and the direction of this relative velocity with respect to the aircraft body axes are strong factors influencing the forces and the moments on the aircraft. Thus, when an aircraft comes across and flies through air masses whose motion has been disturbed by atmospheric processes, the aircraft encounters different wind velocities along its path, and hence experiences variable and unpredictable forces and moments. These forces and moments can disturb the aircraft's static equilibrium and also excite its dynamic or oscillatory modes. Depending on their nature, duration and intensity, the disturbing forces and moments can, in extreme cases, result in loss of control of the aircraft on the part of the pilot, and induce significant departures from the desired trajectory, potentially leading to accidents. In less severe cases, the pilot may retain or recover control, but the temporary trajectory aberrations can cause passenger discomfort, injuries, object dislocations, panic and incidents (i.e. near-accidents).

3.3.1 Wind shear

The best known, and now highly feared, type of air motion associated with many atmospheric processes is *wind shear* (Kessler, 1985; Mahapatra *et al.*, 1982). Aircraft accident statistics consistently show numerous entries against wind shear and other wind-related factors. UK accident analysis for 1977 (Civil Aviation Authority, 1978) shows as many as eight accidents (two of which involved commercial aircraft) as being due to 'unfavourable wind conditions'. US data list many accidents as being due to wind shear and wind-related phenomena such as 'downdraft', 'high wind', 'mountain waves', 'crosswind', 'tail wind', 'unfavorable wind', 'gusts', etc. In 1990, these factors together appeared as causes or factors in 16 accidents involving air carriers of various kinds (National Transportation Safety Board, 1993a) and 345 involving general aviation aircraft (National Transportation Safety Board, 1993b).

A general definition of wind shear refers to spatial as well as temporal rates of variation of wind speed and/or direction. Thus, a certain amount of wind shear exists at all points in the atmosphere at nearly all times. However, wind shear becomes important from the aviation point of view only when it is of certain types and strengths.

The destructive role of wind shear in aviation has been recognised as early as the 1950s (Stewart, 1958; Viemeister, 1961), though the term 'wind shear' was not in use in this context at that time. Through subsequent years, as the phenomenon was better understood and implicated in a growing number of air mishaps, its potential for aviation disaster and disruption became clearly evident. The concern of the scientific community at large regarding the wind shear problem in the aviation context is reflected in a specific study conducted by the US National Academy of Sciences on this topic (National Academy of Sciences, 1983).

44 Aviation weather surveillance systems

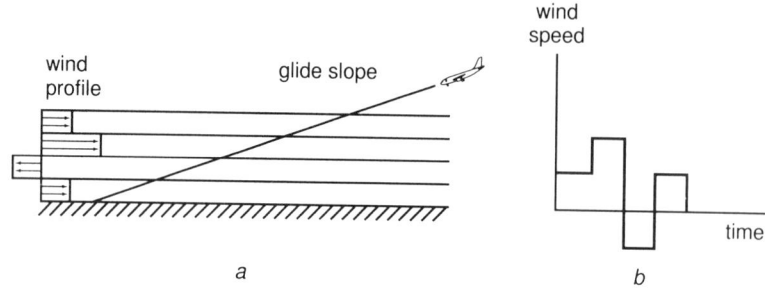

Figure 3.1 A steady but layered wind field (a), and the time-varying wind experienced by an aircraft descending through the layers (b)

Wind shear may be expressed quantitatively in terms of the temporal and spatial derivatives of the wind velocity vector. In studying the effect of wind shear on aircraft, complications are introduced by the spatial motion of the aircraft, which has the effect of coupling the spatial and temporal variations of the wind velocity vector as experienced by the aircraft. As an example, consider an atmosphere with different but constant horizontal wind velocities at each point, as shown in Fig. 3.1a. To all static observers, the wind field would appear steady (as opposed to being gusty or time-varying). However, an aircraft moving through these points would experience a time-varying or unsteady wind (Fig. 3.1b), and such wind variations would affect its flight path and induce oscillations. A practical example of such a situation will be shown later in Fig. 4.15. Since the dynamical equations of aircraft motion are well known (e.g. Etkin, 1982), using appropriate mathematical models of wind shear (e.g. Frost and Camp, 1977; Swolinsky, 1986) it is possible to determine quantitatively the effects of wind fields on aircraft motion (e.g. Brockhaus, 1986; White, 1992).

It is useful to express wind shear in terms of its components. One may define the partial derivative of each component of wind speed with respect to each spatial co-ordinate as a type of wind shear. In three-dimensional space, there would be nine such spatial derivatives. It is convenient to set up a Cartesian co-ordinate system with its x- and y-axes horizontal, oriented along chosen directions (often along and perpendicular to the aircraft heading), and the z-axis vertical. The wind components along the x, y and z directions are designated W_x, W_y and W_z, respectively. Then the wind shear types or components are given by the derivatives:

$$S_{xx} = \frac{\partial W_x}{\partial x} \quad (3.1a)$$

$$S_{yx} = \frac{\partial W_y}{\partial x} \quad (3.1b)$$

$$S_{zx} = \frac{\partial W_z}{\partial x} \quad (3.1c)$$

$$S_{xy} = \frac{\partial W_x}{\partial y} \quad (3.1d)$$

$$S_{yy} = \frac{\partial W_y}{\partial y} \quad (3.1e)$$

$$S_{zy} = \frac{\partial W_z}{\partial y} \quad (3.1f)$$

$$S_{xz} = \frac{\partial W_x}{\partial z} \quad (3.1g)$$

$$S_{yz} = \frac{\partial W_y}{\partial z} \quad (3.1h)$$

$$S_{zz} = \frac{\partial W_z}{\partial z} \quad (3.1i)$$

Each of these shear types or components may be designated by a name. For example, if the x-axis is along the aircraft heading, then S_{xx} is the forward (along-track) shear of forward wind, S_{yx} is the forward shear of lateral wind, S_{xz} is the vertical shear of forward wind, etc. It is often necessary to designate the absolute wind shear at a particular location or over a distance without reference to any aircraft flight. In such a case, terms such as 'horizontal shear of horizontal wind' (or simply 'horizontal shear') and 'vertical shear of horizontal wind' ('vertical shear') are used.

All the types of wind shear listed in eqns. 3.1 do not have the same effect on aircraft. Some of them are far more important than others. Since fixed-wing aircraft, especially those used in civil aviation, fly essentially horizontally (their climb and descent angles do not usually exceed about 6° even during takeoff and landing), the derivatives along the forward direction, given by the first three of eqns. 3.1, are of greater significance for flight. However, since the vertical component of speed is significant during takeoff and landing, the wind shear in the vertical direction is also of importance during these flight phases. The effects of some of the important types of wind shear are discussed in the paragraphs below.

The most serious effect on flight can come from horizontal wind shear in the direction of flight (e.g. Shen *et al.*, 1996) since such shear causes rapid variation of the aircraft's air speed (i.e. speed relative to surrounding air) and,

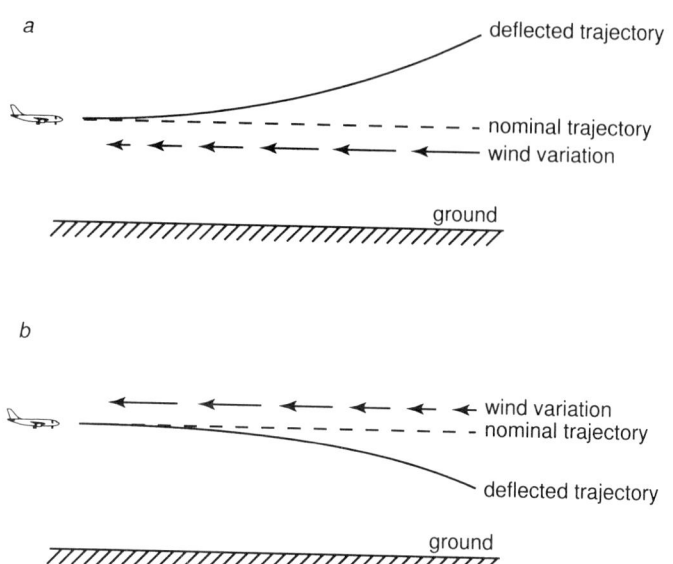

Figure 3.2 *The effect of horizontal wind shear on aircraft flight path. The nominal flight path, shown here to be straight and level, is deflected upward due to increasing headwind (a), and downward due to decreasing headwind (b), along the flight path*

as discussed in Section 2.3, the main forces and moments governing or affecting flight vary nearly as the square of the air speed. The effect of horizontal wind shear on a nominally straight and level flight is depicted graphically in Fig. 3.2. In Fig. 3.2a, the wind is opposite to the direction of flight (i.e. there is *headwind*), and its speed increases along the flight path. Here, if the inertial speed and the angle of attack of the aircraft remain constant, the lift force would increase with time, resulting in an upward deflection of the trajectory. It is, of course, possible to adaptively correct for such deviations by varying the aircraft's angle of attack and/or speed through pilot or autopilot action, but the time lag between the wind disturbance and the corrective measure could still leave a significant uncorrected trajectory deviation.

The trajectory aberration is of an opposite nature if the headwind diminishes along the flight path. In such a case the lift decreases as the flight proceeds, resulting in the drooping trajectory shown in Fig. 3.2b. Similar behaviour is observed if the aircraft experiences a *tailwind* that becomes stronger along the flight path. Finally, a decreasing tailwind, as in the case of the increasing headwind shown in Fig. 3.2a, will cause an upwardly deviated trajectory.

The effect of vertical wind shear, i.e. the vertical variation of horizontal wind, is shown in Fig. 3.3. A rapid change of wind direction from headwind to tailwind, as the aircraft passes through different layers of air during takeoff

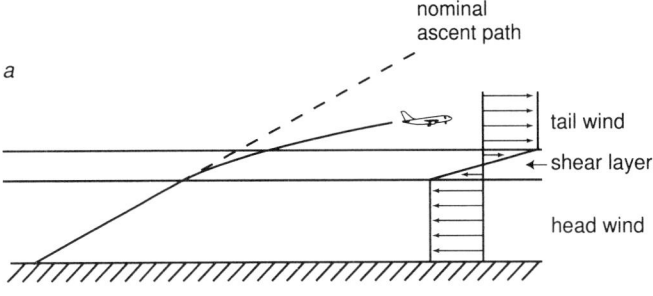

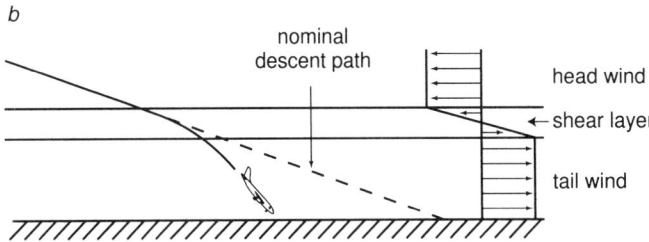

Figure 3.3 Examples of the effect of vertical wind shear on flight. An aircraft taking off from ambient headwind conditions into a layer of tailwind through a shear layer would suffer a trajectory droop (a) and possibly stall, and an aircraft descending from a layer of strong headwind to one of tailwind through a shear layer would experience a steeper-than-expected descent (b) leading to a hard landing or crash

(Fig. 3.3a), results in an equally rapid loss of lift and a corresponding droop in the trajectory relative to the nominal. Similar trajectory perturbations can occur during landing through shear layers. Fig. 3.3b shows a situation in which an aircraft flying in an environment of strong headwind descends through a shear layer into a zone of tailwind existing close to the ground. Here again, the rapid reduction of air speed while flying through the shear layer would result in a loss of lift (or even stall) and consequent steepening of the descent path. The loss of height may be irreversible and catastrophic under certain conditions, e.g. (i) when the wind change is strong and persistent, (ii) when the loss of air speed is strong enough to drive the aircraft to the stall condition, and (iii) if a high level of oscillatory motion is induced by the sudden change in the aircraft's air speed. Tailwind in general was implicated in 68 aircraft accidents in the USA in 1990 (National Transportation Safety Board, 1993a; 1993b), three of these involving commercial air carriers. However, the exact role of the tailwind in these accidents is not mentioned in the statistical data, which probably includes both horizontal and vertical wind shear effects.

Steeper-than-expected encounter with the runway in general causes hard landings. In relatively benign cases, only passenger discomfort results. However, in more severe cases of hard landing, damage to the landing gear and other parts of the aircraft can occur. If the 'undershoot' (shortfall with respect to the landing threshold) due to the trajectory droop is large, the aircraft may touch the ground before the start of the runway. In such cases, damage to the undercarriage is almost certain, with a high probability of more severe damage to the aircraft and passengers. Kayton (1969) mentions that manually controlled and automatic landing systems were typically designed to cope with a vertical wind shear <2 m/s per 100 ft of altitude change, while shear values as high as about 10 m/s per 100 ft altitude were reported by Kramer (1965). However, modern specifications require automatic landing systems to perform satisfactory touchdown with headwinds to 25 knots (~12 m/s), tailwinds to 10 knots (~5 m/s), crosswinds to 15 knots (~7 m/s), moderate turbulence (see Section 3.3.2), and wind shears of 8 knots (~4 m/s) per 100 ft of height from 200 ft to touchdown (Kayton and Fried, 1997, Chapter 13).

Wind shear involving changes in lateral wind is important during touchdown. During this operation, the aircraft is required to maintain its line of flight along the runway centreline, and unexpected strong changes in lateral wind may force the aircraft to veer off this line. Sudden gusts of lateral wind may also excite the oscillatory roll modes of the aircraft, making its control difficult during touchdown. US data show that crosswind was involved as a cause or factor in 115 general aviation accidents (National Transportation Safety Board, 1993b) and three commercial air carrier accidents (National Transportation Safety Board, 1993a) in the year 1990.

The wind shear cases mentioned in this section constitute the simple or basic types of idealised wind shear. Actual wind variations along the path of aircraft in real life are usually much more complex, involving more than one significant component. A type of powerful wind shear field that occurs commonly in nature and is of great hazard value for aviation results from outflows of storms. The nature of such fields, which involve strong horizontal divergence as well as vertical air currents and their gradients, will be discussed in some detail in the next chapter. In particular, dangerous levels of wind shear at low altitudes can arise from microbursts (see Section 4.7), which cause damaging winds over a small patch of ground.

Low-altitude wind shear caused by microbursts and other forms of thunderstorm-induced divergence is said to be of convective origin, and that due to other causes is nonconvective low-level wind shear. Although microbursts are the source of some of the strongest wind shear at low altitudes, a low-level wind shear alert by the control tower personnel does not automatically mean the presence of microburst(s) in the area. However, if a microburst is identified, a microburst alert may be issued irrespective of the actual value of the associated wind shear.

Operationally, low-level wind shear is included in an Aviation Terminal Forecast if there are pilot reports of wind shear that causes air speed gain or loss of 20 knots (~10 m/s) or more within 2000 ft of the surface, vertical shear of 10 knots (~5 m/s) or more per 100 ft within 2000 ft of the surface, or if meteorological conditions for wind shear are satisfied (Jackson, 1991).

The effect of wind shear fields on aircraft performance is quantified through a wind shear hazard index called the *F-factor*. In a simple case of planar wind shear where the wind variation is in a vertical plane and the aircraft also flies in the same plane, the F-factor is given as

$$F = \frac{\dot{W}_x}{g} + \frac{W_z}{V_a} \qquad (3.2)$$

where

$\dot{W}_x$ = time-derivative of the horizontal wind component along the aeroplane ground track
W_z = vertical component of wind, positive downward
V_a = air speed of the aircraft
g = acceleration due to gravity.

In a more general wind shear field the F-factor takes the form

$$F = \frac{\dot{\mathbf{W}} \cdot \mathbf{V}_a}{gV_a} + \frac{W_z}{V} \qquad (3.3)$$

where **W** is the absolute (inertial) wind vector, $\mathbf{V}_a$ is the air speed vector, and $(\mathbf{V}_a/V_a)$ is the unit vector along the air speed direction. It is possible to measure the along-track shear component $\dot{W}_x$ directly with forward-looking sensors mounted on the aircraft itself (see Section 8.5), or the F-factor may be derived from the absolute wind field data obtained from ground-based observations, as discussed in Chapters 6–8.

It is clear from eqns. 3.2 and 3.3 that a positive value of *F* is produced by a performance-decreasing wind shear such as a decreasing headwind or a downdraft or a combination of both. Conversely, a performance-enhancing shear such as an increasing headwind and/or updraft would yield a negative F-factor.

A detailed description of the F-factor may be found in Oseguera *et al.* (1992). Physically, the F-factor refers to the rate at which the wind field changes the energy of the aircraft. The stronger this change the more pronounced would be the effect of the shear field on the aircraft. While moving through a wind shear with a positive F-factor, an aircraft would have to generate an excess thrust (over drag) equal to the F-factor times its own weight to maintain level flight at constant speed. If the shear is so strong that the positive F-factor exceeds the specific (i.e. per unit weight) excess thrust capability of the aircraft, then the aircraft will suffer some loss of air speed and/or altitude in spite of the best possible pilot input. To form an idea of the numbers involved, the maximum specific thrust capability of commercial transport aircraft typically lies in the range from 0.11 to 0.17 depending on the number of engines (Lewis *et al.*, 1994). If the wind shear field produces a negative F-factor, then the aircraft, to maintain its speed and altitude, must cut down its thrust to be lower than the drag by an amount equalling the F-factor times its own weight.

In another, or perhaps equivalent, manner of looking, the F-factor balances with the climb capability of an aircraft. If an aircraft climbing at an angle of θ radians in still air at constant speed encounters a wind shear with a positive F-factor equal to θ then it will cease to climb and fly level if it holds its thrust and speed unchanged. Similarly, if it encounters a wind shear with F-factor equal to $-\theta$ while descending at an angle θ in still air, its path will be restored to level flight if its speed and thrust are held constant.

The F-factor as defined in eqns. 3.2 and 3.3 is a point function and thus can be defined and possibly measured at each point along the flight path of an aircraft. However, since strong wind shear in nature is usually accompanied by significant turbulence (see the end of the following subsection), the instantaneous values of F-factor would contain a corresponding amount of rapidly varying random components. These random variations must be filtered out over an appropriate path length in order for the F-factor to be useful for characterising the wind shear. This aspect will be discussed in some detail in Section 8.5.2 in connection with airborne wind shear detection. Analysis of flight data from the Lockheed L-1011 aircraft, which suffered a fatal accident (National Transportation Safety Board, 1986) due to severe wind shear encounter while attempting to land at the Dallas–Fort Worth airport in Texas, USA, has shown that the F-factor, averaged over a 1 km length, was as high as ~0.30 (Lewis *et al.*, 1994). Comparison of this magnitude of the F-factor with the maximum specific thrust capability of commercial aircraft indicated above would make it appear that the accident was inevitable if the aircraft got into the wind shear without prior warning and evasive action. On the other hand, with sufficient warning, it is possible to escape strong shear fields, as proven by the aircraft that followed the ill-fated L-1011 aircraft in the landing sequence at the Dallas–Fort Worth airport (National Transportation Safety Board, 1986; Fujita, 1986). Visser (1997) discusses optimal guidance strategies for escaping from microburst wind shear encounters by aircraft during final approach. The technique relies on *in situ* (reactive) or short-range forward-look (predictive) wind shear detection (see Section 8.5). More discussion on the methods and benefits of advance warning of wind shear follows in Section 8.5.2.

3.3.2 Turbulence

The phenomenon of turbulence involves random motion of air. This is in contrast to wind shear, which refers to the systematic component of air motion. Clearly, attributes such as randomness and systematism can be defined only with respect to certain scales. The motion of air molecules occurs over a very broad range of scales, spanning from Brownian motion to movement on global scales. The motion also has both temporal and spatial dimensions. Thus, an observer located at a stationary point in the atmosphere will normally experience variation of the wind velocity vector with time. Also, an array of spatially separated observation points will, in general, record different wind velocities at any given point of time.

Of these two types of velocity variation, the spatial variation is the more pertinent one in the context of aircraft flight. It is instructive to view this variation in terms of its spatial frequency spectrum. The passage of an aircraft through a part of the atmosphere harbouring randomly variable velocities induces random forces with a wide frequency spectrum. In relation to the characteristics of the aircraft, the spectrum may be divided into three regions. The high-frequency components are those which have frequencies significantly higher than those of the rigid-body dynamic modes of the aircraft. These components do not have much effect on the motion of the centre of gravity of the aircraft, but can excite structural vibrations (flexible modes) in the aircraft. The mid-frequency range of aerodynamic forces has frequencies comparable to those of the aircraft's rigid dynamics. These forces can induce the aircraft to oscillate as a whole. The results can be instability in attitude and altitude, loss of control, oscillations in the aircraft trajectory, and structural oscillations. Finally, the low-frequency components of aerodynamic force (those with frequencies significantly lower than the aircraft's rigid-body modes) will not induce appreciable oscillations, but will affect the overall lift forces, resulting in trajectory deviations.

In the context of aviation it is reasonable to include under turbulence those scales of random air motion which induce aerodynamic forces of high and medium frequencies. Air motion components limited to low frequencies may be classified as wind shear. Clearly, the dividing line between the two types of air motion is not sharp.

The reason for the fuzzy boundary between the spatial scales of motion defining turbulence and wind shear is partly the imprecise nature of the definition of the three different frequency regions of the wind velocity spectrum. More importantly, the aircraft dynamical frequencies, with respect to which the low-, medium- and high-frequency regions are referenced, themselves vary from aircraft to aircraft. This poses some difficulties for the systems used for sensing and characterising turbulence which are generally required to perform these functions for a wide variety of aircraft. In practice, atmospheric turbulence of concern for aviation systems handling diverse types of aircraft would include spatial scales of wind variation extending from a few metres to several kilometres. This fact has an important bearing on the understanding of the effect of complex wind fields on aircraft in flight, as also on the sensing of turbulence and wind shear using instruments such as the Doppler radar. More details on the characterisation and sensing of turbulence will follow later in this book, particularly in Section 6.6.3.3 and Chapter 11.

The most visible effects of turbulence on aviation are rough flights which are experienced by most frequent air travellers at one time or another. The erratic and rapid vertical and lateral accelerations induced by turbulence may cause dislocation of objects and passengers within the aircraft cabin, resulting in serious passenger injuries (National Transportation Safety Board, 1972a, 1972b, 1972c, 1975) even while the aircraft itself may not be significantly damaged. Besides, the random oscillations forced on the aircraft and its structural members by turbulence can result in high stresses and/or metal fatigue, leading to rupture and structural break-up of the aircraft in flight (National Transportation Safety Board, 1967, 1970). Another major disastrous effect of turbulence on aircraft is through the excitation of strong rigid-body dynamic modes which can lead to difficulties in controlling the aircraft, or even loss of control and consequent accidents (e.g. National Transportation Safety Board, 1974).

Further effects of turbulence manifest themselves through interference with aircraft engine performance. Flight through a highly turbulent atmosphere causes the airflow into jet engines to vary randomly. This causes erratic variation of the thrust developed by the engines, which further accentuates the problems associated with the vibration and control of the aircraft. Overall, turbulence was found to be the single most important cause of air carrier accidents in the US (Lee and Beckwith, 1981).

From considerations of effect on aviation, the level of atmospheric turbulence is often quantified in terms of a *derived gust velocity* U_{de} defined as follows (after Pratt and Walker, 1954):

$$U_{de} = \frac{2(\Delta n)(w/S)}{\rho_\infty (dC_L/d\alpha) K_g V_a} \qquad (3.4)$$

where (see Section 2.3 for explanation regarding parameters)

Δn = turbulence-induced incremental vertical acceleration of the aircraft, as measured at its centre of gravity
w = weight of the aircraft
S = wing planform area
ρ_∞ = density of the undisturbed surrounding air
C_L = lift coefficient of the aircraft
α = angle of attack
K_g = a factor called *gust alleviation factor*
V_a = air speed of the aircraft (equivalent value in the case of fluctuating winds).

The factor (w/S) is the wing loading, and $(dC_L/d\alpha)$ is the slope of the lift curve of the aircraft (see Fig. 2.2). Turbulence of different levels of severity from the point of view of aviation may be defined using either the derived

Table 3.2 Atmospheric turbulence levels in terms of parameters

Turbulence intensity	Derived gust velocity U_{de}, m/s	Incremental vertical acceleration Δn, g	
		RMS	Peak (absolute value)
Light	1.5–6	<0.2	0.2–0.5
Moderate	>6–10	>0.2–0.3	>0.5–1.0
Severe	>10–15	>0.3–0.6	>1.0–2.0
Extreme	>15	>0.6	>2.0

gust velocity or the incremental vertical acceleration. Table 3.2 provides some indicative thresholds for such classification. Turbulence classification from a different consideration is given in Chapter 11.

The nomenclature of different levels of turbulence used in Table 3.2 corresponds to those instinctively felt by pilots and communicated as pilot reports or *pireps*. These reports are an important and reliable source of meteorological information in aviation. In particular, pireps obtained from aircraft landing through nonideal weather conditions can be of great use in alerting the pilots of subsequent aircraft planning to land at the same airport. An example of a pirep-based turbulence map of a region in the USA, covering all flight altitudes, is shown in Fig. 3.4. In Fig. 3.5, the altitude interval is split, with the turbulence reports in the high and low levels (the dividing altitude being at 18 000 feet) mapped separately over the entire conterminous landmass of the country.

The classification of atmospheric turbulence into the intensity levels listed in Table 3.2 and depicted in Figs. 3.4 and 3.5 is done according to its effect on aircraft. A brief description of various levels of turbulence in terms of their effect on aircraft in flight is given below.

Light levels of turbulence may manifest themselves in two different ways. When the effect consists of momentary, slight, erratic changes in the aircraft altitude and/or attitude, the pilot reports a 'light turbulence'. The type of turbulence that causes slight, rapid and somewhat rhythmic bumpiness without appreciable changes in altitude or attitude is reported as 'light chop'. Further, the turbulence is reported as 'occasional' if it occurs over less than one-third of the time, 'intermittent' if occurring between one-third and two-thirds of the time, and 'continuous' if occurring for more than two-thirds of the time. In the interior of the aircraft, the occupants may feel a slight strain against seat-belts or other restrainers and loose objects may be slightly displaced, but walking and food service in the cabin do not pose any significant problem.

Moderate levels of turbulence may also be of two types. The type that is similar to 'light turbulence' but of greater intensity is reported as 'moderate turbulence'. This type of turbulence does induce changes in aircraft altitude

54 Aviation weather surveillance systems

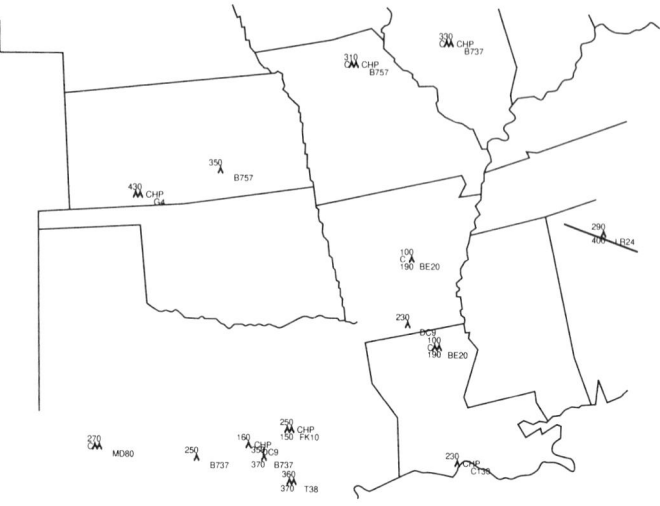

Figure 3.4 Map of a part of the USA with symbolic representation of pilot reports ('pireps') of turbulence encountered during 1700-1900 UTC on 19 March 1997. The Λ symbol represents moderate turbulence, and the ΛΛ symbol indicates light-to-moderate turbulence. The notation CHP to the right of these symbols (if present) indicates 'chop' or turbulence of short duration. The letters C or O to the left of the symbols (if present) refer to continuous or occasional turbulence, respectively. The number above the symbols shows the flight level at which the turbulence was encountered, in hundreds of feet above mean sea level (e.g. Flight Level 230 or FL230 means a flight altitude of 23 000 ft MSL), and where two levels are shown (above and below the symbol) they specify the altitude range over which turbulence was reported to exist. The type of the reporting aircraft is indicated to the lower right of the turbulence symbol (e.g. B757 is Boeing 757). When a pilot report provides a beginning and end position of turbulence encounter, the interval is shown as a line segment. (Courtesy R. J. Williams of the Aviation Weather Center, Kansas City, Missouri, USA)

and/or attitude, but the aircraft remains in positive control throughout the flight. Variations in the indicated air speed are usual. The other type of turbulence, which is a stronger form of the 'light chop' and causes rapid bumps or jolts to the aircraft without appreciable changes in its altitude or attitude, is reported as 'moderate chop'. The occupants of the aircraft feel definite strain against their seat-belts or other restrainers, unsecured objects are dislodged, and difficulties are encountered in food service and walking in the cabin.

Severe turbulence, also reported as 'severe turbulence', causes large and abrupt changes in aircraft altitude and/or attitude and usually causes large variations in the indicated air speed. Under its influence aircraft may go

Atmospheric effects on aviation 55

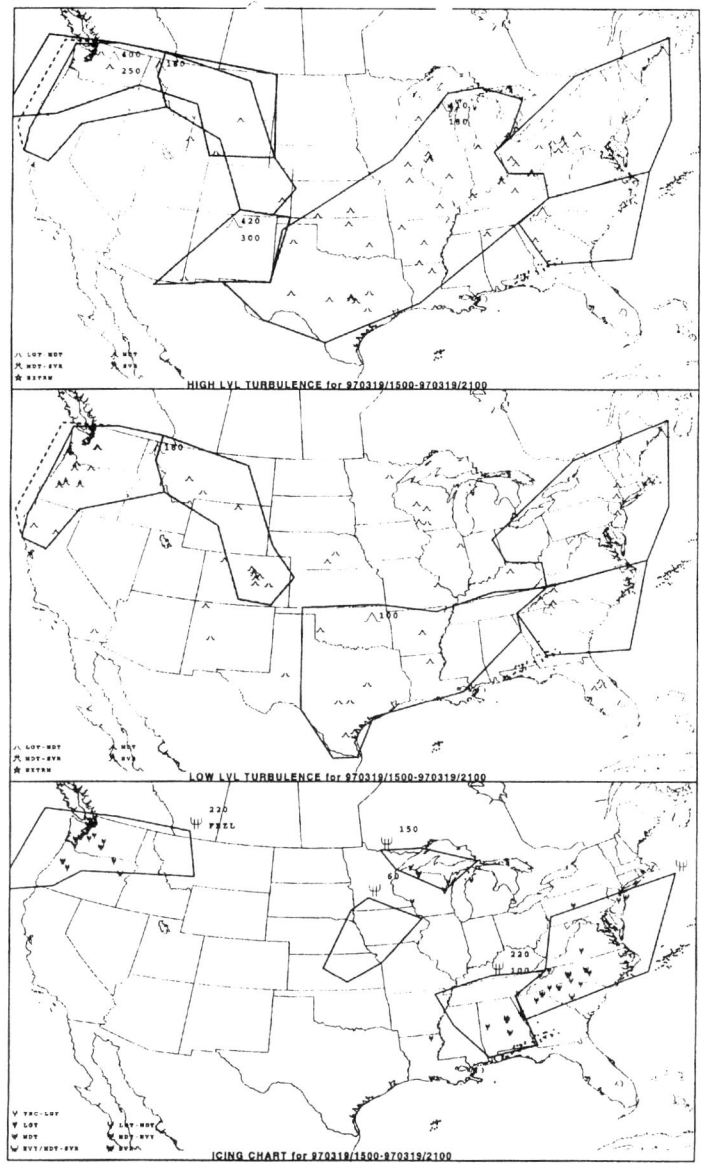

Figure 3.5 Pilot reports of turbulence over conterminous USA during 1500-2100 UTC on 19 March 1997, shown for the high level (> 18 000 ft altitude) (top), and the low level (≤ 18 000 ft altitude) (middle). The numbers inside the map indicate flight levels in hundreds of feet. The symbols for various levels of turbulence intensity are shown on the lower left of each Figure. The bottom figure maps reports of aircraft icing during the same interval. (Courtesy R. J. Williams, Aviation Weather Center, Kansas City, MO, USA)

momentarily out of control. The occupants of the aircraft are forced violently against their seat belts or other restrainers, unsecured objects are tossed about, and food service and walking in the cabin become impossible.

Extreme turbulence (reported as such, i.e. as 'extreme turbulence') is a level of turbulence that tosses the aircraft violently, causing severe loss of control. Besides, it may also cause structural damage to the aircraft. Unsecured objects in the interior of the aircraft may fly around, possibly causing injury to occupants.

The most frequent and strongest source of atmospheric turbulence affecting aviation is associated with thunderstorms, but significant turbulence may arise from strong surface winds, mountain waves, and other topographic factors affecting wind flows. These aspects will be discussed in detail in the following chapter. One term that occurs frequently in the aviation context is *clear air turbulence*, reported by pilots as CAT, which refers to turbulence encountered at relatively high altitudes (normally beyond 15 000 ft above sea level) that is not associated with thunderstorms or cumulus clouds. Pilot reports of CAT are normally qualified by its intensity, and indicate whether it is of the 'chop' type.

Statistically, atmospheric turbulence of various types and origins is responsible for a large number of acccidents involving a variety of aircraft. In the UK, in the year 1977, three aircraft accidents were attributed to turbulence in flight (Civil Aviation Authority, 1978). Table 3.3 shows some relevant data pertaining to the calendar year 1990 in the USA.

A wind phenomenon closely allied with turbulence is that of gust. While turbulence normally imparts to aircraft repetitive but random jolts or a rumbling sensation, gusts consist of individual transitory bursts of wind. A sudden gust can blow an aircraft off its intended position, e.g. the centreline of the runway while landing. It may also induce aerodynamic and structural oscillations in the aircraft, leading to reduction or loss of control. A series of gusts in close succession is akin to turbulence in its effect on flight. Gusts are implicated in numerous aircraft accidents. In 1990 they were found to be the cause or a factor in 94 aircraft accidents in the USA (National Transportation Safety Board, 1993a; 1993b), out of which four involved commercial air carriers.

Although wind shear and turbulence are often treated as two different types of phenomena in terms of their physical nature and effects on aircraft, they are actually two parts of a continuum of scales of air motion, as discussed in the beginning of this subsection. In Chapter 11 we describe how wind shear gives rise to turbulence. In practice, these two hazard factors for aviation are very often found to coexist. Indeed, some authorities consider wind shear to be 'the cause of almost all turbulence' (Buck, 1978, p. 129).

The coexistence of wind shear and turbulence in the atmosphere is illustrated by Fig. 3.6, which shows a time–height surface plot of the wind shear measured by a radar wind profiler (see Section 8.2.2 for a description of this instrument), with pilot reports of turbulence superimposed. It is seen

Table 3.3   Turbulence of various types/origins as a cause or factor in US aviation accidents during 1990

Type of aviation operation[a]	Phenomenon[b]	Number of accidents
14 CFR 121, 125, 127[c]	Turbulence	1
	Turbulence (thunderstorm)	1
	Turbulence (clear air)	2
14 CFR 135 (scheduled)[c]	Turbulence in clouds	1
14 CFR 135 (nonscheduled)[c]	Turbulence	2
General aviation[d]	Turbulence	27
	Turbulence in clouds	1
	Turbulence (thunderstorm)	2
	Turbulence (clear air)	1

[a] For explanation on the type of operation, see Table 2.1
[b] As cited in the sources below
[c] Source: National Transportation Safety Board (1993a)
[d] Source: National Transportation Safety Board (1993b)

that the incidence of high levels of turbulence often coincides with high values of wind shear. This particular observation pertains to the jet stream associated with a long planetary wave, but this is not the only type of phenomenon where strong wind shear and turbulence coexist, as will be clear in the following chapter. An aircraft flying through such a wind field would be exposed to a dangerous combination of the hazards arising from wind shear and turbulence, compounding the chances of disaster. In particular, when an aircraft is driven to a stall or near-stall condition due to wind shear, its ability to recover from such a condition by careful manoeuvres would be impaired by the reduction of the control effectiveness caused by the coexisting turbulence.

3.4 Hydrometeorological phenomena

The atmosphere is an important part of the terrestrial water cycle. By virtue of its contact with oceans, water bodies, and the moist earth, the atmosphere continuously acquires large amounts of moisture. Under certain combinations of temperature, pressure and humidity, the moisture condenses to form droplets, which may undergo further transformation depending on the

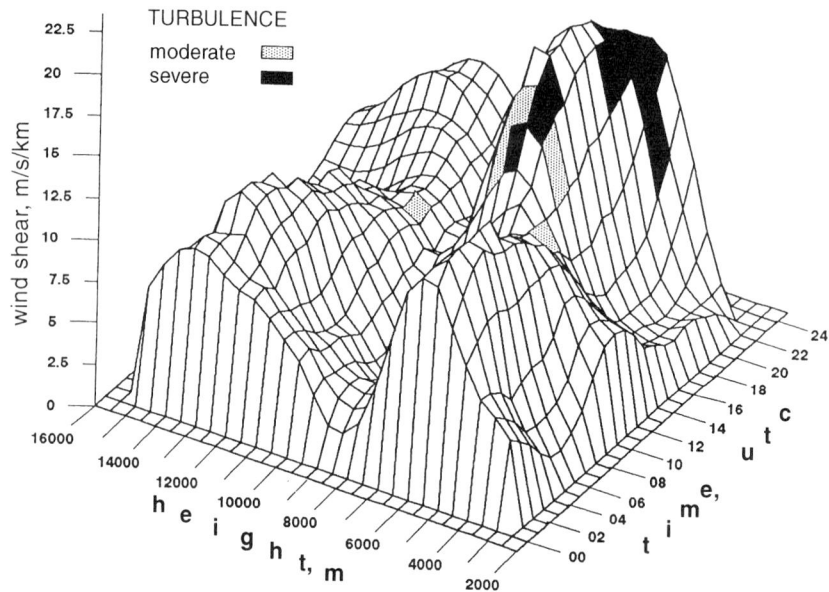

Figure 3.6   *3-D plot of the magnitude of wind shear as a function of height and time, as observed above a wind profiler located at Crown, PA, USA on 21 January 1987. The maximum value of wind shear in the plot is ~18 m/s/km. Pilot reports of observed turbulence are shown shaded at the reported height and time slots. The black patches indicate moderate-to-severe or severe turbulence, and the grey patches signify light-to-moderate or moderate turbulence. Note that significant or high levels of turbulence coincide with the occurrence of strong shear.* (From Syrett, 1991, by author's permission)

ambient conditions. Particles of different forms of water that fall through the atmosphere under the action of gravity are called *hydrometeors*. Hydrometeors may be in the solid, liquid or mixed phase. Their phase, size and shape are determined by the atmospheric conditions in the zone of their origin, as well as those along the path of their travel through air.

The vast majority of hydrometeors are small particles which quickly attain their terminal speeds while falling through the atmosphere. Their fall speeds and direction are governed by their composition, size, shape, orientation, the properties of the ambient air, and the local air currents. As will be discussed later, these attributes of the hydrometeors have a strong bearing on their detectability by remote sensing devices such as the radar. Hydrometeorology is a vast subject of much study. In this section we confine our attention only to the interaction between hydrometeorological phenomena and aviation, and mention those properties of hydrometeors which are relevant from the aviation weather surveillance viewpoint. Discussed below are the three most commonly encountered members of the hydrometeor family: rain, snow and hail.

3.4.1 Rain

Rain is the most visible indicator of bad weather. This is so both for visual observation and for remote sensing by radar. Apart from serving as an indicator of weather activity in a given area, rain also has significant direct effects on aviation.

In low or moderate intensities, rainfall may cause some visibility problems, but modern aircraft design and navigation procedures and aids have made commercial and military aviation quite safe against such levels of rain. But small and general aviation aircraft without sophisticated navigational instrumentation, and largely flying under visual flight rules (VFR), may still be threatened by moderate rain rates. Rainfall of high intensity, however, has high hazard potential for aviation involving all classes of aircraft.

Aside from gross reduction of visibility, and the possibility of hydroplaning (skidding) on very wet runways, a major effect of heavy rain on aircraft in flight is through interference with the combustion process of aircraft engines. Flight through zones with heavy rain causes significant amounts of liquid water to be ingested into jet engines along with the intake air used to support combustion. This liquid water retards engine combustion, resulting in thrust reduction. Further, since regions of heavy rainfall often harbour high levels of turbulence, the water ingestion rate (as also the overall mass flow rate) is erratic, producing randomly variable thrust loss. In extreme cases extinction of the engine combustion (or 'flameout') may occur, leading to total loss of engine power. Such has indeed been found to be the case in certain aircraft accident analyses (National Transportation Safety Board, 1978; 1980).

Heavy rain may also cause significant impairment of aircraft performance by increasing the aerodynamic roughness of the wing surfaces. Heavy rain creates a thick film of water over the wing surfaces which develops ripples due to the high relative velocity of air over the wings. The ripples effectively increase the roughness of the wing surface which disturbs the laminar flow of air over the wings and advances the point of separation of flow from the wing surface forward along the wing cross-section. This change has the effect of reducing the lift generated by the wing and increasing its drag (Haines and Luers, 1983; Luers and Haines, 1983; Phillips, 1989). In high-lift, high-drag conditions of flight, which occur during the takeoff and landing phases, such unfavourable changes in lift and drag forces can appreciably enhance the probability of stall and consequent accidents.

That such a mechanism may be able to best explain performance loss of aircraft in heavy rain environments is illustrated by an analysis of the Pan American Boeing 727 crash that occurred at New Orleans, USA, in 1982. The official investigation of the accident had determined wind shear to be the probable cause (National Transportation Safety Board, 1983). However, detailed analysis by Dietenberger *et al.* (1985) showed that the inclusion of lift loss and drag increase due to heavy rain, in addition to wind shear effects, was able to explain the recorded flight parameters better than wind shear alone.

Thompson and Jang (1996) report that the aerodynamic efficiency of wings in rain depends on the dimensions of the rivulets and droplets formed on the wing surface as a result of rain, and state that flatter and thinner rivulets result in less loss of aerodynamic efficiency as indicated by the lift-to-drag (L/D) ratio. They observe that a wettable wing surface results in thinner water films, flatter and thinner rivulets and flatter droplets than a nonwettable surface, causing a smaller loss of L/D.

Another factor that aggravates the problems due to aerodynamic lift loss and drag increase is the loss of momentum on the part of aircraft by the physical impact of large quantities of hydrometeors, which has the effect of a further increase in the apparent or overall drag.

High levels of rain activity also impact adversely on aviation by impairing the efficacy of navigational aids and instruments. One class of instruments that may be affected by heavy rain are landing guidance systems (Beals, 1969). Radars, especially those operating in the X-band of frequencies (8–12.5 GHz) and above, are also susceptible to rain effects. For example, the rain-induced attenuation of the X-band airborne weather radar signals, and the resultant degradation in the ability to estimate the extent and intensity of the weather disturbances, were determined officially as contributing factors in the case of an airliner accident in Nebraska, USA in 1980 (National Transportation Safety Board, 1980).

For some special aerospace activities any amount of rain is detrimental. An important example is the case of reusable space re-entry vehicles such as the US Space Shuttle which fly in the atmosphere at very high speeds, and whose exterior thermal protective tiles are extremely sensitive to the impact of raindrops at such speeds. Indeed, Space Shuttle landing conditions stipulate that there should be no precipitation forecast at the landing site, and no thunderstorms within 20 or 30 nautical miles (~37 and 55 km, respectively) of the landing site, depending on the type of landing (Rigdon, 1991).

3.4.2 Snow

Dry snow, being light and fluffy, does not directly cause damage to aircraft. However, snowstorms can cause poor visibility and slippery runways. Deposits of thick layers of snow render runways unserviceable for extended periods of time. Snowstorms are frequent causes of airline delays in the higher latitudes in winter.

Wet snow tends to aggregate and accumulate on aircraft surfaces. These deposits (along with ice accumulations) enhance surface roughness, and disturb aerofoil shapes, spoiling the flow over the aerodynamic surfaces and impairing their performance, thus leading to accidents (National Transportation Safety Board, 1981).

3.4.3 Hail

Among all hydrometeors, hail has the highest potential for direct damage to aircraft through impact. Hail can occur in sizes up to a few or even several centimetres, and their population densities can be quite high over localised regions within thunderstorms. Because of the high flight speeds of aircraft, the impact of large-sized hail can cause serious structural damage to the exposed parts of the aircraft such as the leading edges of wings and tail planes, and can crack or shatter wind shields, canopies, radomes, etc. (Fig. 3.7).

In thunderstorms hail very often coexists with heavy rain and turbulence. When ingested into jet engines along with the liquid rain, it produces two distinct types of hazardous effects. First, it contributes further to the erratic combustion, enhancing the possibility of flameout and loss of thrust (National Transportation Safety Board, 1978). Next, hail ingestion and impact with interior components can damage delicate engine parts such as fan and compressor blades which rotate at high speeds with very small clearances from other parts.

Indeed, hailstorms are high on the list of weather phenomena to be avoided by aircraft pilots. As will be shown later, reliable identification of hail zones within weather fields is an important goal of modern weather surveillance systems, and is a topic of active research and implementation.

3.5 Aircraft icing

As the operating heights of aircraft cover a large altitude interval, extending typically up to ~13 km for subsonic airliners and transport planes, and significantly higher for supersonic aircraft, most aircraft encounter freezing temperatures during some part of their operations. The height at which the temperature of the atmosphere equals the freezing point of water under local conditions is called the *freezing height* or *freezing altitude*. The freezing height generally decreases with increase in latitude, and varies cyclically around the year according to seasons, reaching as low as the ground level in winter in the higher latitudes. Under certain atmospheric conditions, mentioned below, the freezing temperatures lead to ice formation and its accretion on exposed parts of the aircraft, especially the leading edges of wings and the vertical and horizontal tailplanes. Accretion of ice on the aerodynamic surfaces of the aircraft changes their aerofoil shapes and increases their roughness. This causes the streamlined flow around the aerofoils to be disturbed (Fig. 3.8), leading to loss of lift and increase in drag, and possibly even impairment of aircraft stability (e.g. Perkins and Reike, 1993). A dramatic picture of severe ice accretion on the leading edge of an aircraft's vertical fin in actual flight may be found in Buck (1978, p. 217).

Considerable research has been conducted to determine the effect of ice deposits on aerofoil performance. An example is a wind tunnel study on an

62 *Aviation weather surveillance systems*

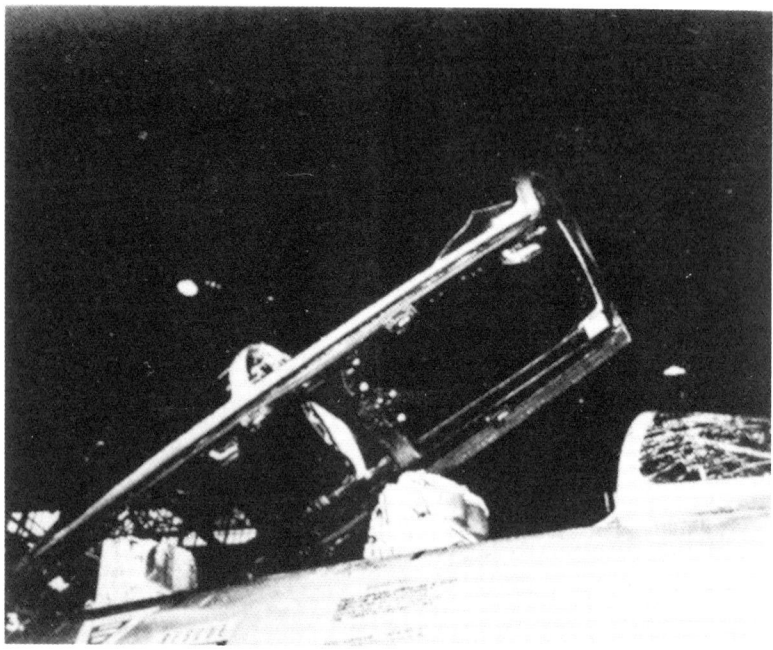

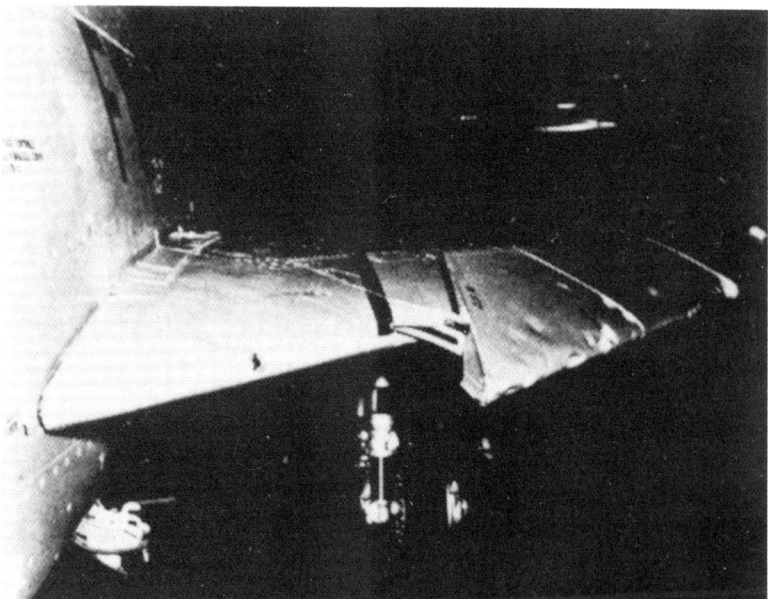

Figure 3.7 The damage to a military aircraft canopy (top) and wing leading edges (bottom) caused by hail impact during flight. (From Mahapatra and Zrnic', 1991. Photographs courtesy J.T. Lee)

NACA 0012 aerofoil reported by Korkan *et al.* (1985). They simulated leading edge outgrowths as well as the roughness on the windward part of the aerofoil similar to those caused by ice deposits occurring in actual flight, and found that both the lift capability and stall characteristic of the aerofoil degraded considerably. For example, at a Reynold's Number of 1.4×10^6 the maximum lift coefficient C_{Lmax} (see Section 2.3) was reduced from ~1.6 for the clean aerofoil to a value less than unity with ice deposits, and the stall angle of attack α_{stall} came down from 18° to ~12°. Such degradation greatly enhances the danger of premature stall under adverse flight conditions.

Icing-related problems constitute a major aviation hazard, and have caused numerous air crashes (National Transportation Safety Board, 1979; 1981; Civil Aeronautics Board, 1963; 1964a). In the UK, icing conditions accounted for four aircraft accidents in 1977, out of which three involved public transport aircraft (Civil Aviation Authority, 1978). According to one estimate (Czekalski, 1983), aircraft icing caused 51 fatal accidents, most often involving general aviation or commuter-class aircraft, and claimed 66 lives on an average in the US in the early 1980s. In 1990, one commercial air carrier accident (National Transportation Safety Board, 1993a) and 18 general aviation accidents (National Transportation Safety Board, 1993b) had icing as a cause or a factor.

In addition to ice accretion on the aircraft surface, icing of certain critical internal components of aircraft may also lead to loss of performance and even serious accidents. One such highly critical component is the carburettor found in a large number of small, piston-engined aircraft used mostly for general aviation purposes. As the frigid ambient air is drawn through the carburettor for mixing with fuel for combustion in the engine, any suspended water droplets freeze on the air passage, causing blockage, leading to improper air–fuel mixing, and consequent loss of engine power or possibly engine failure. Carburettor icing was determined to be a cause or factor in 42 general aviation accidents, of which it was the primary cause in ten cases, in the USA in 1990 alone (National Transportation Safety Board, 1993b), and was involved in at least one commercial carrier accident in the same year (National Transportation Safety Board, 1993a).

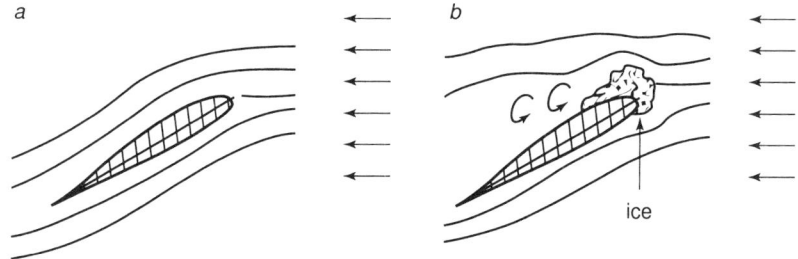

Figure 3.8 Schematic depiction of the streamlined flow around a clean aerofoil (a), and the disturbed flow due to ice accretion around the leading edge (b)

The severity of aircraft icing has a complex dependence on a host of factors. These include the temperature, liquid water content, cloud droplet size distribution, atmospheric turbulence level, and the phase of water in the ambient air (Hansman, 1989). The overall effect of icing on aircraft performance depends on the severity of icing and its distribution over the sensitive parts of the aircraft (wings, tailplanes and control surfaces), and the type and flight condition of the aircraft. The flight conditions include speed, altitude, climb/descent rate, angle of attack, and wing loading.

When aircraft remain parked or are taxiing on the ground at temperatures not much below freezing, liquid or partially frozen rain can freeze over the aircraft skin (at much colder temperatures, there will be little liquid precipitation). Long exposure periods to such conditions can lead to dangerous levels of ice accretion. However, inspection of aircraft for icing is a part of the normal pretakeoff check in cold weather, and there are standard de-icing procedures involving chemical sprays. These procedures have brought the aircraft icing problem on the ground effectively under control, except in cases of extraordinarily rapid ice formation, and of human errors involving improper compliance with stipulated procedures (e.g. National Transportation Safety Board, 1993d).

Another effect of the existence of icing conditions at the ground level is the formation of ice layers on runways, rendering them extremely slippery and causing aircraft to skid and lose control during takeoff and landing.

The icing conditions are quite different in the case of aircraft in flight. Here, unlike the case of aircraft on the ground, any particle impinging on the surface of the aircraft would be subjected to blowing action by the high relative speed of the air on the aircraft skin. The ice accretion process thus occurs in two steps. The first consists of the impact of droplets on to the aircraft skin, and the second involves the freezing of the drops and their adherence to the skin to form ice layers. The ice accretion rate depends on the efficiencies of both these subprocesses.

A body moving through air containing droplets effectively acts as a collector of a certain fraction of the droplets. The collection efficiencies of three cylinders and an aerofoil are shown in Fig. 3.9. The Figure shows that the collection efficiency of a given body shape depends most strongly on the size of the impacting droplets, followed by the collector size and air speed.

After impinging on the aircraft skin at subzero temperatures, a droplet can splash off the surface, freeze rapidly, or flow along the skin for a distance before freezing. The last of the three processes results in glaze or clear ice to be formed. Under favourable conditions such as very low temperatures, small droplet sizes, and a relatively low liquid water mass rate of deposition on the surface, the droplets can release their latent heat rapidly, leading to their freezing at the point of impact. The result is the formation of rime ice. Glaze has a greater surface roughness than rime ice (Politovich, 1993). If liquid water concentration (measured as the weight of water droplets suspended in a unit volume of air) is high and the temperature is also relatively high, then

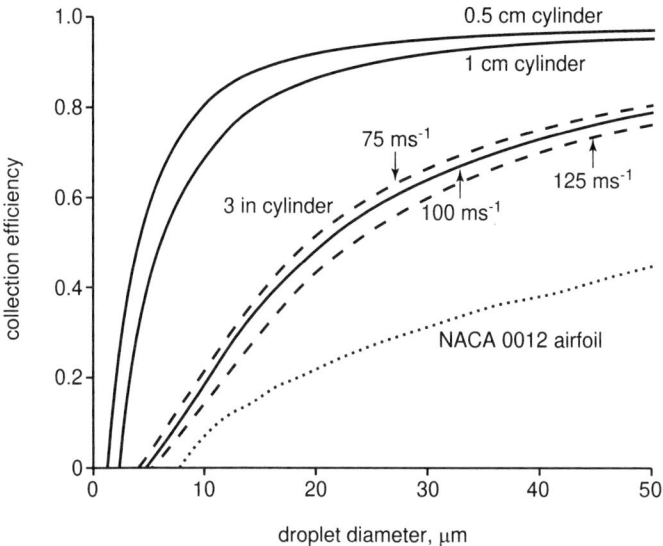

Figure 3.9 Collection efficiency as a function of impacting droplet diameter for the indicated collector sizes and shapes. Air speed is 100 m/s except where indicated otherwise. The efficiencies are based on the calculations of Makkonen and Stallabrass (1987). (From Politovich, 1993, by author's permission)

the latent heat of droplets cannot be transferred quickly enough for them to freeze on impact, and the larger droplets tend to splash off the surface.

One environmental condition considered particularly conducive for aircraft icing is the presence of supercooled water droplets. Under certain conditions, notably in the absence of condensation nuclei, liquid water may exist in the atmosphere in the supercooled state, i.e. at temperatures below the local freezing temperatures. Supercooled water can freeze immediately on contact with the aircraft skin or pre-existing ice layers through the conversion of its negative sensible heat (with respect to freezing temperature) to latent heat of solidification. This facilitates ice formation and accretion by minimising the need for heat removal. The atmospheric temperature range between $-10°$ and $0°C$ is particularly conducive to ice accretion on aircraft (Westwater and Kropfli, 1989), but somewhat colder temperatures may also foster significant icing. At much colder temperatures, progressively larger fractions of cloud droplets turn into ice, and do not readily adhere to the aircraft skin on impact.

It has been found that most serious aircraft icing hazards occur in a surrounding of supercooled liquid droplets, and that larger droplets cause faster build-up of ice deposits on aircraft surfaces. Droplets in the diameter range of 30–400 μm have been found to be highly conducive to aircraft icing (Cooper *et al.*, 1984; Politovich, 1989). In an experimental study, Hoffman

(1991) demonstrated that severe icing was observed on a Do-28 aircraft after flying for 165 km through a cloud with small droplets, having a median volume diameter (MVD)[1] of ~20 μm. However, when the droplets were large, with an MVD >200 μm, only 46 km of flight resulted in heavy icing. In both cases, the ambient temperature and the liquid water concentration were nearly equal, being $\sim -2°C$ and 0.2 g/m^3, respectively, so that the spectacular difference in the icing rates was largely attributable to particle size effect.

The degradation in the basic aerodynamic coefficients of the aircraft such as the lift and drag coefficients, especially the latter, results in the loss of overall performance capabilities of the aircraft. For example, the rate-of-climb capabilities of several research aircraft were found to decline by more than 5 m/s within 10 min of exposure to environments with large supercooled droplets (Sand et al., 1990).

Politovich and Bernstein (1995) provide insightful results of a controlled flight experiment through a winter storm in Colorado conducted in February 1990 under the Winter Icing and Storms Project (WISP). The storm involved clouds of supercooled liquid water that persisted for as long as 36 h and contained regions with large droplets, >30–50 μm (though the mean droplet size was smaller in most parts). There was substantial ice accumulation on the Beechcraft Super King Air 200T aircraft, and it was found that the lift coefficient of the aircraft changed only marginally, while the drag coefficient showed strong degradation. Compared to its value for the clean aircraft, the drag coefficient increased by 38, 58 and 84% during successive passes through the supercooled cloud over a time period of ~1 h. This increase had a strong effect on the gross performance characteristics of the aircraft. For example, after ~20 min of flight, before the drag coefficient had increased by 58%, the maximum climb capacity of the aircraft had been reduced to only ~2.5 m/s from its normal value of 7–10 m/s. The stall characteristic of the aircraft was also degraded. Within somewhat less than an hour of flight (before the drag coefficient increase was 84%) the aircraft, on a stall check, approached the stall condition (as indicated by pronounced buffeting) at an air speed of 115 knots (59 m/s) compared to its normal stall speed of ~90 knots (46 m/s). This means that the aircraft would stall earlier as it slowed down to land, and would be forced to land at a higher speed than normal to maintain adequate stall margin. Further, if such an ice-laden aircraft suffers unexpected loss of air speed (e.g. due to wind shear) during low-speed operations, it would be more prone to stalling and crashing than a clean aircraft. If the ice accumulation occurs while the aircraft is on the ground, stall may occur during takeoff (National Transportation Safety Board, 1993d).

---

[1] In a population of droplets consisting of a range of sizes, if half of the total mass of the droplets is accounted for by the droplets less than (or greater than) a certain diameter, then that diameter is called the median volume diameter (MVD) of the population.

Politovich (1996) reports more results of controlled flight experiments involving the King Air 200T aircraft. These experiments show lift coefficient degradation up to 35%, a drag coefficient increase by as much as 230%, and a maximum loss of climb capability by 6.9 m/s as a result of icing. The largest performance decrease occurred when the liquid water content of the ambient air was $>0.2 \text{ g/m}^3$, the droplet MVD was >30 μm, and the temperature was $>-10°C$. In yet another study, Ashenden *et al.* (1996) report moderate reduction in maximum lift coefficient due to icing caused by supercooled cloud, drizzle and raindrops, but increase in the profile drag of the wing by as much as 56%.

In summary, the important meteorological conditions that facilitate aircraft icing may be listed as (Ellrod and Nelson, 1997): (i) cloud-top temperatures in the range 0 to $-20°C$; (ii) large liquid water droplets, exceeding ~50 μm; (iii) weak upward vertical air motion to replenish the supercooled water supply; and (iv) thick and extensive cloud masses.

It is possible to estimate the icing potential of the ambient air by mounting appropriate instrumentation on aircraft. On sensing entry into a zone with high icing potential, an aircraft may perform the necessary manoeuvres (e.g. change of altitude) to escape from the zone before significant ice deposits are actually formed. The icing potential at any given location can be inferred from three basic measurements regarding the ambient conditions: temperature, liquid water content, and the MVD of the droplet population. Hauf (1993) describes a system for *in situ* measurement and warning of icing conditions which makes use of a hot-film anemometer sensor for drop size measurement. The schematic diagram of the system is shown in Fig. 3.10.

Remote sensing of the icing potentials of the parts of air space along an aeroplane's proposed flight path would, of course, be more useful than *in situ* sensing since it would provide the pilot with advance warning, and help in planning the navigation route more efficiently. Such possibilities are discussed later in this book, especially in Section 8.8 in the context of satellite observations.

3.6 Low visibility

It was mentioned in Section 2.5 that a large number of aircraft flights occur under visual flight rules (VFR). Visual flight is conditioned on the existence of clear visibility conditions. Further, all aircraft utilise a significant amount of visual input during terminal area operations such as approach, landing, taxiing and manoeuvres on the apron. Hence adequate visibility is necessary for this phase of flight operations for all aircraft.

Low visibility in terminal areas is a common source of flight delays, disruptions and accidents. Such conditions are of frequent occurrence in many parts of the world. For example, at Heathrow Airport in London, England, the visibility level is inadequate for Category I landing (see Table 2.3) for an average of 275 h per year, and stays below Category II landing

68 Aviation weather surveillance systems

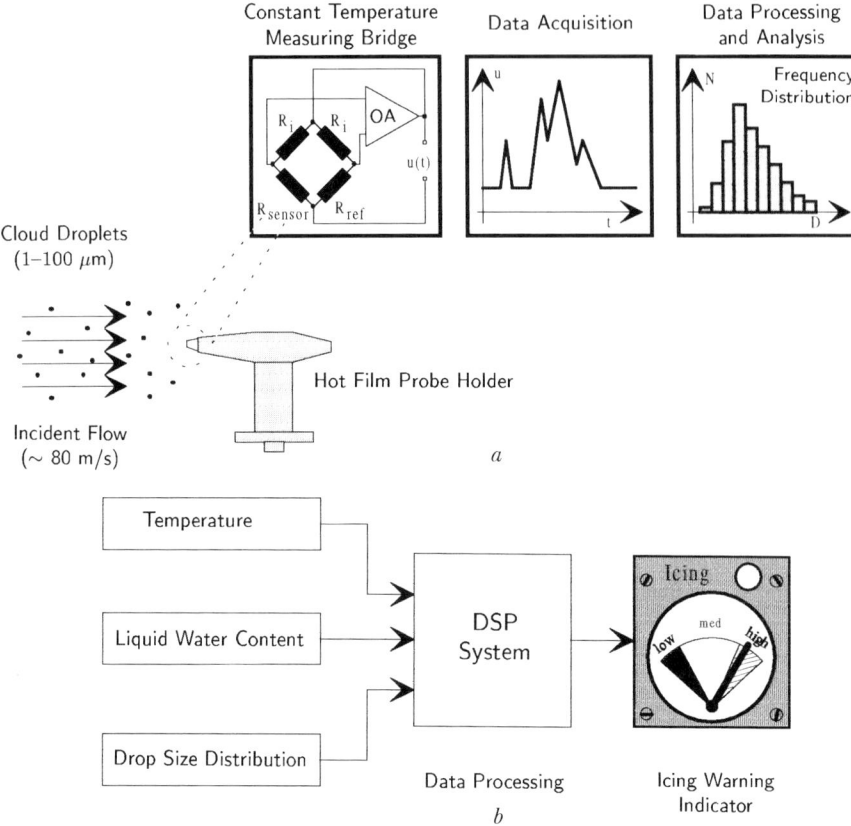

Figure 3.10 *Schematic diagram of an airborne icing warning system: (a) hot-film drop size sensor circuit and principle, and (b) icing warning system structure.* (Courtesy T. Hauf, DLR Institut für Physik der Atmosphäre, Oberpfaffenhofen, Germany)

requirements for 170 h per year (Webber, 1965). In terms of safety, an early study (Jervis, 1964) led to the striking observation that aircraft accidents occur approximately 30 times more often under marginal visibility conditions than under good visibility conditions. Although better instrumentation may have somewhat improved matters since this study, the aviation hazard potential of inadequate visibility remains very high.

Poor visibility may be caused by a number of phenomena. At higher altitudes of flight the most frequent cause of poor visibility is the presence of clouds. Clouds of various types are found in different height bands, some of which may overlap. Cloud heights are also dependent on latitude and local conditions. From an aviation point of view a very important parameter is the height of the *cloud base* which is the bottom surface of a cloud column. Since VFR flight must maintain a specified distance from clouds (see, for example,

Nolan, 1994, Chapter 3), a lowering of the cloud base severely restricts the air space available for such flights. When the cloud base descends to the order of hundreds of feet, it will interfere seriously with the aircraft landing operations, even with instrument aids.

At the ground level, the most frequent visibility-reducing factor is fog, which is essentially a cloud mass in contact with the ground. Fog degrades the efficiency of landings, takeoffs and ground movements of aircraft, reduces the traffic handling capacity of airports, and is the most frequent cause of delayed flights during the winter season at many airports located in the higher latitudes.

At most altitudes of interest to aviation, visibility may be affected seriously due to precipitative phenomena such as rain (Huffman and Haines, 1984) and snow. Other factors sometimes responsible for poor visibility are dust storms and smoke. The former is important in desert and dusty locations and the latter is becoming a significant factor at airports close to big cities and in industrial areas affected by atmospheric pollution. On rare occasions, insect swarms and bird flocks may also interfere with visibility.

In statistical terms, visibility-related problems appear frequently and repeatedly as causes or factors in numerous aircraft accidents and incidents. UK aircraft accident data (Civil Aviation Authority, 1978) show that, during 1977, a total of six accidents were caused by low ceiling (out of which two involved public transport aircraft), and two more due to fog. Table 3.4 shows the distribution of accidents caused by poor visibility or related factors in the USA in 1990.

3.7 Atmospheric electrical phenomena

Under certain conditions, especially those associated with thunderstorms, regions of the atmosphere harbour electrically charged water particles and significant electric field potential gradients. The most visible manifestation of atmospheric electrical activity occurs in the form of lightning flashes, which involve the creation of ionised channels through the atmosphere and the conduction of strong electrical currents through these channels.

Atmospheric electricity is a potent source of hazard to aircraft in flight. Aircraft passing through regions of strongly charged clouds and strong electric fields may encounter lightning naturally occurring in those regions (Mazur et al., 1984). It has also been postulated that aircraft themselves may trigger the occurrence of lightning in certain thunderstorms (Fitzgerald, 1967).

The effect of lightning on aircraft is to cause strong transient electric currents on the aircraft skin and structural members, as well as through the electrical and electronic circuits on board the aircraft. The latter may result in malfunction, possible outage, and even permanent damage to on-board instruments, computers and flight control systems. Lightning can also cause

Table 3.4 Poor visibility and related phenomena appearing as a cause or factor in US aviation accidents during 1990

Type of aviation operation[a]	Visibility-related phenomenon[b]	Number of accidents
14 CFR 121, 125, 127[c]	Fog	2
14 CFR 135 (scheduled)[c]	Clouds	1
	Fog	1
	Low ceiling	3
14 CFR 135 (nonscheduled)[c]	Clouds	2
	Drizzle	1
	Fog	4
	Haze/smoke	1
	Low ceiling	6
General aviation[d]	Clouds	11
	Drizzle	5
	Fog	52
	Haze/smoke	3
	Ice fog	1
	Low ceiling	54

[a] For explanation on the type of operation, see Table 2.1
[b] Does not include phenomena such as rain and snow which also contribute to poor visibility
[c] Source: National Transportation Safety Board (1993a)
[d] Source: National Transportation Safety Board (1993b)

explosions due to the ignition of fuel–air mixtures. Further, strong lightning strokes can cause physical damage to aircraft structures. In a particularly disastrous accident involving a civilian airliner, the fuel in one of its tanks was ignited by lightning, and its left wing was destroyed, killing all the 81 people on board (Civil Aeronautics Board, 1964b). As other examples, the UK Civil Aviation Authority (1978) reported one aircraft accident due to lightning in 1977, and the US National Transportation Safety Board (1993b) listed lightning as a cause or factor in two general aviation accidents in 1990. Less publicised fatal accidents have occurred to military aircraft due to lightning. Other direct damaging effects of lightning on aircraft include pitting, burning and magnetic deformation.

Although the hazardous effects of lightning on aviation are real and varied, the frequency of occurrence of serious accidents and equipment outages due to lightning is not alarming for the present generation of aircraft. However, two unrelated trends in aeronautical engineering have the potential of changing this comfortable situation. The first of these is the increasing use of nonmetallic composite materials in aeroplane construction. Such materials are finding favour with aeroplane designers because of their advantageous strength-to-weight and stiffness-to-weight ratios. In the case of military aircraft, nonmetallic materials have the added advantage of offering low radar reflectivity, greatly reducing the ability of enemy radars to detect them. However, conventional all-metallic aircraft construction is advantageous from the point of view of lightning because of the electromagnetic shielding it provides to the conductors and circuitry located in the aircraft interior. The second line of development in aeronautics that is of concern in the context of lightning is the increasing use of digital electronics to perform critical flight functions such as flight control and even the maintenance of basic aircraft stability. Recall here the mention of fly-by-wire control systems in Section 2.3. The digital systems being used in modern and futuristic aircraft are inherently less robust and tolerant against lightning-induced electrical transients than the conventional analogue circuitry and mechanical and hydraulic controls.

Because of these developments there is a call for greater caution and finer qualitative and quantitative understanding regarding the effects of lightning on aircraft. The most comprehensive and systematic programme to refine such understanding was conducted by the US National Aeronautics and Space Administration (NASA) during the 1980s under the Storm Hazards Program(me). During this programme, extensive electronic and photographic (both movie and still) data were collected by flying a lightning-hardened and custom-instrumented F106B aircraft about 1500 times through thunderstorms at altitudes between 5000 and 40 000 feet. Additional co-ordinated observations were made by multiple ground-based radars. A very good summary of the findings has been presented by Pitts *et al.* (1988).

The multisensor observations recorded and analysed during the Storm Hazards Program corroborated the belief that aircraft themselves trigger a large fraction of the lightning strokes they receive. This conclusion has great significance in the context of lightning avoidance by aircraft. It means that it is not enough to merely sense the existence of naturally occurring lightning zones and instruct the aircraft to avoid them, but it is necessary to identify those parts of the air space which harbour conditions conducive to aircraft-induced lightning, and avoid them as well. Extensive photographic studies also showed that there are distinct differences between the effects of naturally occurring and aircraft-triggered lightning on aircraft in terms of the location and migration pattern of the lightning attachment points.

Quantitatively, the Storm Hazards Program provided the statistics of voltages, currents and fields induced by lightning on aircraft. Each lightning strike on the aircraft produced skin currents between 10 000 and 200 000 A, with a peak rate of change exceeding 200 000 A/μs in some cases. Discontinuities in the flow of these currents, caused by joints and cutouts on the skin, set up electromagnetic fields within the aircraft's interior. Electromagnetic fields, especially the components with relatively low frequencies, also leak through the skin into the interior. The field inside the aircraft, in particular its rate of change, is responsible for inducing voltage and current surges in the electronic and electrical circuits housed in the aircraft. The peak value of the rate of change of electric flux density recorded during the Program was found to be in excess of 100 A/m^2. These high values were found inside the front part of the aircraft fuselage, which is the area that normally houses most of the sensitive circuitry controlling aircraft functions. The time duration of electromagnetic fields set up due to lightning strikes on aircraft bodies may be of the order of a second or more.

A more detailed treatment of the lightning phenomenon from the point of view of aviation is presented in Chapter 12.

3.8 Need for improved aviation weather information

Historically, aviation weather information has been derived from the general pool of meteorological information, obtained through direct sensing as well as forecasting. A system of availability of regular aviation weather data and advisories has been established for most controlled air spaces. However, general weather monitoring, because of the need to cover large areas within limited resources, is generally of a coarse nature in both spatial and temporal dimensions. Unfocused weather advisories and forecasts that warn of average conditions and possible hazards over large areas as single units are only of limited value to aviation users. Knowing that phenomena such as strong precipitation, turbulence or icing conditions are occurring or are forecast to occur somewhere within a certain large area along the flight path is of little use to a pilot except prompting him/her to avoid such areas altogether. This would involve large departures from the scheduled flight path, diversion to an airport different from the original destination, aborting the flight and returning to the starting point, or cancelling the planned flight altogether. Each of these alternatives is costly in terms of money and time, and results in an overall reduction in the traffic capacity of the air space.

The utility of coarse and unfocused weather information for terminal area controllers and managers is even more marginal. Unlike decisions regarding a single flight, the operating status of airports affects a large number of flights, many of which may already be airborne. Just knowing the likelihood of hazardous weather in the general area of the airport would put the managers in a predicament. Shutting down operations over an entire airport is an extreme and highly expensive step that cannot always be taken based on the mere possibility of hazard in a general area, and yet operating airports in a normal manner under such conditions would expose flights to finite probabilities of danger or dislocation. Even shutting down one or more runways in a busy airport leads to considerable loss of traffic capacity. In general, vague forecasts of aviation-hazardous weather over terminal areas would lead to cautious operation, requiring greater vigilance on the part of pilots and controllers and frequently resulting in increased spacing between landings and takeoffs. This again causes loss of traffic capacity of the aviation system.

This situation can be remedied to a great extent by monitoring and forecasting aviation weather information with a greater degree of specificity and precision. The improvement can be in terms of spatial resolution, temporal rapidity of updating, and accuracy of weather parameter estimation. Fine delineation of boundaries of localised hazardous weather features would permit more informed vectoring of aircraft in the controlled air space, especially within the crowded terminal areas. Similarly, accurate prediction (even by a matter of minutes) of the time parameters associated with hazardous weather features such as the starting and ending times of precipitation, poor visibility, etc. would greatly help in anticipating and planning for periods of reduced airport capacity. Further, being able to forecast (again, even in the very short term) winds and wind changes, intensity of visibility-reducing phenomena, etc. would help controllers in maintaining optimal spacing between aircraft in approach and takeoff queues with or without the aid of automated systems.

Another major goal of modern aviation weather surveillance systems is to generate processed weather data, called *weather products*, at a level which would greatly minimise the workload on aviation system operating personnel for assimilation and utilisation. Although many air traffic controllers and pilots are very well informed about weather phenomena that affect aviation, they are usually not professional meteorologists. Presenting weather data to them in the traditional, relatively raw, form, whether alphanumeric or graphical, would require considerable meteorological interpretation. This would introduce subjectivity and delays into the decision chain, and leave room for ambiguity in interpretation and errors in judgment. By sensing accurately and processing the data intelligently, modern aviation weather surveillance systems aim to present hazardous and otherwise undesirable weather phenomena in a form that can be directly utilised by pilots and controllers for decision-making. Because of their finished form, such data products can also be fed directly into automated air traffic control systems without the need for human intervention.

Accurate *en route* weather information can help plan flights optimally in a strategic way by choosing in advance flight paths and timings so as to avoid danger zones. This is greatly facilitated if hazardous weather features can be forecast over timescales of the order of the flight times, usually a few to several hours. In the absence of such information pilots take *ad hoc* or tactical decisions to skirt hazardous phenomena, often after actually encountering them, spending greater amounts of time and fuel in the process and intruding into routes used by other aircraft, some of which may themselves be evading the same danger zones. Strategic route planning, aided by modern aviation weather surveillance systems, will thus make flights safer while promoting economy and schedule-keeping.

Intelligent data processing for high-level weather product generation also helps in compressing the data and making them more manageable. As an example, pilots seeking weather information for a proposed flight from current computer-based information systems often receive a printout of character-based data that is >20 ft long, through which they would have to sift manually to obtain the bits of information that are actually useful for the flight (Sankey, 1993). Even where the data are available in graphical form their inherent accuracy may not be much better. Further, many different graphical data fields may have to be scanned to retrieve the necessary information. This problem is becoming more serious as new independent sensors and systems are being developed and added to the aviation scene (many such are described in Chapters 7 and 8). Accurate data integrated from many sources in an aviator-friendly way, as aimed to be achieved by some of the most modern aviation weather surveillance systems, would make the presentation compact, more informative and reliable simultaneously.

Aviation activity has expanded enormously since many of the world's major airports were designed and commissioned, and these are now operating at or beyond their maximum intended capacities. The capacity is further strained in times of adverse weather. Most of these airports are hemmed in by city areas or other developments, making it difficult to expand their capacity via the addition of new runways. Construction of new airports to add capacity near population centres is even more difficult for reasons of land availability, cost and logistical factors (e.g. providing highways and rapid transit facilities from city centres, transfer between different airports serving a given urban cluster, compatibility with the hub model of the airline operation, etc.). One cost-effective option for increasing airport capacities is to resort to high-level system automation which would minimise the adverse impact of weather and also permit airports to handle more aircraft by reducing their spacing during landing and takeoff. Such systems would require high-grade atmospheric observational and forecast data for effective operation.

The requirement of quality atmospheric data for mitigation of weather effects is self-evident. However, accurate knowledge of atmospheric parameters (e.g. wind vector at various heights, which affects the length of the disturbing wake behind each aircraft) is also necessary for ensuring optimal separation between aircraft during terminal area operations. Sensors and systems for generating such information for modern aviation systems are the subject matter of the later chapters of this work.

3.9 Summary

This chapter provides an overview of the basic types of atmospheric effects on aviation. Atmospheric processes strongly affect both flight and ground operations of aircraft, and have a bearing on most important aspects of aviation such as safety, passenger comfort, flight punctuality, system efficiency and overall economy. Major types of atmospheric effects on aviation are related to wind shear, turbulence, rain, snow, hail, surface icing, visibility impairment and lightning. These phenomena affect aircraft and flight in entirely different ways, and the relative importance of these factors varies among different phases of aircraft operation. The nature of the effects of various significant atmospheric processes on aviation has been the subject matter of this chapter. The details of the sensing of atmospheric phenomena important for aviation follow in much more detail in later chapters.

3.10 References

ASHENDEN, R., LINDBERG, W., MARWITZ, J.D., and HOXIE, B. (1996): 'Airfoil performance degradation by supercooled cloud, drizzle, and rain drop icing', *J. Aircr.*, **33**, pp. 1040–1046

BEALS, G.A. (1969): 'Rain models for landing guidance systems'. US Air Force Report USAF ETAC TN 69-9, November, 1969

BROCKHAUS, R. (1986): 'A global model of the coupled process of aircraft motion and wind field'. Proceedings of the 2nd International Symposium on Aviation Safety, Toulouse, France, November 1986 (Cepad, Toulouse, 1988), pp. 123–137

BROMLEY, E., JR. (1977): 'Aeronautical meteorology: Progress and challenges – today and tomorrow', *Bull. Am. Meteorol. Soc.*, **58**, pp. 1156–1160

BROWNING, K.A., ED. (1982): 'Nowcasting' (Academic Press, London), p. ix

BUCK, R.N. (1978): 'Weather flying' (McMillan, New York, Revised Edition)

CIVIL AERONAUTICS BOARD (1963): 'Lockheed Electra L-188, N6102A, American Airlines, Inc., Municipal Airport, Kansas City, Missouri, January 29, 1963', CAB File No. 1-0018

CIVIL AERONAUTICS BOARD (1964a): 'Slick Airways Division, The Slick Corporation, Douglas C-54B-DC, N384, Castle Island, Boston, Massachusetts, March 10, 1964', CAB File No. 1-003

CIVIL AERONAUTICS BOARD (1964b): 'Boeing 707-121, N709PA, Pan American World Airways, Inc., near Elkton, Maryland, December 8, 1963', CAB File No. 1-0015

CIVIL AVIATION AUTHORITY (1978): 'Accidents to aircraft on the British register, 1977', Civil Aviation Authority, London, Report CAP 419, October 1978

COOPER, W.A., SAND, W.R., POLITOVICH, M.K., and VEAL, D.L. (1984): 'Effects of icing on performance of a research airplane', *J. Aircr.*, **21**, pp. 708–715

CZEKALSKI, L. (1983): 'Overview of FAA's aircraft icing programs'. Proceedings of 7th Annual NASA Workshop, pp. 30–34

DIETENBERGER, M.A., HAINES, P.A., and LUERS, J.K. (1985): 'Reconstruction of Pan Am New Orleans accident', *J. Aircr.*, **22**, pp. 719–728

ELLROD, G.P., and NELSON, J.P. (1997): 'An experimental GOES image product to identify conditions favorable for aircraft icing'. Preprints of 7th Conference on Aviation, Range, and Aerospace Meteorology, Long Beach, CA, 2–7 February 1997 (American Meteorological Society, Boston), pp. 112–115

ETKIN, B. (1982): 'Dynamics of flight: Stability and control' (Wiley, New York)

EVANS, J.E. (1991): 'Integrated Terminal Weather System (ITWS)'. Preprints of 4th International Conference on Aviation Weather Systems, Paris, France, 24–28 June 1991 (American Meteorological Society, Boston), pp. 118–123

FISCHETTI, M.A., and PERRY, T.S. (1986): 'Our burdened skies', *IEEE Spectrum*, **23**, (11), pp. 36–37

FITZGERALD, D.R. (1967): 'Probable aircraft triggering of lightning in certain thunderstorms', *Mon. Weather Rev.*, **95**, pp. 835–842

FROST, W., and CAMP, D.W. (1977): 'Wind shear modeling for aircraft hazard definition'. Federal Aviation Administration Report FAA-RD-77-36

FUJITA, T.T. (1986): 'DFW microburst on August 2, 1985'. The University of Chicago, SMRP Research Paper 217

HAINES, P.A., and LUERS, J.K. (1983): 'Aerodynamic penalties of heavy rain on landing airplanes', *J. Aircr.*, **20**, pp. 111–119

HANSMAN, R.J., JR. (1989): 'The influence of ice accretion physics on the forecasting of aircraft icing conditions'. Proceedings of 3rd International Conference on the Aviation Weather System, Anaheim, CA, 30 January–3 February 1989 (American Meteorological Society, Boston), pp. 154–158

HAUF, T. (1993): 'An in-situ system for warning of icing conditions'. Preprints of 5th International Conference on Aviation Weather Systems, Vienna, VA, 2-6 August 1993 (American Meteorological Society, Boston), pp. 161–162

HOFFMAN, H.E. (1991): 'Some aspects of an icing atmosphere'. SAE Subcommittee AC-9C, Working Group on Aircraft Icing Technology, Orlando, FL

HUFFMAN, P., and HAINES, P.A. (1984): 'Visibility in heavy precipitation and its use in diagnosing high rainfall rates', AIAA Paper 84-0541, January 1984

JACKSON, R.L. (1991): 'Low-level wind shear terminology'. Preprints of 4th International Conference on Aviation Weather Systems, Paris, France, 24–28 June 1991 (American Meteorological Society, Boston), pp. 13–15

JERVIS, E.R. (1964): 'The safety and reliability of an all-weather landing system', ARINC Research Corp. Publication No. 4276-436, Washington, DC, May 1964

KAYTON, M. (1969): 'Landing guidance', Kayton, M. and Fried, W.R., Eds., 'Avionics navigation systems' (John Wiley, New York), Chap. 14

KAYTON, M., and FRIED, W.R. (1997): 'Avionics navigation systems' (Wiley, New York, 2nd edn.)

KESSLER, E. (1988): 'Use, nonuse and abuse of weather radar', *J. Aircr.*, **25**, pp. 448–452

KESSLER, E. (1990): 'On Low-Level Windshear Alert Systems (LLWAS) and Doppler radar in aircraft terminal operations', *J. Aircr.*, **27**, pp. 423–428

KORKAN, K.D., CROSS, E.J., JR., and CORNELL, C.C. (1985): 'Experimental aerodynamic characteristics of an NACA 0012 airfoil with simulated ice', *J. Aircr.*, **22**, pp. 130–134

KRAMER, K.C. (1965): 'An operational all-weather landing system'. Proceedings of 17th National Aerospace Electronics Conference, Dayton, OH (IEEE and AIAA), pp. 489–507

LEE, J.T., and BECKWITH, W.B. (1981): 'Thunderstorms in aviation', in E. Kessler (Ed.): 'Thunderstorms: A social, scientific, and technological documentary, Vol. 1: The thunderstorm in human affairs'. US Department of Commerce, National Oceanic and Atmospheric Administration, Environmental Research Laboratories, Chap. VI, pp. 141–169

LEWIS, M.S., ROBINSON, P.A., HINTON, D.A., and BOWLES, R.L. (1994): 'The relationship of an integral wind shear hazard to aircraft performance limitations'. NASA Technical Memorandum 109080, February 1994

LUERS, J.K., and HAINES, P.A. (1983): 'Experimental measurements of rain effects on aircraft aerodynamics'. AIAA Paper 83-0275, January 1983

MAHAPATRA, P.R., and ZRNIC', D.S. (1991): 'Sensors and systems to enhance aviation safety against weather hazards', *Proc. IEEE*, **79**, pp. 1234–1267

MAKKONEN, L., and STALLABRASS, J.R. (1987): 'Experiments on the cloud droplet collision efficiency of cylinders', *J. Clim. Appl. Meteorol.*, **26**, pp. 1406–1411

MAZUR, V., FISHER, B.D., and GERLACH, J.C. (1984): 'Lightning strikes to an airplane in a thunderstorm', *J. Aircr.*, **21**, pp. 607–611

NATIONAL ACADEMY OF SCIENCES (1983): 'Low altitude wind shear and its hazard to aviation' (National Academy Press, Washington, DC)

NATIONAL TRANSPORTATION SAFETY BOARD (NTSB) (1967): 'Braniff Airways, Inc., BAC 1-11, N1553 near Falls City, Nebraska, August 6, 1966', Docket SA 393

NATIONAL TRANSPORTATION SAFETY BOARD (NTSB) (1970): 'Wien Consolidated Airlines, Inc., Fairchild F-27B, N4905, Pedro Bay, Alaska, December 2, 1968'. Report NTSB-AAR-70-16

NATIONAL TRANSPORTATION SAFETY BOARD (NTSB) (1972a): 'Pan American World Airways, Inc., Boeing 747–121, N739PA near Nantucket, Massachusetts, November 4, 1970,' Report NTSB AAR-72-14, 1972

NATIONAL TRANSPORTATION SAFETY BOARD (NTSB) (1972b): 'National Airlines, Inc., Boeing 747–135, N77772 near Lake Charles, Louisiana, January 4, 1972,' Report NTSB AAR-72-21, 1972

NATIONAL TRANSPORTATION SAFETY BOARD (NTSB) (1972c): 'Northwest Airlines, Inc., Boeing 747–151, N606US over the North Pacific Ocean 105 Nautical Miles West of 150° East Longitude at 36° North Latitude, April 12, 1972,' Report NTSB AAR-72-27, 1972

NATIONAL TRANSPORTATION SAFETY BOARD (NTSB) (1974): 'Skyways International, Inc., Douglas DC-7C, N296 near the Miami International Airport, Dade County, Florida, June 21 1973', Report NTSB-AAR-74-2

NATIONAL TRANSPORTATION SAFETY BOARD (NTSB) (1975): 'Air France Boeing 707B–328B, F-BLCA near O'Neill, Nebraska, May 13, 1974' Report NTSB AAR-75-4, 1975

NATIONAL TRANSPORTATION SAFETY BOARD (1978): 'Southern Airways, Inc., Douglas DC-9-31, N1335U, New Hope, Georgia, April 4, 1977'. Report NTSB AAR-78-3

NATIONAL TRANSPORTATION SAFETY BOARD (1979): 'Rocky Mountain Airways, Inc., DeHavilland DHC-6 Twin Otter, N25RM near Steamboat Springs, Colorado, December 4, 1978'. Report NTSB AAR-79-6

NATIONAL TRANSPORTATION SAFETY BOARD (1980): 'Air Wisconsin, Inc., Swearingen SA-226 Metro, N650S, Valley, Nebraska, June 12, 1980'. Report NTSB AAR-80-15

NATIONAL TRANSPORTATION SAFETY BOARD (1981): 'Redcoat Air Cargo Ltd., Bristol Britannia 253F, G-BRAC, Billerica, Massachusetts, February 16, 1980'. Report NTSB AAR-81-3

NATIONAL TRANSPORTATION SAFETY BOARD (1983): 'Pan American World Airways Clipper 759, N4737, Boeing 727-235, New Orleans International Airport, Kenner, Louisiana, July 9, 1982'. Report NTSB AAR-83-02

NATIONAL TRANSPORTATION SAFETY BOARD (1986): 'Aircraft accident report: Delta Airlines, Inc., Lockheed L-1011-385-1, N726DA, Dallas/Fort Worth International Airport, Texas, August 2, 1985'. Report NTSB/AAR-86/05, August 1986

NATIONAL TRANSPORTATION SAFETY BOARD (1993a): 'Annual review of aircraft accident data, US carrier operations, Calendar Year 1990'. Report NTSB/ARC-93/02, 4 October 1993

NATIONAL TRANSPORTATION SAFETY BOARD (1993b): 'Annual review of aircraft accident data, US general aviation, Calendar Year 1990'. Report NTSB/ARG-93/02, 17 December 1993

NATIONAL TRANSPORTATION SAFETY BOARD (1993c): 'Annual review of aircraft accident data, US carrier operations, Calendar Year 1989'. Report NTSB/ARC-93/01, 7 May 1993

NATIONAL TRANSPORTATION SAFETY BOARD (1993d): 'Takeoff stall in icing conditions, USAIR Flight 405, Fokker F-28, N485US, Laguardia Airport, New York, New York Report NTSB/AAR-93/02, 17 February 1993

NOLAN, M.S. (1994): 'Fundamentals of air traffic control' (Wadsworth Publishing Co., Bellmont, CA, 2nd edn.)

OSEGUERA, R.M., BOWLES, R.L., and ROBINSON, P.A. (1992): 'Airborne *in situ* computation of the wind shear hazard index'. AIAA Paper 92-0291, Proceedings of the 30th Aerospace Sciences Meeting, Reno, NV, 6–9 January 1992 (American Institute of Aeronautics and Astronautics, Washington, DC)

PERKINS, P., and REIKE, W. (1993): 'Aircraft icing problems – after 50 years', AIAA Paper 93-0392, Proceedings of the 31st Aerospace Sciences Meet and Exhibit, Reno, NV, 11–14 January 1993 (American Institute of Aeronautics and Astronautics, Washington, DC)

PHILLIPS, E.H. (1989): 'NASA will study heavy rain effects on wing aerodynamics', *Aviat. Week Space Technol.*, **130**, (7), pp. 38–41

PITTS, F.A., FISHER, B.D., MAZUR, V., and PERALA, R.A. (1988): 'Aircraft jolts from lightning bolts', *IEEE Spectrum*, July 1988, pp. 34–38

POLITOVICH, M.K. (1989): 'Aircraft icing caused by large supercooled droplets', *J. Appl. Meteorol.*, **28**, pp. 856–868

POLITOVICH, M.K. (1993): 'Aircraft icing: Meteorological effects on aircraft performance'. Preprints of 5th International Conference on Aviation Weather Systems, Vienna, VA, 2–6 August 1993 (American Meteorological Society, Boston), pp. 435–439
POLITOVICH, M.K. (1996): 'Response of a research aircraft to icing and evaluation of severity indices', *J. Aircr.*, **33**, pp. 291–297
POLITOVICH, M.K., and BERNSTEIN, B.C. (1995): 'Production and depletion of supercooled liquid water in a Colorado winter storm', *J. Appl. Meteorol.*, **34**, pp. 2631–2648
PRATT, K.G., and WALKER, W.G. (1954): 'A revised gust load formula and reevaluation of V-G data taken on civil transport airplanes from 1933–1950'. Report NACA-1206
PROFS (1979): 'Report of a study to estimate economic and convenience benefits of improved local weather forecasts'. NOAA Technical Memorandum ERL-PROFS-1
RIGDON, G.G. (1991): 'Weather support activities for the Space Shuttle'. Preprints of 4th International Conference on Aviation Weather Systems, Paris, France, 24–28 June 1991 (American Meteorological Society, Boston), pp. 1–6
RUDICH, R.D. (1986): 'Weather-involved U.S. air carrier accidents 1962–1984: A compendium and brief summary'. Proceedings of AIAA 24th Aerospace Sciences Meeting, Reno, NV, Paper AIAA-86-0327
SAND, W.R., POLITOVICH, M.K., and RASMUSSEN, R.M. (1990): 'A program to improve aircraft icing forecasts'. Proceedings of AIAA 28th Aerospace Sciences Meeting, Reno, NV, Paper AIAA-90-0196
SANKEY, D.A. (1993): 'An overview of FAA-sponsored aviation weather research and development'. Preprints of 5th International Conference on Aviation Weather Systems, Vienna, VA, 2–6 August 1993 (American Meteorological Society, Boston), pp. 1–4
SERAFIN, R.J. (1986): 'Nowcasting and aviation safety'. Proceedings of the Workshop on Multiparameter, Doppler Weather Radar for India, Bangalore, India, pp. 15–19
SHEN, J., PARKS, E.K., and BACH, R.E. (1996): 'Comprehensive analysis of two downburst-related aircraft accidents', *J. Aircr.* **33**, pp. 924–930
STEWART, O. (1958): 'Danger in the air' (Philosophical Library, London), pp. 59–62
SWOLINSKY, M. (1986), 'Wind shear models for aircraft hazard investigation'. Proceedings of 2nd International Symposium on Aviation Safety, Toulouse, France, November 1986 (Cepad, Toulouse, 1988), pp. 101–122
SYRETT, W.J. (1991): 'Hourly wind profiler observations of the jet stream: wind shear and pilot reports of turbulence'. Preprints of 25th International Conference on Radar Meteorology, 24–28 June 1991, Paris, France (American Meteorological Society, Boston, MA), pp. 319–322
THOMPSON, B.E., and JANG, J. (1996): 'Aerodynamic efficiency of wings in rain', *J. Aircr.*, **33**, pp. 1047–1053
VIEMEISTER, P.E. (1961): 'The lightning book' (Doubleday, New York)
VISSER, H.G. (1997): 'Lateral escape strategies for microburst windshear encounter', *J. Aircr.*, **34**, pp. 514–521
WEBBER, G.W. (1965): 'Landing aircraft safely in all weather', *Engineering*, 10 September 1965, pp. 333–335
WESTWATER, E.R., and KROPFLI, R.A. (1989): 'Remote sensing techniques of the Wave Propagation Laboratory for the measurement of supercooled liquid water: Application to aircraft icing'. NOAA Technical Memorandum ERL WPL-163, May 1989
WHITE, R.J. (1992): 'Effect of wind shear on airspeed during airplane landing approach', *J. Aircr.*, **29**, pp. 237–242

Chapter 4
Origins of harmful atmospheric effects on aircraft

4.1 General

The basic types of atmospheric effects on aviation were discussed in the preceding chapter. However, these effects seldom occur in isolation or in their pure form. Analysis of aircraft accidents shows most of them to occur because of a combination of circumstances. Meteorological factors may combine with nonmeteorological factors to cause accidents and incidents. Further, where meteorological factors are solely responsible for accidents, a number of such factors may be involved. Again, there may not be a one-to-one correspondence between atmospheric phenomena and the types of effect encountered by aircraft. Thus a single phenomenon may give rise to different types of effects on aircraft. More importantly, certain types of parent phenomena harbour a variety of subprocesses, each with its distinct effects on aviation. It is thus necessary to understand the major types of atmospheric phenomena of significance to aviation before proceeding to discuss the design aspects of the instruments and systems devised to detect and characterise them.

As mentioned before, aviation is much more sensitive to severe local phenomena and local variation of aviation-significant parameters than to large-scale or global climatic processes and trends which are of great significance in many branches of meteorology. The local phenomena may, of course, be driven by the overall climatic conditions. For example, the occurrence of a large low-pressure system over an area would give rise to myriads of thunderstorms. However, the type of instrumentation and modelling that provide information and forecasts about global and regional weather patterns do not provide details on a local scale that are adequate for aviation applications. This chapter concentrates on the types of local phenomena that serve as origins for many of the atmospheric effects on aviation discussed in the preceding chapter.

4.2 Structure of atmosphere

Before proceeding to discuss the origins of atmospheric phenomena of significance to aviation, it is necessary to understand the structure of the

80 Aviation weather surveillance systems

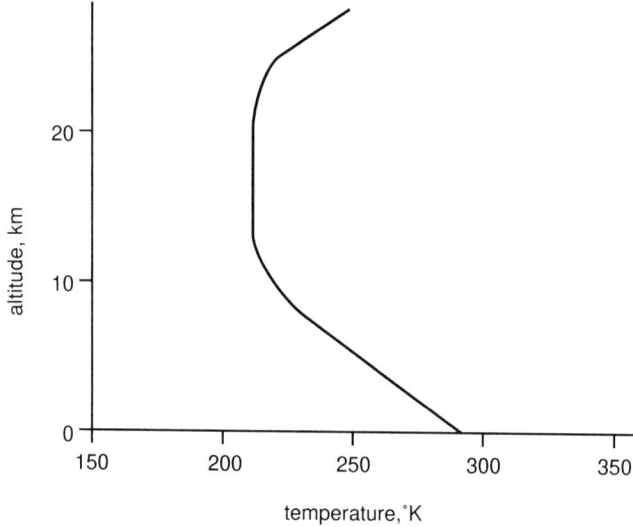

Figure 4.1   The nature of temperature variation in the lower atmosphere

atmosphere which supports flight and also harbours the weather phenomena that affect flight.

Air is a compressible fluid. Its density increases with increase in pressure according to the gas law. Since each layer of the atmosphere bears the weight of the entire column of air above it, the pressure of air is highest near the surface of the earth, falling nearly exponentially with height. The density also shows a corresponding variation. As the pressure and density vanish asymptotically in the outer reaches of the atmosphere, the mean free path of air molecules becomes comparable to the dimensions of flight vehicles or their components, resulting in 'free-molecular flow'. However, at heights covered by normal aircraft flight, the mean free paths are small enough for the air flow to be treated as 'continuum flow'.

Figure 4.1 shows the main layers of the earth's lower atmosphere. These layers are defined according to the nature of temperature variation in a vertical direction. The broad temperature variation of the atmosphere is governed by the energy absorption at different layers of the atmosphere, with the incident and reflected solar radiation, and the earth's self-radiation, serving as inputs. The principal role in the absorption process is played by water vapour, with carbon-dioxide playing a secondary role. Most of the energy in the direct and reflected solar radiation occurs in the visible and near-infra-red frequencies which pass through the atmosphere with little absorption. However, the self-radiation of the earth, with a surface temperature in the region of 300 K, is rich in wavelengths of the order of 10 μm, which are heavily absorbed by water vapour. As the layer of the atmosphere close to the earth's surface has a higher concentration of water vapour because of diffusion from water bodies and the moist ground, most of the earth's radiation is trapped in the lower layers of the atmosphere, raising their temperature. The screening of the earth's radiation by the lower layers, coupled with a decreasing water vapour concentration, results in a steady fall of temperature as height increases, as shown in Fig. 4.1. The lowest layer of atmosphere, exhibiting such a negative temperature gradient with respect to altitude, is called the *troposphere*. Above the troposphere is a layer of air called the *tropopause* where the temperature remains essentially constant, and beyond that lies the *stratosphere*, in which the temperature increases with height. Nearly all of aviation activity occurs within the troposphere and the tropopause, with very little spilling over to the stratosphere.

4.3 Thunderstorms: nature, initiation and evolution

The origin of a vast majority of deleterious atmospheric effects on aviation can be traced to thunderstorms. In the US, during the period from 1962 to 1984, nine out of eleven fatal weather-related air carrier accidents occurred in the area of thunderstorms (Rudich, 1986). Also in the US, an analysis of 51 accidents or incidents caused by wind shear in the extended period between 1959 and 1983 showed about two-thirds of the events to be associated with convective storms, i.e. thunderstorms and rain or snow showers (US Department of Transportation, 1987). A skeletal understanding of thunderstorms is therefore necessary to appreciate the nature of their effects on aviation and to understand the design and operation of aviation weather surveillance systems. The following subsections are intended to provide such a picture of thunderstorms. For more detailed understanding of thunderstorms the reader is referred to many good books that are available on the subject (e.g. Battan, 1961; Magono, 1980; Kessler, 1981–1982).

The ground, heated by solar radiation, sets off convection in the atmosphere. Vertical atmospheric convection may also be initiated by topographic factors such as sloping ground or mountainsides deflecting horizontal wind flows upwards. Yet other factors that may force convection in a vertical direction are related to aerodynamic phenomena such as flow convergence, cold and warm fronts, etc. The moist air near the earth's surface, on rising due to convection, cools down because of adiabatic expansion as well as mixing with the cooler air at higher altitudes. When the temperature of the air mass falls below its saturation temperature, the condensation of a part of the water vapour in the air results in the formation of clouds and possibly rain.

Relatively weak convection results in *cumulus clouds* which are scattered cloud masses, often occurring in fair weather over large areas. Cumulus clouds may have dimensions up to a couple of kilometres across, and are separated several kilometres from one another on average. Such clouds are usually too small to yield precipitation. Relatively hot spots or other strong topographical features may yield stronger convection that rises significantly higher than the surrounding cumulus clouds.

Cumulonimbus convection resulting in the formation of thunderstorms is facilitated in environments where a deep *conditionally stable atmospheric layer* exists from the vicinity of the ground upward. A conditionally stable layer is one in which the temperature *lapse rate* (the rate of decrease of temperature with height) exceeds the adiabatic rate of cooling of water-vapour-saturated air during its ascent through the atmosphere. This so-called *saturated adiabatic lapse rate* is lower than the lapse rate for unsaturated adiabatic cooling because of the condensation of water and the consequent release of the latent heat of condensation. In such a conditionally stable layer a rising parcel of saturated air is unstable, i.e. its rise through the layer leaves it warmer than the surrounding air at any given height, subjecting it to positive (upward) buoyancy forces which further accelerate its ascent. Successive masses of air going through the same process establish a continuing column of upward moving air, leading to the formation of a thunderstorm cell. The buoyancy force driving the air column and the resulting vertical acceleration are further reinforced by the reduction in the mass of the air parcels due to the loss of water in the form of precipitation.

Although the *parcel theory of convection* offers a convenient way of visualising and modelling the formation of thunderstorms, they are actually rather complex processes with many factors governing their energetics (Darkow, 1982). In particular, while the parcel theory assumes the convected air to rise without energy exchange with the surrounding air mass, in practice there are varying levels of mixing between the two air masses. Among other factors, this mixing lends a high degree of variability to the evolution of thunderstorms under apparently similar atmospheric stability conditions. Indeed, in a large fraction of the cases of convection in the conditionally stable troposphere, the clouds that form fail to grow beyond the small cumulus stage. A deep conditionally stable tropospheric layer is thus a necessary, but not sufficient, condition for the formation of thunderstorms (Ludlam, 1982). This unpredictability of the detailed aspects of the location, growth, intensity and structure of individual thunderstorms makes their aviation hazard potential assessment and alleviation more dependent on observation than on modelling and prediction.

Returning to the thunderstorm process itself, the rising air mass experiences positive buoyancy, and hence upward acceleration, till it reaches the equilibrium level at the *height of neutral density* where the density of the rising air parcel equals that of the ambient air. Because of the nonexistence of density differential, the buoyancy force vanishes at that height, which is therefore also called the *height of neutral buoyancy*. As a result of the integrated effect of the upward acceleration up to the equilibrium level, the ascending air mass attains the highest vertical speed at this height. Beyond this height the air still continues to rise by its momentum, but the rising air parcel, being cooler and denser than the surrounding air, experiences negative buoyancy (i.e. the weight of the air mass exceeds the upward buoyancy force) and hence undergoes deceleration. The height at which the vertical air velocity is

Origins of harmful atmospheric effects on aircraft 83

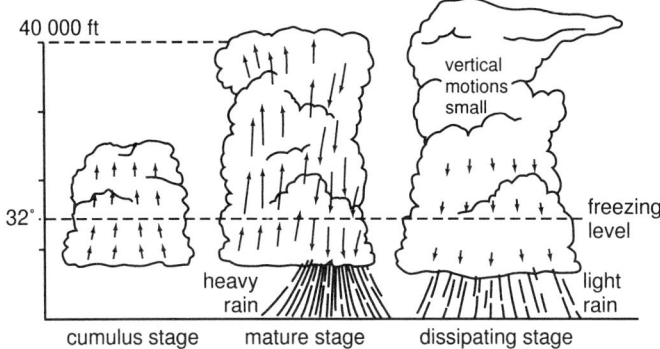

Figure 4.2 The three main stages in the life-cycle of a thunderstorm (adapted from Byers and Braham, 1949)

completely nullified, i.e. the air mass stops rising, corresponds to the *storm top*. When the vertical speed of the rising air is arrested, it spreads out horizontally, often in an asymmetric manner due to the winds prevailing at the altitude, forming an *anvil*.

The above evolutionary picture shows thunderstorms to be highly dynamic phenomena with well defined life-cycles. The life-cycle of a typical thunderstorm may be divided into three phases, as shown schematically in Fig. 4.2. The first phase of thunderstorm growth, characterised by the presence of updraft throughout the cell, and the formation of cloud columns due to the condensation of moisture, is called the *cumulus stage*. At a certain point in this stage the water or ice particles in the cloud grow to sufficiently large sizes to start falling down. During the fall they grow further in size due to the coalescence of colliding droplets. The growing particle mass causes the fall speeds to increase against atmospheric drag. The drag force between the falling particles and the surrounding air causes a part of the air mass to descend. The air mass is further cooled by evaporation of the droplets or melting ice particles. The resulting increase of density accelerates the descent of the air mass. Thus the mature thunderstorm contains both rising and descending shafts of air, i.e. *updraft* and *downdraft*, side by side. The overall structure of a mature thunderstorm is shown schematically in Fig. 4.3.

The falling water droplets constitute rain. At high altitudes, where the temperatures are below the freezing temperature of water, the water drops freeze to form hail. During their fall, hail pellets are often retarded strongly by rising air currents which may keep them lifted within the frigid zone for considerable lengths of time. This allows the hail to accrete more water mass and grow to large sizes.

After an active *mature stage*, the thunderstorm reaches its final or *dissipating stage* during which the factors forcing the vigorous updraft are weakened. This happens due to a combination of rain-quenching of the hot spot initiating the updraft, the descent of the cool air from the storm core to the lower levels,

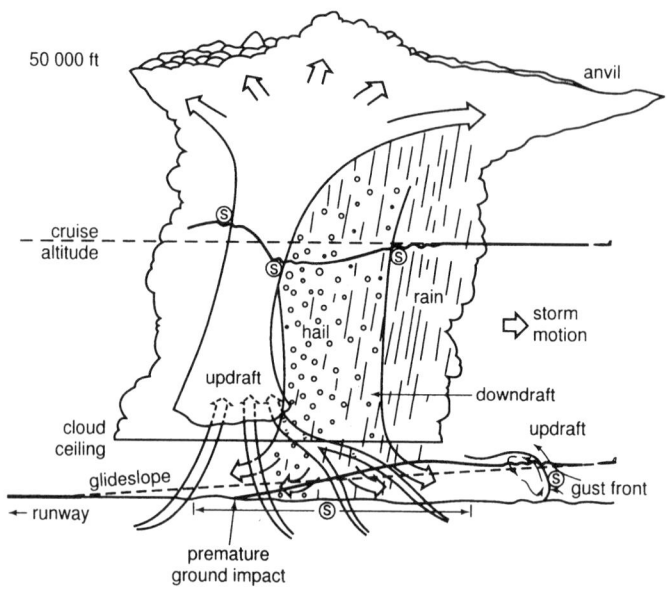

Figure 4.3 The overall structure of a typical thunderstorm, and its possible effects on aircraft during cruise and landing. Dotted lines represent expected flight paths and the solid lines actual flight paths. The notation (S) indicates the likely zones of wind shear, and the wavy segments of flight paths result from turbulence. The updraft may have a superimposed rotation, resulting in helical air motion

the counteracting of the rising air mass by the descending precipitation and air mass, and by the mixing of the rising shaft of air with the surrounding air mass and the cool descending air current. The dissipating stage is characterised by relatively weak downdrafts throughout, and the dying down of the turbulence, precipitation and lightning activity that are strongly manifest during the mature stage. The cloud debris left after the dynamic phases of the thunderstorm eventually evaporates away.

An individual thunderstorm cell, to a fair approximation, is a closed system in the sense that the air masses maintaining the ascending and descending flows within the thunderstorm do not significantly interact with the ambient air. A class of severe thunderstorms, called *supercells*, are open systems believed to organise a steady flow to propagate through the ambient air, interacting with the large-scale environment of the storm. Supercells are found in environments characterised by strong wind shear at the lower levels, rotation of the wind vector with height (Colour Plate 1), and substantial convective instability (Ray, 1990). These thunderstorms are of relatively infrequent occurrence, but are generally more violent than ordinary single-cell storms, often producing giant hail, strong surface winds, strong and persistent tornadoes, and intense updrafts coexisting with strong downdrafts for long periods of time (Burgess and Lemon, 1990).

Thunderstorm cells often occur in groups or complexes containing several individual cells in different stages of evolution. *Thunderstorm complexes* naturally have a larger spatial and temporal extent than individual thunderstorm cells. A linear arrangement of thunderstorms is called a *squall line*. Squall lines may extend over hundreds of kilometres and often generate violent levels of rain, hail and strong winds.

4.4 Thunderstorm parameters

Following the phenomenological description of the thunderstorm process, it would be instructive to quantify the various parameters associated with thunderstorms that are of importance to aviation. A single thunderstorm cell is typically a few to several kilometres across, and has a life-cycle spanning the order of an hour. Supercells have longer lifetimes on the average, lasting up to 6 h.

The updraft in thunderstorms often reaches speeds of the order of 25 m/s (~50 knots) at heights of the order of 25 000 ft. Vertical speeds of 50–70 m/s are possible in the upper levels of thunderstorms. These figures may be viewed in the context of normal cruising altitudes of aircraft (see Table 2.2), which are typically between 30 000 and 40 000 ft for subsonic jet airliners. Supercells are known for their strong and sustained updrafts, of the order of 25–50 m/s.

Another parameter of great importance for aviation is the height of thunderstorms. As mentioned before, the thunderstorm updraft often accelerates all the way up to the tropopause, with its momentum helping it penetrate into the lower layers of the stratosphere. In middle and mid-high latitudes thunderstorm tops frequently reach heights of the order of 40 000 ft above sea level. Particularly energetic thunderstorms may reach significantly higher levels. For example, the tops of the Big Thompson thunderstorms in Colorado, USA, on 31 July 1976, which caused heavy flash floods, were estimated to be as high as 57 000 ft (~18 km) (Maddox *et al.*, 1977). In tropical regions, where the tropopause is higher, storm tops generally reach greater heights than in the mid-latitudes, exceeding 63 000 ft (~20 km) in a significant percentage of cases (see Fig. 4.4). Such a height distribution means that nearly all types of aviation activity are subject to the vagaries of thunderstorms.

Also of interest from the point of view of aviation is the global distribution of thunderstorms. Since solar energy provides the ultimate driving force for atmospheric processes, thunderstorm activity is the highest in the belt surrounding the earth's equator, which receives the highest average insolation. Within this belt land areas generally receive more rainfall than those covered by the oceans. Thunderstorm activity is particularly frequent in the

86 Aviation weather surveillance systems

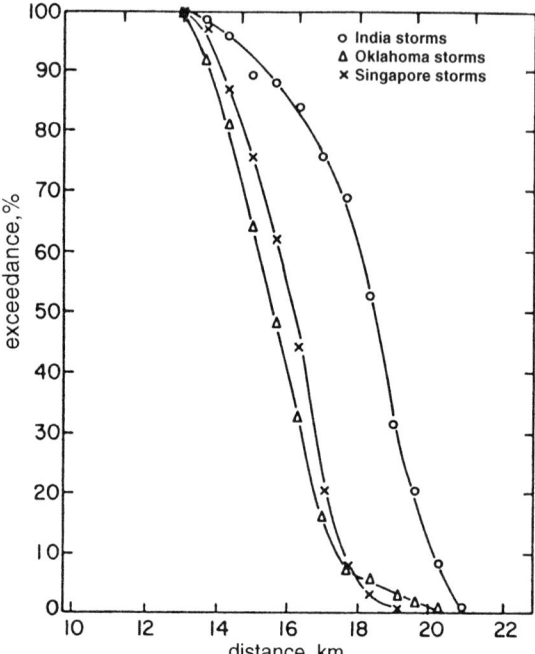

Figure 4.4   *Percentage of storm tops exceeding a given height at three different geographical locations on the earth* (from Lee and Beckwith, 1981, by editor's permission)

tropical parts of South America, coastal regions of equatorial Africa, and the East Indies (Court and Griffiths, 1982). Over the seas thunderstorm frequency is high in the region of the Gulf of Panama, Caribbean Sea, Gulf of Guinea, and the part of the Indian Ocean around the Indonesian archipelago (Sanders and Freeman, 1982). It must, however, be noted that the frequency of thunderstorm occurrence does not accurately indicate the strength of individual thunderstorms; individual thunderstorms and storm complexes in the mid-latitudes are often more violent than those anywhere else.

The parameters given above pertain to thunderstorms as a whole. The quantitative aspects of the various individual types of phenomena associated with thunderstorms are discussed in the following subsections.

4.5 Phenomena associated with thunderstorms

The effects of different types of atmospheric phenomena on aviation were discussed in Chapter 3. However, the origin of such phenomena was not traced. We are now in a position to understand that the thunderstorm is the causative factor for many of these phenomena. Although the thunderstorm is

Origins of harmful atmospheric effects on aircraft 87

a single entity in an evolutionary sense, it is a highly complex process that spawns and/or harbours many different types of individual phenomena that are of concern to aviation. The significant ones among these are discussed below.

4.5.1 Divergence and convergence

The outward flow of air from a localised zone constitutes *divergence*. Its opposite, i.e. the inward flow of air towards a common region, is termed *convergence*. Divergent and convergent flows may occur from or towards a common point (or small area/volume), in which case the flow is said to involve *radial divergence* or *radial convergence*. Divergence or convergence from or towards a line (or narrow linear zone) is called *linear divergence* or *linear convergence*. These possibilities are shown schematically in Fig. 4.5a–d. In a more general sense divergence or convergence may be said to occur if there is a sudden jump (increase or decrease) in air speed along the flow path, though the direction of flow may not change (Fig. 4.5e,f).

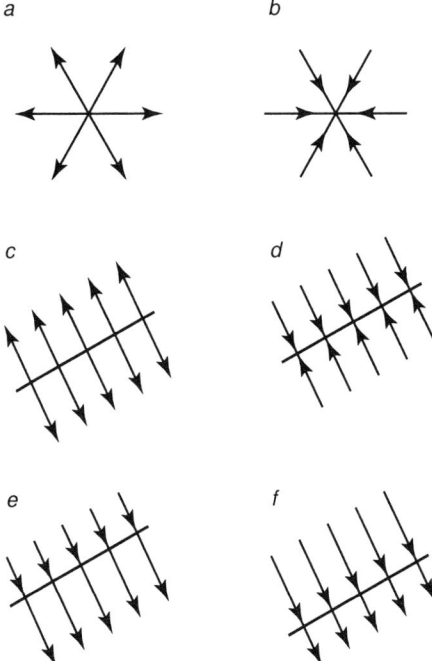

Figure 4.5 Types of two-dimensional divergent and convergent flow fields: (a) radial divergence, (b) radial convergence, (c) linear divergence, (d) linear convergence, (e) generalised divergence and (f) generalised convergence. In the last two cases the arrow lengths represent the flow speeds.

88 *Aviation weather surveillance systems*

Radial divergence is associated with thunderstorms at two altitude levels. At higher altitudes, near the tops of storms, divergence is caused when the vertical speed of the rising column of air in the unstable layer is arrested by the stable layer above. The thunderstorm also gives rise to divergence at low altitudes when the descending shaft of cold air encounters the ground and spreads outwards. Both these types of divergence may be visualised by referring to Fig. 4.3. Thunderstorm-generated divergence may be modelled as two-dimensional flow with good accuracy for practical purposes. As the ground is a hard boundary, low-level divergence associated with thunderstorms usually occurs over a thinner layer and hence packs higher horizontal speeds and speed gradients than the divergence at storm tops. The low-level divergence is very often radially symmetric, to a first approximation, though significant asymmetry is not uncommon (Eilts and Doviak, 1987).

In a radially divergent field, the air speed is the highest near the source of divergence, and decays outwards as the air expands in ever-enlarging circles. The resulting spatial gradient of wind speed constitutes wind shear. An aircraft entering a divergent flow field first encounters head wind, which increases as the aircraft flies towards the source of divergence. While flying through or close by the source, the high head wind turns rapidly into high tail wind, which then decays as the aircraft flies out of the divergence (Fig. 4.6).

Wind shear fields due to the low-altitude divergence caused by thunderstorms are particularly hazardous to aviation. The spatial scale of peak wind reversal is typically a few kilometres, and aircraft attempting to land through such a field may experience the reversal within a timescale of the order of 1 min. To minimise the landing speed, and hence runway length requirement, aircraft have low stall margins during approach to landing. The usual

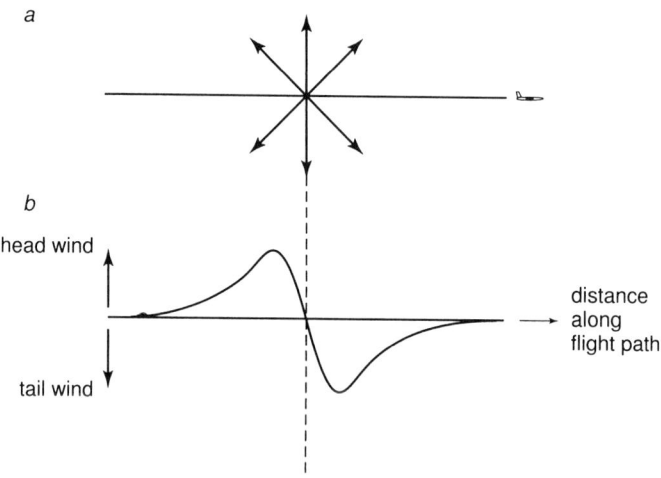

Figure 4.6 *The passage of an aircraft through a radially divergent flow field (a), and the nature of the resulting wind variation experienced by the aircraft (b)*

stall margins are of the order of 30%, i.e. the landing speed is only about 1.3 times the stall speed of the aircraft (Kayton and Fried, 1969). Under such flight conditions, sudden or rapid loss of air speed caused by flow reversal from headwind to tailwind as shown in Fig. 4.6 can cause high rates of descent (called *sink rate*), leading to premature and steep ground impact. The discussion in Section 3.3.1 and reference to the lower flight trajectory in Fig. 4.3 would be instructive here. The sink rate is further enhanced by the downward velocity component of the thunderstorm downdraft. Downdraft was held as being responsible for four aircraft accidents (including one commercial and one executive aircraft) in the UK in 1977 (Civil Aviation Authority, 1978), and 29 accidents (including one involving a scheduled commercial air carrier) in the USA in 1990 (National Transportation Safety Board, 1993a; 1993b).

Wind shear due to low-altitude thunderstorm divergence, and the resulting high sink rate, have been blamed for a large number of aircraft accidents during landing (National Transportation Safety Board, 1974a; 1974b; 1974c; 1976a; 1977; 1978a; 1983; 1985). Loss of air speed, and hence altitude, can also occur during aircraft takeoff through such divergence, leading to accidents (National Transportation Safety Board, 1976b; 1978b).

As already mentioned, because of the gradual braking action due to negative buoyancy, the divergence corresponding to storm tops occurs over a larger depth than that near the ground. The wind shear at storm tops is therefore, in general, weaker than in the case of low-altitude divergence. Further, a relatively small fraction of aviation activity occurs in the height band normally occupied by storm tops. Finally, aircraft have a long time interval and height reserve to recover from any loss of height or transient loss of control induced by the divergence at storm tops. Hence storm top divergence is a matter of less concern to aviation safety than low-level divergence, though the former can cause considerable passenger discomfort.

The convergent flow fields feeding air to the rising and descending shafts within thunderstorms can have effects on aviation that are qualitatively similar in the sense that horizontal wind velocity reversal is encountered during passage through the field. However, thunderstorm-induced convergence normally occurs over relatively large depths, and therefore involves weaker velocity gradients than divergence.

4.5.2 Turbulence

The wind fields associated with thunderstorms are not in general streamlined but contain significant gusty components. Gusts, which are irregular, local and transitory variations in velocity fields, are a matter of common experience in thunderstorm surroundings. The random, small-scale temporal and spatial variations in a wind field constitute turbulence which, as discussed in Section 3.3.2, is a strong source of hazard and discomfort for aviation.

Strong turbulence occurs at many locations in a thunderstorm. The strong updraft that initiates the thunderstorm and maintains its mature stage produces eddy-like pockets in its leading edge due to encounter and mixing with the relatively still surrounding air. These turbulent eddies give a cauliform appearance to the tops of cumulonimbus clouds and the anvils of developed thunderstorms. As mentioned in Section 4.3, a mature thunderstorm has the updraft and downdraft existing side by side. The air in the interface region between the two oppositely moving shafts is subjected to strong shearing effects and breaks up into turbulent eddies. At the bottom of the thunderstorm, the downdraft air spreads out, forming a front at its interface with the ambient air. The air at this interface is subject to mixing, resulting in the formation of turbulent eddies and gustiness.

From experience, certain characteristic levels of turbulence are found to be frequently associated with different types of convective clouds. Light turbulence (see Section 3.3.2 for classification of turbulence) is usually encountered within or in the immediate vicinity of fair-weather cumuli and altocumulus clouds. Moderate levels of turbulence are found in the environment of cumulonimbus clouds and towering cumuli, which may, occasionally, harbour severe turbulence. All thunderstorms generally contain regions with moderate levels of turbulence, and may contain some regions of severe turbulence. Severe turbulence is, however, found more frequently in rapidly growing or mature thunderstorms. Again, such storms can sometimes harbour extreme turbulence, which in turn is frequently associated with severe thunderstorms. Overall, about two-thirds of the severe and extreme turbulence events encountered by aircraft are of thunderstorm origin.

There is evidence of some correlation between the maximum level of turbulence in a thunderstorm and the strongest rain occurring within it. In a joint experimental programme called Project Rough Rider in the mid-1960s and 1970s, involving the British Royal Aeronautical Establishment, and the National Aeronautics and Space Administration, Air Force, Federal Aviation Administration, and the National Severe Storms Laboratory of the US, several structurally strong and instrumented aircraft were flown over a thousand times through thunderstorms to study the related turbulence phenomena. Among the numerous findings of the programme was an observation that storms with maximum rain intensity that measured <40 dBZ on the radar reflectivity scale (see Section 6.2.3), which corresponds to a rain rate of ~10 mm/h, did not produce any extreme or severe turbulence. Another significant finding was that high turbulence levels in a thunderstorm did not necessarily coincide with regions of high reflectivity (signifying rain activity) or reflectivity gradients, but were frequently encountered outside the thunderstorm core, sometimes more than 15 km from the storm centre (Lee and Beckwith, 1981). This observation is of high significance for aviation since avoiding high rainfall areas in a storm would not always mean avoiding the storm-related turbulence.

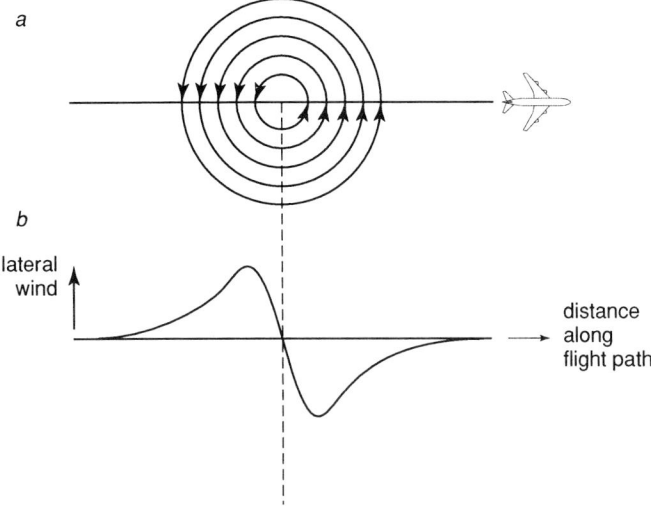

Figure 4.7 *The passage of an aircraft through a cyclonic (rotating) wind field top view causes a reversal of the direction of the lateral wind (a), as well as variation of the magnitude of the wind (b)*

4.5.3 Downburst

A thunderstorm downdraft which causes a harmful burst of outward winds at very low altitudes is called a *downburst*. Downbursts are of great interest to aviation because of their high hazard potential. A particularly dangerous type of downburst is the *microburst*. Because of its special significance in the context of aviation hazard, this phenomenon is covered separately in Section 4.7.

4.5.4 Cyclonic motion and tornadoes

Another type of air motion that can act as a significant source of wind shear involves cyclonic or rotational motion, normally about a near-vertical axis. Passage of aircraft through cyclonic fields causes variation and reversal of winds, but the winds in this case are lateral (Fig. 4.7), unlike the fore-and-aft winds in the case of convergent and divergent fields.

Mesocyclones are rotating masses of air with characteristic dimensions of the order of 10 km. The large-scale wind shear produced by these phenomena is usually not strong enough to be of serious concern to aviation. Also, their relatively large size gives aircraft sufficient time to adjust to the wind field, and recover from perturbations. Further, mesocyclones are relatively persistent, providing adequate opportunity for their detection and warning. However, relatively strong and spatially small features within mesocyclones may excite the dynamical modes of large aircraft. Mesocyclones are of particular concern to aviation because they are the source of the most violent tornadoes.

Cyclonic motion of much larger spatial scales occurs in tropical cyclones, hurricanes, and similar phenomena which appear very distinctly in satellite photographs. The total energy associated with the cyclonic motion in these phenomena is, of course, enormous. The wind fields within these phenomena consist of a smoothly varying large-scale (hundreds of kilometres) circulating component, superimposed with strong local variations. While steady strong winds are of concern to flight, the strong and rapid local variations of wind are of a much greater hazard potential.

The rotational air motion that is most destructive on a small spatial scale arises from *tornadoes*, which are spinning shafts of air usually spawned by mesocyclones (tornadoes of relatively low strength and duration, called by some 'gustnadoes', may arise in the absence of parent mesocyclones). Tornadoes have an average diameter of the order of 100 m (Davies-Jones, 1982), with the vast majority having diameters <1.5 km, and a significant number <30 m (Battan, 1961). The maximum wind speeds in intense tornadoes have been estimated to lie in the range 110–125 m s^{-1} (~220–250 knots) (Davies-Jones, 1982; Zrnic' et al, 1985). The distribution of speeds of dust aggregates in a violent tornado in Ohio, USA, obtained by photogrammetric tracking (Abbey and Fujita, 1981), shows a maximum speed of ~125 m s^{-1}. Such extreme speeds, reversing direction within the small distance corresponding to the tornado diameter, generate tremendous wind shear which endows tornadoes with devastating destructive power capable of physically destroying aircraft on encounter. A further complicating factor is that tornadoes form rapidly and move in apparently random directions, making their prediction difficult.

In spite of their awesome hazard potential, and regular and considerable toll of houses and other terrestrial structures, tornadoes have been associated with few hazards to flying aircraft. This is because of the small size and short life of tornadoes, which makes their probability of encounter with flying aircraft very small. For example, of the ~700–1200 tornadoes reported in the USA each year, a majority have diameters <100 m and travel <2 km before dissipating (Abbey and Fujita, 1981). The probability is further reduced due to the highly uneven distribution of tornado occurrences over the globe. A vast majority of the tornado events in the world occur in the central and midwestern parts of the USA (particularly the state of Oklahoma) which are not among the zones of heaviest aviation activity. Among other tornado-prone areas on the globe are the north-eastern parts of India and northern Argentina, where aviation activity is at a low level. There are, however, some areas of the world such as the UK, Western Europe, South Africa, Japan, New Zealand, and the eastern and western coastal areas of Australia which experience considerable tornadic activity (Fujita, 1973) while supporting moderate to high levels of aviation.

Another reason that flying aircraft have low probability of tornado encounter is that good flying practice prompts pilots to avoid severe thunderstorms which are generally the ones that spawn a majority of tornadoes. In spite of these saving factors tornadoes are known to have destroyed aircraft parked on the ground, where they are exposed to the chance of tornado encounter for extended periods. Tornadoes have also been suspected to be the causative factor in some unexplained accidents involving flying aircraft. As aviation activity and density continues to increase, the threat of the tornado as a hazard factor cannot be taken lightly.

4.5.5 Rain

Rainfall is a natural outcome associated with thunderstorms. In a mature thunderstorm heavy sustained (>5–10 min periods) rain rates >25 mm h^{-1} at the ground level are of frequent occurrence, and those exceeding 100 mm h^{-1} are not uncommon. Even higher rates of rainfall can occur occasionally. As an example, in the case of the Big Thompson thunderstorms already mentioned, radar data indicated incremental rainfall of >150 mm within a half-hour interval, yielding a rainfall rate of ~300 mm h^{-1} (Hoxit *et al.*, 1981). Similar rates have been observed elsewhere (Kessler, 1981). The maximum instantaneous rain rates (e.g. over 1 min periods) within each of the sustained rainfall periods would naturally be higher than their average rain rates.

Rainfall within the thunderstorm is usually confined to a well-defined volume called the *rain cell*. Because of its reflecting power, rain cells normally appear quite clearly in radar pictures of thunderstorms. The contours of the rain cell vary during the evolution of the thunderstorm, dependent on the stage of its maturity and the ambient winds.

The vertical profile of rainfall within a thunderstorm is, in general, not uniform. Starting from the heights where raindrop formation is initiated, the rainfall rate tends to increase downwards as the drops grow in size and fall faster. Simultaneously, the evaporation of a part of the rain during its fall tends to reduce the rain rate. These opposing effects, coupled with the effects of vertical air currents, produce a variability in the vertical distribution of rain within a thunderstorm. An important corollary of this fact is that the local rain rates at various altitudes within thunderstorms can be quite different from those observed on the ground. This fact is important for aviation because an aircraft can encounter heavy rain at its flight altitude even though ground-based observers and instruments such as rain-gauges may experience little or no rain. Remote sensing instruments such as radars can detect the precipitation aloft even when there is no rain at the ground level.

4.5.6 Hail

Thunderstorms produce hailstones of a wide spectrum of sizes. The common designation for different sizes of hailstones (Court and Griffiths, 1982) is shown in Table 4.1. The terminology given therein often appears in the aviation context in the pilots' reports and other description of hail encounters.

Table 4.1 Common designation of hail according to size (length of the longest axis)

Designation	Typical diameter, mm
Shot	<5
Pea	5–10
Grape	10–20
Walnut	20–30
Golf ball	30–50
Hen egg	50–60
Tennis ball	>60

Hail sizes up to a few centimetres across are encountered frequently, and diameters of several centimetres are possible. Individual hailstones with equivalent diameters of 10 cm and weighing >700 g have been found, but their occurrence is rare. The average distribution of hailstone sizes aloft in thunderstorms follows an exponential variation with respect to diameter, the value of the coefficient of the diameter in the exponent lying between ~ -2 and -4 (Morgan and Summers, 1982). The exact value of the coefficient within this range is believed, at least in part, to depend on the climatic region in which the storm occurs. For example, Douglas (1963, 1965) obtained the following distribution for Alberta (Canada) hailstorms:

$$N_d = 10e^{-3.1d} \text{ m}^{-3} \text{ per mm size interval} \tag{4.1}$$

while Federer and Waldvogel (1975) obtained the distribution

$$N_d = 12.1e^{-4.2d} \text{ m}^{-3} \text{ per mm size interval} \tag{4.2}$$

in a multicell storm in Switzerland. In eqns. 4.1 and 4.2, d is the hailstone diameter in centimetres, and N_d is the number of hailstones per cubic metre of spatial volume per millimetre size interval at that diameter. A lower magnitude of the negative exponent in the distribution signifies a higher probability of finding larger hail within a population; such storms are more hazardous from the aviation point of view.

Hail fall is very often organised along well defined columns called *hailshafts*. Hailshafts usually cover an area <1 km² in cross-section, and do not appear to have any definite location relative to the rain cell in a thunderstorm. When the hailshaft is within the rain cell, rainfall and hail fall occur simultaneously. However, in certain locations large hail may typically fall entirely outside the rain cell area, followed by rain and small hail.

Hail of large sizes is usually associated with the strongest thunderstorms, which also produce other severe phenomena such as tornadoes, severe winds and strong localised rain, all of which are of serious concern for aviation.

The geographical distribution of hail occurrences is of great interest to aviation. Because of its inherent difficulties the global coverage of hail measurements is patchy. However, it is known that frequent hail falls occur in some part of every continent, except Antarctica (Gokhale, 1975). Thus hail hazard to aviation is widespread globally. Large hail is believed to occur most frequently in India and, in North America, in the lee of the Rocky Mountains (Flora, 1956). Even considering the overall frequency of hail occurrence, the Indian land mass is the most hail-prone in the world, with large tracts of northern India recording an average of 9 days or more of hail fall in a year (Frisby and Samson, 1967; Williams, 1973). Other hail-prone zones of the earth are the eastern parts of South Africa, the central areas of the USA, Central America, the Andes Mountains zone in South America, the Caucasus, Central Asia, and France.

In terms of height, the maximum frequency of hail encounters has been found, during the US Thunderstorm Project (Byers and Braham, 1949), to be between 10 000 and 15 000 ft (~3–5 km) in Ohio, USA, though there is some probability of hail encounter at all altitudes up to 25 000 ft. The height of maximum hail encounter in Florida, USA, was 16 000 ft. In general hail is believed to be most frequent at the middle levels of thunderstorms (Morgan and Summers, 1982). During formation hail has been found to be carried by thunderstorm updrafts to heights as much as 12 or 13 km, and moved horizontally by as much as 20 km (Browning and Foote, 1976). Battan (1975) has also reported evidence of hail all the way up to the top of a mountain hailstorm at 12 km height. Hail growth regions have been reported to be in the altitude band of 4–6.5 km above sea level in the Georgian Republic in Europe (Sulakvelidze, 1967). Thus, altitudewise, all aviation activity, except some rare extra-high-altitude flights, is subject to the hazards of hail encounter in the area of thunderstorms.

4.5.7 Lightning, electric fields and atmospherics

Lightning is also an inevitable product of thunderstorms. Indeed, almost all lightning, and consequently thunder, originates from cumulonimbus clouds or thunderclouds. The upper portions of thunderclouds are normally positively charged and their lower parts are negatively charged. A part of the negative charge from the latter parts is transferred to the ground via cloud-to-ground lightning flashes, conduction currents and charge transfer via raindrops being the other electricity-conveying mechanisms.

Under fair weather conditions the atmosphere has a positive electric potential relative to the earth which increases with height. The atmosphere therefore has a positive potential gradient with respect to altitude, its value being ~ 100 V m^{-1} at the surface of the earth. However, with the approach of a thunderstorm the potential gradient usually goes negative (because of the negative charge near the cloud base), and fluctuates in magnitude; its peak value is often of the order of 10 kV m^{-1} (Pierce, 1982). Lightning flashes produce rapid and violent field changes which may exceed 50 kV m^{-1} and may even momentarily reverse the direction of the field (potential gradient).

The fields within thunderclouds (Winn et al., 1974) are often higher than those outside and are, for the most part, in the range of ~50–200 kV m^{-1}. There are, however, localised regions extending a few hundreds of metres where the fields may be as strong as 400 kV m^{-1}. These high-field foci, through a chain of complex processes not completely understood, serve as the starting points of ionised conducting paths called *leaders* which, on average, carry currents of the order of 100 A, advance at speeds of ~2×10^5 m s^{-1}, and persist from tens to hundreds of milliseconds. These leaders provide the path for the potential differences within clouds to be neutralised. When a leader encounters a pocket of opposite charge along its path the charge flashes through the conducting path in the form of a current surge of several kiloamperes, which produces the lightning effect. This process constitutes *intracloud lightning*.

Some of the leaders originating in the clouds head towards the earth, and generally have branches on emergence from the cloud base. The leaders to the ground are usually stronger, carrying currents of a few kiloamperes, and progress in an apparently stepped fashion. When the main branch of the conducting leader short-circuits the charge on the ground, the charge transfer sets up a surge current with values averaging ~20 kA (extreme value >140 kA), progressing typically at speeds of 5×10^7 m s^{-1} (Pierce, 1982). The high current momentarily produces temperatures of 2×10^4 to 3×10^4 K and, consequently, brilliant light flashes. Such high surge currents also produce strong field fluctuations. The ionised channel often supports several recoil strokes (median number 2–3, maximum 10–11), giving *lightning-to-ground* its familiar flickering effect. The total flash duration averages 180 ms, with its maximum value being of the order of 1 s. Flashes to the ground are, as a rule, stronger than intracloud lightning, though the latter are far more numerous than the former.

The US NASA Storm Hazards Program to collect data on aviation hazards of lightning was mentioned in Section 3.7. A very important finding from the Storm Hazards Program was that, contrary to prior assumption, lightning strike rates on aircraft penetrating thunderstorms were low in areas of strong precipitation and turbulence. This negates the instinctive and readily implemented guideline for lightning avoidance on the part of pilots and air traffic controllers, since areas of strong weather activity, involving heavy rain, hail and turbulence, are easily sensed both visually and by airborne and ground-based radars.

The Program also showed that aircraft are found to be most susceptible to lightning when flying high within thunderstorms at ambient temperatures of −40°C or less, the other conditions being the absence of strong precipitation and turbulence as mentioned above. At altitudes <20 000 ft (6 km) the greatest threat of lightning comes from decaying storm cells. This condition, again, is counterintuitive, since decaying storms are normally accompanied by a falling rate of natural lightning flashes, while their effect on aircraft is a higher strike rate. Although the two conditions stated above enhance the probability of lightning strikes to aircraft, there remains a finite probability of lightning strikes to aircraft at all temperatures and altitudes of flight within thunderstorms (within the 5000–40 000 ft interval covered in the study).

The transient electromagnetic fields generated by lightning are called *atmospherics* (or *spherics*). These are of particular interest to aviation because they interfere with communication signals and induce random, possibly damaging, voltages in avionic circuits. The lightning process is rich in high-frequency impulses.

Lightning discharges to earth produce strong very-low-frequency (VLF) transients which are isolated, but may repeat a few times. Intracloud flashes produce isolated VLF signals, but these are considerably weaker. Both types of discharge, however, produce comparable levels of high-frequency (HF) and very-high-frequency (VHF) signals, at a rate often >10 000 pulses per event. These pulses are ~1 μs in duration and are spaced 20–100 μs apart. The signal strength of atmospherics has an approximately inverse dependence on frequency.

For more detailed reading on the complex electrical phenomena associated with thunderstorms, a number of excellent reference texts are available, e.g. Chalmers (1967), Uman (1969) and Golde (1977). Chapter 12 provides more details on the lightning phenomenon.

4.5.8 Icing

It was mentioned in Section 3.5 that the presence of supercooled liquid water droplets, particularly in the diameter range of 30–400 μm, in the flight environment fosters ice accretion on aircraft surfaces. Such conditions are often encountered in thunderstorm clouds in a height band above the freezing altitude level, but normally the flight time of aircraft through single thunderstorm cells is not long enough to cause dangerous levels of icing. However, when there is a high concentration of supercooled liquid water, and the flight through the storm zone is of considerable duration (such as in flying through multicell storms and squall lines), the ice accumulation can be high enough to affect flight adversely.

4.5.9 Poor visibility

Thunderstorms are the major source of many factors responsible for poor visibility. The tall columns of dense cumulus and cumulonimbus cloud in a thunderstorm impair visibility at flight altitudes. Thunderstorms usually have a well defined cloud base which corresponds to the condensation height of the rising air mass. In a highly moist ambience the cloud base descends to very low altitudes, often restricting visibility to <1 km above the ground.

Rain and hail are the other visibility-reducing factors produced by thunderstorms. In arid and dusty regions the flying particulate matter and larger objects raised by thunderstorm outflows are a main cause of poor atmospheric visibility affecting aviation.

4.5.10 Overall thunderstorm scenario

The preceding subsections discuss the different thunderstorm effects on aviation individually. However, the overall picture in the area of a thunderstorm is highly complicated, with more than one of these subphenomena and effects existing simultaneously or sequentially along an aircraft's path. For example, aircraft in the process of landing or takeoff through a thunderstorm may experience icing, wind shear due to low-level divergence, severe turbulence due to the gustiness of the descending and ascending air masses, heavy rain and hail, and poor visibility, all at the same time (e.g. National Transportation Safety Board, 1981). Similarly, aircraft entering a storm outflow region may first encounter a sharp discontinuity in air mass characteristics (temperature, density) at the gust front, followed by a zone of turbulence, and finally the wind shear due to divergence (see the lower flight path in Fig. 4.3). In such rapid sequencing of hazard factors, the deleterious effects of each factor on the aircraft are felt even before the effects of the preceding ones have died down. The combined effect of multiple hazard factors is, in general, more detrimental than isolated factors. For this reason, intelligent detection, characterisation and hazard potential estimation of thunderstorms and their related phenomena form the prime focus of the modern electronic systems designed for the enhancement of aviation safety and management.

4.6 Gust fronts and related phenomena

In the preceding section mention was made of the radially divergent flow of air resulting from the horizontal deflection of the thunderstorm downdraft by the ground. The leading edge of this diverging air mass is called the *gust front* (see the bottom right part of Fig. 4.3). The gust front thus forms the interface between the warm and moist ambient air close to the ground in the environs of the thunderstorm, and the cool air originating from the core of the storm. The colder storm air, being denser, tends to undercut and displace the ambient air, which is lifted and flows over the advancing layer of divergent flow. The nature of air flow on both sides of the gust front is quite different, and involves a velocity discontinuity. A gust front is therefore an approximation to a line of flow convergence, as shown in Fig. 4.5d. To an aircraft crossing the gust front, the sharp discontinuity in the wind velocity appears as a strong localised wind shear. Further, it also experiences a jump in air density across the gust front because of the considerable temperature difference between the cold storm-outflow and the ambient air. As seen from eqns. 2.1 and 2.2, both velocity (appearing through speed and angle of attack) and density are strong parameters in determining the forces (and therefore also moments) acting on aircraft, and sudden changes in these quantities would trigger the dynamic modes of the aircraft.

Table 4.2 Gust front strength categories[a]

Category	Velocity difference ΔV (m s^{-1})
Weak	$5 \leqslant \Delta V < 10$
Moderate	$10 \leqslant \Delta V < 15$
Strong	$15 \leqslant \Delta V < 25$
Severe	$\Delta V \geqslant 25$

[a] From Troxel and Delanoy (1995). Reprinted with permission of MIT Lincoln Laboratory, Lexington, MA

4.6.1 Characteristics

The horizontal speed of the air behind the gust front sometimes exceeds 25 m s^{-1}, and occasionally even approaches speeds of the order of 50 m s^{-1} (Sasaki and Baxter, 1982). Such speeds lend the gust front high hazard potential. Because of a certain degree of mixing, the transition from the ambient air to the cold outflow air occurs over a distance of ~500 m which corresponds to a flight time of a few to several seconds for an aircraft approaching to land. The gust front is a low-altitude phenomenon, and air speed changes of the magnitude of a few or even several tens of metres per second occurring over such short durations have a high hazard value for aircraft flying at low levels.

The strength of a gust front is given by the difference ΔV in the wind speeds (perpendicular to the front) across the front. An indicative classification of gust fronts according to strength may be made by using differential velocity thresholds as given in Table 4.2.

In addition to the sharp wind gradients, aircraft passing through gust fronts are also suddenly exposed to a high level of turbulence behind the front. The turbulence arises from the complex pattern of eddies that form when the divergent flow is impeded by the nearly static ambient air, and mixing, undercutting and lifting take place. Fig. 4.8 shows a magnified view of a gust front, with a schematic representation of the associated flow pattern. The

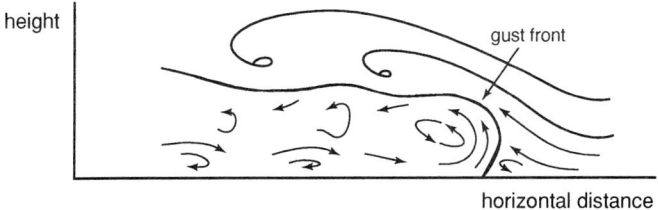

Figure 4.8 Schematic diagram of flow pattern in a gust front, showing zones of turbulence

100 *Aviation weather surveillance systems*

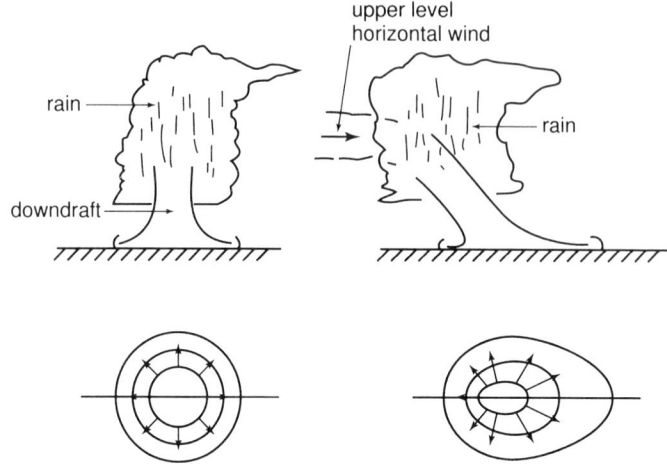

Figure 4.9 Vertical structure of thunderstorm downdraft (top) and progression of gust front in plan view (bottom) for (a) symmetric and (b) asymmetric gust fronts

pattern may be more complex in the presence of ground features and roughness, and significant ambient winds, and may possess more than one bulge and mixing zone.

The wind at high speeds close to the ground behind the gust front picks up large quantities of dust, fine sand, dry leaves and other light debris which rise up in the eddy motion at the head of the expanding flow. The characteristic arcual or curvilinear shape of this line of flying debris, as seen from the cockpit of aircraft aloft, is an important, often the only, clue to the existence of the gust front. Another clear and important visual manifestation of the presence of a gust front is the formation of *arcus* or *roll cloud* due to the lifting and cooling of the warm and moist ambient air by the storm outflow air. In a plan view the roll cloud pattern follows the shape of the gust front, though it may not be continuous.

As there is a sharp discontinuity in the radial (i.e. lateral) velocity at the arc representing the gust front, the front constitutes a line of convergence according to the generalised definition given in the preceding section. As will be seen later, this observation is very important for human-aided as well as automatic detection and delineation of gust fronts from Doppler radar pictures and data.

Under ideal conditions the gust front resulting from a vertical thunderstorm downdraft should spread out in an expanding circle (Fig. 4.9a), with the substorm point as the centre. However, the actual conditions of gust front formation and propagation very often differ from the ideal. One major distorting factor is the presence of significant ambient winds at the ground level which makes the arc of the gust front asymmetric relative to the thunderstorm, and also makes the strength of the front vary along the arc.

Further, in the presence of strong winds aloft, the shaft of downdraft air does not remain vertical. A horizontally moving layer of air, on receiving rainfall from the upper reaches of the thunderstorm and cooling down because of its evaporation, descends to the ground level (further aided by the momentum and mass loading of the rain), but retains its horizontal momentum. This momentum causes the gust front to progress preferentially in the direction of the original upper-level air current (Fig. 4.9b). It also imparts to the outflow winds and to the gust front a high destructive potential.

Although gust fronts commonly originate from thunderstorms, they can propagate long distances away from the parent storms (Colour Plate 8), loaded by the cold and dense air, and propelled by any residual horizontal momentum. Gust fronts may be found several tens of kilometres away from their generating thunderstorms (Sasaki and Baxter, 1982) and indeed may continue to propagate for some time after the parent storm has dissipated or moved far away. Colour Plate 3 shows a gust front, marked by the waves it generates, occurring far from any rain activity. Thus, unlike many other thunderstorm-related aviation hazards, it is not always possible for aircraft to avoid gust fronts by merely skirting visible thunderstorms.

Except when they are still in the vicinity of thunderstorms, gust fronts may be dry, i.e. the air mass on either side of the front may not contain raindrops. This fact is important from the point of view of radar detection of gust fronts, since raindrops are a strong source of radar energy reflection.

Gust fronts are important agents in the initiation of fresh thunderstorms. It may be recalled (Section 4.3) that thunderstorm processes are started when warm and moist air parcels from the lowest layers of the atmosphere are lifted into a thick conditionally stable layer within which the parcel will continue to rise on its own. When the conditionally stable layer is widespread, the ambient air near the ground, lifted by the cool outflows from existing thunderstorms, enters the layer, leading to the initiation of fresh thunderstorms. The process can repeat itself as long as conducive conditions persist.

4.6.2 Outflow-induced waves and bores

The outward flow of the storm air, after its deflection by the ground and intrusion into the ambient air mass, is propelled essentially by gravity force resulting from the density differential between the two types of air. The gust front is therefore the front end of what is essentially a *gravity current* or *density current*.

The disturbance produced in the ambient air mass due to the *ram effect* of the diverging cold air mass has its own natural speed of propagation. As long as the speed of the advancing gust front is more than the natural propagation speed of the disturbance, the disturbance is pushed forward at the head of the gravity current of the outflow. However, as the gravity current weakens due to the expanding circle of divergence and the momentum loss caused by friction with the ground and the ambient air mass, its speed may fall below that of the

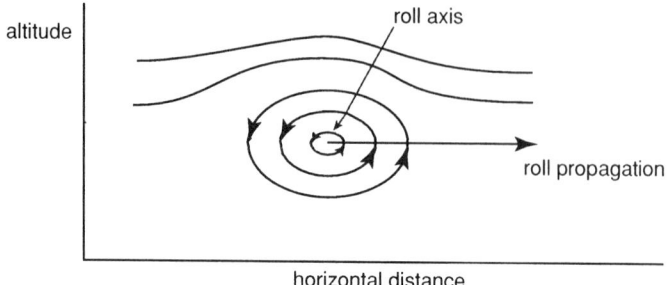

Figure 4.10   Transverse vertical section through an idealised horizontal roll

propagation speed of the disturbance. The disturbance may then detach itself from the outflow air mass and propagate forward on its own. Such propagating disturbances may be in the form of either *rolls* or *buoyancy waves*. The latter were earlier often called *gravity waves*, a term used by some even today. The advancing wave often traps a parcel of the outflow air, which gradually leaks out of its lee end, leading to a weakening of the wave.

The undercut–overflow effect in the region of the gust front produces a significant amount of vorticity about a horizontal axis. When this vorticity detaches itself and propagates forward as a horizontal rotor, it forms a *solitary roll*, an idealised form of which is shown in Fig. 4.10. Aeroplane flight through a roll would cause it to experience incremental vertical-plane air currents of opposite sense, on either side of the roll axis, which would be manifested as a rapid change in the angle of attack and, consequently, in the lift forces acting on the aircraft. The roll may also carry in its wake a few cycles of undulating air motion (Fulton *et al.*, 1990), more about which is discussed in the paragraphs below.

Another type of disturbance in the atmosphere that produces low-altitude wind shear is the *solitary wave*. The solitary wave, also called *soliton*, is a large-amplitude, short-duration single-crested wave. It is a nonlinear disturbance which sustains itself through a balance between the differential propagation speeds of the different amplitude levels of the wave on the one hand, and those of its different frequency components on the other. Atmospheric solitary waves commonly have amplitudes between 300 and 1000 m (~900–3000 ft), effective horizontal lengths between 0.5 and 7 km (~0.3–4 miles) along the propagation direction, and propagation speeds ranging from 6 to 16 m s^{-1} (~12–32 knots); occasionally exceeding 30 m s^{-1} (~60 knots) (Doviak and Christie, 1989). The amplitude of solitary waves can grow as their width decreases; both these effects tend to reinforce each other in enhancing the wind shear generated by these waves along the flight paths of landing aircraft (Doviak and Chen, 1988).

Atmospheric conditions characterised by the presence of a strongly stable layer close to the earth's surface, and near-neutral stability of the air above it, are conducive to solitary wave propagation (Crook, 1986). Such conditions may occur during late nights or early mornings due to radiative cooling of the air. The cold air generated by thunderstorms can also form stable layers over the ground at any time of the day. Under such conditions the solitary wave can propagate large distances from its origin. In the case of a solitary wave studied in detail by Doviak and Chen (1988) the wave had travelled a distance of 60 km from its parent storm and still retained significant wind shear. At a height of 90 m above the ground the horizontal rate of change of the horizontal wind was found to be 4.4×10^{-3} s^{-1} and the altitude rate of change of the horizontal wind was as high as 60×10^{-3} s^{-1}. The horizontal shear was felt over a distance of ~1500 m. Shorter waves riding over this solitary wave had an even higher horizontal shear of 7.6×10^{-3} s^{-1} over a distance of 700 m. An aircraft attempting to land through such a solitary wave along a standard 3° glideslope would be subject to a temporal headwind change of 19 m s^{-1} (~38 knots). This is worse than the headwind decrease of 17.5 m s^{-1} (~35 knots) which caused a Boeing 747 aircraft to drop through a height of 50 m during its approach to land at Melbourne, Australia (Woodfield, 1983). The wave carried a zone of turbulence behind itself, with a wind shear value in the region of 50×10^{-3} s^{-1}. In addition to generating velocity disturbances, solitary waves often trap a parcel of cool storm outflow air in their core, as this wave did, and a penetrating aircraft will additionally be subject to density difference effects on its lift and drag forces.

Given that the above values were observed when the solitary wave had already travelled 60 km from its originating storm before passing over the measurement station, worse could be expected closer to the storm. This is corroborated by radar measurements which showed the amplitude of the wave to be almost twice at a point just 15 km ahead, i.e. after propagating as much as 45 km from its origin. The wind shear values are expected to be correspondingly higher for such wave amplitudes.

It is not just the numerical value of the maximum wind shear, but also the nature of wind variation along solitary waves that is of great concern for aircraft flight. The quasi-sinusoidal variation of wind across the wave can cause resonance with the aircraft's dynamic modes, especially the phugoid (Section 2.3), amplifying the disturbances to the aircraft flight path. The effect can be quite significant even when the forcing function is only about a half-cycle.

The lifting and consequent cooling caused by solitary waves may create characteristic cloud bands, but these may be absent in relatively dry atmospheric conditions. Since aircraft do not always avoid thunderstorms by such large distances as tens of kilometres, strong outflow-induced waves are a source of insidious aviation hazard. Doviak and Chen (1988) discuss a number of cases suggesting the involvement of rolls and waves in aircraft accidents and incidents.

An atmospheric *bore* is a long-period disturbance, akin to a hydraulic jump (Clarke, 1972), which propagates on a stable layer of air. There is a discontinuity in the level of the layer across the propagating front. However, unlike a gust front, it may not be propelled by a contiguous mass of outflow air stretching all the way back to the source, usually a thunderstorm.

Quite often bores are not flat-topped, but have quasi-sinusoidal waves superimposed on their tops. Such bores are called *undular bores*. The fine structure may be present on the bore right from its origin, or it may develop while the bore propagates as a nonlinear disturbance, as shown through numerical simulations by Christie and Muirhead (1983).

It is common for atmospheric solitary waves to occur as groups or 'wave packets' of ordered (normally decreasing) amplitude. Mahapatra *et al.* (1991) reported a multisensor observation of one such very well developed phenomenon propagating through central Oklahoma in the USA. Fig. 4.11 shows a satellite photograph of the alternating cloud bands formed by the lifting of air due to the waves. The variation of the radial velocity (relative to the observing radar) along the wave packet is shown in Colour Plate 2. A

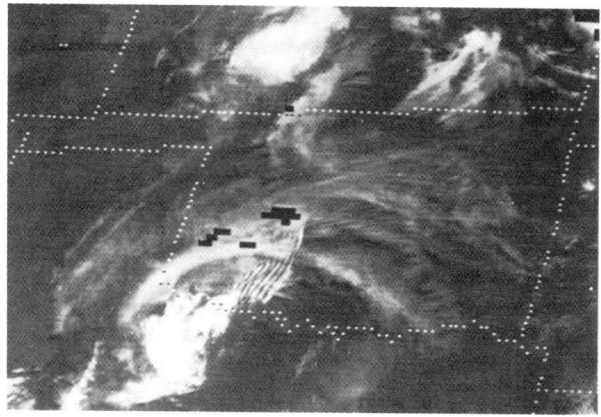

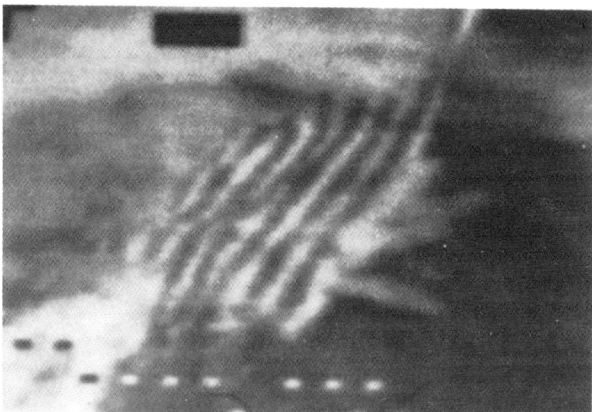

Figure 4.11 Satellite photograph of cloud bands formed by an ordered wave packet propagating through central Oklahoma, USA, on 22 June 1987: the larger scene (top), and close-up view showing details (bottom)

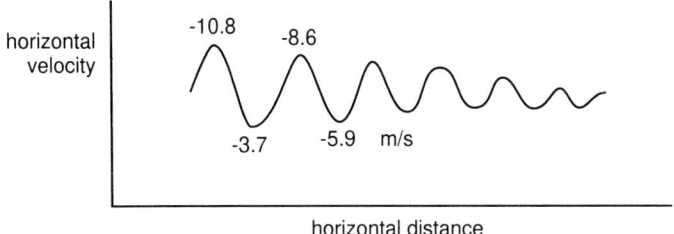

Figure 4.12 *Variation of horizontal velocity along the wave packet shown in Fig. 4.11 and Colour Plate 1*

better appreciation of the horizontal velocity variation through the waves can be had from the plot in Fig. 4.12, which clearly shows the amplitude decay of the successive wavelets. A periodic horizontal velocity perturbation of this type is particularly worrisome for aircraft, especially in low flight, because its periodicity is of the same order as the phugoid periodicity (see Section 2.3) of larger aircraft, and can therefore induce violent altitude oscillations through resonance.

4.7 Macrobursts and microbursts

The downburst was briefly mentioned in Section 4.5.3 as a downdraft which induces a divergent outburst of damaging winds on or near the ground. The spatial extent of such winds is of great importance for aviation. A downburst with its outburst wind zone exceeding 4 km in horizontal dimension is called a *macroburst*. Smaller downbursts, with damaging winds extending 4 km or less, are called *microbursts* (Fujita, 1985). More precisely, a microburst may be defined (Wilson *et al.*, 1984) as a divergent outflow for which the differential radial velocity between maxima is 10 m/s or more, and the distance between the maxima is $\leqslant 4$ km (yielding a radial divergence of 2.5×10^{-3} s^{-1} or more). When the outflow has a radial divergence $\geqslant 2.5 \times 10^{-3}$ s^{-1}, but the distance between the radial velocity maxima exceeds 4 km, the event is classified as a macroburst. A macroburst can produce winds as high as 60 m s^{-1} (120 knots or 134 mph), with the damaging winds lasting 5–30 min. A microburst may cause winds up to 75 m s^{-1} (150 knots or 168 mph), but the intense winds often last only 2–5 min. A Doppler radar view of a microburst is shown in Colour Plate 16.

The microburst as a hazardous atmospheric phenomenon is of relatively recent research focus, yet it is recognised as a prime hazard to aviation because of its high wind shear content and its transitory nature. It has been implicated in many aircraft accidents and incidents, most notably the well-studied accidents at New York's John F. Kennedy International Airport in 1975 (Fujita and Byers, 1977; Fujita and Caracena, 1977), New Orleans in

1982 (Fujita, 1983a; Dietenberger *et al.*, 1985), and Dallas-Fort Worth in 1985 (Fujita, 1986), all in the USA. Posterior analysis, discussed quite graphically by Fujita (1985), has also brought out a clear possibility of microbursts having been responsible for airliner accidents elsewhere in the world, e.g. at Kano, Nigeria, as early as 1956, Pago Pago, American Samoa, in 1974, and Doha, Qatar, in 1976.

Microbursts are phenomena of frequent occurrence in many parts of the world, and their spatial and temporal density can be quite high. In a systematic observational programme in the vicinity of Darwin in northern Australia, Potts (1991) reports a daily average of 5 and maximum of 16 microburst events within a circle of 40 km radius during a period of 15 days. Given that the influence zones of individual microbursts can be tens of square kilometres, such a high density and frequency of occurrence of microbursts makes it distinctly probable for aircraft to encounter them while operating in the area.

4.7.1 Microburst types

A microburst is difficult to detect because of its small spatial dimensions and short lifetime. A number of studies have been conducted to generate a knowledge base regarding microbursts to facilitate their detection. These include dedicated major co-ordinated projects such as the Northern Illinois Meteorological Research on Downbursts (NIMROD) (Fujita, 1979), the Joint Airport Weather Studies (JAWS) (Fujita and Wakimoto, 1983), as well as numerous other studies (e.g. Eilts and Doviak, 1987).

Microbursts do not always coincide with significant rainfall at the ground level. Those which produce appreciable rain on the ground are called *wet microbursts*; others are termed *dry microbursts*. A comparison between the two types of microburst is made in Table 4.3.

Wet microbursts can produce intense rain which may exceed rates of 200 mm h^{-1} over a few minutes. The microburst blamed for the 1982 airliner crash at New Orleans (Fujita, 1983a) was accompanied by heavy rain. Fujita (1983b) studied a microburst which hit the Andrews Air Force Base airport along with extremely intense rain a few minutes after the US Air Force One, the plane used by the US president, landed on 1 August 1983. Although wet microbursts can be as hazardous as, or possibly more hazardous than, the dry ones, the latter are of greater concern as they do not offer significant visual clues to deter pilots from entering their area. This is corroborated by accident investigations. For example, in 1990, dry microbursts were held as the cause of four general aviation accidents in the USA, while only one accident was attributed to a wet microburst (National Transportation Safety Board, 1993b). As will be discussed later, the absence of precipitation also makes dry microbursts much more difficult to detect using radars.

Because of the projects and studies mentioned above, a fairly detailed knowledge base has been generated regarding microbursts within the US

Table 4.3 Wet and dry microbursts[a]

Characteristics of dry microbursts	Characteristics of wet microbursts
• Little or no rain reaches the surface. Often associated with virga shafts containing lightly rimed snowflakes that completely evaporate before reaching the surface (i.e. generally small particles)	• Associated with heavy rainfall, and precipitation core is mainly in the form of ice, e.g. melting hail (i.e. generally larger particles)
• Strong surface winds caused by negative buoyancy due to evaporation, melting and sublimation of precipitation below cloud base	• Strong surface winds caused by precipitation loading in addition to negative buoyancy. Downward momentum transfer and/or dynamically induced pressure gradients may also contribute, especially in strong events
• Strong synoptic-scale forcing not necessary	• Strong synoptic-scale forcing not necessary
• Downdraft entrainment considered minimal	• Downdraft entrainment of environmental air at level of minimum equivalent to potential temperature considered important
• Dry or nearly dry adiabatic subcloud lapse rate	• Pseudo-adiabatic subcloud lapse rate early in the day, becoming dry adiabatic by the time of maximum solar heating
• Dry in the subcloud layer	• Relatively moist in the subcloud layer
• Moist at mid-levels	• Dry at mid-levels
• Relatively weak convection/updrafts	• Relatively strong convection/updrafts
• Relatively high cloud bases	• Relatively low cloud bases
• Are a function of solar heating. Thus the time of maximum occurrence is mid-afternoon, local time	• Are a function of solar heating. Thus the time of maximum occurrence is mid-afternoon, local time
• Exhibit relatively small lowering of surface temperature during the event	• Exhibit relatively large lowering of surface temperature during the event

[a] Adapted from Nelson and Ellrod (1997) by permission of the author

landmass. A good review of the characteristics of microbursts in the US has been made by Wolfson (1988). It has been found that the proportion of different types of microbursts has a geographical variability. For example, the JAWS programme showed that 83% of the microbursts around the city of Denver in the state of Colorado were dry, while the results of the NIMROD programme indicated only 36% of the microbursts in the northern parts of the state of Illinois to be dry (Fujita and McCarthy, 1990).

4.7.2 Characteristics

As wind shear is the main danger arising from microbursts, the horizontal and vertical structure of air flow within microbursts is important in the aviation context. The wind structure is also of great importance for designing and optimising sensors for microburst detection. Also of crucial importance from the detection point of view is the evolutionary lifecycle of the microburst, especially the timescales associated with its rise, persistence and decay. Wilson *et al.* (1984), using Doppler radar data from the JAWS project, determined that the shaft of downdraft air associated with microbursts has a typical diameter of ~1 km, and it begins to spread horizontally at a height that is normally <1 km from the ground. Temporally, from the initiation of divergence on the ground it takes a median time of 5 min for the microburst to develop its maximum horizontal wind shear, as measured by the differential of wind velocity across the divergent flow field of the microburst. The maximum velocity differential in the study had a median value of 22 m s^{-1}, and occurred over an average distance of 3.1 km. Further, the maximum wind shear was found to occur at a height of ~75 m above the ground. However, this height was the lowest height of data collection for horizontal wind speeds; hence the actual height of occurrence of the maximum wind shear could possibly be somewhat less than 75 m.

The average and median spatial parameters are shown schematically in a hypothetical microburst in Fig. 4.13 for visualisation purposes. All these microburst parameters are of great significance for aviation weather

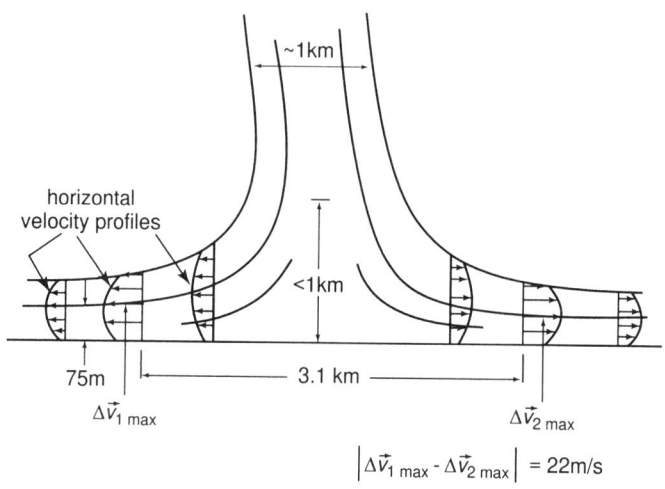

Figure 4.13   A hypothetical microburst showing the average parameters

surveillance. However, particular attention is drawn to the fact that the highest horizontal wind shear in a microburst occurs within the lowest 100 m of height. This means that an aircraft will encounter maximum microburst wind shear just prior to landing, at heights of only tens of metres above the ground, where it is most susceptible to disturbances. Further, the occurrence of high wind speeds so close to the ground creates a thin layer of high vertical wind shear (i.e. vertical rate of change of horizontal wind speed) as the horizontal wind speed decreases to zero at the ground level due to friction with the ground. The drop in horizontal wind speed above the height of maximum differential also causes vertical shear, but its magnitude is lower than that in the layer below the maximum. Finally, the detection of winds at such low levels requires very special considerations regarding radar location, scanning strategy, sensitivity and clutter filtering for their detection within the time frame of their rapid evolution. These aspects will be considered later in the appropriate context.

In another detailed study, also based on JAWS data, Hjelmfelt (1988) arrived at similar microburst parameters. He noted that microburst outflows have depths varying from 300 to 1200 metres. In addition, he observed that the average time from the microburst reaching its maximum horizontal wind shear to its decay is ~8 min. Together with the 5 min build-up time of the microburst mentioned above, this gives an average total lifetime of 13 min for the JAWS microbursts. In this study, downdraft diameters were in the range of 1.5 to 3 km, and the maximum downdraft speeds varied in the range of 6 to 22 m s^{-1}. Another important observation made from this study was that the outflow morphology of the microbursts was independent of their associated precipitation rates. Thus the rainfall intensity, as observed by rain-gauges or many current and older generations of weather radars, cannot offer any significant clue to the occurrence of microbursts. Indeed, the study noted that some of the strongest microbursts (maximum differential velocity >25 m s^{-1}) occurred with very low radar reflectivities (<0 dBZ[1]). Similar lack of correlation has also been reported by Wilson *et al.* (1984).

The low altitude of occurrence of dangerous wind shear due to microbursts is also borne out from observations and investigations carried out in connection with aircraft accidents and incidents caused by microbursts. In the case of the Dallas-Fort Worth accident (National Transportation Safety Board, 1986) the Lockheed L-1011 aircraft involved entered the heavy shear zone of a microburst at a height of ~750 ft (~250 m) from the ground, from which it could not recover. In an incident at the Denver Stapleton Airport, a Boeing 737 aircraft encountered a microburst during the final approach to land, and was driven down to a height <100 ft (~30 m) above the ground by the wind shear before managing to recover (Schlickenmaier, 1989). Another incident, at Atlanta Hartsfield airport, involved a Boeing 767 aircraft that penetrated a microburst just prior to landing and descended uncontrollably down to as low

---
[1] See Chapter 6 for definition

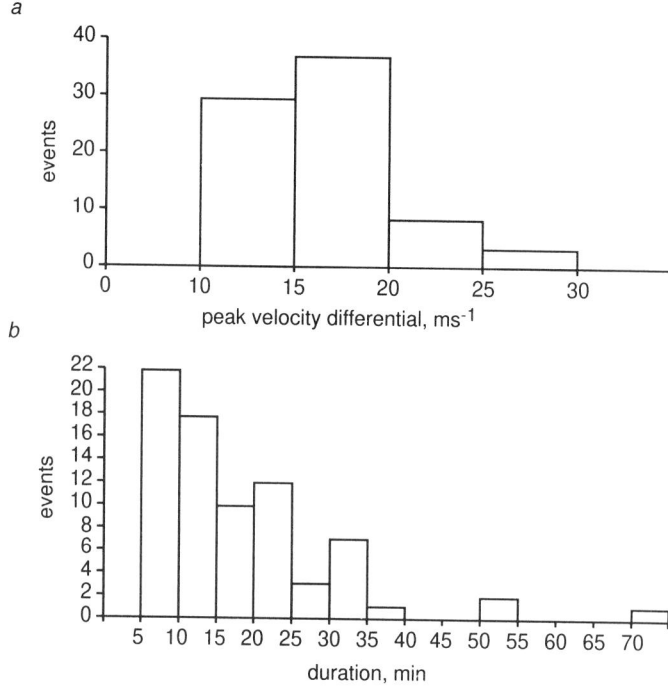

Figure 4.14 Parametric distribution of 76 microburst events observed over a 15-day period in February 1989 within 40 km of a 5.33 cm Doppler weather radar located near Darwin in northern Australia: (a) peak velocity differential, (b) lifetime (from Potts, 1991, by author's permission)

as ~70 ft (~20 m) above ground, short of the runway threshold, before it could recover and execute a missed approach (Lewis *et al.*, 1994).

Some studies have also been conducted on microbursts outside the USA. Two important parameters of microbursts observed in northern Australia (Potts, 1991) are presented in Fig. 4.14. The peak velocity differential of a vast majority of the 76 microbursts detected over 15 days lay between 10 and 20 m/s, with the maximum reaching 27 m/s and the median value being 17 m/s. This median is less than the corresponding value of 22 m/s found from JAWS as mentioned above, but the author ascribes the difference largely to the difference in the range resolution of the observing radars (this effect is discussed in Chapter 6). The lifetime of the phenomenon (the interval over which the radial divergence $\geq 2.5 \times 10^{-3} \mathrm{s}^{-1}$) was found to vary from 5 to 55 min (mean value 15 min), with the shorter lifetimes being more probable. The results show that even in the tropical regions the microbursts have major characteristics similar to those studied in the USA under the JAWS programme. Other important observations during the study were that all the microburst events were associated with thunderstorms, coincided with reflectivity maxima (which correspond to the zones of maximum rain, as will be discussed in Chapter 6) which were of moderate to high intensity, and occurred during the afternoon or evening, peaking between 1500 and 1700 h.

4.7.3 Asymmetry

A characteristic of considerable importance in the description and detection of a microburst is its *asymmetry*. The asymmetry of divergent fields was briefly discussed in the preceding section. Microbursts very often display an asymmetry, with radial outflows being stronger and more spread out in certain directions from the microburst centre (the point at which the axis of the downdraft shaft encounters the ground) than in others. Accordingly, the *strength asymmetry* of a microburst is the ratio of its maximum to its minimum strength ('strength' is the highest differential velocity) over all aspect angles or viewing directions, and the *shape asymmetry* is the ratio of the longest to the shortest spatial extent of the outflow field over all such directions (Hallowell, 1990). Hjelmfelt (1988) obtained an average strength asymmetry of ~2, and shape asymmetry values of the same order. The study by Wilson *et al.* (1984) deduced strength asymmetry ratios up to ~6, with an average of ~3, over distances of 3 km, which is a spatial scale of crucial importance when considering wind shear effects on aircraft. Hallowell (1993) analysed a large number (859) of microburst observations and obtained generally lower values of asymmetry (range 1.0 to 3.0, median value 1.34) after removing the effects of radar configuration used for data gathering, temporal difference between radar observations, and residual errors of apparent asymmetry. He also found that the difference in microburst asymmetry between widely separated geographical areas (within the US) such as Orlando, FL, and Denver, CO, is minimal. As will be discussed in Section 10.6, the extent of asymmetry has a strong bearing on the automatic detection of microbursts and their hazard estimation based on data from single radar installations, which is the normal mode of data generation and processing.

4.8 Other sources of atmospheric hazard

Although thunderstorms and phenomena deriving from them, such as gust fronts and microbursts, are responsible for a vast majority of weather hazards to aviation, there are other significant sources of atmospheric hazards. These are related to atmospheric air currents of relatively longer scales, and flow aberrations caused by topographic factors.

One significant source of wind shear consists of *low-level jets* or *nocturnal jets* which can produce wind speeds of the order of 10–30 m s^{-1} (20–60 knots) at a height of ~600 ft above the ground while the winds at the ground may be calm (Kessler, 1974; Stull, 1988; Hoecker, 1963). As an illustration of the nature and severity of wind shear hazards posed by such jets, Fig. 4.15a shows the wind profile of a low level jet, and Fig. 4.15b indicates the wind variations

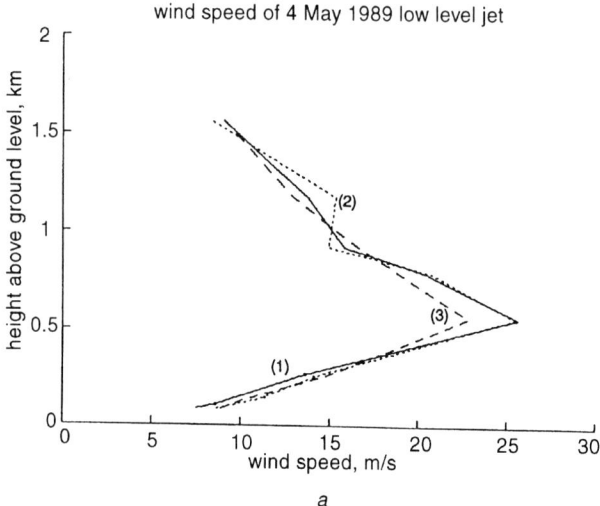

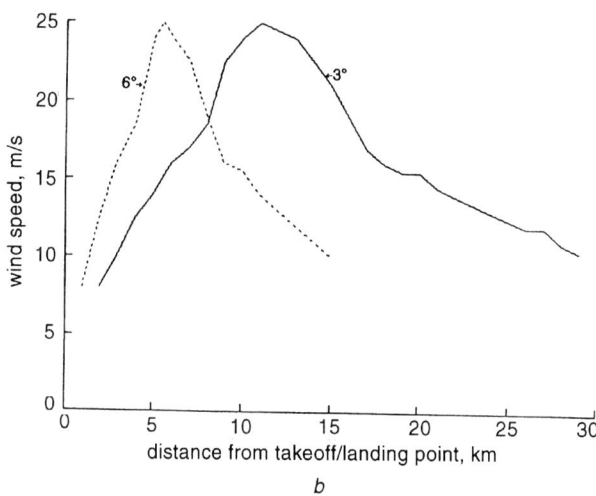

Figure 4.15 Profiles of wind speed due to a low level jet observed on 4 May 1989 in the vicinity of Kansas City International Airport, USA during the FAA evaluation of Terminal Doppler Weather Radar (see Section 7.5). The vertical profile (a) shows a sharp peak at ~1500 ft (500 m) altitude, with strong wind shear above and below this height. An aircraft attempting to take off (at 6° angle) or land (along a 3° glideslope) through the jet would experience the wind variation shown in (b). (Courtesy D.S. Zrnic', NOAA National Severe Storms Laboratory, Norman, Oklahoma)

that may be experienced by aircraft executing typical landings and takeoffs through the jet.

At higher levels, the *jet stream* can produce significant wind shear. Jet streams are layers of strong winds occurring at certain altitudes and in certain latitude belts which show seasonal variation. Pilots often take advantage of jet streams by adjusting their flight altitudes (within allowed limits) to maximise favourable wind (or minimise opposing wind). However, variations in the jet stream winds give rise to wind shear. Fig. 4.16 shows an example of the strong winds and wind shear caused by a jet stream phenomenon.

The passage of cold or warm fronts can generate appreciable wind shear because of a differential vector velocity occurring across the frontal surface. The *frontal shear* usually has both vertical and horizontal components because of the cold air displacing the warm air horizontally as well as vertically through a process of undercutting.

Significant wind shear may also be caused by the land breeze and sea breeze phenomena in regions close to the sea. These winds, as also surface winds of other origins, can reach hazardous levels due to the 'wind tunnelling effect' caused by the presence of topographic features such as mountains, mountain ranges and valleys. Topographically induced updrafts are common in hilly regions, as also are mountain waves and turbulence induced by air currents in the lee of mountains. Wind shear induced by topographic factors is often more significant for aircraft takeoff than landing due to limitations imposed by operational requirements (Konig *et al.*, 1980).

Of particular significance in the aviation context are mountain waves. Strong winds blowing generally normal to a mountain ridge can produce moderate or stronger turbulence. Mountain-induced turbulence is likely to be found at levels near the ridge top, in quite stable layers which may be up to ~5000 ft thick and may extend as much as 150 km downstream. Mountain wave development may be reinforced by the presence of strong wind currents such as jet streams. Also, mountain waves may interact with turbulence caused by wind shear (see Sections 3.3.2 and 11.2) to produce variations in turbulence over a wide range of altitudes. The pressure changes associated with the flow past mountains often cause the formation of characteristic lenticular or rotor clouds which serve as visual indicators of the presence of mountain waves to pilots. Such clouds may not form in the absence of sufficient moisture in the air, but the mountain wave could still harbour significant turbulence. Mountain waves often form the source of the clear air turbulence referred to in Section 3.3.2.

Close to the ground, significant turbulence can be created by winds blowing over rough and uneven terrain. Such mechanical turbulence can be reinforced due to interaction with thermal convection caused by warm ground. As an indicator, light turbulence may be expected when surface winds are >15 knots (~7.5 m s^{-1}) and the air is colder than the underlying ground. When the ground–air temperature differential is high, possibly due to strong solar radiation and/or cold air moving in over warm ground, the

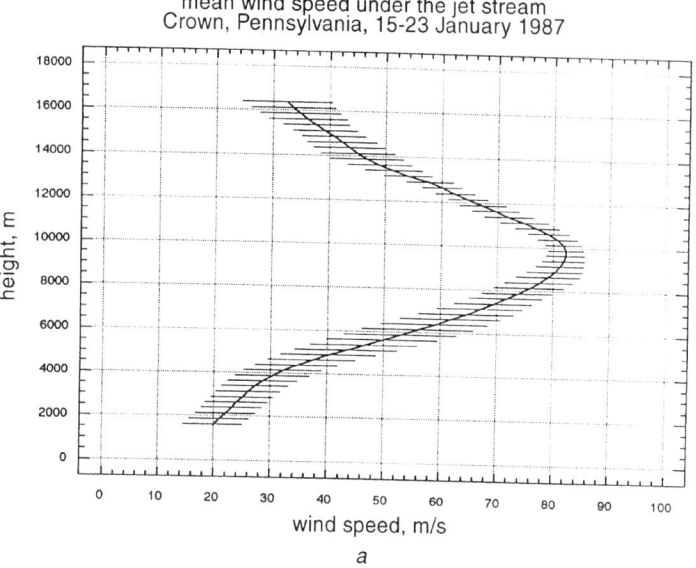

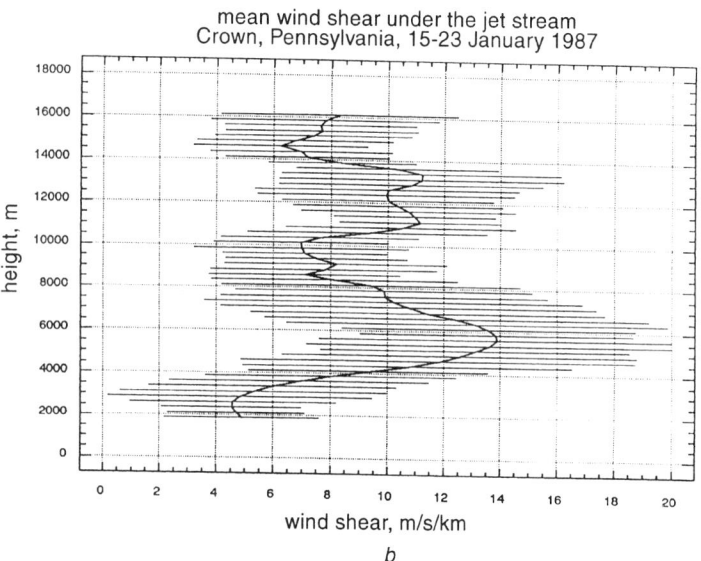

Figure 4.16 Mean wind speed (a), and mean wind shear (b) during a strong jet stream observed by the 50 MHz wind profiler (see Section 8.2.2) at Crown, PA, USA during mid-January 1987. The horizontal lines are standard error bars. Note that curve (b) is not the derivative of curve (a) as both represent averaged quantities (from Syrett, 1991, by author's permission)

lower layers of the atmosphere may become unstable, producing moderate levels of turbulence. Such turbulence may also be generated by surface winds >25 knots (~13 m/s).

4.9 Summary

The origins and characteristics of atmospheric phenomena that affect aviation adversely have been described in this chapter. Special emphasis has been laid on those aspects of the phenomena which have a bearing on the design and operation of the sensors and systems for their detection, which will be addressed in subsequent chapters.

The thunderstorm is shown to be the primary source of most atmospheric phenomena hazardous to aviation. Heavy rain, hail, lightning, turbulence, and wind shear deriving from divergent and stratified flows are the principal hazard-causing factors associated with thunderstorms. The evolutionary process of the thunderstorm is responsible for the creation of these phenomena, each of which has its own associated spatial and temporal scales.

Two phenomena of great hazard potential for aviation, the gust front and the microburst, also owe their origin to thunderstorms, but may occur in conditions where their association with the visible parts of the parent storms may not be readily apparent. This makes their evasion by aircraft difficult. In the case of gust fronts this may happen when the front propagates far away from the storm, and is devoid of significant rain. Bores and solitary waves (or wave packets) driven by the front may travel even further while retaining sufficient destructive power. Microbursts become less noticeable when they occur outside the wet zones of thunderstorms. Both gust fronts and microbursts produce wind shear at very low altitudes which makes them particularly dangerous for aviation, and which is more difficult to detect. Detection of the effects of these phenomena at very low altitudes, of the order of tens of metres, is a major determinant of the design philosophy of the electronic weather surveillance systems described in the following chapters.

4.10 References

ABBEY, R.F., and FUJITA, T.T. (1981): 'Tornadoes as represented by the tornado outbreak of 3–4 April 1974', in E. Kessler (Ed.): 'Thunderstorms: A social, scientific, and technological documentary, Vol. 1: The thunderstorm in human affairs', US Department of Commerce, National Oceanic and Atmospheric Administration, Environmental Research Laboratories, Chap. III, pp. 47–84

BATTAN, L.J. (1961): 'The nature of violent storms' (New York: Doubleday)

BATTAN, L.J. (1975): 'Doppler radar observations of a hailstorm', *J. Appl. Meteorol.*, **14**, pp. 198–208

BROWNING, K.A., and FOOTE, G.B. (1976): 'Airflow and hail growth in supercell storms and some implications for hail suppression', *Q. J. R. Meteorol. Soc.*, **102**, pp. 499–533

BURGESS, D.W., and LEMON, L.R. (1990): 'Severe thunderstorm detection by radar', in D. Atlas (Ed.): 'Radar in meteorology' (American Meteorological Society, Boston, MA), Chap. 30a, pp. 619–647

BYERS, H.R., and BRAHAM, R.R., JR. (1949): 'The thunderstorm' (US Government Printing Office, Washington, DC), 287 pp.

CHALMERS, J.A. (1967): 'Atmospheric electricity' (Pergamon Press, New York), 515 pp.

CHRISTIE, D.R., and MUIRHEAD, K.J. (1983): 'Solitary waves, a hazard to aircraft operating at low altitudes', *Aust. Meteorol. Mag.*, **31**, pp. 97–109

CIVIL AVIATION AUTHORITY (1978): 'Accidents to aircraft on the British register, 1977', Civil Aviation Authority, London, Report CAP 419, October 1978

CLARKE, R.H. (1972): 'The morning glory: an atmospheric hydraulic jump', *J. Appl. Meteorol.*, **11**, pp. 304–311

COURT, A., and GRIFFITHS, J.F. (1982): 'Thunderstorm climatology', in E. Kessler (Ed.): 'Thunderstorms: A social, scientific, and technological documentary, Vol. 2: Thunderstorm morphology and dynamics', US Department of Commerce, National Oceanic and Atmospheric Administration, Environmental Research Laboratories, Chap. II, pp. 11–52

CROOK, N.A. (1986): 'The effect of ambient stratification and moisture on the motion of atmospheric undular bores', *J. Atmos. Sci.*, **43**, pp. 171–181

DARKOW, G.L. (1982): 'Thunderstorm energetics', in E. Kessler (Ed.): 'Thunderstorms: A social, scientific, and technological documentary, Vol. 2: Thunderstorm morphology and dynamics', US Department of Commerce, National Oceanic and Atmospheric Administration, Environmental Research Laboratories, Chap. IV, pp. 79–108

DAVIES-JONES, R.P. (1982): 'Tornado Dynamics', in E. Kessler (Ed.): 'Thunderstorms: A social, scientific, and technological documentary, Vol. 2: Thunderstorm morphology and dynamics', US Department of Commerce, National Oceanic and Atmospheric Administration, Environmental Research Laboratories, Chap. X, pp. 297–361

DIETENBERGER, M.A., HAINES, P.A., and LUERS, J.K. (1985): 'Reconstruction of Pan Am New Orleans accident', *J. Aircr.*, **22**, pp. 719–728

DOUGLAS, R.H. (1963): 'Recent hail research: A review', in 'Severe local storms', *Meteorological Monograph* 5 (American Meteorological Society, Boston, MA) pp. 157–167

DOUGLAS, R.H. (1965): 'Intermittency in western Canadian hailfall'. Proceedings of the International Conference on Cloud Physics, Tokyo, Japan, May 1965, pp. 291–295

DOVIAK, R.J., and CHEN, S (1988): 'Observations of a thunderstorm generated gust compared with solitary wave theory'. US Department of Transportation, Federal Aviation Administration Report DOT/FAA/SA-88/1, Washington, DC

DOVIAK, R.J., and CHRISTIE, D.R. (1989): 'Thunderstorm-generated solitary waves: A wind shear hazard', *J. Aircr.*, **26**, pp. 423–431

EILTS, M.D., and DOVIAK, R.J. (1987): 'Oklahoma downbursts and their asymmetry', *J. Clim. Appl. Meteorol.*, **26**, pp. 69–78

FEDERER, B., and WALDVOGEL, A. (1975): 'Hail and raindrop size distribution from a Swiss multicell storm', *J. Appl. Meteorol.*, **14**, pp. 91–97

FLORA, S.D. (1956): 'Hailstorms of the United States' (University of Oklahoma Press, Norman, OK), 201 pp.

FRISBY, E.M., and SAMSON, H.W. (1967): 'Hail incidence in the tropics', *J. Appl. Meteorol.*, **6**, pp. 339–354

FUJITA, T.T. (1973): 'Tornadoes around the world', *Weatherwise*, **57**, pp. 56–62

FUJITA, T.T. (1979): 'Objectives, operation and results of Project NIMROD'. Preprints of 11th Conference on Severe Local Storms, Kansas City, KS, pp. 259–266

FUJITA, T.T. (1983a): 'Microburst wind shear at New Orleans International Airport, Kenner, Louisiana on July 9, 1982'. SMRP Research Paper 199, University of Chicago

FUJITA, T.T. (1983b): 'Andrews AFB microburst'. SMRP Research Paper 205, University of Chicago, Chicago, IL

FUJITA, T.T. (1985): 'The downburst' (The University of Chicago, Chicago, IL)

FUJITA, T.T. (1986): 'DFW microburst' (The University of Chicago, Chicago, IL)
FUJITA, T.T., and BYERS, H.R. (1977): 'Spearhead echo and downbursts in the crash of an airliner', *Mon. Weather Rev.*, **105**, pp. 129–146
FUJITA, T.T., and CARACENA, F. (1977): 'An analysis of three weather-related aircraft accidents', *Bull. Am. Meteorol. Soc.*, **58**, pp. 1164–1181
FUJITA, T.T., and McCARTHY, J. (1990): 'The application of weather radar to aviation meteorology', in D. Atlas, (Ed.): 'Radar in meteorology'. American Meteorological Society, Boston, MA, Chap. 31a, pp. 657–681
FUJITA, T.T., and WAKIMOTO, R.M. (1983): 'Microbursts in JAWS depicted by Doppler radars, PAM and aerial photographs'. Preprints of 21st Conference on Radar Meteorology, Edmonton, Canada, pp. 638–645
FULTON, R., ZRNIC', D.S., and DOVIAK, R.J. (1990): 'Initiation of a solitary wave family in the demise of a nocturnal thunderstorm density current', *J. Atmos. Sci.*, **47**, pp. 319–337
GOKHALE, N.R. (1975): 'Hailstorms and hailstone growth' (State University of New York Press, Albany, NY), 465 pp.
GOLDE, R.H. (Ed.) (1977): 'Lightning' (Academic Press, London, Vols. 1 & 2), 848 pp.
HALLOWELL, R.G. (1990): 'Aspect angle dependence of outflow strength in Denver microbursts: spatial and temporal variations'. Preprints of the 16th Conference on Severe Local Storms and the Conference on Atmospheric Electricity, 22–26 October 1990, Kananaskis Park, Alberta, Canada (American Meteorological Society, Boston, MA), pp. 397–402
HALLOWELL, R.G. (1993): 'Dual-Doppler measurements of microburst outflow strength asymmetry'. Preprints of 26th International Conference on Radar Meteorology, 24–28 May 1993, Norman, OK (American Meteorological Society, Boston, MA), pp. 664–666
HJELMFELT, M.R. (1988): 'Structure and life cycle of microburst outflows observed in Colorado', *J. Appl. Meteorol.*, **27**, pp. 900–927
HOECKER, W.H., JR. (1963): 'Three southerly low-level jet systems delineated by the Weather Bureau special pibal network of 1961', *Mon. Weather Rev.*, **91**, pp. 573–582
HOXIT, L.R., LIEB, H.S., CHAPPELL, C.F., and MOGIL, H.M. (1981): 'Disaster by flood', in E. Kessler (Ed.): 'Thunderstorms: A social, scientific, and technological documentary, Vol. 1: The thunderstorm in human affairs, US Department of Commerce, National Oceanic and Atmospheric Administration, Environmental Research Laboratories, Chap. II, pp. 23–45
KAYTON, M., and FRIED, W.R. (1969): 'Avionics navigation systems' (Wiley, New York), Chap. 14
KESSLER, E. (1974): 'Survey of boundary layer winds with special reference to extreme values', Proceedings of AIAA 7th Fluid and Plasma Dynamics Conference, Palo Alto, CA, Paper 74-586
KESSLER, E. (1981): 'Thunderstorms in a social context', in E. Kessler (Ed.): 'Thunderstorms: A social, scientific, and technological documentary, Vol. 1: The thunderstorm in human affairs'. US Department of Commerce, National Oceanic and Atmospheric Administration, Environmental Research Laboratories, Chap. I, pp. 1–22
KESSLER, E. (Ed.) (1981–1982): 'Thunderstorms: A social, scientific, and technological documentary', Vols. 1, 2 and 3. US Department of Commerce, National Oceanic and Atmospheric Administration, Environmental Research Laboratories
KESSLER, E. (1985): 'Wind shear and aviation safety', *Nature*, **315**, pp. 179–180
KONIG, R., KRAUSPE, P., and SCHANZER, G. (1980): 'Procedure to improve flight safety in wind shear conditions'. 'Proceedings of 12th Congress of the International Council of the Aeronautical Sciences, Munich, FRG, pp. 744–757
LEE, J.T., and BECKWITH, W.B. (1981): 'Thunderstorms in aviation', in E. Kessler (Ed.): 'Thunderstorms: A social, scientific, and technological documentary, Vol. 1: The thunderstorm in human affairs'. US Department of Commerce, National Oceanic and Atmospheric Administration, Environmental Research Laboratories, Chap. VI, pp. 141–169

LEWIS, M.S., ROBINSON, P.A., HINTON, D.A., and BOWLES, R.L. (1994): 'The relationship of an integral wind shear hazard to aircraft performance limitations'. NASA Technical Memorandum 109080, February 1994

LUDLAM, F.H. (1982): 'Introduction', in E. Kessler (Ed.) 'Thunderstorms: A social, scientific, and technological documentary, Vol. 2: Thunderstorm morphology and dynamics'. US Department of Commerce, National Oceanic and Atmospheric Administration, Environmental Research Laboratories, Chap. I, pp. 1–9

MADDOX, R.A., CARACENA, F., HOXIT, L.R., and CHAPPELL, C.F. (1977): 'Meteorological aspects of the Big Thompson flash flood of 31 July 1976'. NOAA Tech. Rep. ERL 388-APCL41 (NOAA Environmental Research Laboratories, Boulder, CO), 87 pp.

MAGONO, C (1980): 'Thunderstorms' (Elsevier, Amsterdam)

MAHAPATRA, P.R., DOVIAK, R.J., and ZRNIC, D.S. (1982): 'Radar detection of low level wind shear affecting aircraft terminal navigation'. Proceedings of the National Aerospace Meeting of the Institute of Navigation, Moffett Field, CA, pp. 52—59

MAHAPATRA, P.R., DOVIAK, R.J., and ZRNIC, D.S. (1991): 'Multisensor observation of an atmospheric undular bore', *Bull. Am. Meteorol. Soc.*, **72**, pp. 1468–1480

MORGAN, G.M., and SUMMERS, P.W. (1982): 'Hailfall and hailstorm characteristics', in E. Kessler (Ed.): 'Thunderstorms: A social, scientific, and technological documentary, Vol. 2: Thunderstorm morphology and dynamics'. US Department of Commerce, National Oceanic and Atmospheric Administration, Environmental Research Laboratories, Chap. XI, pp. 363–408

NATIONAL TRANSPORTATION SAFETY BOARD (1974a): 'Aircraft accident report: Ozark Airlines, Inc., Fairchild Hiller FH-227B, N421S near the Lambert–St. Louis International Airport, St. Louis, Missouri, July 23, 1973'. Rep. NTSB AAR-74-5

NATIONAL TRANSPORTATION SAFETY BOARD (1974b): 'Aircraft accident report: Delta Air Lines, Inc., McDonnell Douglas DC-9-32, N3323L, Chattanooga Municipal Airport, Chattanooga, Tennessee, November 27, 1973'. Rep. NTSB-AAR-74-13

NATIONAL TRANSPORTATION SAFETY BOARD (1974c): 'Aircraft accident report: Iberia Lineas Aereas de Espana (Iberian Airlines) McDonnell Douglas DC-10-30 EC-CBN, Logan International Airport, Boston, Massachusetts, December 17, 1973'. Rep. NTSB AAR-74-14

NATIONAL TRANSPORTATION SAFETY BOARD (1976a), 'Aircraft accident report: Eastern Airlines, Inc., Boeing 727-225, N8845E, John F. Kennedy International Airport, Jamaica, New York, June 24, 1975'. Rep. NTSB AAR-76-8

NATIONAL TRANSPORTATION SAFETY BOARD (1976b): 'Aircraft accident report: Continental Airlines, Inc., Boeing 727-224, N88777, Stapleton International Airport, Denver, Colorado, August 7, 1975'. Rep. NTSB AAR-76-14

NATIONAL TRANSPORTATION SAFETY BOARD (1977), 'Aircraft accident report: Pan American World Airways, Inc., Boeing 707-321B, N454PA, Pago Pago, American Samoa, January 30, 1974'. Rep. NTSB AAR-77-7 (Revised)

NATIONAL TRANSPORTATION SAFETY BOARD (1978a): 'Aircraft accident report: Allegheny Airlines, Inc., Douglas DC-9, N994VJ, Philadelphia, Pennsylvania, June 23, 1976'. Rep. NTSB AAR-78-2

NATIONAL TRANSPORTATION SAFETY BOARD (1978b): 'Aircraft accident report: Continental Air Lines, Inc., Boeing 727-224, N32725, Tucson, Arizona, June 3, 1977'. Rep. NTSB AAR-78-9

NATIONAL TRANSPORTATION SAFETY BOARD (1981), 'Aircraft accident report: Redcoat Air Cargo Ltd., Bristol Britannia 253F, G-BRAC, Billerica, Massachusetts, February 16, 1980'. Rep. NTSB AAR-81-3

NATIONAL TRANSPORTATION SAFETY BOARD (1983), 'Aircraft accident report: Eastern Airlines, Inc., Boeing 727, N8838E, Raleigh, North Carolina, November 12, 1975'. Rep. NTSB AAR-83/06

NATIONAL TRANSPORTATION SAFETY BOARD (1985), 'US Air Inc., Flight 183, McDonnel Douglas DC-9-31, N964VJ, Detroit Metropolitan Airport, Detroit, Michigan, June 13, 1984'. Report NTSB AAR-85/01

NATIONAL TRANSPORTATION SAFETY BOARD (1986): 'Aircraft accident report: Delta Airlines, Inc., Lockheed L-1011-385-1, N726DA, Dallas/Fort Worth International Airport, Texas, August 2, 1985'. Rep. NTSB/AAR-86/05, August 1986

NATIONAL TRANSPORTATION SAFETY BOARD (1993a): 'Annual review of aircraft accident data, US air carrier operations, Calendar Year 1990'. National Transportation Safety Board, Washington, DC, Report NTSB/ARC-93/02, 4 October 1993

NATIONAL TRANSPORTATION SAFETY BOARD (1993b): 'Annual review of aircraft accident data, US general aviation, Calendar Year 1990'. National Transportation Safety Board, Washington, DC, Report NTSB/ARG-93/02, 17 December 1993

NATIONAL TRANSPORTATION SAFETY BOARD (1993c), 'Annual review of aircraft accident data, US air carrier operations, Calendar Year 1989'. Report NTSB/ARC-93/01, 7 May 1993

NELSON, J.P. III and ELLROD, G.P. (1997): 'Recent developments in a microburst risk image product derived from GOES I-M satellite sounder data'. Preprints of 7th Conference on Aviation, Range, and Aerospace Meteorology, Long Beach, CA, 2–7 February 1997 (American Meteorological Society, Boston), pp. 262–265

PIERCE, E.T. (1982): 'Storm electricity and lightning', in E. Kessler (Ed.): 'Thunderstorms: A social, scientific, and technological documentary, Vol. 2: Thunderstorm morphology and dynamics'. US Department of Commerce, National Oceanic and Atmospheric Administration, Environmental Research Laboratories, Chap. XIII, pp. 447–466

POTTS, R.J. (1991): 'Microburst observations in tropical Australia'. Proceedings of 4th International Conference on Aviation Weather Systems, Paris, France, 24–28 June 1991 (American Meteorological Society, Boston), pp. J67-J72

RAY, P.S. (1990): 'Convective dynamics', in D. Atlas (Ed.): 'Radar in meteorology' (American Meteorological Society, Boston, MA), Chap. 24a, pp. 348–390

RUDICH, R.D. (1986); 'Weather-involved U.S. air carrier accidents 1962–84: A compendium and brief summary'. Proceedings of the AIAA 24th Aerospace Sciences Meeting, Reno, NV, Paper AIAA-86-0327

SANDERS, F., and FREEMAN, J.C. (1982): 'Thunderstorms at sea', in E. Kessler (Ed.) 'Thunderstorms: A social, scientific, and technological documentary, Vol. 2: Thunderstorm morphology and dynamics'. US Department of Commerce, National Oceanic and Atmospheric Administration, Environmental Research Laboratories, Chap. III, pp. 53–77

SASAKI, Y.K., and BAXTER, T.L. (1982): 'The gust front', in E. Kessler (Ed.): 'Thunderstorms: A social, scientific, and technological documentary, Vol. 2: Thunderstorm morphology and dynamics'. US Department of Commerce, National Oceanic and Atmospheric Administration, Environmental Research Laboratories, Chap. IX, pp. 281–296

SCHLICKENMAIER, H.W. (1989), 'Windshear case study: Denver, Colorado, July 11, 1988'. Report DOT/FAA/DS-89/19, November 1989

STULL, R.B. (1988): 'An introduction to boundary layer meteorology' (Kluwer Academic, Dordrecht, The Netherlands), Chap. 1, p. 15

SULAKVELIDZE, G.K. (1967): 'Rainstorm and hail'. English Translation: National Technical Information Service, Springfield, VA, NTIS-TT-68-50466, 1969

TROXEL, S.W., and DELANOY, R.L. (1995): 'Machine intelligent gust front detection for the Integrated Terminal Weather System (ITWS)'. Preprints of 6th Conference on Aviation Weather Systems, Dallas, TX, 15–20 January 1995 (American Meteorological Society, Boston), pp. 378–383

UMAN, M.A.(1969): 'Lightning' (McGraw-Hill, New York), 264 pp.

US DEPARTMENT OF TRANSPORTATION (1987): 'Windshear Training Aid'. Prepared for FAA under Contract DTFA 01-86-C-00005 by the Boeing Co., Lockheed Corp., United Airlines, American Weather Associates, Inc., and Helliwell, Inc.

WILLIAMS, L. (1973): 'Hail and distribution'. Special Report of the Engineer Topographic Laboratories, Ft. Belvoir, VA, ETL-SR-73-3, 6 pp.

WILSON, J.W., ROBERTS, R.D., KESSINGER, C. and McCARTHY, J. (1984): 'Microburst wind structure and evaluation of Doppler radar for airport wind shear detection', *J. Clim. Appl. Meteorol.*, **23**, pp. 898–915

WINN, W.P., SCHWEDE, G.W., and MOORE, C.B. (1974): 'Measurement of electric fields in thunderclouds, *J. Geophys. Res.*, **79**, pp. 1761–1767.

WOLFSON, M.M. (1988); 'Characteristics of microbursts in the continental United States', *Lincoln Lab. J.*, **1**, pp. 49–74.

WOODFIELD, A.A. (1983): 'Wind shear and vortex wake research in U.K. 1982'. Proceedings of 6th Annual Workshop on Meteorological and Environmental Inputs to Aviation Systems, W. Frost and D.W. Kamp (Eds.), NASA Rep. CP-2274 and Dept. of Transportation Rep. DOT/FAA/RD-82/72, 66–83

ZRNIC', D.S., BURGESS, D.W., and HENNINGTON, L. (1985): 'Doppler spectra and estimated wind speed of a violent tornado', *J. Clim. Appl. Meteorol.*, **24**, pp. 1068–1081

Chapter 5
Requirements of systems for aviation weather surveillance

5.1 General

The preceding discussions should have served to convey a picture of the extreme diversity of the aviation weather scenario in terms of the types of phenomena encountered and their spatial and temporal characteristics. There is also a considerable degree of geographical variability of these phenomena in respect of their characteristics, intensity, frequency of occurrence, and their diurnal and seasonal distribution. Further, there is a gradation of atmospheric phenomena from the consideration of the seriousness of their hazardous effects on aviation. Yet again, the cost of surveillance equipment and systems is a highly variable parameter, depending not only on the generic type, but also on the desired performance level. Finally, airports and airways vary widely in their traffic patterns and densities, and the equipment configuration that is optimum for one situation will, in general, not be so for another.

All these variable factors make the specification, design and selection of electronic equipment for aviation surveillance rather complex. It is therefore necessary to evolve the requirements of aviation weather surveillance systems of different types and match them, as optimally as possible, with the requirements of modern aviation and the needs of the particular installation under consideration. This chapter is devoted to evolving the attributes and characteristics desired of electronic weather surveillance systems designed in support of modern aviation. In addition to technical parameters, there are considerations of human factors such as the skill level and availability of human operators and data interpreters, environmental factors such as restrictions on electromagnetic spectrum and radiated power, and infrastructural factors such as communication facilities, quality of electric power, and engineering support base, in the design and choice of surveillance equipment. Finally, the global and local regulatory and legal framework governing aviation activity, and the allocation of responsibility and liability in this field has a strong bearing on the development and deployment of weather surveillance systems in support of aviation.

The technical requirements of aviation weather monitoring systems must be discussed from the spatial as well as temporal points of view. The spatial requirements include the volume coverage and spatial resolution parameters of the surveillance systems. These must be matched with the corresponding requirements over the air space covered by the air traffic service of which the weather information dissemination system is a part. The density and traffic handling capacity of different parts of the air space are also important considerations in determining the spatial parameters.

The temporal agility and data refreshment rate of the surveillance systems are required to be compatible with the rapidity of evolution of the weather phenomena of interest, the frequency of aircraft operations (e.g. landing and takeoff) and also with other elements of the overall air traffic service system.

5.2 Types of weather surveillance systems for aviation

5.2.1 In situ and remote sensing

There are two basically different philosophies for sensing of aviation weather: *in situ* sensing and remote sensing. In *in situ* sensing the sensing element is in direct contact with the medium whose parameters are being sensed to help deduce the qualitative and quantitative characteristics of the phenomena under surveillance. An example of *in situ* sensors is the anemometer used for measuring air speed. In contrast, in remote sensing, the sensing device is spatially removed from the scene being observed, and the parameters of the phenomena in the scene are measured by detecting and suitably processing the radiation, reflection or absorption characteristics of the medium which supports these phenomena. The radiation involved is usually electromagnetic, but acoustic waves are also sometimes used. The radar is a prime example of a remote sensing device for atmospheric phenomena.

The output of sensing devices usually has to be processed further to render it intelligible and usable. In both *in situ* and remote sensing such processing may be done at the location of the sensor itself. Alternatively, the sensed data may be transmitted to one or more different locations, either in raw form or after a certain amount of preprocessing, for final processing before display and for interpretation. Such transmission is usually done to improve human access to the data and/or to make available more sophisticated processing and display equipment than is available or desirable at the sensor site. When multiple sensors are necessary to cover a phenomenon or region, the pooling of data from all these sensors to a common processing station becomes essential. This is frequently the case in aviation weather surveillance systems.

As *in situ* instruments make measurements by 'contact' with the medium of interest, they are essentially point measuring devices, i.e. their measurements are true (within their inherent accuracy) only at the spot where they are located. Thus a bucket-type rain-gauge or a vane-type anemometer would provide measures of the rainfall or wind speed at its location only. This amounts to a spatial sampling of the parameter being measured. In using such observations for inferring the distribution of the parameter over an area or volume, the assumption is made that the parameter value measured by an instrument is valid over its immediate vicinity. Such an assumption would naturally be less accurate as one moves farther from the point of measurement.

Extended coverage of measurement can be achieved by using an array of spatially separated *in situ* sensors, and assuming a smooth variation of the measured parameters between individual sensors. For this to be valid, at least to a first approximation, the spacing between adjacent sensors should be equal to, or closer than, half of the fastest Fourier component of the spatial frequency of the phenomenon of interest. Thus, while *in situ* meteorological instruments have served well in the area of large scale (global or regional) weather monitoring and modelling, their use in aviation weather surveillance does not ensure adequate coverage. The reasons are given below.

First, as shown in the preceding chapter, the phenomena of most significance for aviation often have high spatial gradients, frequently showing significant parameter variation over distances as small as hundreds of metres, or even tens of metres in some cases. This would call for an impractically close packing of sensors in the array, resulting not only in very high cost, but also problems of their siting, installation and maintenance. Aviation being a wide-area activity, anything like full coverage of the controlled air space through *in situ* weather sensing is clearly impossible.

Aviation is not well served by *in situ* sensing for another important reason. The bulk of aviation activity occurs at significant height above the ground, whereas *in situ* sensors are mounted either on the ground itself or on masts or towers which are at the most a few tens of metres tall. As was clearly observed in the preceding chapter, for the severe local phenomena that form the source of most aviation hazards, the measurement of atmospheric parameters near the ground alone serves a very limited purpose in the assessment of hazardous weather conditions aloft.

For these reasons *in situ* weather sensing plays only a limited role in the modern and futuristic scheme of aviation weather surveillance. However, it is effective in monitoring hazards to flight at very low altitudes and over localised areas. The best example of such an application is the Low Level Wind Shear Alert System, which is discussed in Section 8.4. *In situ* instruments can complement the measurements made by remote sensors such as radars in areas which fall significantly below the radar horizon. Another useful application for *in situ* sensors in the modern aviation weather scene is in the measurement of atmospheric parameters at discrete spots to act as ground truth, calibration or initialisation bases for more complex and automated hazard parameter estimation algorithms. Some of these algorithms will be discussed in Chapter 10. *In situ* sensors will, of course, remain important for measuring parameters such as temperature and pressure, which are difficult to measure accurately by remote sensing, and which by themselves are not hazard factors for aviation.

5.2.2 Ground-based, airborne and spaceborne sensors

The limitations of *in situ* sensing due to coarse spatial sampling as well as those imposed by near-ground location of sensors, may apparently be minimised by mounting the sensors on a moving platform such as an aircraft or a balloon. This provides a spatial continuity of measurements along the flight path of the platform, but not in other dimensions. Even along the flight path, since the sensor is at a given point only at a particular instant, the measurement involves a temporal sampling, usually of a nonrepetitive nature.

In spite of these difficulties airborne *in situ* atmospheric sensing is useful in certain situations of importance to aviation. The airborne measurement of aircraft icing potential mentioned in Section 3.5 is one such application. Balloons have been used traditionally for routine monitoring of altitude profiles of atmospheric parameters. Among other data, these can provide a definite indication of the existence of near-ground stable layers which can support buoyancy waves, as mentioned in Section 4.6.2. Radar tracking of balloons provides the vertical profile of wind, from which the vertical wind shear can be deduced directly. However, balloon measurements are reliable only for detecting relatively large phenomena and measuring the slow-varying parameters of the atmosphere. The spacing of routine balloon launches is not close enough in space and time to be of great use in alerting aircraft against the dynamic and localised phenomena which are of most significance to aviation.

In situ instrumentation borne on aircraft has special advantages in several situations. Observations from aircraft landing on a particular runway can be used to alert closely following aircraft if hazardous features are noticed. The role of pilot reports or pireps as a good source of turbulence information was illustrated in Section 3.3.2. With proper instrumentation and algorithms aircraft may also infer the presence of wind shear when they begin to enter into such flow fields (McLean, 1988).

Aircraft-borne *in situ* instruments provide the most direct measurement of thunderstorm phenomena aloft, such as rain, hail, lightning and turbulence. However, such airborne measurement is used essentially for research and data generation, and not in an operational mode in support of aviation, for obvious reasons.

The vantage point provided by aircraft platforms also offers unique advantages to remote sensing devices such as radars. Storm warning radars are standard equipment in most airliners and other large aircraft. For aviation weather research applications radars carried on aircraft are used to map the interiors of phenomena such as thunderstorms and cyclonic storms. Such a procedure provides the advantages of fine resolution because of the proximity of the mapping radar, ability to observe horizontal cross-sections of phenomena, and the facility of looking at the phenomena from various aspect angles by flying around the phenomenon.

A far more common and useful location for remote sensors for aviation weather applications, however, is on the ground. Ground-based sensor systems can be bigger and more sophisticated, permitting accurate parameter sensing over large volumes of space. More importantly, in aviation operations it is required to cover given volumes of air space rather than cover individual aircraft paths or weather phenomena. Ground installations permit their attachment to designated air traffic monitoring and control centres, and efficient use of their data in air traffic services. Modern and futuristic aviation weather surveillance systems derive the bulk of their data from ground-based remote sensors.

Data from space-based remote sensors is also of value in aviation weather monitoring. Pictures and data from satellite-borne instruments now have an established place in the visualisation, monitoring, analysis and prediction of global and regional weather patterns. With improvement in the accuracy and fineness of these data, satellite-based observations are becoming a useful source of data for direct use in aviation. Satellite observation of aviation weather is a topic of discussion later in this book.

5.3 Spatial coverage

The spatial coverage of aviation weather surveillance systems should include all the designated flight corridors within the air space, as well as the space above terminal areas. Further, it should cover areas of commercial, general and military aviation activities.

The height coverage of aviation weather surveillance systems should extend over the altitude slab occupied by all types of aviation activity. As mentioned before, normal aviation activity, including the flight of large subsonic jet airliners and transport aircraft, occurs from ground up to ~40 000 ft in height. Less frequent flights, including those of supersonic transport and high-altitude military missions, may occur up to twice this height.

When we talk of height coverage, it clearly pertains to remote sensing systems, since locating *in situ* instruments at all positions except those on or close to the ground is impractical. Remote sensing systems, of which radar is the best example, are inherently well suited for collecting data from higher elevations. However, there are difficulties in collecting data from the lower elevations due to the earth's curvature effects and the interfering effect of the ground.

If a straight line propagation for radiation is assumed, the spherical surface of the earth drops off from the propagation path by a height h at a distance d from a ground-based sensor (see Fig. 5.1a). The two parameters are related as

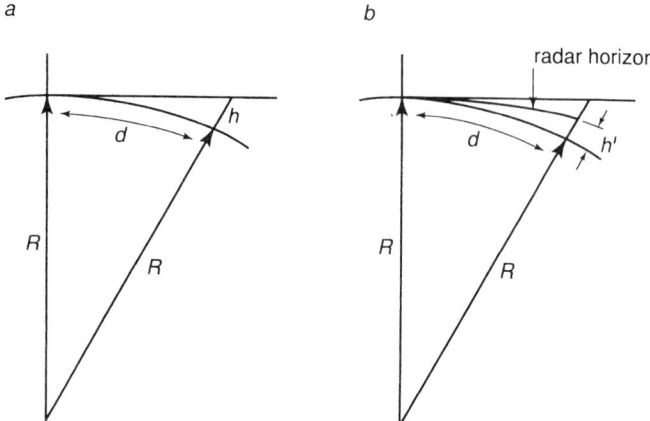

Figure 5.1 *Geometry of optical (or electromagnetic) horizon for rectilinear propagation (a), and for curvilinear propagation ('ray bending') due to atmospheric refraction (b)*

$$h = R\left[\sec\left(\frac{d}{R}\right) - 1\right] \qquad (5.1)$$

where R is the radius of the earth (mean value ~6367.5 km). For small values of the ratio d/R, eqn. 5.1 may be approximated as

$$h \simeq \frac{d^2}{2R} \qquad (5.2)$$

In reality, however, electromagnetic radiation does not, in general, propagate in a straight line in the atmosphere because of the density gradient and the resulting refractive index variation of the atmosphere with height. The consequent bending of the rays is beneficial for the remote sensing of atmospheric phenomena because it enhances the field of view of the sensor by helping the rays to partially follow the curvature of the earth, as shown in Fig. 5.1*b*. The path of the ray starting horizontally from the sensor is called the *radar horizon*. Tracing the true radar horizon is rather complex. However, if the atmospheric refractive index is assumed to vary linearly with respect to height, then either of eqns. 5.1 or 5.2 may still be used to obtain the clearance h' of the radar horizon from the ground if the earth's radius R is replaced by a larger fictitious radius $R' = kR$. A commonly used value for k is 4/3. Thus,

$$h' \simeq \frac{d^2}{2R'} = \frac{3d^2}{8R} \qquad (5.3)$$

for small d/R. Geometrically, eqn. 5.3 means that the clearance of the radar horizon from the spherical earth, considering atmospheric refraction, approximately equals the clearance of the straight-line (unrefracted) horizon from a sphere with diameter four-thirds of that of the earth. The height given by eqn. 5.3 is less than that obtained from eqn. 5.2. Thus, the atmospheric refraction helps electromagnetic remote sensors such as weather radars 'see' closer to the ground than is possible without considering the refraction effect. More accurate estimation of the refraction effect may be obtained by assuming an exponential variation of the atmospheric refractive index with respect to height. The resulting expressions are complex and would require numerical computation for evaluation, but standard charts are available relating the various parameters such as range, height and elevation angle (Blake, 1968a; 1968b).

Geometrically, an electromagnetic remote sensor is blind to the parts of the atmosphere falling below the radar horizon. To be able to 'see' all points at a height h_{min} and above at a given location, such a sensor must be positioned within a maximum distance d_{max} from that location, obtained by inverting eqn. 5.3 as

$$d_{max} = \left[\left(\frac{8R}{3}\right)h_{min}\right]^{1/2} \qquad (5.4)$$

This maximum distance is higher than the corresponding distance $(2Rh_{min})^{1/2}$ in the case of straight line propagation obtained from eqn. 5.2. Eqn. 5.4 is very useful for determining the coverage and location of ground-based remote sensors such as radars in the aviation context.

As mentioned at the beginning of this section, a vast majority of normal aviation activity occurs within a height of 40 000 ft or ~12 km. Also, most of the significant hazardous weather factors occur within this height at the middle and the high latitudes (in the tropics the tropopause is higher and thunderstorms may extend to heights >12 km, see Fig. 4.4). As shown in Fig. 5.2, a remote sensor must be located within a certain maximum distance to be

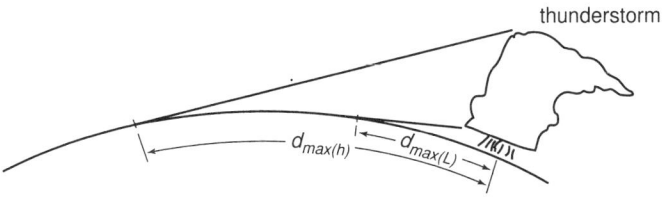

Figure 5.2 Geometry of horizon limitation for viewing different parts of a thunderstorm. $d_{max(h)}$ is the farthest range at which the highest part of the thunderstorm can be observed, and $d_{max(l)}$ is the maximum distance for 'viewing' the lowest point requiring surveillance

able to observe at least a part of these phenomena. For a 12 km altitude limit, the radar-horizon-limited maximum surveillance range works out to ~390 km for straight-line propagation, and ~450 km when atmospheric refraction is taken into account. This amounts to an increase of ~15% in the surveillance range limit, and 33% in the surveillance area, due to refraction. An important fact to be noted here is that these geometric range limits for observation of thunderstorms would override even if a radar has no limits on other range-determining resources such as power and sensitivity.

In the vicinity of the runway complex of an airport, where flights occur at low levels, a surveillance system should provide coverage down to very low altitudes. The minimum altitude of surveillance, called the *observation ceiling*, is determined by the lowest heights from the ground at which atmospheric hazards to aviation may be found. The phenomenon that is of the highest significance in this context is the outflow from microbursts which is the strongest source of wind shear at low altitudes. As mentioned in Section 4.7.2, microbursts produce their strongest wind outflows at heights of ~75 m or possibly even lower. If a radar is to be so located as to observe atmospheric phenomena at heights as low as 75 m (Fig. 5.2), its horizontal distance from the point of observation should not exceed ~35 km as given by eqn. 5.4. Equivalently, a given radar would not be able to observe the part of the atmosphere lower than 75 m outside a circle of radius 35 km centred at its location. This is a very important consideration for determining the location of terminal area weather surveillance radars, and will be referred to in Section 7.4.2.

As one goes farther out from the runway area towards the *en route* flight regions, the minimum altitude of flight operations also increases, and hence the need for low-altitude weather surveillance may be correspondingly relaxed. For convenience, such relaxation may be effected in a stepped manner. One such scheme is presented in Section 5.5, along with spatial resolution requirements.

It is worth pointing out that the discussions on horizon effects in this section are based on the assumption of a smooth spherical earth. The presence of ground undulations, surface features and structures, and other obstructions can alter the horizon line considerably. These effects are site-dependent and must be evaluated for each sensor location. Further, the horizon limits are also altered if the sensor is mounted at a height from the ground. The additional geometric range capability obtained due to the sensor height can be evaluated by appropriate use of eqns. 5.2–5.4. More discussion on refraction and coverage aspects may be found in Blake (1990).

Spatial coverage requirements may sometimes be phenomenon-specific, and may also have associated reliability criteria. An example is provided by microburst detection in terminal areas. As per the requirements of the US Federal Aviation Administration, the critical area for microburst detection covers runways themselves, and includes 3 miles (~5 km) on the approach side and 2 miles (~3 km) on the departure side of each runway (Turnbull,

1995). These are areas where the aircraft is likely to be below 1000 ft in altitude and hence subject to microburst-induced wind shear. Further, at least 90% of the microbursts in the vicinity of the airport must be detected with a false alarm rate not exceeding 10%.

5.4 Data update rates

In the context of aviation weather surveillance systems, the data update rate is among the most important specifications. The optimum data update rate is arrived at as a compromise between the need to reliably capture the fastest of the phenomena that are hazardous to aviation, and to keep the requirements of system resources such as communication bandwidth, computing power, memory and operator involvement at an affordable level. In view of the strong influence of this parameter on the aviation weather surveillance system design, systematic studies have been conducted to determine the most desirable data update rates.

The *agility* of an aviation weather surveillance system is indicated by the speed or rate at which it refreshes or updates its data regarding its volume of observation. The agility required of the system is decided primarily by the timescales over which aviation-hazardous phenomena build up, decay or change significantly.

In general, atmospheric phenomena of larger spatial scales are temporally more stable than smaller ones, providing more time for their detection. Based on a systematic study of several Oklahoma storms, Mahapatra and Zrnic' (1982; 1984) noted that mesocyclonic features associated with thunderstorms have a characteristic timescale of the order of several tens of minutes. This is true of the general activity level of thunderstorms (e.g. updraft, rain, lightning frequency).

Individual subphenomena within thunderstorms may, however, build up or decay much more rapidly than the thunderstorm as a whole. They may also shift position to occupy more hazardous locations, such as aircraft approach paths and landing thresholds, within relatively short intervals of time and in an unpredictable manner. Modern, high-performance aviation weather surveillance systems, intended to provide fine guidance against localised hazards in dense-traffic environments, must be agile enough to track such fast-evolving phenomena.

Perhaps the most stringent requirement in terms of detection system agility is demanded by the needs of microburst detection. In Section 4.7 it was mentioned that microbursts build up to their maximum wind shear in ~5 min. The same data also showed (Hjelmfelt, 1988) that the storm outflow reached microburst intensity in 2.5 min on average, after low-altitude divergence was first observed by radar. To minimise false alarms, aviation management procedures usually require an observation to be confirmed in at least two consecutive observation cycles before warnings can be issued. Thus, the data update intervals of surveillance systems should be at least half of the build-up periods of hazardous phenomena. Considering that many microbursts would build up faster than the ensemble average of 2.5 min, and allowing time for processing, communication, display and interpretation, such a criterion sets an upper limit of ~1 min on the data update interval of surveillance systems for microburst detection. Indeed, the US Federal Aviation Administration requires that microburst warnings must be updated every minute, and that they cannot be more than 20 s old when first displayed (Turnbull, 1995).

5.5 Spatial resolution

The spatial resolution of the data acquired and processed by aviation weather surveillance systems must be consistent with the requirements of the different regimes of flight operations. Intuitively, one would expect that in the *en route* phase of aircraft flight, where the aircraft are dispersed sparsely in the air space, and have more time and space reserve for recovering from errors or deviations, the requirement for detail in the spatial picture of the flight situation would not be as stringent as in the case of terminal-area operations where aircraft are densely packed in a limited air space and where the margin for error is small.

Refer to Section 2.5, wherein the elements of air traffic control were introduced. From the point of view of collision avoidance, which is the prime focus of the current air traffic control systems, a greater separation between aircraft is required during *en route* flight, but a closer spacing is permitted in the terminal areas where aircraft converge. A similar consideration is applicable to weather avoidance during the different phases of flight. Presenting a fine spatial picture of the weather situation within terminal areas permits (i) the depiction of the detailed structures of hazardous weather features within the areas, facilitating their recognition and precise delineation, and (ii) the accurate location of zones of high gradients of precipitation and wind velocities, and reliable estimation of these gradients. This facilitates more efficient use of the limited air space and expensive airport facilities. However, stipulating a highly detailed picture for the entire air space would entail the handling of too much information in terms of storage, transmission, retrieval and display. This is better appreciated by realising that a vast proportion of the navigated air space is covered by *en route* traffic. Hence, any reduction in data density for *en route* areas has a large contribution to the reduction of overall data handling and storage requirements. On the other hand, terminal air spaces are localised areas constituting a small fraction of the overall air space. Hence, a detailed coverage of the weather pattern in these areas is possible without an inordinate increase in the overall data volume.

Requirements of systems for aviation weather surveillance 131

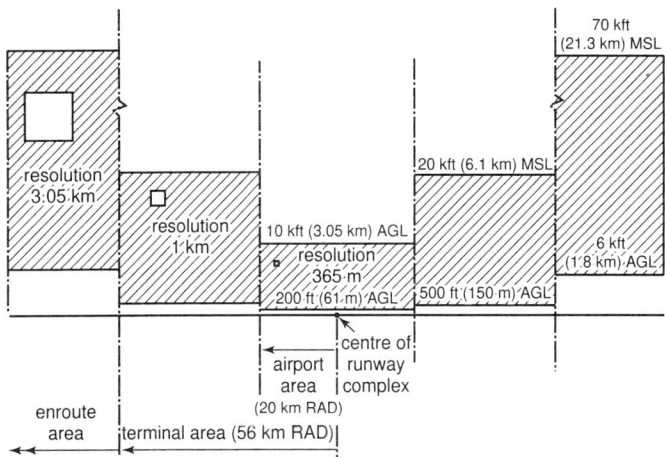

Figure 5.3 Schematic representation of the FAA requirements for weather radar coverage (maximum and minimum heights of observation) and resolution in different flight areas

Such a scheme of variable spatial resolution requirement for weather data has been recommended by the US Federal Aviation Administration (US Department of Transportation, 1981) in the context of airport weather surveillance by radars. The scheme, represented graphically in Fig. 5.3, demarcates three different zones and assigns different levels of spatial resolution to each of them, as given below:

1. *Airport area* covers a radius of 20 km (12 miles) from the centre of the runway complex. The spatial resolution requirement in this area is 365 m. Within this area, the radar-based weather surveillance system is required to cover the height interval between 200 ft (61 m) and 10 000 ft (3.05 km) above the local ground level.
2. *Terminal area* extends over a radius of 56 km (35 miles) from the centre of the runway complex. Over the annular part of the terminal area that is outside the airport area, the spatial resolution requirement of the aviation weather surveillance system is 1 km, and the altitude coverage is between 500 ft (150 m) and 20 000 ft (6.1 km) above ground level.
3. *En route area* covers the zone outside the terminal area. The spatial resolution here is stipulated at 3.05 km, and the altitude coverage is specified to be from 6000 ft (1.8 km) to 70 000 ft (21.3 km) above ground level.

The US Terminal Doppler Weather Radar system, which will be discussed in detail in Section 7.5, takes into account these progressively increasing resolution and altitude coverage specifications.

Note that in the above scheme the resolution requirement in the *en route* air space is over eight times in linear terms, and about 583 times in volumetric terms, compared to that in the airport area. These numbers give an idea

132 *Aviation weather surveillance systems*

about the great saving in data handling and storage requirements of aviation weather surveillance systems by adopting a stepped resolution requirement rather than a uniformly fine data grid.

5.6 Data processing and display systems

Sensors form the front end of any surveillance system. In the case of weather surveillance, the sensors measure the parameters of the atmosphere (e.g. pressure, temperature and humidity) in the region of surveillance, and also sense the nature and intensity of other indicators of weather condition, such as precipitation. The output of these sensors is called *raw data*. Earlier generations of sensors produced *analogue outputs* in the form of voltages or currents, which were either indicated on corresponding measuring instruments or recorded on strip chart recorders. Modern sensors generally produce *digital outputs*, i.e. their data are in digitally coded form, which can be fed directly to computers for further processing and/or logging.

Indeed, the ability to perform sophisticated data processing using computers is responsible in a major way for the development of the modern aviation weather surveillance systems. Raw data from sensors require human intervention for interpretation and hazard potential evaluation. In modern, data-intensive, systems the volume of raw data is so large that human observers would be overwhelmed. This is especially true in the case of highly dynamic situations such as those associated with aviation, where actions may have to be taken based on the sensed data within a matter of minutes. Modern aviation weather surveillance therefore relies heavily on computer-based data processing systems not only for such mechanical functions as the recording, retrieval and display of data, but also for intelligent functions such as weather feature identification, hazard potential evaluation, hazardous weather prediction, and warning generation. The role of human operators in such systems is intended to be essentially supervisory. This is the only way that modern aviation weather surveillance systems, with their multifarious functions, high resolution, and agility can be cost-effective, affordable and reliable.

The processing of the raw weather data generated by sensors is a very involved and diversified topic, which is discussed in some detail in Chapters 9 and 10. Here the attention is focused only on the generic aspects of data processing and display systems for aviation weather surveillance, and their specification and requirements.

5.6.1 Stages in data processing

Modern aviation weather surveillance systems are based on a multiplicity of sensors, some of which may be of generically different types. Data processing and utilisation in the context of such multisensor surveillance systems has the following three main stages, as depicted schematically in Fig. 5.4:

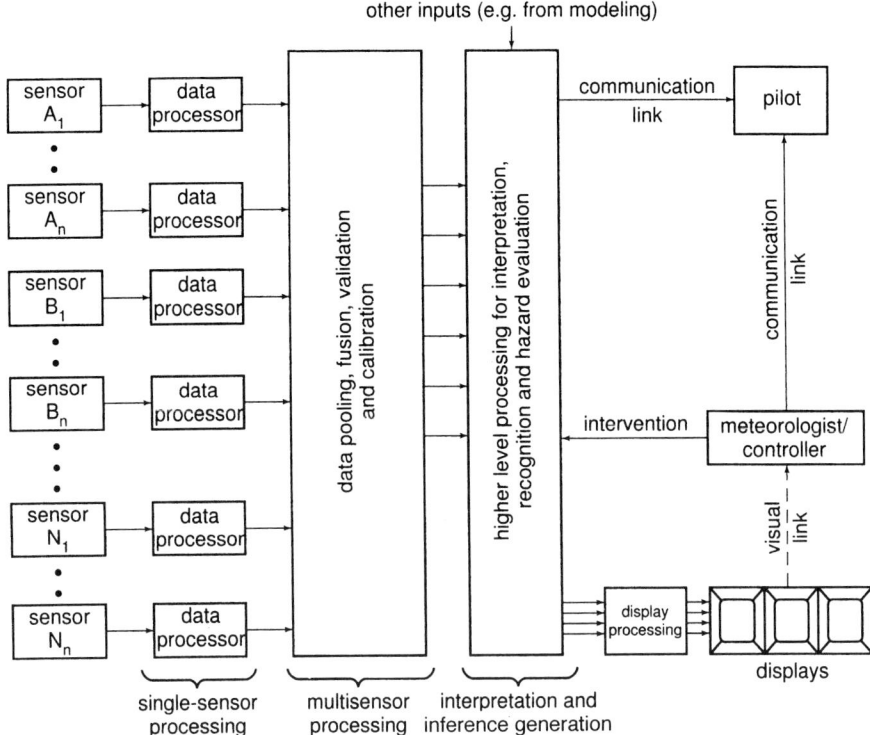

Figure 5.4 Levels of data processing in multisensor aviation weather surveillance systems

1. **Single-sensor processing:** This consists of processing the raw or 'low-level' data of each sensor to generate 'high-level' data which would make more direct sense to a human interpreter if displayed, or can be easily combined with data from other sensors in an automatic processing system. Such processing is readily demonstrated in the case of weather radars where, for example, the microwave echo power in watts received from raindrops constitutes the raw data, but it must be converted into equivalent rainfall intensity (see Section 6.3.2) before it becomes meteorologically meaningful, and can be compared or combined with rain data from another sensor, such as a rain-gauge. In reality, of course, the processing of modern Doppler weather radar data has far more aspects, which are covered in later chapters.

2. **Multisensor processing:** In this stage of processing, data from several sensors of similar and/or different types are combined or fused to generate a more

comprehensive picture of the weather scene. Combining data from multiple sensors serves one or more of the following purposes:

(i) enlarging the field of surveillance, e.g. combining the observation fields of several radars to cover a larger area than is possible with a single radar
(ii) refining data, i.e. enhancing the accuracy of data by processing multiple observations, e.g. by averaging
(iii) observation from different aspects to get a more true picture of weather features, which is especially beneficial when individual observations are aspect-sensitive
(iv) enhancing data reliability in the case of individual sensor malfunction, temporary loss of data from individual sensors, etc.
(v) data validation and calibration, whereby the measurements with indirect sensors such as radars are calibrated at discrete points by using data from direct sensors such as the rain-gauge.

3. *Interpretation and inference generation:* In this final phase of processing, the data field derived from different sensors is subjected to analysis to determine the presence of features of various descriptions, and to evaluate the strength of the different types of feature(s). Feature recognition is generally done through a process of signature analysis and template matching, and may involve the use of techniques based on artificial intelligence. For example, the low-altitude divergent field associated with a microburst has a characteristic signature in terms of the velocity distribution, and can be recognised if a picture of the distribution of wind velocities is available. Such data fields may be provided by Doppler weather radars or anemometer arrays. Once the nature of a feature is recognised, the next stage of processing involves the estimation of the parameters associated with the feature. For example, in the case of a microburst, some of the important parameters are the maximum outflow wind speed, the maximum wind shear, the location of the centre of the divergent field, and the diameter of the outflow region. Once these parameters are estimated, the next step of processing consists of converting the parameters into an aviation hazard index or severity indicator. This information is then communicated to airports, air traffic control centres, and pilots for warning purposes, and for taking the necessary follow-up action. More on automatic feature detection and recognition follows in Chapter 10.

5.6.2 Display of aviation weather data

Aviation weather scenarios being multidimensional and complex, very often a simple warning message of the 'fly' or 'no-fly' type does not adequately convey the reality of the situation. For example, such a message cannot take into account the differences between the hazard levels in different parts of the zone covered by the surveillance system, and the susceptibility of different types of aircraft to varying levels of the hazard index. To take care of such variabilities, the warning threshold may have to be set at a level that is considered safe under the worst assumptions, resulting in inefficient utilisation of premium air space and other aviation resources.

A better appreciation of hazardous weather situations is obtained by putting further information at the operator's disposal, in addition to the automatically assessed hazard index. Indeed, enabling the pilot in the cockpit to take more and more of the flight-related decisions in an informed manner based on accurate and timely data forms the basis of evolution of the futuristic aviation philosophy called *free flight*. Properly processed and presented weather data would be very helpful for the success of the free flight regime (Evans, 1997).

In the case of sensing a single simple phenomenon, such as horizontal wind shear at a specific location along the extended runway centreline, it may be adequate to transmit and display the parameters of the phenomenon in an alphanumeric format besides the warning signal. However, in more complex situations covering multiple phenomena over extended ground areas or air space volumes, it becomes necessary to present the information in a pictorial format to make quick sense to the operator (airport manager, air traffic controller or aircraft pilot). The format of such pictorial presentation must be carefully designed, keeping in mind several factors:

(i) minimising data transmission bandwidth, which permits the data to be conveyed via low-capacity channels and even piggyback on other data in pre-existing channels (this is very important for ground-to-pilot data transmission)

(ii) making realistic or maximally informative projections to display three-dimensional data fields on two-dimensional surfaces such as computer displays or paper printouts

(iii) providing relatively uncluttered and stable displays, i.e. displays that do not contain too many details or change too rapidly

(iv) using colour schemes and symbol sets that make instinctive sense to operators as far as possible, which minimises operator training and response time

(v) relating the weather field information with the physical and navaid-related environment of the surveillance area by using appropriate overlays.

This calls for a further level of data processing to convert the multisensor and topographic data into composite *situation displays*.

5.6.3 Requirements of data processing and display systems

In addition to the functions mentioned above, the data processing and display systems for modern aviation weather surveillance systems must satisfy fairly stringent time constraints and requirements of simultaneous data accessibility for multiple users requiring different levels of processed data.

The constraints on data processing time arise from system agility considerations. As mentioned in Section 5.4, the requirement of microburst detection and warning sets the limit on the system data update rate to ~1 min. Since much of this time is taken for raw data collection through a mechanical scanning of the space of interest, only a small fraction of a minute is available for completing all the stages of data processing, including transmission and display. Indeed, the time is inadequate for a full volumetric scanning of the space of interest, and inferences usually have to be drawn based on limited scan data (e.g. scanning at only one or two elevations). Of course, many other phenomena are not as fast-evolving as microbursts, and information about them can be derived from a more complete scanning, using elaborate processing algorithms. Such multilevel time constraints call for innovative processing schemes which are optimised simultaneously for each type of data requirement.

One point that emerges from the above discussion is that the data processing schemes of modern aviation weather surveillance systems are intimately interrelated with other system aspects such as scan strategy. In general, the more elaborate and complete the scan cycle for data collection, the more involved and time-consuming the data processing cycle. It is shown later (Section 7.4.3, Chapter 10) that the scanning strategies of modern remote sensing devices for aviation weather, and the processing algorithms, are jointly governed strongly by the types of phenomena required to be monitored and the nature of data products required.

Accessibility to data is also an important consideration for designing data processing and retrieval systems for modern aviation weather surveillance systems. The data products generated by the processing system are mandatorily transmitted to certain locations as demanded by operational requirements, but because of the high current value of the data they may also be in demand by many other users at the same time. However, not all users would require data in the same form, level or detail. Their data requirements would depend on their purpose, and on the hardware and software they use for further processing or utilising the data. For example, meteorologists or general aviation pilots unburdened with too much operational load may be satisfied with the raw weather picture of the surroundings, based on which they make their own judgments. Similarly, independent operators with their own processing software may prefer raw data which gives them maximum flexibility to process the data in their own preferred ways. On the other hand, operators (pilots, controllers) who need data in finished form for ready use may request data products of a higher level. The central data storage and retrieval mechanism of the aviation weather surveillance system should be able to cater to as many of such diverse requests in real time as possible.

Many of the considerations for the design or choice of display systems for aviation weather surveillance systems were listed in the preceding subsection. In addition, the display should be so designed as to provide the necessary weather information to the operator in the simplest way possible, i.e. with the least additional demand on his attention and requiring a minimum of his intervention. This is necessary because airport managers, air traffic controllers and aircraft pilots are normally quite busy with other tasks related to flight safety such as ensuring the health of the aircraft and support systems, and performing mandatory flying and air traffic control functions. Further, the format of the aviation weather system display should be matched to the data sources in the system, and to the intended use of the displayed data. For example, the data from an anemometer array which provides indication of ground-level wind shear may be displayed in the alphanumeric format or use low-resolution graphic displays. In contrast, radar-derived data usually require high-resolution multicolour displays. More about these types of sensors and their displays follows in later chapters.

5.7 Automated operation

It should be clear from the above discussion that the special requirements of aviation support and the complex nature of data processing involved preclude large-scale human involvement in the routine operation of modern aviation weather surveillance systems. Indeed, to meet the stringent requirements of aviation, not only has the data processing to be automated, but even the other elements of the system such as sensors should operate to the maximum extent without requiring human support for routine and normal functioning. It has been accepted by the US Federal Aviation Administration that the hazard to aircraft due to low-altitude wind shear, such as that produced by microbursts, can only be addressed by a fully automatic detection and warning system (McCarthy, 1989).

The philosophy of automated operation has guided the design of the modern aviation weather surveillance systems. Each sensor used in such systems is provided with its data encoder to generate digital output of its sensed parameter(s). Discrete *in situ* sensors are connected to a central site through wire or radio links so that data may be pooled at the central site and logged on to a computer automatically. In the case of remote sensing instruments such as the radar, which collect data over an extended volume from a single location, the installation is connected to the central site with two-way links. The power-on and -off operations of the installation and its various operating modes can be remotely controlled from the control computer at the central site, and the data from the installation are transmitted to the logging computer(s) at the central site through cable or microwave links. This obviates the need for routine operator presence at the actual instrument location except for periodic maintenance and repair functions. Such automated operation reduces system delays, helps in meeting the agility needs of aviation operations, minimises human errors induced by fatigue and boredom, and makes the surveillance systems affordable.

5.8 Selection of primary sensors

Sensors are the devices that form the origin of the observational data which feed the aviation weather surveillance system. Their characteristics are therefore of utmost importance in designing and assessing the performance of such systems. The sensors that provide direct data regarding the weather parameters of interest to aviation are called *primary sensors* in the context of aviation weather surveillance. A host of primary sensors is available for use in modern aviation weather surveillance systems. They vary greatly among themselves in regard to the parameters they measure, and their coverage, speed, accuracy and sensitivity. Keeping all this diversity in mind, a proper choice of the complement of primary sensors, and assigning appropriate roles to them within the system architecture, are very important parts of the design of modern aviation weather surveillance systems.

5.8.1 Atmospheric parameters monitored for aviation

In Chapter 4 we discussed the atmospheric phenomena that are of concern from the point of view of aviation safety and efficiency. However, the presence of these phenomena in the air space of interest cannot be directly inferred by sensing instruments. Even the most modern instruments do not have the cognitive power of the human eye–brain combination, but the human eye is grossly limited in observation distance compared to modern instruments, especially in bad weather conditions, which are precisely those of interest for aviation. The only way out of this bind is to sense the sensible parameters of the atmosphere, and infer the presence of deleterious phenomena by appropriately processing these parameters. These parameters constitute the measurands for the primary sensors used in aviation weather surveillance systems.

The earlier chapters of this book have made it abundantly clear that although aviation is affected by the overall climatic conditions and large-scale weather patterns, flights of individual aircraft are most seriously endangered by violent local phenomena such as the thunderstorm and its attendant subphenomena. Emphasis is therefore placed on the sensing of those atmospheric measurands that can yield an accurate qualitative and quantitative picture of the weather features that affect aviation in the most serious way.

Precipitation is the most visible indicator of inclement weather. It affects flight directly, and its presence and pattern are reasonable guides to the nature of the weather features harbouring it. In the absence of any other information, the precipitation pattern alone can be used to warn aircraft pilots regarding the types of dangers to expect and the areas to avoid. Indeed, this has been the case operationally for a few decades now. Precipitation also happens to be one of the most readily sensible and measurable quantities using *in situ* as well as remote sensing instruments. The intensity of precipitation is therefore among the most important measurands for modern aviation weather surveillance systems.

Other atmospheric parameters that affect flight directly and seriously are the localised air motions which manifest themselves as wind shear and turbulence. The effects of these on flight have been discussed before. Hence, the determination of atmospheric wind fields and their parameters constitutes another important measurement function of modern aviation weather surveillance systems.

Many other atmospheric variables serve to indicate the state of the atmosphere that supports flight. The most important and basic among these are temperature, pressure and humidity. These have traditionally been measured directly by *in situ* instruments, and continue to be so measured. However, these quantities can also be measured indirectly by remote sensors, and such measurement is the only way when localised and current measurements are needed at flight altitudes, such as when determining, nowcasting, or trying to forecast aircraft icing conditions. Quantities such as air density at altitudes are normally computed from these three basic measurements.

5.8.2 *Primary sensors for modern aviation weather surveillance*

The desired attributes of the primary sensor(s) for a modern aviation weather surveillance system should be clear from the preceding discussions. For a comprehensive surveillance system, these may be summarised as follows. The sensor(s) must:

(i) be able to measure the necessary atmospheric parameters, especially important being the precipitation intensity and air motion parameters
(ii) comprise the minimum number of different types to perform measurement of all the required parameters
(iii) be quantitative and accurate
(iv) provide volume coverage, extending from ground up to the maximum height of aviation activity in the area
(v) observe weather up to the longest possible range, in order to serve a large air space and to minimise the number of sensors required to cover a given geographical area
(vi) provide data with high spatial resolution, to depict the detailed structure of hazardous weather features, recognise and delineate zones of high precipitation and velocity gradients, and permit subsequent sophisticated data processing leading to weather feature recognition and hazard estimation
(vii) provide for locally enhanced spatial resolution, such as near the centre of terminal areas, and progressively decreasing resolution away from these areas
(viii) have the required agility to provide rapid updates of the weather scenario, as necessary for following fast-evolving hazardous phenomena

(ix) make available the data for the entire surveillance volume at a single point for easy handling, processing and dissemination, and minimising communication and networking requirements (with the attendant costs, delays and failures)
(x) provide data in digital format for direct input to computer-based data storage, retrieval, processing and communication systems
(xi) permit all-weather operation, i.e. its observing and measuring abilities should not degrade significantly in the presence of the very weather conditions that it is intended to observe.

These requirements are, of course, besides the normal considerations of cost, reliability and simplicity of operation.

A matching of these requirements with the characteristics of various atmospheric sensors would show that the most suitable primary sensor for a modern aviation weather surveillance system would be a Doppler radar (or a combination or network of such radars) specifically designed for weather observation. Such a sensor can accurately measure both precipitation and air motion parameters under all weather conditions, provide fast coverage of volumes of the atmosphere, match the resolution requirement of aviation weather surveillance, observe precipitative weather phenomena up to hundreds of kilometres away, and provide data in digital form.

In the context of versatile atmospheric observation, the Doppler weather radar has an important advantage as a primary aviation weather sensor. Hazardous levels of atmospheric air motion very often occur in an environment of precipitation, as is clear from the discussion on thunderstorms in the preceding chapter. In the presence of precipitation, air motion is readily detectable by radars. However, requirements for aviation weather surveillance also include detection of hazardous air motion in clear air, which is associated with phenomena such as gust fronts and dry microbursts. This requirement is more important for terminal areas than for *en route* flight. The modern Doppler weather radars are capable of detecting clear air motion, albeit over a reduced range than in precipitation-carrying air.

There are other modern sensors that provide useful data for aviation weather surveillance. These include simpler instruments based on the Doppler radar principle, as well as automated anemometer arrays.

5.9 Summary

Modern aviation weather surveillance systems are intended to serve current and future aviation activity, which is characterised by a high level of dynamism and traffic density. The features required of such systems are therefore dictated by the needs of the evolving and projected aviation scenario. Further, these systems must also incorporate other features suggested by the deeper understanding, derived in recent years, of the behaviour of the atmospheric phenomena that are hazardous to aviation. Some of the characteristics of such phenomena that drive the important design features of the modern aviation weather surveillance systems include the maximum values of the different atmospheric parameters associated with these phenomena, the sharpest gradients of these parameters, the fastest evolution and decay rates of the phenomena, and the extremes of their spatial extent.

Microbursts, as the fastest evolving among the highly hazardous atmospheric phenomena, set the speed of data updating in modern aviation weather surveillance systems to the order of 1 min. This includes the time required for both sensing as well as data processing. To meet such stringent speed requirements, both these aspects of aviation weather surveillance operations need to be performed automatically. Automated operation also has other important advantages such as large-scale coverage at lower cost and less human-dependent error. Such operation is a much favoured norm for modern aviation weather surveillance systems.

The primary sensors of the atmospheric parameters act as the principal source of the vital data necessary for the successful operation of such systems. Considering the demanding data requirements of aviation weather surveillance systems, the modern Doppler weather radar has emerged as the most favoured primary sensor. Because of the central role of the Doppler weather radar in modern aviation weather surveillance systems, we devote the following two chapters entirely to an understanding and survey of this very potent weather sensor. Other important modern sensors of aviation weather are described in a subsequent chapter.

5.10 References

BLAKE, L.V. (1968a): 'Ray height computation for a continuous nonlinear atmospheric refractive-index profile', *Radio Sci.*, **3** (New Series), pp. 85-92

BLAKE, L.V. (1968b): 'Radio ray (radar) range-height-angle charts', *Microw. J.*, **4**

BLAKE, L.V. (1990): 'Prediction of radar range', in M.I. Skolnik (Ed.): 'Radar handbook, 2nd edn.' (McGraw-Hill, Singapore), Chap. 2 pp. 2.81–2.96

EVANS, J.E. (1997): 'Coupling terminal weather information to next generation automation, traffic flow management, and "free flight" systems'. Proceedings of 7th Conference on Aviation, Range and Aerospace Meteorology, Long Beach, CA, 2–7 February 1997 (American Meteorological Society, Boston), pp. 11–15

HJELMFELT, M.R. (1988): 'Structure and life cycle of microburst outflows observed in Colorado', *J. Appl. Meteorol.*, **27**, pp. 900–927

MAHAPATRA, P.R., and ZRNIC', D.S. (1982): 'Lifetimes of convective atmospheric phenomena hazardous to aviation: A study with application to the design of scan strategies for NEXRAD'. Federal Aviation Administration Report DOT/FAA/RD-82/69

MAHAPATRA, P.R., and ZRNIC', D.S. (1984): 'A physical basis for NEXRAD data update rates', *J. Aircr.*, **21**, pp. 840–850

McCARTHY, J. (1989): 'Advances in weather technology for the aviation system', *Proc. IEEE*, **77**, pp. 1728–1734

McLEAN, D. (1988), 'Airborne detection of wind shear'. Proceedings of 2nd International Symposium on Aviation Safety, Toulouse, France, November 1986 (Cepad, Toulouse), pp. 227–244

TURNBULL, D.H. (1995): 'Aviation weather radar'. Proceedings of IEEE International Radar Conference, Alexandria, VA, 8–11 May 1995, pp. 748–751

US DEPARTMENT OF TRANSPORTATION (1981): 'Terminal area weather radar detection and convective prediction development'. Interagency Agreement DTFA01-81-Y-10521 between DoT/FAA and DoC/NOAA, 7 January 1981

Chapter 6
Doppler weather radar as a primary aviation weather sensor

6.1 General

Radar is a highly versatile sensor. It is an active device which beams electromagnetic energy into the space it observes, and detects the presence of objects by sensing a tiny part of the energy reflected by them. Radar technology has come a long way since the days of its early development preceding and during the Second World War, and the radar has been put to multifarious uses during the intervening decades. Currently radar systems range in size and complexity from huge ones that probe hundreds of millions of kilometres into space to small hand-held sets used in sports and traffic monitoring. The radar was first used for detecting incoming hostile aircraft during World War II, and it has subsequently been perfected in this role through sustained research and development. Aircraft and other radar targets of relatively small size are called *point targets*, and a majority of radars are used for detecting such targets.

Radars used for observing point targets in the atmosphere also receive the echo power scattered by any raindrops within their volume of observation. Rain echoes generally interfere with the radar's ability to observe the point targets, and are dismissed as *clutter*. In fact, the signal processors of such radars are specifically designed to maximally eliminate *weather clutter* and make the point target(s) stand out clearly against the background of the interfering weather echo signals.

It was, however, the observation of weather echoes in aircraft-detecting and ship-detecting radars that first suggested the idea of using radars for detecting and observing weather *per se*. This idea gave birth to *weather radars*. Weather radars are similar to point-target-detecting radars in their basic principle and major components. However, the former are designed specifically to optimally detect and estimate the parameters of weather fields, and provide the relevant data in conveniently usable forms.

Starting quite early in the developmental history of the radar, the Doppler phenomenon, whereby moving scatterers return echo signals to the radar at a frequency slightly different from that of the transmitted signal, has been used extensively to enhance the detection of point targets against a background of stationary or slowly moving clutter. The Doppler frequency shift has also been used to accurately estimate the closing or receding speed of point targets (i.e. the component of the target's relative speed vector along the radar–target radial line). More recently, with the availability of fast digital processing equipment, these techniques have been extended to weather radars to map the wind velocity fields in the atmosphere, besides determining the distribution of precipitation. The resulting *Doppler weather radars* are the principal primary sensors for modern aviation weather surveillance systems.

Weather radars have many distinguishing features relative to those used in other roles, even within the field of aviation (see Section 2.6). This chapter focuses on the special features of weather radars, assuming a fair acquaintance with the principles, construction, and working of radars in general. For readers interested in learning more about radar systems, many excellent books can be recommended. The handbook edited by Skolnik (1990a) provides broad and fairly detailed coverage of most aspects of radars, including meteorological radars. There are only a few books specifically devoted to weather radars. Among early works dedicated to the subject were those by Atlas (1964) and Battan (1973). The most comprehensive book on the subject of Doppler weather radar is by Doviak and Zrnic' (1993), and a lucid treatment of radar for meteorologists has been made by Rinehart (1991). Collier (1989) focuses on the application of weather radars. Compact coverage of the essentials of the topic may be found in Doviak and Zrnic' (1988) and Serafin (1990).

6.2 Basic aspects

In its simplest form radar senses only the energy backscattered by its target(s) and reaching its receiver *via* its antenna (Colour Plate 5). Any inference that can be drawn about the nature of target(s) is based only on the amount of this energy, taking into account the parameters of the radar system.

In the more common role of the radar as a point-target observer, the direction of the target relative to the radar can be inferred by noting the pointing direction of the antenna for which the echo power received by the radar is maximum. The distance of the target from the radar is ascertained from the time taken by the electromagnetic waves to travel from the antenna to the target and back, measured as the time difference between the instant of the waves leaving the antenna, and the instant of the corresponding echo arriving at the antenna. Such correspondence is difficult to establish without time marks (or modulation) on the transmitted signal. To provide the time marks, and for other important advantages, radar signals are most commonly transmitted in the form of short, high-power pulses, typically of the order of 1 μs in duration, spaced relatively wide apart in time, the interpulse interval often being of the order of 1 ms. Such radars are called *pulsed radars*. The structure of pulsed transmitted signals is shown schematically in Fig. 6.1, and the important parameters of pulsed radars are listed in Table 6.1 along with their brief definitions.

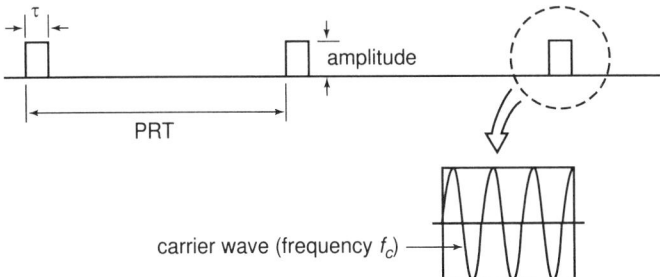

Figure 6.1 Structure of a simple transmitted pulse train showing the main parameters

Weather radars are also pulsed radars, but their operation differs from point target detection in very remarkable ways. The most significant difference is that the targets of weather radars are scattered over extended parts of their field of view, sometimes covering nearly all of it. The basic weather radar therefore does not estimate the range and angular co-ordinates of its target(s); instead, it observes *all parts* of its field of view, estimates the parameters associated with the scatterers everywhere in the field, and provides a pictorial presentation (or whole-field data) of the distribution of the parameters in the field. The radar antenna may scan at a fixed elevation to present a two-dimensional picture, or scan a volume of space to provide three-dimensional data.

It is now imperative to comment on the nature of parameter(s) that a weather radar measures. Radars can most naturally and readily measure the strength of the echo power returned from scatterers, and the initial applications of radars have been based on this ability. This also holds true for weather radars. The first generation of weather radars based their observations only on the sensing of the power reflected by the hydrometeoric scatterers, principally raindrops, occurring in weather phenomena. Such radars are called *conventional weather radars*, and form the subject of discussion of the following section. The more sophisticated *Doppler weather radars* are discussed in subsequent sections. In the following part of this section we discuss some of the common aspects of most types of weather radars.

6.2.1 Weather radar resolution

The schematic diagram of a weather radar observing an ensemble of raindrops is shown in Fig. 6.2. Weather radars normally employ *pencil beams* which are conical beams of narrow width. At a distance r from the radar, a beam of angular width θ_b has a linear width of $r\theta_b$. As the radar illuminates and receives echoes from all scatterers within this width, it cannot distinguish or *resolve* these scatterers. Thus the *angular resolution* of a weather radar

146 *Aviation weather surveillance systems*

Table 6.1 Important parameters of pulsed radars

No.	Parameter	Symbol[a]	Usual units	Brief definition or description
1	Range coverage (*also* maximum range)	r_m	km	The distance up to which a radar can detect or make meaningful observation of its intended targets
2	Azimuth coverage		deg rad	The sector over which the radar antenna can scan in a horizontal plane, expressed as the angle of the sector
3	Elevation coverage		deg rad	The sector over which the radar antenna can scan in a vertical plane, expressed as the angle of the sector
4	Operating frequency (*also* carrier frequency *or* radio frequency)	f_c	MHz GHz	Frequency of the electromagnetic wave within the pulses radiated from the radar antenna
5	Peak [transmitted][b] power	P_t	MW	Average value of the power within the narrow pulse transmitted by the radar
6	Average [transmitted] power	P_a	kW	Average value of the transmitted power over a time period much longer than the pulse repetition time
7	[Transmitted] pulse width	τ	μs	Width of the radar transmitted pulse
8	Pulse repetition time (PRT) *or* pulse repetition interval (PRI)	T_r	μs ms	The time interval between the corresponding points of successive transmitted pulses
9	Pulse repetition frequency (PRF)	f_r	Hz kHz	The number of pulses transmitted by the radar per unit time
10	Duty ratio	ρ_d	dB	Ratio of the width of the transmitted pulse to the pulse repetition time

Table 6.1 Continued

No.	Parameter	Symbol[a]	Usual units	Brief definition or description
11	[Maximum] unambiguous range	r_u	km	The maximum distance of the target from the radar for which the two-way signal transit time does not exceed the pulse repetition time
12	Nyquist frequency	f_N	Hz kHz	The maximum Doppler frequency shift due to the target which is not aliased due to the effective sampling caused by the transmitted pulses of a given pulse repetition frequency
13	[Maximum] unambiguous velocity	v_u	m s^{-1}	The target radial speed which produces a Doppler shift equal to the Nyquist frequency
14	[Antenna] beam width[c]	θ_b	deg rad	Angular width of the antenna beam, measured between the points that are 3 dB lower than the peak of the one-way radiation pattern of the beam
15	Antenna gain	G	dB	The ratio of the highest value of the power density (per unit solid angle) radiated by the radar antenna to the power density around an omnidirectional antenna radiating the same amount of power
16	Resolution volume (*also* resolution cell)	V V_6		The smallest spatial element that the radar can resolve. Individual scatterers within this element cannot be distinguished from one another by the radar

148 *Aviation weather surveillance systems*

Table 6.1   Continued

No.	Parameter	Symbol[a]	Usual units	Brief definition or description
17	[Antenna] scan rate	θ	deg s^{-1} rad s^{-1}	The angular rate of scanning motion of the radar antenna
18	Receiver bandwidth	B	kHz MHz	Bandwidth of the radar receiver filter, measured between the two points 3 dB below the peak of its transmission characteristic curve
19	Signal-to-noise ratio (SNR)	ρ_{sn}	dB	The ratio of the target echo signal to the unwanted noise signals at specified points in the radar signal processing chain
20	Receiver sensitivity (*also* minimum detectable signal)	S_m	W dB$_W$ dB$_m$	The minimum echo power that can be detected by the receiver, while providing the required signal-to-noise ratio
21	[Receiver] dynamic range		dB	The ratio of the maximum signal power that the receiver can handle without getting saturated or distorting the signal shape, to the minimum detectable signal of the receiver

[a] As used in this book
[b] Words in square brackets are optional (often omitted unless the context demands them)
[c] The beamwidth may in general be different in the vertical and horizontal planes. However, weather radars normally employ conical beams which have equal width in all planes through the beam axis

generally equals its beamwidth. Further, because of the finite width τ of the transmitted pulse, at any instant, the radar receiver receives echo signals from all particles within a finite radial distance Δr. For monotone transmitted pulses, i.e. pulses which are not frequency-modulated (pulse compression is generally not used in weather radars, since peak power is seldom a limitation), and a receiver with its bandwidth matched to the pulses, as is usually the case with weather radars, this radial distance corresponds to half

Doppler weather radar as a primary aviation weather sensor 149

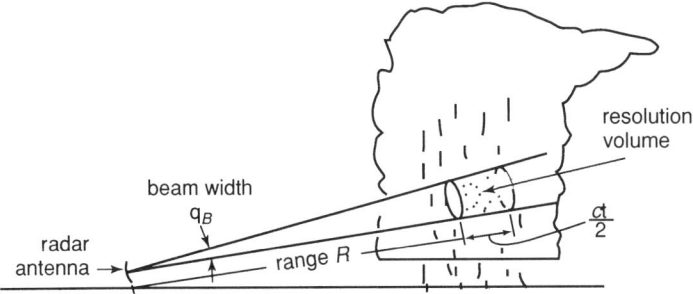

Figure 6.2 Schematic diagram of a pulsed weather radar observing an ensemble of raindrops

of the spatial extent of the transmitted pulse, i.e. the *range resolution*

$$\Delta r = c\tau/2 \qquad (6.1)$$

where c is the speed of light. The factor of half arises from the two-way transit of the radar signals. The receiver cannot resolve the particles occurring within this distance. The volume of space extending radially over this distance, and bounded laterally by the beamwidth, is called a *resolution volume*, denoted V.

The radar antenna beamwidth is normally measured between the 3 dB points of the one-way radiation pattern. By virtue of the two-way propagation involved in radar signal reception, a scatterer at the edge of the beam would be weighted twice by the pattern factor, and hence contribute echo signal energy that is 6 dB lower than what the same scatterer would return when located at the centre of the resolution volume. Similarly, the finite bandwidth of the radar receiver causes the edges of the echo pulse to roll-off gently, necessitating a clear definition of the radial depth of the resolution volume. It is customary to threshold the radial response function also at the 6 dB level to demarcate the resolution volume. In fact, the resolution volume is bounded in three dimensions by a contour on which the product of the two-way beam pattern function and the radial response function is 6 dB below its peak value (which normally occurs at the centre of the volume). A resolution volume defined this way is sometimes designated as V_6 to explicitly show the 6 dB thresholding. The weighting of the scatterers within the resolution volume influences the estimation of parameters by weather radars.

Resolution volume is so named because it is the smallest element of space that a radar can resolve. A smaller radar resolution volume is considered more desirable because it results in more detailed pictures of the weather pattern. The radial depth of the resolution volume, which is determined by the transmitted pulsewidth and receiver bandwidth, remains constant throughout the field of view of the radar. However, its lateral extent $r\theta_b$ increases linearly away from the radar. Consequently, the volumetric resolution degrades as the square of the range to the point of observation. Therefore, weather radars can provide much more detailed pictures of the weather scene close to their location, but the picture quality degrades at faraway ranges.

150 *Aviation weather surveillance systems*

   This point may be illustrated by assuming certain realistic radar parameters. For a transmitted pulsewidth of 1 μs and an antenna beamwidth of 1°, typical of modern weather radars, the radial resolution would be ~150 m at all ranges, but the lateral (azimuth and elevation) resolution would vary from ~175 m at a distance of 10 km to almost 8 km at 450 km, which is the order of the highest range of observation of weather radars (see the spatial coverage aspects discussed in Section 5.3). Volumetrically, the resolution volume degrades (i.e. increases) by a factor of >2000 between 10 and 450 km.

   The effects of such resolution degradation over range are easy to see. The most important among these are as follows:

(i) The absolute loss of lateral resolution becomes detrimental to detailed weather feature observation and parameter estimation at long ranges. For example, the 8 km resolution volume width mentioned above would cover the greater part of the height of a typical thunderstorm cell, providing only a grossly averaged picture of the entire cell visible above the radar horizon. Note that when the radar beam is directed nearly horizontally, as is the case for long-range observation, the vertical dimension of a storm falls along a lateral dimension of the resolution volume.

(ii) The degradation in lateral resolution alone causes a distortion in the shape of the resolution volume, from a fairly symmetric one (i.e. with comparable dimensions in radial and lateral directions) at closer ranges to a highly asymmetric one at far ranges. This makes the radar-observed weather picture appear uniformly sharp in a radial direction, but increasingly blurred in a lateral or circumferential direction at longer ranges. Further, when the radar data are used as inputs for further sophisticated processing, significant asymmetry of resolution volumes introduces difficulties in such processing.

Range-dependent variability and asymmetry of the basic resolution volume is a fundamental characteristic of weather radars, and must be borne in mind while designing and utilising radars for weather surveillance.

6.2.2 Mapping of weather fields

Only one value of echo power is obtained from each resolution volume during each measurement, and it corresponds to the weighted sum of the scattered signals from each of the particles within the volume. A weather radar maps larger areas within its field of view, or the entire field of view, by covering it densely with resolution volumes. The resolution volumes are stacked rangewise to cover the slender space within the antenna beam, and the beam is scanned angularly to provide a two-dimensional picture of a slice through the atmosphere. Such scanning at multiple elevation angles is necessary to cover a volume in space. By assigning the echo power received from each resolution volume to the spatial location of that volume, a complete distribution of the echo strength over the radar field of view may be obtained.

The conical space contained within the radar beam staring along a fixed direction is called a *radial* along that direction. After a powerful microwave pulse is transmitted by the radar, the echo signal is received continuously from all ranges along the radial, though the echoes from certain range intervals may be weak or even undetectable. The signal waveform, sampled at a discrete instant, contains contributions from the scatterers lying within a corresponding resolution volume. The location of each resolution volume in space is determined by the time instant of sampling of the echo signal and the antenna pointing direction at the sampling instant (neglecting the small shift of the antenna position during the signal transit time). Figure 6.3 shows the geometry of range–azimuth scanning and the relationship between range and time sampling. If t_s is the instant of sampling, measured with respect to the radar transmitted pulse, the radial distance r of the corresponding resolution volume from the radar antenna is given as

$$r = ct_s/2 \tag{6.2}$$

To observe weather scatterers out to the maximum range r_m of the radar, the echo waveform must be sampled up to the time instant t_m given by

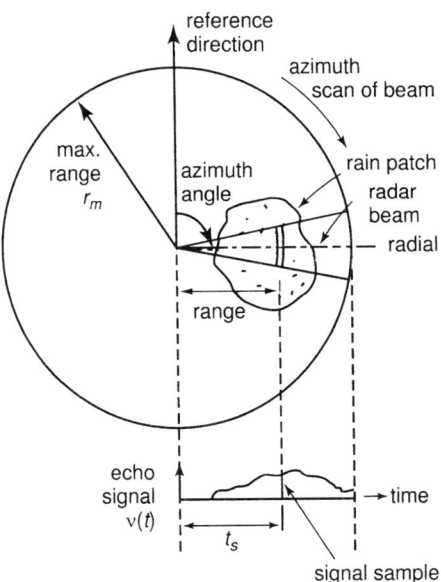

Figure 6.3 In a pulsed weather radar, scanning in range along a radial is performed by sampling the echo signal at different time instants, and the 'radial' scans in azimuth angle to cover a circle or sector

152 Aviation weather surveillance systems

$$t_m = 2r_m/c \qquad (6.3)$$

The range interval is most densely or completely covered by sampling the echo waveform at a uniform interval equal to τ, the width of the transmitted pulse. At such a sampling rate the resolution volumes are contiguous in the range direction, and a maximum amount of information is obtained about the scatterer population. This yields the highest range resolution for the given transmitted pulsewidth.

It may appear that a higher resolution should be obtainable by oversampling the echo waveform, i.e. by sampling at intervals shorter than the pulsewidth. However, this is not so. Oversampling merely makes the resolution volumes overlap one another. To the extent that overlapping resolution volumes share some scatterers, their echoes are not entirely independent in terms of information content. Thus, if the echo waveform is optimally sampled, i.e. sampled at intervals of τ, and the samples are recorded, then any other sample of the waveform can be obtained by interpolation of these samples.

To achieve contiguity of resolution volumes in the azimuthal scan direction (in a direction orthogonal to the radial), the successive antenna beam positions must be 'stacked' or displaced angularly in such a way that each beam position differs from the previous one by one beamwidth, i.e. the successive beam positions overlap one another along the 3 dB points of their one-way patterns. Such discrete positioning is possible with electronically steered radars. However, in actual weather radars, which are hitherto mechanically scanned, the antenna position varies continuously. In such systems, the echo signals from successive beam positions or radials (at given ranges) are considered to be independent only when the angular displacement between the beam positions reaches a value equal to the beamwidth. Data may, of course, be received during the intermediate beam positions, but such data are essentially the interpolation of the independent samples obtained at radials one beamwidth apart.

Each resolution volume in the physical space being observed by the radar is represented by one *pixel* in the display or data array. The parameters displayed or recorded are assumed to be constant within each pixel. From eqns. 6.1 and 6.3 it is evident that there would be $r_m/\Delta r$ range samples along each radial for full range coverage, and there would be $2\pi/\theta_b$ independent radials in a full rotation of the antenna in a horizontal plane. The total number n_r of resolution volumes or pixels involved in a full-circle maximum-range coverage is therefore given as

$$n_r = \left(\frac{2\pi}{\theta_b}\right)\left(\frac{r_m}{\Delta r}\right) \qquad (6.4)$$

If a surveillance volume is to be covered by multiple scans of the antenna at different elevation angles, the total number of pixels would be higher by a

Doppler weather radar as a primary aviation weather sensor 153

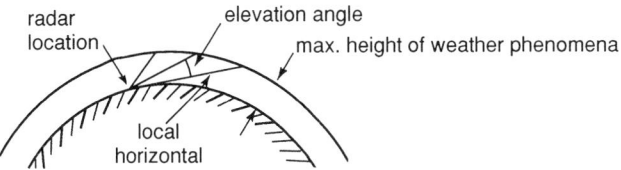

Figure 6.4 Geometry showing the reduction of 'useful' range (range interval over which significant weather information may be found) with increase in radar beam elevation angle

corresponding factor. The maximum range of observation at each elevation may, however, be different. Generally the maximum weather radar surveillance range decreases at higher elevation angles. This is so because significant weather phenomena are confined to an altitude of 15 to 20 km (depending on latitude) from the ground, and the radar beam crosses this altitude ceiling within shorter ranges at higher elevation angles. Figure 6.4 illustrates this point.

In the discussion hitherto the resolution of the weather radar has been assumed to be the same as its basic resolution volume. However, the two are often somewhat different, though the former always depends intimately on the latter. Radar echoes are normally feeble, and the returns from a number of transmitted pulses are summed or *integrated* to enhance the signal power in most types of pulsed radars. In the case of weather radars, where the signals of interest and their parameters to be estimated are themselves statistical in nature (this point will be elaborated later) the integration process also improves the accuracy and robustness of the estimates. The continuously scanning antenna of a weather radar would move over a finite angle during the time that the number of pulses required for integration are transmitted, converting each resolution volume along the radial to a moving window in the azimuthal scan direction. Integration over such a moving window results in an effective widening of the resolution volume and a corresponding degradation of resolution along the azimuth.

The radar resolution is also sometimes deliberately degraded to achieve some desired objectives. The main reasons for effecting such degradation are:

(i) *Data smoothing*: the parameters measured by radar for each resolution volume have random or noise-like components due to the statistical uncertainty inherent in the measurement process. The scatter in radar-derived parameters can be reduced by averaging the measurements over a number of adjacent resolution volumes. The average is more commonly done in the range direction for the reason given below.

(ii) *Symmetrising the resolution volume in range and azimuth*: the asymmetry in the resolution volume, because of the crossradial (i.e. along azimuth and elevation) resolution of radars increasing with range while the radial

154 Aviation weather surveillance systems

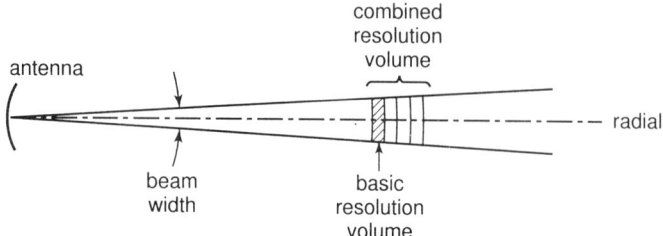

Figure 6.5   *If the basic resolution volume of a weather radar is highly asymmetric along range and azimuth directions, multiple volumes may be combined (resulting in loss of resolution) to achieve greater symmetry*

resolution remains constant, can be remedied to a certain extent by summing the echo samples from a number of radially stacked resolution volumes, as shown in Fig. 6.5.

(iii) *Reduction of data rates*: combining the data from a number of adjacent resolution volumes to produce larger pixels makes the weather picture appear coarser but correspondingly reduces the requirement for data storage and handling because of a reduction in the number of pixels used for mapping a given weather scene.

6.2.3 *Scattering by raindrops and radar reflectivity of weather*

Since weather radars monitor atmospheric phenomena primarily by detecting the backscattered energy from raindrops, it is necessary to understand the radar scattering characteristics of raindrops to be able to interpret the echo signals in meteorological terms. In a weather phenomenon involving rain, many of the radar resolution volumes will contain large numbers of raindrops. Such resolution volumes would generally return stronger radar echoes than those volumes which do not contain raindrops. Further, the stronger the rain activity at a given location, the larger should be the echo power. In other words, echo power from resolution volumes forms the basis of radar detection of weather. More sophisticated sensors such as Doppler weather radars would derive more information from the echo signals backscattered by the raindrop population.

The nature of scattering of electromagnetic waves by particles has a strong dependence on the relative values of the dimension D of the particles and the wavelength λ of the waves (Knott, 1990) — particles for which $D/\lambda \ll 1$ are called *Rayleigh scatterers*. Raindrops may be up to several millimetres in diameter, with those of smaller diameters being the most numerous. At weather radar wavelengths, which are typically several centimetres, raindrops behave as Rayleigh scatterers.

The most important parameter of a target that determines how well it can be detected by a radar is its *radar cross-section*, a hypothetical area which, when placed at the target location and oriented normal to the radial line from the radar, would intercept an amount of energy which, in turn, if radiated omnidirectionally, would provide at the receiving antenna location a power density equal to that produced by the actual target. In simple terms, the radar cross-section indicates the 'size' of the targets as the radar 'sees' them. The radar cross-section of a target depends most significantly on its physical size, shape and orientation (relative to the direction to the antenna), material (more specifically, its dielectric constant) and the frequency of operation of the radar. In the case of weather radars, which usually use a single antenna for transmission and reception, one is interested in the *backscattering cross-section* of ensembles of raindrops.

The backscattering cross-section σ_b of small spherical drops of water, in the Rayleigh scattering domain, varies directly as the sixth power of the drop diameter, and inversely as the fourth power of the radar wavelength, i.e.

$$\sigma_b = k \frac{D^6}{\lambda^4} \qquad (6.5a)$$

where k is a constant dependent on the complex refractive index m of water as

$$k = \pi^5 |K_w|^2 \qquad (6.5b)$$

with

$$K_w = \frac{m^2 - 1}{m^2 + 2} \qquad (6.5c)$$

For a given radar, the wavelength λ is constant, so that the backscattering cross-section of individual raindrops depends essentially only on the drop diameter, raised to the 6th power.

In practice, however, the radar would not look at a raindrop in isolation, but at an ensemble of raindrops contained within a volume, which is typically the radar resolution volume. The ensemble would consist of numerous drops having a range of diameters. In such a case the pertinent parameter is the backscattering cross-section per unit volume of the atmosphere, called the *reflectivity* of the rain process. The reflectivity η thus defined has the form

$$\eta = \left(\frac{k}{\lambda^4}\right) Z \qquad (6.6)$$

where Z is the *reflectivity factor*, defined as

$$Z \equiv \frac{1}{\Delta V} \sum_i D_i^6 \qquad (6.7)$$

In this definition D_i is the diameter of the ith raindrop occurring in a volume element ΔV, and the summation is carried out over all the drops within the volume. It is easy to see that eqn. 6.6 has a form similar to eqn. 6.5a, with the single scatterer diameter replaced by a sixth-power-averaged diameter on a volumetric basis.

The definition in eqn. 6.6 is a neat one intuitively, but is not useful for the actual determination of the reflectivity factor, since it requires the diameter of each individual scatterer, which is usually not available. A more practical expression is based on the statistical distribution of drop sizes, as will be discussed later.

The volume element ΔV in eqn. 6.7 is a general one, assumed to be small enough that the nature and intensity of the rain activity may be considered to be uniform over it. However, in the context of radar detection, this volume element is usually the same as the radar resolution volume V or V_6.

Equation 6.5a, which is the basis for the definition of the reflectivity factor, is accurate for spherical drops with diameter $D \leq \lambda/16$ (Doviak and Zrnic', 1993). For larger drops, errors creep into the reflectivity factor, and corrections may be necessary to get accurate results. Large raindrops also tend to be oblate in shape, which causes differential scattering behaviour in the vertical and horizontal planes. This aspect, and its beneficial use in radar observation of weather, will be considered in the context of polarisation diversity radars in Chapter 13.

A volume of atmospheric air may contain scatterers other than raindrops. Some of these are hydrometeors directly related to weather, such as ice pellets or frozen rain, hail, graupel or soft hail, and snow. There may also be mixed-phase scatterers such as partially melted ice pellets or wet snow. Further, the atmosphere may also contain particles and flying matter only indirectly related to weather, such as the dust and debris raised by storm outflows. Finally, the atmosphere, especially at lower altitudes, often harbours populations of insects which serve to scatter radar energy (Vaughn, 1985). These scatterers not only differ from raindrops in their material and hence dielectric constant, but are very often highly nonspherical in shape. For such scatterers the definition in eqn. 6.7 is not applicable. Instead, an *equivalent reflectivity factor* Z_e is used which is the value of Z obtained from eqn. 6.6 by using for η the actual radar backscattering cross-section per unit volume, and the value of k for water as in the case of rain.

Looking at eqn. 6.7, the MKS unit of reflectivity factor should be m^6/m^3 (dimensionally equivalent to m^3), and this unit is used where dimensional consistency with other parameters is required. However, the drop diameters D_i are more conveniently expressed in millimetres because that is the order of these diameters, and hence a commonly accepted unit for the reflectivity factor Z or Z_e is mm^6/m^3 or $mm^6\ m^{-3}$. The MKS and practical units differ by a factor of 10^{18}.

The values of reflectivity factor corresponding to the extremes of naturally occurring precipitation span over several orders of magnitude. To avoid

Table 6.2 US National Weather Service standard reflectivity levels for different rainfall intensities

Level	Reflectivity interval (dBZ)	Rainfall category
1	18–30	Light (mist)
2	30–41	Moderate
3	41–46	Heavy
4	46–50	Very heavy
5	50–57	Intense
6	>57	Extreme (with hail)

dealing with such large numbers, it is customary to express the reflectivity factor on a decibel scale and designate the resulting value in dBZ. Thus,

reflectivity factor in dBZ = $10 \log_{10} Z$, with Z expressed in mm^6/m^3 (6.8)

To form an idea of practical values, the reflectivity factor associated with cumulus congestus clouds is of the order of 0 dBZ, while that of very heavy rainfall accompanied by hail can exceed 60 dBZ.

The US National Weather Service has specified six reflectivity slabs that correspond to rainfall of different intensity levels. These are listed in Table 6.2.

6.2.4 Radar echoes from clear air

The above discussion throws light on the mechanism of radar echo generation from weather phenomena with particulate scatterers. However, it is often necessary to monitor weather phenomena or regions of the atmosphere which do not harbour significant solid or liquid particulate matter. Many phenomena of serious import for aviation, such as gust fronts and dry microbursts, occur in immediate environments devoid of rain or hail. Fortunately, atmosphere without such hydrometeoric scatterers, called *clear air*, does return radar echoes, albeit feebly. These echoes are detectable at relatively close ranges, making the observation of clear-air weather phenomena possible.

The echoes from nonionised clear air masses are caused by Bragg scattering due to refractive index irregularities or inhomogeneities of certain scales. In particular, the Fourier component of spatial variation of refractive index corresponding to half of the radar wavelength (the '$\lambda/2$ scale') is primarily responsible for the backscattering by clear air masses.

The refractive index irregularities or inhomogeneities responsible for clear-air backscattering are caused by the turbulent transport and intermixing of air streaks of different temperatures and moisture content. Since the earth's surface is the chief source of heat and moisture for clear air, the strongest irregularities in refractive index normally occur closest to the ground. The strength of these irregularities, and consequently the radar reflectivity of clear air, generally decreases with height from the ground. However, there may be enhanced reflectivity from atmospheric regions where high levels of moisture and refractive index gradients are present. These include the mixing boundaries between air masses of different temperature and moisture levels. Similarly, near the ground, stronger clear-air returns may be observed over hot spots such as large rocks or cities with their black roads and concrete structures.

Unlike the case of particulate scatterers, in which the radar reflectivity depends inversely on the fourth power of the wavelength as shown in eqn. 6.6, the backward reflectivity of clear air due to the Bragg scatter has only an inverse one-third-power dependence on the wavelength:

$$\eta \propto \lambda^{-1/3} C_n^2 \tag{6.9}$$

where C_n^2 is called the *turbulent structure parameter of refractive index*, defined as (Gossard and Strauch, 1983)

$$C_n^2 = k[\Delta n(\mathbf{r}) - \Delta n(\mathbf{r} - \Delta \mathbf{r})]^2 / |\Delta \mathbf{r}|^{-2/3} \tag{6.10}$$

Here $\mathbf{r}$ and $\mathbf{r} - \Delta \mathbf{r}$ refer to the position vectors of two points close to each other within the scale of a wavelength, Δn is the perturbation in refractive index at a point, and $k\bullet/$ denotes time averaging. The definition in eqn. 6.10 is valid in the inertial subrange of homogeneous, isotropic turbulence. The meaning of these terms, and a more detailed treatment of atmospheric turbulence, is given in Chapter 11.

Equation 6.9 shows that the higher the value of the structure constant the more would be the radar backscattering from clear air, resulting in a greater detectability. Because of stronger temperature and moisture gradients close to the ground, C_n^2 is usually high in the lowest layers of the atmosphere. The boundaries between moist and drier air masses are often rendered clearly visible to radars as the strong moisture gradients there may drive the C_n^2 to values as high as 10^{-11} m$^{-2/3}$ (Gossard and Strauch, 1983, p. 25). C_n^2 is highly variable spatially, and may change by orders of magnitude over a few metres of altitude (Doviak and Zrnic', 1993, p. 471). Its value is also temporally migratory. Chadwick and Moran (1980) have shown C_n^2 at a fixed height of 805 m to be varying over three orders of magnitude, between 10^{-16} and 10^{-13}, through timescales of the order of 12 h. The variations can be faster in the presence of strong convective or wind-driven phenomena.

6.2.5 Weather attenuation of radar signals

The part of the radiated microwave energy returning to the radar from atmospheric scatterers is responsible for the detection and visualisation of weather phenomena. In addition to the properties of these scatterers, the nature of the intervening medium also strongly influences the echoes reaching the receiver. In particular, the presence of heavy weather (strong precipitation) along the propagation path alters the echo signals from the target weather field, and thereby adversely affects the weather observation process. Because of the great importance of weather effects on radar operation, they have been the subject of much study (e.g. Battan, 1973).

Heavy rain has very pronounced effects on microwave propagation. The most important effect of rain on microwave signals is the absorption of the power associated with the signals. *Rain attenuation* is an important consideration in the design and operation of most radars with significant range coverage through the atmosphere.

When microwave radiation of an initial power level p_0 traverses through a distance r in a uniformly attenuating atmosphere, its power p decreases exponentially according to the law

$$p = p_0 e^{-\varepsilon r} \tag{6.11}$$

where ε is called the *extinction coefficient* or *specific attenuation*. The fraction of power lost during propagation through a unit distance of atmosphere is called the *attenuation coefficient*, commonly and conveniently expressed in dB/km and designated K. In this form, the attenuation due to successive layers of atmosphere becomes additive.

The attenuation coefficient K would, in general, have components due to absorption and scattering. However, for normally used wavelengths for weather radars, which are in the centimetres range, absorption is the predominant attenuation mechanism for all rain rates. The absorption due to droplets depends on their *absorption cross-section*. An important point to note is that while the backscatter cross-section of a small drop (relative to radar wavelength) is proportional to the sixth power of its diameter, its absorption cross-section depends only on the cube of the diameter.

The relationship of the attenuation coefficient K with the rain activity in a given atmospheric volume is a complex one, difficult to determine analytically. However, good empirical relations of K with rain activity are available. One set of relations quoted by Doviak and Zrnic' (1993, p. 42) based on data from Burrows and Attwood (1949) is of the form

$$K = \alpha R^\beta \text{ dB km}^{-1} \tag{6.12}$$

where the rainfall rate R is expressed in millimetres per hour, and the constants α and β are as given in Table 6.3. It is apparent from the Table and

Table 6.3   Rain attenuation parameters

Radar wavelength, cm	α	β
3.2	1×10^{-2}	1.21
5.0	1.8×10^{-3}	1.05
10.0	3.43×10^{-4}	0.97

eqn. 6.12 that rain attenuation increases rapidly with increase in the frequency of the radar (i.e. decrease in wavelength) within the normally used range.

It is worth reiterating that relations such as eqn. 6.12 are only approximations meant for analytical usage or quick and rough estimation of rain attenuation values. There are a number of variables that such simple relations cannot take into account. Important among them are the actual distribution of drop sizes in different types of rain in different geographical and climatological regions, the phase composition (proportion of frozen and liquid water) of the precipitation, and the temperature of the hydrometeors. For more accurate determination of attenuation, detailed data incorporating such effects must be made use of. Another fact to remember is that relations such as eqn. 6.12 and attenuation tables are based on the assumption of uniform hydrometeor properties. Relatively long paths in the atmosphere would very often not satisfy this condition. In such cases the path may be segmented into stretches of (approximately) uniform properties, or a path-integral form of eqn. 6.12 may be used.

Care must also be exercised in using the appropriate value of the rain rate either in eqn. 6.12 or another equivalent relationship, or in look-up tables. In particular, the rain rate sensed at the ground level, as done with rain-gauges, may not be the same as the rate aloft. When the radar beam clears the ground by a great height, such as for high antenna elevation angles or for long-range observation (due to the earth's curvature; see Figs. 5.1 and 5.2), the use of ground-level rain rate for estimating rain attenuation may lead to significant errors. Improvement in the estimates is possible by using a model of rain rate variation with height. For widespread rain, such as that due to frontal systems or monsoon, the rain rate aloft is found to decrease approximately exponentially with the square of the height, starting from its value at ground level, when the rainfall is continuous (Atlas and Kessler, 1957). Widespread rain is very important from the attenuation point of view because radar waves would have to penetrate large distances through such rain to map its farther reaches or phenomena beyond its boundaries. Local rains such as thunderstorms, however, have a much less predictable variation with altitude. One way to overcome this problem is to estimate the rain intensity and hence attenuation aloft from the radar reflectivity data itself, but because of the mutual dependence between path attenuation and reflectivity, such methods must be used very carefully, and may not yield reliable results in many situations.

To form an idea of the values of the attenuation coefficient associated with normally occurring rain rates, at a radar wavelength of 10 cm the coefficient in dB/km varies from 0.000416 in light rain at a rate of 1.25 mm/h to 0.0149 for heavy rain at 50 mm/h, and 0.0481 for extremely heavy rain at 150 mm/h (Burrows and Attwood, 1949). These values assume a raindrop temperature of 18°C. The corresponding values are much higher for a 3 cm radar, being 0.0161, 1.46 and 4.97 dB/km, respectively. Taking the last of these numbers as an extreme but possible case, the power of a 3 cm radar beam could be drastically reduced to about a tenth of its original value within a path length of 2 km, or two-way propagation over only 1 km, through rain of extreme intensity.

The fine droplets of liquid water present in clouds and fog return little backscattered microwave power at normal weather radar frequencies because of the D^6 dependence, but the attenuating effects of these media can be appreciable under certain conditions, as the absorption varies only as the cube of the droplet diameter. Because the volume of droplets also has a cubic dependence on their diameters, the attenuation rate of radar signals in air with a suspension of fine water droplets can be expressed directly in terms of the liquid water density (i.e. the total mass of liquid water suspended in a unit volume of the air), independent of the drop size distribution. The attenuation coefficient may also be tabulated as a function of the visibility distance through the aerosol, which is a fair indicator of the density of the aerosol, and is more readily determined than the liquid water density. At a temperature of 0°C and a wavelength of 3.2 cm, which is close to the shortest wavelength normally used for weather surveillance, the attenuation in very dense fog or cloud with a visibility distance of 30 m is 0.2 dB/km, while the attenuation in thin fog with 300 m visibility is 0.007 dB/km (Saxton and Hopkins, 1951). The corresponding figures for a 10 cm radar are almost an order of magnitude lower, at 0.02 and 0.001 dB/km, respectively. Battan (1973) provides attenuation data for different temperatures and wavelengths.

Clouds composed of tiny ice crystals cause very little attenuation at radar frequencies, and the attenuation of radar signals by hail is about two orders of magnitude lower than rain attenuation (Ryde, 1946). The attenuation coefficient K_s due to dry snow at 0°C is given by an empirical relation (Battan 1973):

$$K_s = 3.5 \times 10^{-2} \frac{R_s^2}{\lambda^4} + 2.2 \times 10^{-3} \frac{R_s}{\lambda} \text{ dB/km} \qquad (6.13)$$

where the wavelength λ is in centimetres, and R_s is the rate of snowfall, expressed as millimetres per hour of water obtained by melting the snow. The presence of a layer of liquid water on the surface enhances the attenuating effect of ice pellets significantly. Pellets with <10% melted may exhibit as much attenuation as fully melted pellets. Further, at certain melt fractions, biphase particles may attenuate signals significantly more than fully melted particles. The attenuation due to water-coated ice spheres may be obtained from tabulated data (e.g. Battan, 1973). Wet snowflakes and ice aggregates of highly irregular shapes have complex attenuation behaviour which cannot be readily described through simple models.

The deleterious effect of rain attenuation on weather radar performance can be very serious. When a radar looks at an extended zone of heavy precipitation, its leading edge would receive the full power as predicted by the inverse-square law, and the reflectivity there can be accurately determined. However, the points farther from the leading edge would receive progressively less power than predicted, and return significantly less echo than expected because of the two-way attenuation. This would result in gross underestimation of the rain activity in the farther reaches of the storm zone. In extreme cases, the radar may show little or light apparent rain activity deep in the storm zone while indeed there may be heavy rain in progress. Such underestimation is more serious with weather radars of relatively short wavelengths such as 3 cm, which are used frequently as airborne weather-detection radars.

Although this section primarily deals with attenuation due to hydrometeors, it is pertinent to touch briefly on types of atmospheric attenuation that are not necessarily or directly caused by particles of water in different phases. Apart from absorption and scattering due to other types of particulate matter such as insects and debris raised by storms, which may be appreciable over localised areas but are difficult to predict and model, significant microwave power may be lost due to absorption by certain atmospheric gases, notably oxygen and water vapour. Because of a greater concentration of the absorbing gases near the ground and in the lower levels of the atmosphere, the absorption in the case of ground-based radars is the highest when the beam is horizontal, and decreases progressively for higher beam elevation angles. The gaseous absorption phenomenon can be responsible for a few to several dB of two-way loss over long ranges at the normal weather radar wavelengths. For example, for a zero beam elevation, the calculated two-way gaseous absorption over a range of 450 km is ~4.4 dB for a radar wavelength of 10 cm, and ~6.6 dB for 3 cm (Blake, 1990, p. 2.50). As pointed out in Section 5.3, a range of the order of 450 km is the highest range of surveillance by weather radars. Further down the wavelength scale, at 1 cm the corresponding absorption shoots up to ~34 dB. The value is significantly higher at the intermediate wavelength of 1.35 cm, the resonance wavelength of water vapour, and blows up to hundreds of dB at a wavelength of 0.5 cm, the resonance wavelength of oxygen, which is the stronger absorber. While these shorter wavelengths are not used for weather radar application, atmospheric absorption, along with other attenuating factors, is in fact a strong reason that weather radars cannot exploit the advantages of the shorter wavelengths.

6.2.6 Operating frequencies of weather radars

The operating frequency is a fundamental parameter to be decided while designing or selecting weather radars for a given application. The choice of radar frequency is usually arrived at as a compromise between various conflicting requirements. The principal factors influencing the choice of frequency in the case of weather radars are briefly discussed below.

Doppler weather radar as a primary aviation weather sensor 163

The primary requirement is to maximise the backscattered power from hydrometeors. Because of the fourth-power dependence of the reflectivity of droplet populations on the operating frequency, as evident from eqn. 5.6 noting that the wavelength λ is the inverse of the frequency f_c, it is preferable to employ as high a frequency as possible to maximise the echo power for improved detection of weather.

A higher operating frequency also results in a more compact radar system. The largest physical component of a radar is usually its antenna. For a given beamwidth, as necessary for obtaining a specified angular resolution, the antenna diameter decreases in direct proportion to an increase in frequency. The antenna drive, pedestal, support structure and radome also become correspondingly smaller and lighter, as also the feed structure and waveguides, resulting in a compact radar system.

The strongest argument against the use of higher operating frequencies or shorter wavelengths comes from attenuation considerations. As seen in the preceding subsection, microwave attenuation through rain increases rapidly with frequency in the centimetre and subcentimetre wavelengths, making the higher frequencies less desirable for weather radar applications. Considering this fact, together with the other frequency-determining factors mentioned above, the following preferential frequency allocations emerge:

1. Weather radars intended for surveillance over long ranges of hundreds of kilometres cannot work reliably with atmospheric attenuation coefficients which are more than a few hundredths of a decibel per kilometre. This precludes wavelengths shorter than ~10 cm for such radars.
2. Radars with a more localised operation, such as surveillance over the terminal area of an airport, can possibly tolerate somewhat higher attenuation rates, permitting a correspondingly lower wavelength of operation.
3. Where compactness of the radar system is an overriding consideration, even shorter wavelengths, perhaps down to ~3 cm, may be used, taking cognisance of the possible loss of system performance due to the high weather attenuation at such frequencies and taking the steps necessary to mitigate its effects as far as possible. Severe size constraints arise typically in the case of airborne and other transportable weather radars.
4. Microwave wavelengths shorter than ~3 cm are seldom used for weather radars. A special application of optical wavelengths occurs in the form of *lidars* or laser radars which have the capacity to detect fine aerosols of atmospheric pollutants and dust. Laser beams are heavily attenuated by rain, clouds and dense fog, but can provide information on cloud edge location. This property is useful for determining the height of cloud bases (which often decides the ceiling of visibility for aviation operations) from ground-based installations, and cloud tops from space-borne platforms.

Table 6.4 Common frequency bands used for atmospheric radars

Radar frequency band*	Representative wavelength(s)	Atmospheric radar application	Aviation weather surveillance application
VHF	6 m	Mesospheric, stratospheric, tropospheric (MST) radar	None
UHF	75 cm, 33 cm	Atmospheric wind profiling	Obtaining atmospheric wind profiles for *en route* and terminal area flight
S	10 cm	Long-range weather observation	Weather surveillance of air space over wide areas for *en route* flight
C	5 cm	Localised weather observation	Weather surveillance of air space over terminal areas
X	3 cm	Short-range weather observation with ground-based, airborne and transportable radars, also as a component of multifrequency weather observation	Aircraft-borne weather radars
Laser	Infra-red	Observation of atmospheric aerosols and dust	Possible use for visibility ceiling (cloud base height) and fog and visibility quantification

* The letter designations are from IEEE Standard 521-1984 (Skolnik, 1990b, Table 1.1, p. 1.14)

A summary of the significance of various common radar frequency bands in the context of atmospheric radars and aviation weather surveillance is given in Table 6.4.

6.3 Conventional weather radar

We recall from the discussions of the preceding section that, reflectivity being the most readily measurable parameter of weather phenomena, the first generation of weather radars based their observation essentially on mapping the reflectivity at various points within their surveillance zone, and that such radars are called conventional weather radars. In spite of such a simplistic approach to weather surveillance, the conventional or reflectivity-only weather radars have rendered yeoman service to the detailed observation of weather in general, and to aviation operations in particular, for over four decades. Such radars have hitherto formed the mainstay of quantitative and real-time weather mapping in many countries and meteorologically important areas of the world. Significant areas of the USA and Europe have been covered by such systems. As examples, the weather radar coverage of the UK and France are shown in Figs. 6.6 and 6.7, respectively. India Meteorology Department has been operating a chain of conventional weather radars along the Indian coastline for the purpose of monitoring and warning against monsoon-driven cyclones that ravage large tracts of South Asia and surrounding regions with frightening regularity.

Although the reflectivity of hydrometeors is the only variable measured by the conventional weather radars, the reason that these radars have been so successful for so long is that very important information can be gleaned from the patterns of reflectivity associated with different phenomena. Experienced radar meteorologists can draw insightful inferences about a host of different types of phenomena and their possible causative factors by visually analysing the display of reflectivity distribution over and around these phenomena. Conventional weather radars have therefore been used primarily in a human-centred mode.

6.3.1 *Reflectivity measurement: radar range equation*

The radar echo power received from a given resolution volume depends not only on the backscattering characteristics of the scatterers populating the volume element but also on its distance from the radar, in addition to a number of parameters relating to the radar. The closer that a group of scatterers is to the radar, the more echo power they are expected to return to the radar receiver. Similarly, the received signal power is expected to increase if the level of the power transmitted by the radar is raised. However, to be able to characterise phenomena based on their backscattering characteristics, it would be far preferable to measure an absolute reflectivity parameter associated with the scatterer population. The reflectivity factor Z happens to be one such parameter, as seen from eqn. 6.7. Weather radars determine the reflectivity factor from the measured echo power by using a relationship between the two, called the *radar range equation*, or simply *radar equation*.

Radar detection of distributed targets such as weather fields differs in important ways from that of point targets which may be familiar to those with an understanding of the working of aerospace surveillance systems. In the case of point targets, the intensity of the power transmitted along the radar antenna beam decreases with distance according to an inverse square law, and hence a target of a given radar cross-section would intercept an amount of

166 *Aviation weather surveillance systems*

Figure 6.6 *Weather radar coverage of the UK area. The radii of the smaller and the larger circles are 75 and 210 km, respectively. Two of the radars in England, in the counties of Shropshire and Devon, provide Doppler data. The three radars in Scotland have Doppler capability which is not used, and the remaining radars in the map are conventional (as of February 1997).* (Map supplied by the UK Meteorological Office based on the Ordnance Survey Map with the permission of HMSO. Crown Copyright. Licence No. AL850306)

Doppler weather radar as a primary aviation weather sensor 167

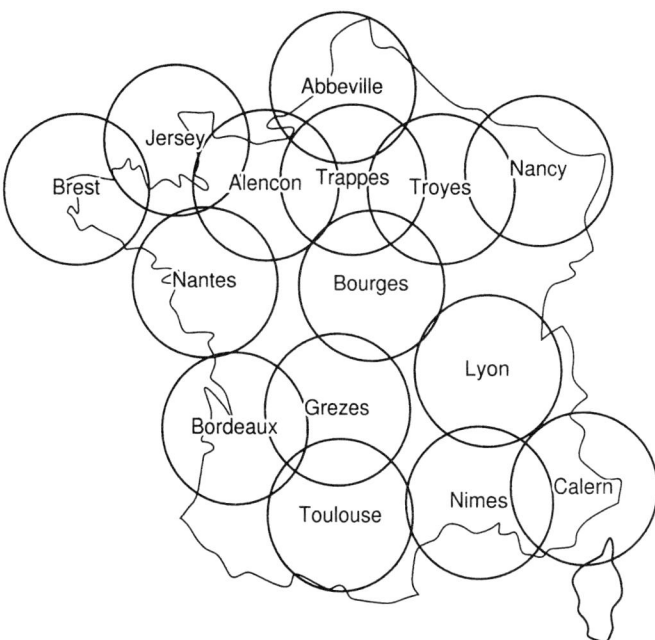

Figure 6.7 Conventional weather radar coverage of France as in 1990. A majority of the radars operate in the C-band and a few (near Bordeaux, Brest and Nimes) in the S-band. The range circles have a radius of 125 km. (Courtesy J.-L. Cheze, Service Central d'Exploitation de la Météorologie, Toulouse, France)

energy that is inversely proportional to the square of its range from the radar. Further, the small amount of energy intercepted and scattered by the target again spreads out on a spherical front, resulting in a second square-law diminution of the echo power intensity with distance back to the radar. The net result is that the echo power received by the radar from a point target ideally has an inverse fourth power dependence on the range to the target (Blake, 1990).

In the case of a target space filled with dispersed scatterers, the transmitted power intensity along the antenna beam does vary inversely with range squared, but this is compensated by an equivalent increase in the size of the range resolution volume and a corresponding increase in the number of scatterers simultaneously receiving the radiated power and contributing to the echo sample (see Figs. 6.2 and 6.3). The echo signal strength received by the radar therefore decreases only as the square of the range to a resolution volume, which is caused by the inverse-square decay of the scattered power intensity on the return path. This is the situation obtaining in the case of weather radars. Comparing this with the fourth-power-of-range decay of echo signal strength for point target detection, the weather radar has a strong advantage in terms of range performance.

The weather radar range equation for a given pixel naturally involves the inherent scattering characteristics of the population of scatterers in the pixel and the range to the pixel, besides a number of parameters of the radar. However, since many of the radar parameters are interrelated, and since the signal levels can be referenced to and measured at different points in the radar, the radar equation can take many variant forms (Serafin, 1990; Rinehart, 1991; Doviak and Zrnic', 1993). A common and convenient form of the ideal weather radar range equation that contains the reflectivity factor explicitly is derived as follows. In consistent SI units, the echo power P_r received by the radar antenna from a given resolution volume at a range r from the radar is

$$P_r = [P_t G]\left[\frac{1}{4\pi r^2}\right]\left[\pi\left(\frac{r\theta_b}{2}\right)^2\left(\frac{c\tau}{2}\right)\right][\eta]\left[\frac{1}{4\pi r^2}\right]\left[\frac{G\lambda^2}{4\pi}\right] \qquad (6.14a)$$

On the right side, the first term in square brackets is the effective power radiated along the beam (in the direction of the resolution volume), the second is the spherical spreading loss out to the target, the third is the volume measure of the resolution volume (assuming an ideal conical beam), the fourth is the reflectivity, i.e. the backscattering cross-section of the population of droplets per unit volume, th fifth is the spherical spreading loss back to the antenna, and the sixth term is the collecting cross-section of the receiving antenna. Substituting eqn. 6.5b in eqn. 6.6 and noting the subsequent concept of equivalent reflectivity factor Z_e, η can be written as

$$\eta = \pi^5 |K_w|^2 \left(\frac{1}{\lambda^4}\right) Z_e \qquad (6.14b)$$

In consistent SI units, λ is in m and Z_e is in m^6/m^3, giving the reflectivity η in m^2/m^3. Substituting eqn. 6.14b in eqn. 6.14a and rearranging,

$$P_r = \frac{\pi^3 P_t G^2 \theta_b^2 c\tau |K_w|^2 Z_e}{2^9 (2\ln 2)\lambda^2 r^2} \qquad (6.14c)$$

Note that the denominator of eqn. 6.14c contains an additional term $2\ln 2$ (≈ 1.39), which is a shape factor arising from the use of a nonideal antenna. While an ideal conical beam would have a uniform radiated power density within a narrow cone and zero power outside it, practical beam shapes are smooth functions of angle. The factor $2\ln 2$ corresponds to the one-way half-power (3 dB) beamwidth of a Gaussian-shaped radar beam. Combining the numerical constants of eqn. (6.14c), we obtain the final form of the range equation as:

$$P_r = \frac{\pi^3 P_t G^2 \theta_b^2 c\tau |K_w|^2 Z_e}{2^{10}(\ln 2)\lambda^2 r^2} \qquad (6.14d)$$

The location of the resolution volume is decided by the antenna beam pointing direction and the echo signal sampling instant as per eqn. 6.2. The dielectric parameter K_w of water is as given by eqn. 6.5c, and the equivalent reflectivity factor Z_e was defined in the discussion following eqn. 6.7. The velocity of light is c, the radar wavelength is λ, and the other symbols have been defined in Table 6.1. This form of the radar equation (eqn. 6.14d) assumes a pencil beam of circular cross-section, as is common with weather radars. However, in special cases where the beam may not be circular, the factor θ_b^2 in the numerator may be replaced by the product $\theta_{ba}\theta_{be}$ of the azimuth and elevation beamwidths.

In eqn. 6.14d P_r is the received echo power appearing at the antenna output terminals. However, it is only at the receiver output that the echo signal is normally sensed for detection, display, recording and/or further processing. It is therefore more practical to write the radar range equation corresponding to the receiver output. This would involve the receiver signal power gain G_s, usually obtained by calibration with a continuous wave (CW) or monotone sinusoidal test signal.

The form of the radar equation (eqn. 6.14d) is an ideal one in the sense that it does not take into account the losses occurring in the radar system or the propagation path. Energy losses may be accounted for in the radar equation by incorporating appropriate *loss factors*. The loss factor of a device or path is the ratio of the power entering it to the power exiting. Loss of energy in the weather radar context occurs in a variety of elements and for a variety of reasons (Blake, 1990, Section 2.7), but three main types of losses may be considered:

(i) *System losses*: This refers to the fraction of energy lost in transit through the various elements of the radar system along the signal path, between the point where the transmitted power P_t is specified and the point where the received echo power P_r is measured. The *system loss factor* is designated L_s, and covers all losses that are not included in any other variable of the radar equation. For example, the power losses in the antenna are usually included in the specification of the antenna gain G, and the losses in the receiving transmission line are a part of the definition of the receiver gain G_s; hence these are excluded from the definition of system loss factors.

(ii) *Atmospheric losses*: Atmospheric attenuation of radar signals due to weather and other factors was discussed in Section 6.2.5. The *atmospheric loss factor* takes into account the fraction of power lost due to absorption and scattering in the atmosphere. If L_a is the loss factor for a one-way transit through the atmospheric medium between the radar and the resolution volume of interest, then the two-way atmospheric loss factor, taking into account the energy lost from the transmitted beam as well as during the return passage of the echo signal, is L_a^2. The atmospheric loss factor may also include a *lens effect loss* component (Weil, 1973) caused by the differential atmospheric refraction of the rays from the antenna at varying elevation angles. A zero elevation angle is the worst case, resulting in a two-way lens effect loss of about 1 dB for distributed targets (~2 dB for point targets) at a range of 450 km (Shrader and Weil, 1987), which is the order of the longest surveillance range for ground-based weather radars (see Section 5.3).

(iii) *Receiver filtering losses:* The third loss factor arises from the finite bandwidth of the processing filter in the radar receiver (Doviak and Zrnic', 1993, p. 79-80). Depending on the frequency characteristics (or range weighting function in the time domain) of the receiver, the echo signal will lose a part of its energy during the filtering process, relative to a continuous wave (CW) or monotone sinusoidal test signal passing through the same receiver. The *finite bandwidth loss factor* is denoted as L_f.

After including the receiver gain and the various loss factors, a more realistic form of the weather radar range equation emerges:

$$P_{ro} = \frac{\pi^3 P_t G^2 G_s \theta_b^2 c\tau |K_w|^2 Z_e}{2^{10} (\ln 2) L_s L_a^2 L_f \lambda^2 r^2} \tag{6.15}$$

where P_{ro} is the echo signal power at the output of the radar receiver.

The range eqns. 6.14 and 6.15 have been derived analytically and hence require the variables to have consistent scientific units such as those in the SI system. To use the more familiar and convenient units, eqn. 6.15 may be recast as

$$P_{ro} \cdot 4.37 \times 10^{-25} \frac{P_t G^2 G_s \theta_b^2 c\tau |K_w|^2 Z_e}{L_s L_a^2 L_f \lambda^2 r^2} \tag{6.16}$$

where the dimensioned variables must be expressed in the following units:

P_{ro}, mW; P_t, W; τ, μs; θ_b, degrees; Z_e, mm^6/m^3; λ, cm; r, km.

The equivalent reflectivity of rain or other types of scatterers can be obtained from the range equation. The receiver output is sampled at regular intervals (usually τ). The range r to the resolution volume under consideration corresponds to the sampling time t_s as given by eqn. 6.2, and P_{ro} is equated to the power associated with this sample. By using these two quantities in either eqn. 6.15 or eqn. 6.16 with appropriate units, the equivalent reflectivity Z_e can be obtained, assuming the radar system parameters and the constants to be known. It may be noted that all the quantities on the right-hand side of the range equation, except Z_e, r and L_a,

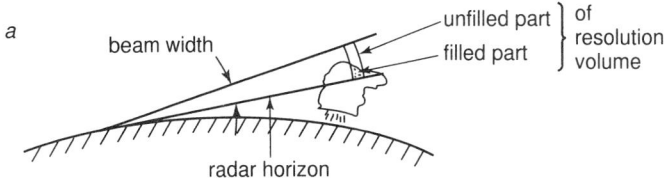

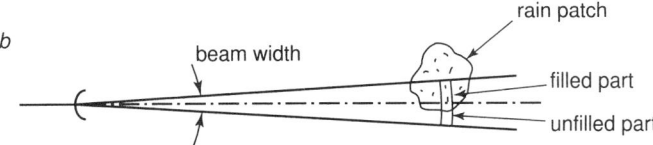

Figure 6.8 *Schematic diagram showing partial beam filling on top of a storm in frontal view (a), and on the edge of a precipitation patch in plan view (b)*

remain invariant for each resolution volume in the observation field. The system parameters may vary slowly with time, requiring calibration at intervals. If the transmitter power P_t is variable, it may be sensed at a convenient point and used to normalise the receiver output P_{ro}. The only unknown is the atmospheric loss factor L_a, which can be a source of significant inaccuracy in reflectivity estimation in the case of conventional weather radars.

The weather radar range equation is usually derived on the assumption that each resolution volume to which it is applied is completely filled with a population of scatterers having homogeneous properties over the entire volume. If the scattering properties of the population vary over the volume, then the radar equation will yield an average value of the reflectivity factor, weighted by the antenna pattern function and the receiver range weighting function. A special case of scatterer nonuniformity in the resolution volume arises from incomplete beam filling, which occurs when the radar beam illuminates the edge of a weather phenomenon. In such a case only a part of the relevant resolution volumes would be filled with weather scatterers, while the remaining parts would have none of them. The situation is shown schematically in Fig. 6.8. Averaging over a partially filled resolution volume would show a lower value of reflectivity than the actual value present, leading to an underestimation of the rainfall intensity.

Nonuniform or partial beam filling is a more frequent and serious problem at longer ranges, because the lateral width of the beam is higher at such ranges. As pointed out in Section 6.2.1, the lateral extent of practical weather radar beams can reach dimensions of the order of 10 km at the maximum operating ranges of the modern generation of weather radars, and significantly higher for older ones. There can be significant variation of local reflectivity over such large spatial scales. Indeed, even entire thunderstorm cells may not fill the beam completely, and partial illumination of larger phenomena is common at such long ranges.

As the radar beam scans past patches of reflectivity, the effective reflectivity sensed by the radar according to the range equation corresponds to the convolution of the true reflectivity distribution and the radiation pattern function of the antenna. This has the effects of (a) smoothing the details of reflectivity variations over the beamwidth, (b) softening or blurring the sharp edges of weather phenomena, and (c) diluting or underestimating reflectivities within high-reflectivity patches which may be as large as a few or even several kilometres across. Obviously, the larger the range and the corresponding linear beamwidth, the more severe will be the blurring effect.

Convolution also occurs in the radial direction with respect to the range weighting function, but this effect is usually not serious because of the relatively small and constant depth (range extent) of the resolution volume. However, when the range extent of the resolution volume is increased to improve the signal-to-noise ratio and/or to render the resolution volume more symmetric (see Section 6.2.1), the effect of range-wise convolution can be significant.

The processing involved in obtaining the reflectivity factor from the range equation was performed in conventional weather radars using analogue elements and simple digital circuitry. The range equation remains basically the same for modern Doppler weather radars, but the use of improved radar components and more sophisticated and extensive digital processing and calibration techniques allow much more accurate determination of the reflectivity factor.

6.3.2 *Estimation of rain rates*

The primary use of the radar reflectivity data is to serve as a measure of the rain activity. Rain is most readily sensed by radar, and rain activity is a fair indicator of foul weather. Although the raw radar reflectivity value itself may serve as a crude indicator of the local rain activity, accurate quantitative estimation of the level of rainfall rate from the reflectivity data is fairly complicated.

Radar reflectivity is a good indicator of rain rate because these two quantities are highly correlated: a heavier rainfall rate generally results in a higher radar reflectivity and, conversely, strong reflectivity from a given volume of the atmosphere indicates a high level of precipitation activity there. The high correlation, in turn, is because both the quantities depend on the size and density of water drop populations. However, the exact nature of dependence is different for the two quantities, so that the relationship between them is rather complex and somewhat uncertain.

In a volume of air containing water drops or droplets, the *liquid water content* is defined as the amount of suspended liquid water per unit volume, normally expressed in units of kg m^{-3}. This quantity depends on the sum of the cubes of the diameters of the water drops suspended in a unit volume. However, the liquid water content, by itself, does not constitute rain. Rain rate, as we normally understand, is the volume of water that passes during a unit time interval through a unit horizontal area at the location of interest. The rain rate, therefore, depends on the fall speed of drops in addition to the diameters and the numerical density of drops in the volume of observation.

Raindrops, being a few millimetres to less than a millimetre in diameter, quickly attain terminal speed during fall through air. Since raindrops grow in size gradually due to the acquisition of water from saturated or supersaturated ambient air, or in small steps through collision–coalescence, it is a fair assumption that the drops always fall at their respective terminal speeds. The terminal speed of a spherical drop of water is a monotonically increasing function of its diameter, and the volume of the drop depends on the cube of its diameter. Therefore the contribution of a drop to the local rain rate, which is a product of its volume and its fall speed, varies with drop diameter as a power higher than the third. A good approximation to the terminal velocity of raindrops in the most useful diameter range between 0.5 and 5 mm is given as (Atlas and Ulbrich, 1977)

$$w_t(D) = 386.6 D^{0.67} \tag{6.17}$$

where $w_t(D)$ is the terminal speed of fall[1] of a spherical raindrop of diameter D in still air. With such an expression for terminal velocity, the rain rate would depend on the 3.67th power of the drop diameter.[2]

Natural rain, however, does not consist of drops of uniform size, but contains a continuum of drop diameters. The rain rate expression therefore becomes more complex, being a function of the proportion of drops of different sizes in the population. Such a proportion is given by the *drop size distribution*, normally expressed through the functional notation $N(D)$, where N is the average number density (i.e. number per unit volume) of drops per unit diameter interval at the diameter D. Thus, if ΔN is the number of drops per unit volume that have diameters lying between D and $D + \Delta D$, then

$$N(D) = \lim_{\Delta D \to 0} \frac{\Delta N}{\Delta D} \tag{6.18}$$

which is usually expressed in units of (number)/(m^3 mm) or m^{-3} mm^{-1}.

---

[1] The notation w_t is used here for the terminal velocity in preference to a more instinctive symbol such as v_t because the fall speed is along the vertical or z-axis. Since velocities along all the three axes are of concern in Doppler radar and atmospheric studies, the symbols u, v, w normally refer to the velocity components along the x, y, z axes, respectively

[2] It is to be noted that a given raindrop will have a higher terminal speed at an elevated altitude because of the lower resistance offered by the rarer air; the factor of increase relative to its sea level value is somewhat less than the square root of the ratio of air density at sea level to that at altitude. This effect is not very important, however, for operational conventional weather radars, which normally employ a nearly horizontal scanning so that the beam stays close to the ground. For volume scanning radars, which may scan at relatively high elevation angles, appropriate altitude correction for the terminal speed variation should be applied for accurate results

174 *Aviation weather surveillance systems*

The drop size distribution is a very important function because it not only determines the rain rate in terms of drop diameters, but also serves to relate the rain rate to the observed radar reflectivity. The relationship between the two quantities is not inherently unique because the quantities depend differently on the diameters of raindrops. While the radar reflectivity depends on the 6th power of drop diameters, as in eqn. 6.7, the rain rate varies with diameter according to a lower power, typically lying in the vicinity of the 3.67th. Thus different combinations of drop sizes and numbers can result in the same rainfall rate but different reflectivities. For example, it is possible for a smaller number of larger drops to produce the same rain rate as a larger number of smaller drops, but the former type of rain would show a larger radar reflectivity than the latter because of the stronger dependence of reflectivity on diameter.

If, however, the drop size distribution in the rain is known or assumed, then the size–number uncertainty vanishes, and the relationship between the rain rate R and radar reflectivity Z becomes unique. The two quantities are expressed in terms of the drop size distribution as follows:

$$R = \frac{\pi}{6} \int_0^\infty D^3 N(D) w_t(D) \, dD \tag{6.19}$$

and

$$Z = \int_0^\infty N(D) D^6 \, dD \tag{6.20}$$

Note that eqn. 6.20 deals with the same quantity, the reflectivity factor, as eqn. 6.7. While the latter expresses the reflectivity in terms of the diameters of individual drops, which are difficult to ascertain, the former is based on the statistical distribution of drop sizes, which may be more readily estimated, assumed or approximated by mathematical functions.

Because of the crucial importance of drop size distribution in the radar determination of rain, extensive research has been carried out to determine this distribution. The landmark study by Marshall and Palmer (1948) resulted in a distribution, now called the *Marshall–Palmer distribution*, which has been extensively used in weather radar work. This study, as well as those by others (e.g. Laws and Parsons, 1943; Sekhon and Srivastava, 1971; Srivastava, 1971) point to an exponential type of variation of the number density of raindrops with diameter, with the exponent dependent on the intensity of rain. In particular, the Marshall–Palmer distribution function relates the drop size distribution $N(D)$ to the rain rate R through the following two-step exponential relationship:

$$N(D) = N_0 e^{-\Lambda D} \tag{6.21a}$$

where

$N_0 = 8000$ m^{-3} mm^{-1}, D is the drop diameter in mm, and

$$\Lambda = 4.1 R^{-0.21} \tag{6.21b}$$

with R expressed in mm h^{-1}.

Although the Marshall–Palmer distribution is simple and widely used, the actual distribution of raindrop sizes can vary considerably from it. The drop size distribution function varies significantly among various types of rain such as thunderstorm rain, stratiform rain[3] and orographic rain.[4] It may also vary according to the intensities of rain in a way that is different from what is indicated by eqn. 6.21. Further, the distribution function may also depend on the geographical and climatic region of occurrence of the rain, local conditions such as temperature, and the altitude of sampling in a thunderstorm. Because the drop size distribution is a strong parameter in relating rain rate to the observed radar reflectivity, any variation of the distribution function from the assumed one would introduce a corresponding uncertainty in the relationship between rain rate and reflectivity.

The relationship between radar reflectivity factor Z and rain rate R is often referred to as the Z–R relationship. As discussed above, this relationship can be established through the drop size distribution. However, based on the large body of observational data accumulated over the decades since the first attempt at rain measurement by radar, it is possible to relate the two quantities empirically in a direct way. Such a direct relationship is not only simple and straightforward to use, but can be postulated even in the case of precipitation other than rain, such as snowfall.

The Z–R relationship is typically given in the form

$$Z = aR^b \tag{6.22}$$

where a and b are constants, and R and Z are usually specified in units of mm h^{-1} and mm^6 m^{-3}, respectively. Because of the variability of drop size distribution among various rain phenomena, a and b can take diverse values, and widely differing values of each constant have been reported. Battan (1973) has listed a large number of Z–R relationships for rain of different types and in different locations, and many more such relationships have been reported since his list was compiled. Table 6.5 provides a sample set of Z–R relationships, showing their variability and claimed or implied applicability.

There is clearly a lot of overlap between the applicability of the various Z–R relationships. In a given application the choice of the appropriate formula to be used is made from a knowledge of the type of precipitation and the climatic and geographic region involved. Information regarding the type of precipitation is not inherently available from the radar observation, but usually has to be provided by the user. It is also possible to divide the precipitation scale into two or more intervals and use a different law for each to achieve better accuracy. Such domain-splitting is especially desirable when

---

[3] Spatially widespread and relatively uniform rain, usually of considerable duration
[4] Rain that is induced or influenced by the presence of mountains or hills

Table 6.5 Representative Z–R relationships and their applicability in terms of precipitation type and/or geographical area

S. no.	Z–R relationship	Applicability	Source
1	$Z = 200 R^{1.6}$	Stratiform rain	Marshall et al. (1955)
2	$Z = 31 R^{1.71}$	Orographic rain, Hawaii	Blanchard (1953)
3	$Z = 486 R^{1.37}$	Thunderstorm rain	Jones (1955)
4	$Z = 155 R^{1.88}$	Rain, North Dakota	Cain and Smith (1976)
5	$Z = 300 R^{1.4}$	Rain, Miami, FL	Woodley et al. (1975)
6	$Z = 300 R^{1.5}$	Rain, Switzerland	Joss and Waldvogel (1970)
7	$Z = 2000 R^{2}$	Snow	Gunn and Marshall (1958)

Notes:
(a) The relationship in the first row, referred to as the Marshall–Palmer formula, is the most commonly used Z–R relationship
(b) In the Z–R relationship for snow in the last row, the precipitation rate R corresponds to that of the melted snow, and is called *water equivalent precipitation rate*

each interval corresponds to a different type of precipitation. For example, very low reflectivity factors of the order of 10 dBZ or less are often due to snow, while regions showing very high reflectivity factors of the order of 55 dBZ and beyond harbour hail with a high probability. Even rain events may be classified according to their intensity, and different Z–R relationships utilised for each intensity interval. Such domain-splitting may be made based on the radar-derived reflectivity value. To a certain extent, the user's preference and past experience also play a part in the choice of the Z–R relationship to be applied. Therefore, good estimation of rain rate from reflectivity data alone is, at least partially, an art.

In spite of the limited number of sample Z–R relationships quoted in Table 6.5, the most striking observation is that the two constants used in the relationships vary over wide ranges. A part of this variation may be due to radar system calibration errors resulting in inaccurate measurement of reflectivity values for any given rain rate. The Z–R relationship is an absolute one, i.e. it relates the absolute values of the two parameters. Therefore any systematic error in one or both of the measured parameters, such as those due to poor calibration of the rain-measuring radar or the validating rain-gauge, would translate into errors in the Z–R relationship derived from their readings. Since many of the proposed Z–R relationships are based on measurements in localised areas with single radar systems and one or a few rain-gauges, and since many of the older conventional weather radars had difficulties with accurate calibration, the possibility of the contamination of some of the Z–R relationships due to poor measurement accuracy cannot be ruled out.

In addition to poor calibration, a number of other factors can affect the accuracy of the Z–R relationships. Several such factors are listed by Doviak and Zrnic' (1993, p. 225). These include atmospheric and weather attenuation of radar signals, differences between ground-level precipitation and precipitation aloft (as radars do not normally 'see' all the way down to the ground level except at close ranges on flat terrain) due to horizontal winds and evaporation, incomplete or nonuniform filling of the radar beam by precipitation, and nonzero vertical air speeds (which makes the fall speed of hydrometeors different from their terminal speeds in still air, thus altering the rain rate for a given drop size distribution).

It is to be borne in mind that measurement errors affect the rain rate estimate in two stages. First, the inaccuracy of the radar and the rain-gauge(s) used for establishing the Z–R relationship affects the correctness of the relationship itself. At the next level, errors in the measurement of the reflectivity factor would translate directly into a corresponding error in deriving the rain rate from the Z–R relationship.

Fortunately, in spite of the large spread among the different Z–R relationships, a fair number of them provide comparable results between 2 and 200 mm h^{-1} of rainfall (Doviak and Zrnic', 1993, p. 226). This covers the most useful range from light to heavy rain. A plot of the available Z–R relationships would show a corridor where a majority of the traces are bunched, as attempted by Doviak and Zrnic' (1993, p. 224) for the 69 such relationships quoted by Battan (1973). Some of the outliers on such a plot would appear to be extremely wayward relative to this corridor, but even the corridor itself is not too well focused, yielding variable estimates of rain rates for given measured values of the reflectivity factor.

Indeed, the uncertainty or errors in rain rate estimation using a single parameter such as reflectivity remains a major weakness of the conventional weather radar. Errors as high as a few hundred percent in deducing the rain rate from radar reflectivity are possible, and 50 to 100% errors are frequently encountered. The inadequacy of the conventional weather radar for accurate precipitation rate measurement, together with the need to incorporate certain useful features it does not have, has been a powerful motivating factor for the development of the modern generation of Doppler weather radars. It is shown in Chapter 13 that the use of additional parameters, such as echo signal amplitudes and phases at different polarisations, can greatly enhance the ability of radars to estimate and characterise precipitation parameters.

The inaccuracy in rain rate estimation by conventional weather radars is more detrimental to aviation interests than to many other fields of use of precipitation data. Applications such as hydrology, flood forecasting and agriculture usually require precipitation data integrated over relatively large geographical areas and over extended periods of time, of the order of weeks or months. Because of the averaging effect of such integration, the standard deviation of total rainfall estimates derived from radar data is generally much lower than that of instantaneous rain rates. Aviation, on the other hand, is

178 Aviation weather surveillance systems

Table 6.6 Approximate basic parameters of the WSR-57 radar

Unit	Parameter/facility	Value/description
System	Operating frequency	3 GHz
	Operating wavelength	10 cm
	Sensitivity[a]	−6.84 dBZ
Antenna	Diameter	3.7 m
	Beamwidth	1.8°
	Gain	38.5 dB
	Polarisation	Horizontal
	Environmental protection	Spherical rigid radome
Transmitter	Peak power	500 kW
	Pulsewidth	0.25 μs, 4 μs
	Pulse repetition frequency	200–800 Hz
Receiver	Bandwidth	750 kHz
	Sensitivity (minimum detectable signal)	−110 dBm

[a] System sensitivity is defined here as the minimum detectable reflectivity factor at a distance of 50 km from the radar

sensitive to the nature and rate of the precipitation obtaining at the location and instant of flight, the estimate of which is relatively unreliable in the case of conventional weather radars.

6.3.3 WSR-57 radar

A good example of a conventional weather radar is the WSR-57 (the number indicates the year of its acceptance for introduction into service), which has been the principal weather-mapping radar system operated by the US National Weather Service for decades. Its successor radar, the WSR-74, inherits many of its basic features. Similar radars have been produced and/or operated by a number of countries for general weather monitoring as well as for specific purposes such as cyclone tracking and warning. When located near airports, the systems provide a fairly detailed picture of the precipitation pattern in and around the terminal area. These conventional radars measure and display only the reflectivity associated with weather phenomena. The basic parameters of the WSR-57 radar, as a representative of this class, are listed in Table 6.6.

The WSR-57 radar normally performs a circular scan and displays the reflectivity distribution over the scan circle on a *plan position indicator* or PPI (Fig. 6.9). The display device is a cathode ray tube with a circular face and a radial trace. The trace is modulated with a reflectivity signal, and the scale is so adjusted that the radial corresponds to the maximum range of observation. The trace is rotated about the centre in synchronism with the antenna rotation, so that the PPI display refreshes itself once for every rotation of the

Doppler weather radar as a primary aviation weather sensor 179

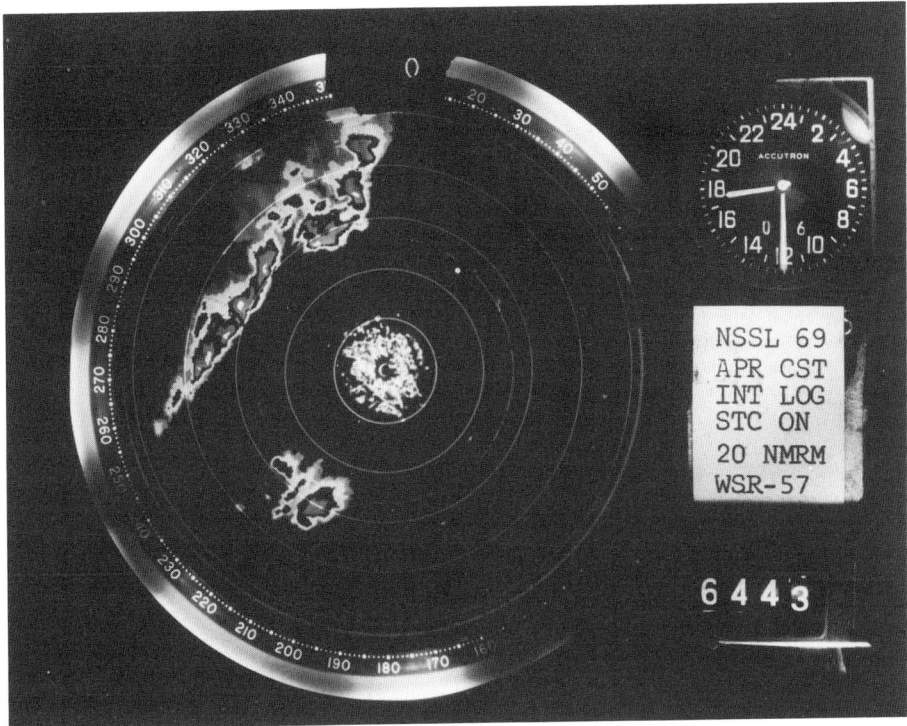

Figure 6.9 Example of PPI display of WSR-57 radar showing the reflectivity distribution associated with two storm phenomena. The elongated patch extending from the west to the north of the centre shows a squall line with multiple storm cells, giving it a bead-like appearance. The reflectivities are intensity modulated, and are thresholded so that when the value exceeds a preset level, the display returns to a dark shade, and brightens again with increase in rain intensity. The resulting contours depict rainfall intensities as well as spatial gradients, with narrower contours indicating higher gradients. The circles indicate range from the radar, set at 20 nmi in this case. The assemblage of bright spots inside the innermost circle represents strong returns due to ground clutter close to the radar. (Courtesy D.S. Zrnic', National Severe Storms Laboratory, Norman, Oklahoma)

antenna. An optional movie camera attachment can take photographs of the display at the rate of one frame for every antenna rotation. A replay of the film at a faster speed provides an animation effect and helps visualise the evolution of dynamic phenomena.

Because of the inherently low dynamic range of the WSR-57 display, the entire range of reflectivity values encountered in weather observation cannot be differentiated well on the display. To overcome this difficulty, and to provide a quantitative indication of the reflectivity gradients, a thresholding and contouring facility is provided. The reflectivity values crossing given

180 *Aviation weather surveillance systems*

thresholds are depicted by abruptly changing grey shades, resulting in a contouring effect. A closer spacing of these contours indicates a sharper spatial gradient of reflectivity.

6.4 Motivation for developing modern weather sensors

An important parameter to note from Table 6.6 is the beamwidth. At a distance as close as 60 km, the beamwidth of $1.8°$ would result in a lateral spread of ~ 2 km of the beam, which is enough to engulf the lateral dimension of many thunderstorm rain shafts entirely. At further ranges, the lateral resolution of the WSR-57 radar would be proportionately worse, becoming almost 15 km at a range of 450 km. This results in poor spatial resolution and enhanced blurring effects. There is clearly a need for better angular resolution of the primary sensor, and the modern weather radars incorporate such an improvement.

Conventional weather radars including the WSR-57 provide a picture of the reflectivity distribution in their fields of view by applying the radar equation to each pixel in their observation fields. However, in analysing and utilising the weather scene depicted by the conventional weather radar for aviation and other purposes, it must be remembered that the raw reflectivity distribution provides only a crude picture of the weather scenario, and further *meteorological interpretation* is necessary to derive full benefit from such data. Such interpretation utilises, in addition to the reflectivity values, the patterns of distribution of reflectivity, information on the synoptic setting in which each phenomenon occurs, and data from sensors other than the radar. Because of the high level of intelligence and familiarity required for such pattern recognition and interpretation, relative to the level of technology that went into the development of such radars, meteorological interpretation of conventional weather radar data is heavily dependent on the involvement of experienced human interpreters. This fact is particularly significant from the point of view of aviation since aviation requires the products of data interpretation to be available in real time to support operational decision making. It is clear that extensive use of conventional weather radars in support of modern aviation systems would entail the deployment of a correspondingly large complement of experienced personnel on a routine, round-the-clock basis for data observation and interpretation. This would not only add significantly to the cost of operation, but also enhance the probability of errors of judgment resulting from human fallibility, fatigue, and boredom from repetitive tasks. As discussed at several points in the remaining part of this book, a major emphasis of the design philosophy of modern aviation weather surveillance systems is to automate as much of the functioning of the sensors and systems as possible within affordable costs and without compromising reliability.

Modern aviation weather surveillance systems, therefore, not only aim to incorporate improved technology and specifications for their individual

Doppler weather radar as a primary aviation weather sensor 181

elements, but also to shift the paradigm from human-intensive operation to one of emphasis on automation and autonomous intelligence, based on high-quality multiparameter data of high volumes. The following are the principal motivating factors for the development of modern sensors and systems for aviation weather surveillance:

1. acquisition of more accurate basic data, through improved system sensitivity and other specifications, to better identify aviation-hazardous phenomena and hazard parameters
2. improvement of resolution through narrower beamwidth
3. acquisition of data on more parameters associated with weather phenomena than what is provided by conventional weather radars, most importantly data on wind velocities and velocity-derived parameters
4. automated system operation, data collection and data recording
5. generation of higher-level data products using the basic data from sensors
6. use of complex and versatile signal and data processing methods to improve system performance and weather parameter estimation under the widest possible range of operating conditions
7. use of advanced computing hardware and software for automated and autonomous data processing, interpretation and inference generation
8. versatile and information-rich display systems for functionally efficient interface with the human elements in the system
9. automated multi-user data access
10. compatibility with other elements of modern air traffic control systems and services.

Of particular interest in the list above, from the aviation point of view, is the third item, which deals with multiparameter data acquisition. As discussed in Section 3.3, aircraft flight is most directly affected by strong air currents and their gradients in the atmosphere, but the conventional weather radar has no way of sensing these important parameters. This shortcoming of the conventional weather radars was the strongest motivating factor for the development of the modern Doppler weather radars capable of directly measuring parameters associated with wind velocities in the air space supporting flight.

6.5 Doppler weather radar: basics

The modern Doppler weather radar has evolved by utilising the vast experience gained in the development and use of conventional weather radars, but differs in the important and fundamental aspect that it can directly measure many parameters associated with air motion, which the conventional weather radars cannot. Radars utilising the Doppler principle for velocity estimation had been in use in the context of point target observation long before they were developed for dedicated weather observation. This experience with Doppler radars has also contributed significantly to the development of the Doppler weather radar, especially because many of the components and subsystems were readily available for the latter programme. However, like the conventional weather radar, the Doppler weather radar also differs in important ways from its counterparts used for point target detection functions. The differences owe their origin to the need for data mapping over extended fields, requiring specialised signal and data processing and display techniques. The development of Doppler weather radars in recent decades, culminating in their operational deployment, has been motivated by a general need for obtaining a more complete weather picture than that provided by conventional weather radars, and has been particularly spurred by the demands of modern aviation support. This section is dedicated to an understanding of the principles and techniques of wind velocity parameter estimation using the modern Doppler weather radars, and the design and development philosophy of such radars.

6.5.1 Basic principle and limitation

The Doppler radar is a *coherent radar*, i.e. it measures the phase of the echo signals in addition to their amplitude. According to the Doppler principle, the motion of the scattering target(s) results in a phase shift of the scattered signal. The rate of change of phase of the echo signal relative to that of the transmitted signal appears as a frequency shift called the *Doppler shift* or *Doppler frequency*. It is therefore possible to estimate motion-related parameters of the target(s) by appropriately processing the phase information in the echo signal.

The Doppler frequency f_d due to a scatterer, however, depends only on the radial component v_r of the scatterer velocity relative to the radar:

$$f_d = -\left(\frac{2}{\lambda}\right) v_r \qquad (6.23)$$

where λ is the radar wavelength. The negative sign in eqn. 6.23 indicates that a positive range rate of the scatterer, corresponding to a radially outward or receding scatterer velocity, causes a negative Doppler shift relative to the carrier frequency. The Doppler shift due to a moving scatterer is insensitive to the components of scatterer motion perpendicular to the radial from the radar. A Doppler weather radar is therefore inherently capable of determining only the radial component of the motion parameters associated with atmospheric phenomena. This fact constitutes a basic and major limitation of Doppler weather radars, and any system designed with such radars as sensors must take this limitation into account. Such a drawback notwithstanding, the

Doppler weather radar is still a very useful sensor for atmospheric motion parameters, and is of great value in aviation weather surveillance. In due course we comment on the effects of this limitation on the choice and design of data processing algorithms, and on some methods of overcoming this limitation.

6.5.2 Atmospheric wind tracers

Wind velocity measurement using radars requires the existence of *tracers* of air motion in the atmosphere. Air molecules *per se* are not sensible by weather radars. It is therefore necessary that there be a significant number of radar scatterers which are airborne and move at the same speed as the local air, so that estimating the velocity parameters of such scatterers amounts to the estimation of the velocity parameters of the air that carries them. Such scatterers are called tracers of wind speed.

As discussed in Section 6.2, hydrometeoric particles, airborne insects, dust and debris, and eddies with refractive index irregularities or inhomogeneities are the common sources of radar scattering in the atmosphere. It is instructive to comment on the behaviour of each of these types of scatterers as tracers of atmospheric air motion.

Atmospheric eddies are part of the air motion itself, and are carried along velocity fields. The eddies faithfully follow all local air motion of scales significantly larger than themselves. Since eddies of characteristic dimension equal to a half of the radar wavelength are responsible for most backscattering, and this works out to a few centimetres in the case of weather radars, eddies can act as faithful tracers for all air motion of interest in weather radar observation. A major difficulty with atmosphere eddies, however, is their weak backscattering strength, which limits the range of their usefulness to a few tens of kilometres from the radar. Even over much of this zone, the backscatter from eddies is plagued by poor signal-to-noise ratios or ground clutter contamination. In spite of such difficulties, the eddies with refractive index irregularities remain the primary tracers for radar estimation of winds in the absence of particulate scatterers.

Aerosols of fine dust are poor radar scatterers at the commonly used radar wavelengths of 3–10 cm, but many types of debris raised by strong air currents such as those induced by tornadoes and storm outflows including microbursts can serve as air motion tracers in the lower atmosphere. Airborne debris is highly localised (confined to zones harbouring strong winds at the ground level), short-lived, unpredictable, and cannot be calibrated in terms of reflectivity. However, in the absence of weather scatterers, such as while detecting and estimating the velocity parameters of tornadoes and microbursts in rain-free surroundings, airborne debris can act as a useful tracer.

Airborne insects form an important class of air motion tracers in the absence of weather scatterers. Insects are particularly abundant in the lower layers of the atmosphere and in warmer temperatures, and may be found in radar-detectable strengths up to a few or even several kilometres of altitude above the ground. The occurrence of insects in the atmosphere may be banded, i.e. particular species of insects may fly more densely within certain height slabs, and their occurrence and distribution usually shows a significant seasonal and diurnal variation. The type and density of insects also varies according to the geographical location and the type of terrain and vegetation. From the point of view of Doppler radars, weakly flying small insects or 'atmospheric zooplankton' are of particular interest as wind tracers because they have little flight speed of their own and hence follow the local air motions quite faithfully. Fortunately, the zooplankton form the majority of the insect population found in the atmosphere. More energetic insects such as swarms of locusts and grasshoppers are also carried by the atmospheric winds, but their speeds can differ significantly from these winds and hence provide misleading velocity information to the radar. The same is true in a stronger way for flocks of flying birds. Birds and strong insects therefore act as sources of disturbance or clutter for Doppler weather radars. The saving feature is their localised and well delineated zones and/or time slots of occurrence, and their relative infrequency.

Of the highest interest from the point of view of wind determination by Doppler weather radars are the wind tracing characteristics of hydrometeoric scatterers. This is so because hydrometeors are present in the zones of severe weather, and because the wind fields associated with severe weather are of prime concern for aviation safety and operations. Another advantage of hydrometeoric scatterers as tracers is their high reflectivity and strong echo signal, resulting in high signal-to-noise ratios which facilitate accurate estimation of velocity parameters.

Very small water droplets such as those constituting clouds and fog follow air motions faithfully, but these are of little interest to normal weather radars because of their negligible reflectivity. As already mentioned, raindrops, which are the most common weather scatterers, have definite fall speeds relative to still air. The equilibrium or terminal fall speed of each drop in still air is determined primarily by its diameter. In the presence of air motion in three dimensions, the raindrops may be assumed to follow the horizontal wind components faithfully. The vertical component of the local wind would modulate the fall speed of each drop with respect to its still-air fall speed. Thus, when the radar beam is narrow and points horizontally, the raindrops crossing the beam act as good tracers of the radial components of the wind field.

However, when the beam has a significant elevation from the horizontal, a component of the fall speed of the drops also contributes a Doppler signal (Fig. 6.10), causing bias in the wind speed estimates. This bias should be removed before the wind field can be deduced from the radar-sensed velocity parameters of the tracers. The bias due to beam elevation increases with increase in the elevation angle, and is maximum for a vertically pointing radar which will see the full terminal fall speed of the raindrops superimposed on

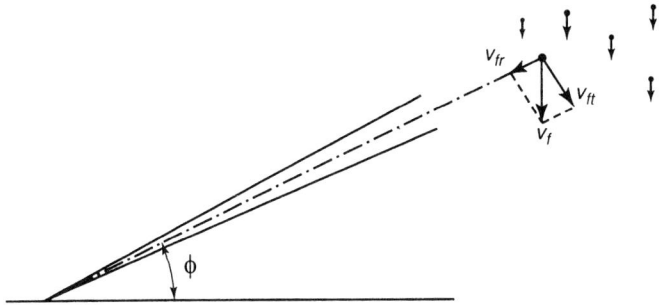

Figure 6.10 Schematic diagram showing Doppler signal generation due to falling raindrops. Even in the absence of winds, the vertical fall velocity v_f of a scattering raindrop would have a radial component $v_{fr} = v_f \sin \phi$ along the beam axis of elevation ϕ, yielding a Doppler frequency of $2v_f \sin \phi / \lambda$

the vertical component of the wind speed. The velocity estimation is further complicated by the fact that the drop sizes in natural rain processes are nonuniform, generating a spectrum of Doppler frequencies rather than a single one. Notwithstanding such difficulties, raindrops are the natural wind tracers for Doppler weather radars in weather zones for the reasons cited in the last paragraph. In regions of heavy weather, backscatter from rain (including hail) overshadows that from all other scatterers, making it difficult to detect and distinguish signals from any other type of tracer. Further, since rainfall intensity itself is a parameter of interest in radar observation, analysis of the echo signals from raindrops can provide both intensity and velocity information simultaneously.

Among other types of hydrometeors, snowflakes are good wind tracers because of their high-drag shapes and low terminal speeds. However, the reflectivity of dry snow is low, resulting in weak echo signals, poor signal-to-noise ratios, and inaccurate velocity estimation at relatively long ranges. Hail particles up to a few millimetres in diameter have wind tracing characteristics similar to raindrops. Large hailstones have relatively high terminal speed and can cause errors in wind velocity estimation at significant elevation angles. Because of their high inertia, large hailstones smooth out the effects of small-scale wind variations and respond only to the relatively slowly varying components of the wind field.

6.6 Doppler weather radar: primary data products

As already emphasised, modern Doppler weather radars constitute a significant improvement over conventional weather radars because of the larger number of dimensions of information they provide, in addition to making more accurate measurements. Thus, the primary data products of the

186 *Aviation weather surveillance systems*

Doppler weather radar include wind-velocity-related data regarding each resolution volume in its field of view, besides data on the reflectivity distribution. Typically, the velocity-related data generated by weather radars consist of the mean and the spread of the radial wind speeds within each resolution volume. These data, together with reflectivity, are obtained by suitably processing the amplitudes and phases associated with batches of echo pulses received by the radar. The nature of such signal processing is outlined in this section, as also the structure of Doppler weather radars that make such processing possible.

6.6.1 *Spectral moments of weather echo signals*

Refer to Fig. 6.2, which shows a weather radar observing an ensemble of scatterers in the atmosphere. Each scatterer in the resolution volume contributes an echo signal component which appears at the receiver as a complex quantity, with an amplitude and a phase. Since the radar cannot resolve individual scatterers within a resolution volume, the instantaneous signal value sensed or sampled by its receiver is the sum of the echo returns from all the scatterers in the volume, weighted by the antenna pattern function and the receiver response function. Successive samples of the return from a given resolution volume form a complex time series, which can be processed coherently to generate the basic data products of the Doppler weather radar.

The nature of Doppler weather radar data products and their correspondence with weather parameters are readily visualised by considering the signal structure in the frequency domain. The frequency spectrum of the echo signal corresponding to a given resolution volume is obtained by taking the Fourier transform of the complex time series representing the successive returns from the volume. The spectrum has contributions from the ensemble of scatterers in the resolution volume, and its shape will depend on the attributes of these scatterers.

Schematically shown in Fig. 6.11 are a series of spectra for different types of distribution of scatterer velocities. If the radar is looking at only one uniformly moving scatterer for a long period of time, the idealised spectrum of the Doppler-shifted echo signal would be a single spike located at a point on the frequency axis whose value is given by eqn. 6.23. This is depicted in Fig. 6.11*a*. The same spike-shaped spectrum would result if one considers a hypothetical case in which there are multiple scatterers in the resolution volume, but they are static relative to one another, and are moving radially relative to the radar as a frozen block. Such a set of scatterers would behave as a single point scatterer. When a single scatterer or a frozen set of scatterers is observed by a radar with a scanning beam, the observation period becomes limited, resulting in a finite width of the spectrum, as shown in Fig. 6.11*b*.

In real situations, the scatterers within a resolution volume are not bound together rigidly, but move around independently. To the extent that they may all be carried by a common air current, they may be translated *en bloc*, but they

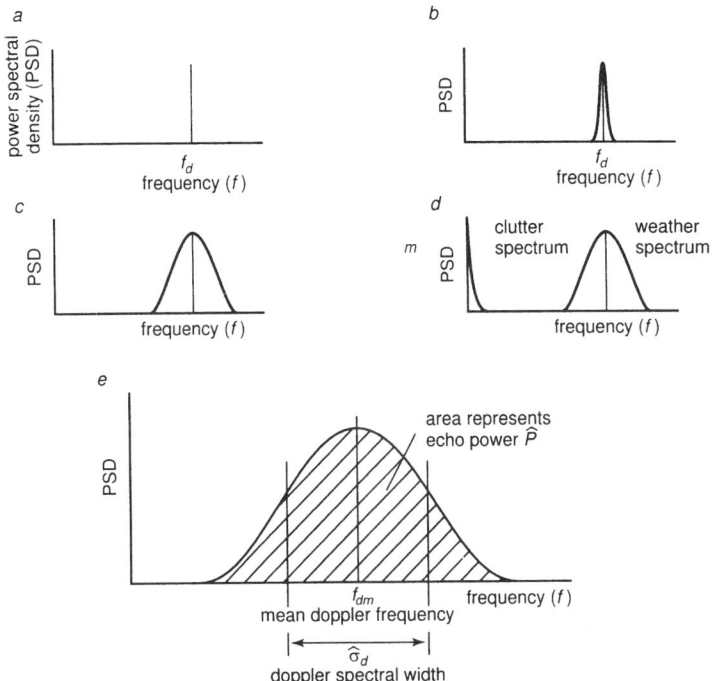

Figure 6.11   *The representation of weather radar echo signals in the spectral domain: (a) ideal return (of infinite length) from a point target moving at a uniform radial speed relative to the radar, (b) signal from a point target having finite spectral width due to antenna rotation and other spectrum-broadening effects, (c) ideal return from weather targets, (d) composite return due to weather and ground clutter, and (e) magnified view of weather spectrum alone, showing the three principal parameters*

also usually have motion relative to one another. Such differential motion may be due to random or turbulent air motion within the resolution volume, systematic velocity gradients across the volume, velocity differentials among tracers (e.g. differing fall speeds of raindrops of varying sizes, and independent flight of individual insects), and other factors. Each independently moving scatterer in general produces a different Doppler shift, causing the spectrum of the composite signal to broaden, as shown in Fig. 6.11c.

Besides echoes from atmospheric scatterers, a weather radar also receives returns from *ground clutter*. Ground clutter is the signal backscattered from objects or features on the earth's surface or those anchored to it, and interferes with the detection of weather, especially at close ranges from the radar. The ground clutter signal corresponding to any given resolution volume consists of the composite echo returned from the scatterers on the ground which arrives at the receiver at the same time as the echo from the volume. For very low antenna elevation angles the main lobe of the pattern

188 *Aviation weather surveillance systems*

grazes the ground, and the major ground clutter contribution comes from the strong main beam illumination. At higher elevation angles, ground scatterer illumination and reception through the antenna sidelobes are responsible for ground clutter. Ideally, for a ground-based radar, the Doppler spectrum of returns from stationary ground clutter should be a spike located at the origin, i.e. over the zero Doppler frequency. In practice, because of wind-driven foliage motion (and wave motion on water surfaces) and various spectral broadening factors (e.g. finite length of the echo signal time series), the ground clutter spectrum would have a finite width. Figure 6.11d shows the combined signal due to weather and ground clutter.

The spectral moments of weather echo signals, which are the primary data products generated by Doppler weather radars, can be readily defined with the aid of the signal spectrum in Fig. 6.11c. This spectrum is shown magnified in Fig. 6.11e for clarity, and the basic data products derived from the first three moments of the spectrum are marked therein. The three products are as follows:

1. Reflectivity: The radar reflectivity of the scatterers in each resolution volume is derived directly from the power associated with the spectrum of echo signals from that volume. The power is simply given by the area under the spectral density plot, which corresponds to the zeroth moment of the spectrum. The reflectivity can be calculated from the echo power by using the radar range equation. The reflectivity measured by Doppler weather radars is physically the same quantity as the reflectivity obtained from conventional weather radars, though different signal processing circuitry and algorithms may be employed in the two systems, and the accuracy of the estimates may vary between them. Therefore the reflectivity factor derived from Doppler weather radars can be used to estimate rainfall rates and draw inferences regarding the nature of weather scatterers, as discussed in Section 6.3.2 for conventional weather radars.

2. Mean radial velocity: The mean Doppler frequency f_{dm} of the weather spectrum in Fig. 6.11e can be obtained from the first moment of the spectrum by normalising it with respect to the area under the spectrum curve, i.e. the power associated with the spectrum. The mean radial velocity v_{rm} of the scatterers in the volume is then obtained by inverting eqn. 6.23 as

$$v_{rm} = -\left(\frac{\lambda}{2}\right)f_{dm} \qquad (6.24)$$

Physically, the mean radial velocity represents the motion of the scatterers due to the average radial component of the wind occurring over the resolution volume. Assuming the scatterers to be faithful tracers of the wind motion, the mean Doppler frequency provides a good measure of the radial wind speed itself. Following the sign convention explained for eqn. 6.24, a positive or higher-than-carrier Doppler frequency signifies a radially inward, i.e. towards-the-radar, wind, and negative Doppler frequencies mean away-from-the-radar winds. The mean radial wind velocity is a very important data output of the Doppler weather radar. A pixel-by-pixel display of the radial wind speeds provides a very good picture of the overall flow field in the surveillance zone of the radar.

3. Doppler velocity spectrum width: The width of the Doppler frequency spectrum provides a measure of the differential motion of the scatterers within a resolution volume. The more the differences in the radial velocities of the individual scatterers in the resolution volume, the more spread out would be their Doppler frequency contributions, resulting in a broader overall spectrum of the echo signal received from the volume. A measure of this spread is given by the Doppler spectrum width σ_d, which is the square root of the second moment of the power-normalised spectrum about the mean Doppler frequency.

Several factors may contribute to the broadening of the Doppler velocity spectrum. Some of them arise from the distribution of hydrometeor density and velocity within the resolution volume, and others are related to radar parameters such as antenna beam width and scan rate, transmitted pulsewidth and receiver bandwidth. The following are the significant causes of Doppler spectral broadening:

(a) turbulence within the resolution volume, which imparts to the scatterers random motion relative to one another in three-dimensional space, having a random radial component with respect to the radar
(b) systematic wind gradients or wind shear over the resolution volume, which cause scatterers across the volume to move, on average, at different radial speeds relative to the radar
(c) differing fall speeds of raindrops and/or other hydrometeors of varying sizes, shapes and densities, to the extent that these fall speeds have a radial component when the radar beam is not horizontal
(d) vibration of hydrometeors, involving oscillatory fall paths and/or pulsating or random change in shape and orientation of nonspherical particles, which causes the radar cross-sections of the particles to vary
(e) the scanning or sweeping motion of the radar beam, which causes each resolution volume to shift continuously during the sample collection period, resulting in a variable weighting of the scatterers from pulse to pulse.

As the spectrum-broadening effects of these factors are independent of one another, they can be added in a root-mean-square fashion to yield the overall Doppler velocity spectrum width σ_v, i.e.

$$\sigma_v^2 = \sigma_t^2 + \sigma_s^2 + \sigma_f^2 + \sigma_o^2 + \sigma_r^2 \tag{6.25}$$

where

σ_t = spectral broadening due to turbulence
σ_s = spectral broadening due to shear
σ_f = spectral broadening due to fall speed differences between hydrometeors
σ_o = spectral broadening due to oscillation or vibration of hydrometeors
σ_r = spectral broadening due to rotation of the radar antenna beam.

Equation 6.25 does not include processing-dependent effects such as that due to the window function, and special effects for particular types of scatterers, such as the spectral broadening caused by the independent flight of individual insects in a population acting as wind tracers.

In a majority of cases, especially when the beam elevation is not high and hence the radial component of hydrometeor fall speeds is small, it would be possible to neglect all but the effects of turbulence and wind shear in eqn. 6.25 without sacrificing much accuracy. Between turbulence and wind shear, the effect of the former is often more dominant, especially at close ranges where the resolution volume is small and hence the wind shear across the volume is not high. For this reason, in the absence of more sophisticated processing, the Doppler velocity spectrum width is often attributed to the level of turbulence alone. However, it is possible to estimate the radial wind shear at a given location from the mean Doppler velocities in the adjacent resolution volumes, and removing the shear contribution from the overall spectrum width (in a root-mean-square sense) would yield the contribution of turbulence. Such additional processing is worthwhile if an accurate estimation of the turbulence level is desired, especially in the presence of strong wind gradients.

6.6.2 Doppler weather radar system features and architecture

We now consider some specific hardware features of the Doppler weather radar, especially those which make the generation of basic data products possible. The Doppler weather radar has all the principal subsystems of the conventional weather radar (as indeed most other types of radar) such as the transmitter, antenna, receiver, signal processing and display. However, being a coherent radar, it has special requirements with regard to the phase shift and frequency characteristics of its subsystems. Further, its signal processing circuitry has the capability of handling signals in complex form, preserving their amplitude and phase information. Complex signals are represented in coherent radar receivers through their in-phase and quadrature components (the so-called *I* and *Q* components). The subsystems of noncoherent and coherent radars are discussed in detail in the various chapters of Skolnik (1990a).

An important requirement of a modern weather radar is a high dynamic range. Dynamic range is an important system parameter of a radar which indicates the spread between the strongest and the weakest signals it can detect. Because of the inherent nature of atmospheric scatterers, the dynamic range requirements of modern Doppler weather radars are very demanding. As mentioned before, clear-air phenomena of meteorological interest may exhibit reflectivity factors of the order of -15 dBZ or less, while the reflectivity factor of intense rain with hail may reach 65 dBZ. This represents a spread of >80 dB in the target reflectivity, calling for a high dynamic range of the receiver.

It is normally desirable that radar receiver-amplifiers be linear to preserve the relative strengths of targets in the signal domain. However, logarithmic receivers are often preferred where the dynamic range requirement is high. Receivers with logarithmic response characteristics provide high gain, and hence high sensitivity, at low signal levels, and yet do not get easily saturated as the input signal strength increases. It is possible to achieve dynamic ranges of the order of 80 dB with well designed logarithmic amplifiers (Taylor, 1990).

It is possible to use logarithmic amplifiers in conventional weather radars to avoid receiver saturation and obtain the required dynamic range. However, in the case of Doppler weather radars the use of a nonlinear device such as the logarithmic receiver-amplifier causes distortion of the spectrum, which affects the accuracy of spectral moment estimation. Modern Doppler weather radars therefore commonly employ linear receivers and utilise *automatic gain control* (AGC) circuitry to achieve high dynamic ranges. However, since an estimate of the original signal power is necessary for deciding the gain setting of the AGC, the control signal for the AGC circuit itself is often derived from a logarithmic amplifier. The block diagram of a generic Doppler weather radar receiver incorporating this principle is shown in Fig. 6.12.

To estimate the spectral moments accurately, the transmitted signals of Doppler weather radars must be coherent from pulse to pulse. This means that the phase angles associated with successive transmitted pulses must be either the same or controlled in a known and precise manner. The requirement of *pulse-to-pulse phase coherence* makes klystron amplifiers the preferred device for the high-power elements of the transmitters of modern Doppler weather radars.

The antenna is typically a centre-fed paraboloidal reflector type, generating a narrow conical or pencil beam. A fan beam may occasionally be used if the weather function is combined with other functions, such as in the case of airport surveillance radars. A full-circular azimuthal scan capability of the antenna is common, with maximum scan rates of several revolutions per minute. The antenna pedestal and drive mechanisms usually permit scan along the elevation angle as well as varying the elevation tilt for performing the azimuthal scan. Programmable antenna scan cycles for conducting volume scans are also a common feature of modern Doppler weather radars.

The wide variability of the scattering strength of weather phenomena imposes severe limitations on the permissible sidelobe levels of the antenna. The effects of sidelobes become particularly apparent or disturbing while observing regions of high-reflectivity gradients. If the antenna has significant

192   Aviation weather surveillance systems

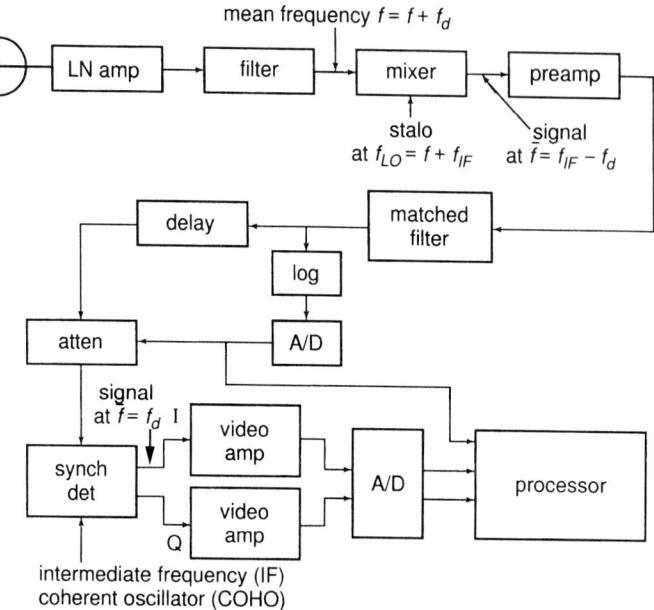

Figure 6.12   *Simplified block diagram of a generic Doppler weather radar (LN Amp: low-noise amplifier, STALO: stable local oscillator, Preamp: preamplifier, Atten: attenuator, Log: logarithmic amplifier, A/D: analog-digital converter, Synch Det: synchronous detector, IF: intermediate frequency, I: in-phase signal, Q: quadrature signal).* (Courtesy D.S. Zrnic', National Severe Storms Laboratory, Norman, Oklahoma)

sidelobes, areas of strong reflectivity viewed through a sidelobe may be comparable to, or even stronger than, the feature in the main lobe at the same range. When the two are comparable, the sidelobe feature will be superimposed on the mainlobe feature, interfering with the observation of the latter (Fig. 6.13a). It may be noted here that all signals received by a radar are assumed to be from targets along its boresight (or main beam) and are so treated by the radar processor and display. When the feature in the sidelobe is much stronger than that observed in the main lobe, the radar processor or observer will assume the former to be in the position of the latter (Fig. 6.13b). The scanning of the radar beam will cause attenuated replicas of strong features to be formed at locations separated by angular intervals equal to the angular separation between the main lobe and the strong sidelobe(s). This effect is shown schematically in Fig. 6.14. The factor of attenuation equals the sidelobe ratio; hence the need to keep the sidelobe ratios low. The requirement of a 'clean' pattern results in a significant tapering of the illumination function of the antenna feed over the reflector.

For fixed-polarisation radars, a horizontal polarisation is more common, but a circular polarisation is employed, at least as an option, in some cases.

Doppler weather radar as a primary aviation weather sensor 193

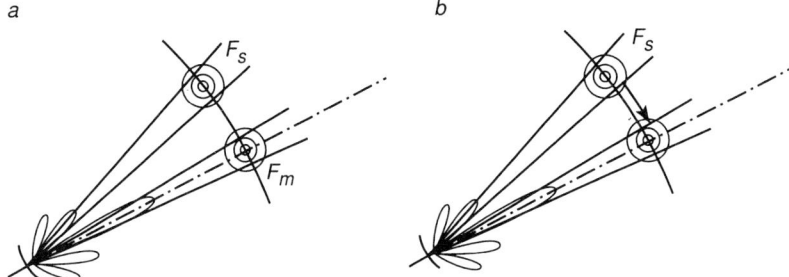

Figure 6.13   Schematic diagram showing effects of antenna pattern sidelobes: (a) the weather feature F_m on the main lobe and the feature F_s seen by a sidelobe cannot be distinguished by the radar, and (b) a feature present in a sidelobe would appear to the radar as being in the main beam

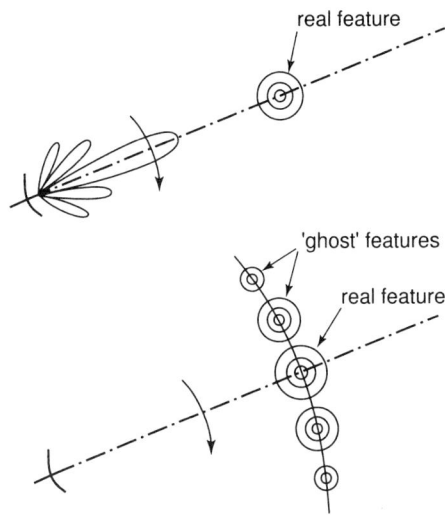

Figure 6.14   A single strong weather feature is displayed as a series of features of graded strengths because of interception by the main lobe and the sidelobes

For more complex Doppler weather radars employing polarisation diversity, special types of feed and related switches must be provided, and the requirements of pattern and sidelobe matching between different polarisation states are stringent. More discussion about polarisation diversity radars follows in Chapter 13.

A major component of modern Doppler weather radars that distinguishes them from conventional weather radars is the sophisticated radar processor that computes the primary data products. In addition to performing the numerical calculations for determining the spectral moments, incorporating operations such as calibration and clutter filtering (see Colour Plate 7), the processor usually does a host of other functions, e.g. cleaning up the data to remove the effects of artefacts such as ambiguity (see Section 7.3), and formatting the data for display and storage. Radar data processing for aviation use may also involve the generation of higher-level products such as estimates of wind shear, detection and extrapolation of gust fronts, etc. Such processing may be performed either at the radar site or at a central location where the basic radar data products are pooled. The computation of spectral moments and the corresponding basic data products is a common function of Doppler weather radars, and the nature of such processing is discussed below.

6.6.3 Computation of basic data products

As introduced in Section 6.6.1 the basic data products of Doppler weather radars are reflectivity, mean radial velocity and Doppler velocity spectrum width, and these are derived from the first three moments of the Doppler spectrum of the echo signal. As the weather spectrum is derived from a large number of independently moving scatterers, it may, to a first approximation, be assumed to be Gaussian in shape. The idealised spectrum shown in Fig. 6.11c has been sketched with this assumption.

In practice, however, very often the weather spectrum would not be so clean. This is especially true for echo signals with high noise contamination, i.e. when the signal-to-noise ratio is low. In such cases, the spectrum would present a jagged appearance. Besides, the spectrum may not be unimodal, i.e. have a single peak as in Fig. 6.11c. For example, more peaks than one may result when a resolution volume straddles air streams with different velocities, separated by a shear layer. The probability of such occurrence is higher at longer ranges from the radar where the size of the resolution volume is larger.

A special case of occurrence of an additional peak in the spectrum, other than that due to atmospheric scatterers, is due to *ground clutter* as mentioned in Section 6.6.1 and shown in Fig. 6.11d. For static radars, the ground clutter is concentrated in the vicinity of the zero-Doppler frequency. If the radar is airborne, the centre frequency would be shifted to a non-zero value determined by the speed of the aircraft and the inclination of the radar beam with respect to the aircraft heading.

Realistic overall spectra for a ground-based radar are shown schematically in Fig. 6.15. Each comprises a weather spectrum that departs from a Gaussian shape, and a peak representing ground clutter. The method employed to estimate Doppler spectral moments in a weather radar must be robust enough to provide accurate estimates for a wide variety of spectrum shapes, and down to as low values of the signal-to-noise ratio as possible. Further, the moments must be computed for each of the large number of resolution volumes in a

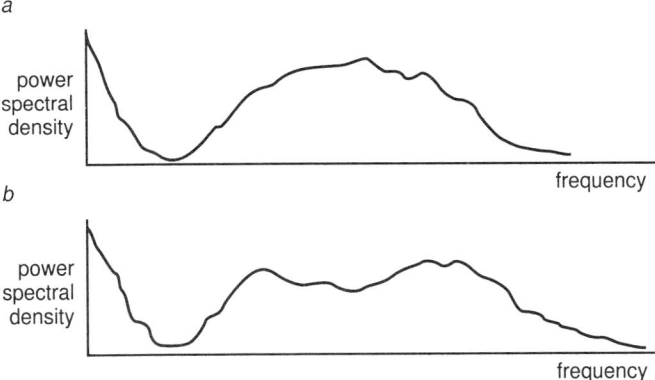

Figure 6.15 Examples of realistic composite weather radar spectra showing (a) unimodal and (b) bimodal weather spectra

scan volume in real time, i.e. within the time required to perform the scan. Considering a realistic example, for a radar with a 1° beamwidth, 1 μs transmitted pulsewidth (and echo signal sampling time), 130 km range of coverage, and 3 rev/min of circular scan rate, the processor may be required to compute ~10^6 sets of spectral moments per minute. The computational load would be proportionately higher for faster scan rates and longer observational ranges which are within the capability of modern Doppler weather radars. The algorithms for spectral moment estimation should therefore be computationally simple and fast.

6.6.3.1 Reflectivity

The determination of the power (i.e. the zeroth moment of the spectrum) of the signal at the receiver output, which is used as a measure of rainfall rate, is straightforward. The power P of each signal sample is obtained by summing the squares of the magnitudes of the in-phase and quadrature signals (I and Q) derived from the detector:

$$P_i = |I_i|^2 + |Q_i|^2 \tag{6.26}$$

where i is the sample number. The samples of the receiver output are randomly varying quantities because of the effects of noise and independent scatterer motion within the resolution volume. It is therefore necessary to average the signal power over a large number of samples to obtain reliable estimates of reflectivity. The actual number depends primarily on the signal-to-noise ratio, Doppler spectrum width, and the detection characteristic of the receiver. The same is true of the other spectral moments as well. If M samples of the receiver output, numbered 0 to $M-1$, are averaged, then the estimated power is

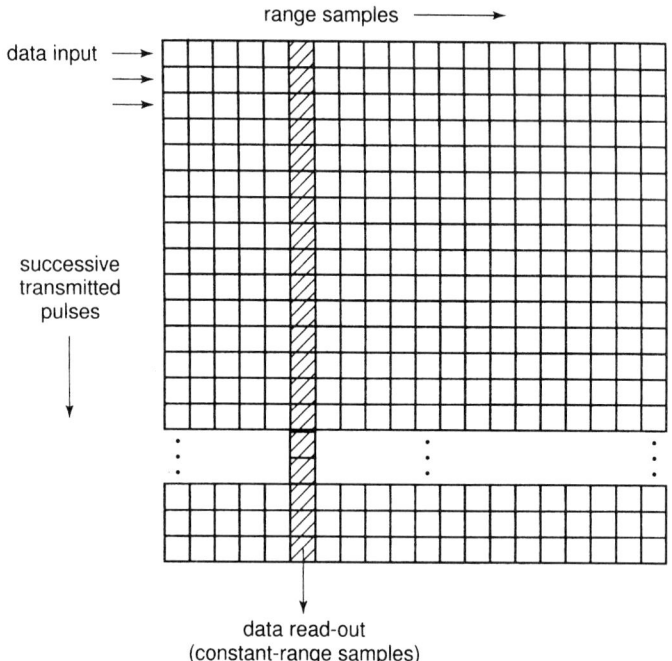

Figure 6.16 Schematic diagram of the organisation of memory matrix for range-Doppler data storage and retrieval. Each row corresponds to the successive range samples obtained from a single transmitted pulse. Data are serially loaded along rows and retrieved along columns for Doppler processing

$$\hat{P} = \frac{1}{M} \sum_{k=0}^{M-1} P_i \qquad (6.27)$$

Note that the samples power-summed in eqn. 6.27 are derived from successive transmitted pulses, and correspond to a given resolution volume which is determined by fixing the sampling instant relative to the transmitted pulse. To obtain such a sample set, the successive range samples of the receiver output from each of M transmitted pulses are stored as a row in a memory matrix, and a column of the matrix corresponding to an assigned range delay is read out. The arrangement is shown in Fig. 6.16. Normally all the columns in the memory matrix are retrieved and processed according to eqn. 6.27, either sequentially or simultaneously. This provides estimates of the echo power in all the resolution volumes along a radial stretch of interest.

The signal appearing at the receiver output is actually a sum of weather echo and noise, and their associated powers are additive. The power estimate (eqn. 6.27) may be assumed to be due to weather only if the noise power is small enough to be neglected. To obtain an accurate estimate of weather echo power, the best estimate of the noise power level at the receiver output must be subtracted from the estimate given by eqn. 6.27. Noise power estimation is a normal part of the calibration process of modern weather radars.

This power estimate, thus obtained for each resolution volume of interest, may be used as the receiver output power P_{ro} in the radar range equation (eqn. 6.15 or 6.16), with appropriate units, to derive the reflectivity factor of the scatterers in the volume, and thence the rain rate.

6.6.3.2 Mean radial velocity

There are many methods to compute the mean Doppler frequency of the scatterers. The spectral parameters of the echo signals may be computed either in the time domain or in the frequency domain. One important member of each class of methods is outlined here.

Time-domain computation: pulse-pair processing: The basis for *pulse-pair processing*, also called *covariance processing*, was laid by Rummler (1968a, 1968b) and Miller and Rochwarger (1972). The method is so called because it derives the Doppler spectral moments from the single-lag autocovariance function of the signal sample set, which involves the multiplication of signal samples or 'pulses' pairwise.

The pulse-pair method utilises the time series formed by $M+1$ successive complex samples x_i of the receiver voltage output corresponding to a given resolution volume, which is determined by fixing the sampling instant relative to the transmitted pulse:

$$x_i = I_i + jQ_i, \ i = 0, 1, \ldots, M \tag{6.28}$$

where $j = \sqrt{-1}$. As discussed in the case of echo power (reflectivity) estimation, the samples x_i are derived from successive transmitted pulses, and hence their spacing equals the pulse repetition time T_r of the radar at the time of sample collection. The first step in the pulse-pair method is to estimate the autocorrelation R of the sample sequence corresponding to a time lag equal to one sample interval, T_r. The estimate is given as

$$\hat{R}(T_r) = \frac{1}{M} \sum_{i=0}^{M-1} x_i^* x_{i+1} \tag{6.29}$$

where * denotes the complex conjugate. The mean Doppler frequency estimate is then obtained from the argument of the complex autocorrelation as

$$\hat{f}_{dm} = \frac{1}{2\pi T_r} \arg\{\hat{R}(T_r)\} \tag{6.30}$$

Finally, the estimated value of the mean radial velocity over the resolution volume is found by substituting the mean Doppler frequency estimate from eqn. 6.30 for f_{dm} in eqn. 6.24.

Although the signal samples have been assumed here to be uniformly spaced, such an assumption is not necessary for spectral parameter estimation by the pulse-pair method. The use of the pulse-pair method for staggered pulse pairs has been dealt with by Zrnic' (1977) and Zrnic' and Mahapatra (1985).

Frequency-domain computation: spectral processing: The first step in the spectral method of moment estimation consists of taking a series of M uniformly spaced complex samples of the receiver output corresponding to a given resolution volume, and forming the *periodogram*

$$\hat{S}_k = \left| \frac{1}{M} \sum_{i=0}^{M-1} x_i e^{-j2\pi ik/M} \right|^2 \tag{6.31}$$

where $\hat{S}_k$ is the estimate of the kth sample of the periodogram. The estimate $\hat{f}_{dm}$ of the mean Doppler frequency shift due to the scatterers in the resolution volume may then be obtained from the relation

$$MT_r \hat{f}_{dm} = \frac{\sum_{k=-M/2}^{M/2} k \hat{S}_k}{\sum_{k=-M/2}^{M/2} \hat{S}_k} \tag{6.32}$$

It is easy to recognise the numerator of the right-hand side of eqn. 6.32 as the first moment of the spectrum about the zero-frequency line (i.e. $k=0$), and the denominator, which is the sum of the spectral components, as the power associated with the spectrum.

The basic relation (eqn. 6.32) becomes erroneous, producing a biased estimate, when a part of the original Doppler spectrum gets displaced due to aliasing (see Section 7.3 on ambiguities). The following improved estimator of mean Doppler frequency minimises the bias (Zrnic', 1979):

$$MT_r \hat{f}_{dm} = k_m + \frac{1}{\hat{P}} \sum_{k_m - (M/2)}^{k_m + (M/2)} (k - k_m) \hat{S}_{\text{mod}_M(k)} \tag{6.33}$$

where $\hat{P}$ is the total power in the periodogram, which is equal to the denominator of eqn. 6.32, and $\text{mod}_M(k)$ represents the remainder when k is divided by M. The spectral moment calculation represented by the second term on the right-hand side of eqn. 6.33 is not about $k=0$ but about an estimated or approximated mean frequency estimate k_m. In the absence of more elaborate algorithms for determining its value, k_m may, for example, be taken as the index of the largest Fourier coefficient in the periodogram. This is because, in a symmetric unimodal spectrum, the mean frequency coincides with the tallest spectral component.

Again, the mean radial velocity may be found from the Doppler spectral mean by using eqn. 6.24.

Comparison between pulse-pair and spectral processing: Pulse-pair or autocovariance processing is generally significantly less computation-intensive than the spectral method, which requires explicit calculation of the Fourier coefficients. The former works best when the echo signal spectrum is symmetric about the mean Doppler frequency, but is robust with respect to noise effects and spectral distortions, i.e. departure of the spectrum relative to a best-fitting Gaussian shape. The spectral method permits straightforward noise subtraction and other spectral operation such as velocity alias detection and dealiasing, and estimation of parameters associated with each mode in a bimodal or multimodal spectrum. On balance, the pulse-pair method has more advantages in normal weather radar applications, and is used more frequently as the processing scheme in such radars.

6.6.3.3 Doppler velocity spectrum width

Like the mean value of the Doppler frequency spectrum, its width can also be calculated either from the autocorrelation coefficient or from the spectral coefficients.

Pulse-pair processing: The calculation of the width of the Doppler spectrum using the pulse-pair method is greatly simplified by assuming the spectrum to be Gaussian in shape, which is a fair assumption for weather spectra under a wide variety of conditions. With this assumption, the width of the radial or Doppler velocity spectrum is calculated as

$$\hat{\sigma}_v = \left(\frac{\lambda}{2\sqrt{2\pi}\,T_r}\right)|L|^{1/2} \qquad (6.34a)$$

where

$$L = \ln\frac{\hat{S}}{|\hat{R}(T_r)|} \qquad (6.34b)$$

and $\hat{S}$ is the estimate of the signal power, obtained by removing the noise power N (assumed known from the radar calibration data) from the estimated receiver output power, which is the sum of signal and noise powers. Thus,

$$\hat{S} = \hat{P} - N \qquad (6.34c)$$

where $\hat{P}$ is given by eqn. 6.27.

Since echo pulses are expected to be progressively less correlated as their time separation increases, the estimated magnitude of the single-lag correlation, $|\hat{R}(T_r)|$, should normally be smaller than the echo power estimate $\hat{S}$ which is the zero-lag autocorrelation. Thus, from eqn. 6.34b, L should be positive and so should the spectrum width $\hat{\sigma}_v$ from eqn. 6.34a. However, in practice, L occasionally does turn out to be a negative number because of noise effects, numerical problems or data corruption, and this would cause problems in computing $\hat{\sigma}_v$ from eqn. 6.34a. To guard against such possibilities, a common form of practical spectrum width estimator is obtained by modifying eqn. 6.34a as follows (Doviak and Zrnic', 1993, p. 136):

$$\hat{\sigma}_v = \left(\frac{\lambda}{2\sqrt{2}\pi T_r} \right) |L|^{1/2} \text{sgn } L \qquad (6.34d)$$

The sign indicator sgn L here is a flag used to tag those data points for which L turns out to be negative. These negative values are then usually treated in one of three ways: (i) set to zero, (ii) eliminated from the data field, or (iii) averaged with positive values to reduce the bias in the estimate. It is worth noting here that sgn L is not used in eqn. 6.34d with its usual mathematical meaning of being +1 when L is positive and −1 when L is negative.

Spectral processing: Using the same notation as in eqns. 6.24 and 6.33, the width of the Doppler velocity spectrum can be estimated as (Doviak and Zrnic', 1993, p. 140)

$$\hat{\sigma}_v^2 = \frac{\lambda^2}{4\hat{P}T_r^2} \sum_{k_m - (M/2)}^{k_m + (M/2)} \left(\frac{k}{M} + \frac{2\hat{v}_{rm}T_r}{\lambda} \right)^2 \hat{S}_{\text{mod}_M(k)} \qquad (6.35)$$

which minimises the bias due to aliasing.

6.6.3.4 Some general aspects of Doppler moment estimation

Pulse-pair and spectral methods constitute two of the most commonly used approaches to the estimation of the spectral moments in modern Doppler weather radars. These are, however, not the only ones available for the purpose. Many other processing methods have been devised which offer special advantages. Among them are several variants of poly-pulse-pair processing, autoregressive processing and periodogram maximisation (Mahapatra and Zrnic', 1983; Sirmans and Bumgarner, 1975; Strauch et al., 1978; Lee, 1978).

An important first step in the processing of weather echo signal time series to extract spectral moments is the choice of the number M of samples used for processing. For simplicity of hardware design and software implementation, the processing is often based on 2^n samples, where the integer index n is chosen depending on the conditions of signal processing and the operating mode of the radar. The choice of the sample number M may be automatically made by the signal processor based on preprogrammed criteria, and many processors allow the number to be set by the operator.

Many factors have a bearing on the processing batch length for spectral parameter estimation. Processing a longer sample set provides more accurate parameter estimation, i.e. reduces the variances of the estimates, and minimises noise effects. It also enhances the efficacy of the clutter rejection algorithms. For this reason, a relatively large number of samples (usually 256 or 128) is usually used for processing signals received from resolution volumes close to the ground, which have a stronger contamination due to ground clutter returns. Long sample trains, however, take a correspondingly long time to be collected, during which the antenna scans over a significant angle, causing blurring and a degradation in angular resolution. To minimise this effect, antenna scan rates are usually reduced at lower elevations and in other situations which require longer sample trains for processing. As we shall see later, the tradeoff between parameter estimation accuracy and angular resolution plays a major role in deciding the scan strategy or scan cycle planning in weather radars.

Modern weather radars, especially those intended for aviation weather surveillance, are required to scan at a fast rate to provide fast updates of rapidly evolving aviation-significant phenomena. At such high scan rates dependence on long sample batches for processing is not practical without seriously compromising angular resolution. For such radars it is necessary to be able to perform accurate spectral parameter estimation based on short sample trains. Mahapatra and Zrnic' (1983) have evolved several practical algorithms for the estimation of mean radial velocities in pulse Doppler weather radars using a small number of samples, and have studied their performance for different levels of signal quality (as given by the signal-to-noise ratio) and varying Doppler spectral width. Their study has shown that reliable estimation of mean velocities is possible by using as low as 16 samples through a proper choice of the processing algorithm.

6.6.4 Display of basic products

As in the case of conventional weather radars, to obtain area coverage of the geographical sector under weather surveillance, the plan position indicator (PPI) is the most commonly used display in modern weather radars. However, there are profound differences between the PPI displays of conventional and modern radars. An example of a PPI display of a basic weather parameter is shown in the top frame of Colour Plate 6.

The most significant difference arises from the fact that the modern Doppler weather radar generates and displays multiple parameters for each resolution volume, while the conventional radar displays only the reflectivity. Thus while the conventional weather radar has only one PPI display, the Doppler radar typically needs three PPI scopes to display the basic products or spectral moments (reflectivity, radial velocity and Doppler velocity spectrum width) discussed in Sections 6.6.1 and 6.6.3. Pictorial views of such displays appear later in this book in the context of specific phenomena displayed, e.g. Colour Plate 18 (bottom frame). Some simpler installations use only two PPIs, showing the reflectivity and radial velocity distribution, while more sophisticated radars may display more fields than the basic three.

Another major characteristic of modern Doppler weather radars that influences the nature of displays is that the displayed parameters are the product of significant amounts of computation. The displays are therefore naturally of a computer-compatible type, in contrast with the circular-faced cathode ray tube displays of older generations of conventional radars with centre-originating radial traces. Computer-compatibility permits a wide variety of commercial and special-purpose display devices to be used, depending on the desired size, resolution, reliability and cost considerations. Computer control also provides a high level of flexibility in the display format. Typical facilities provided on modern weather radar displays include zooming, choice of full-circle or sector displays, centre relocation on the screen, and colour scheme changes including a grey-scale display.

Modern weather radar displays are generally in full colour to efficiently convey the most possible information regarding the phenomenon being displayed. This is especially important in aviation weather surveillance systems where the time available for data assimilation and inference generation is very short.

In parametric displays such as those showing the three basic Doppler weather radar data products and a variety of higher-level products, the parameter being displayed is quantised into a number of levels, usually a power of two, and a uniform colour is assigned to each level. The parametric displays are therefore essentially colour-coded contour plots, but by choosing a sufficiently high number of quantisation levels and assigning incrementally varying shades of colour to neighbouring quantisation levels, an effect approximating a continuous picture of the parameter variation may be created. To make instinctive sense, higher values of a given parameter, say reflectivity, are often assigned increasingly darker shades while areas with low values of the parameter are depicted in light shades. However, threshold values of the parameter which correspond to caution, warning or danger levels, or which must be otherwise emphasised, may be marked by abrupt changes in the assigned shade or intensity. The number of quantisation levels and the display colour scheme for each parameter are very often kept operator-selectable (with suitable default options and built-in schemes for fast set-up) to maximise the visual impact of the display.

Aviation weather surveillance systems may also utilise nonparametric displays for specific purposes. Typically these are used to depict complex scenes with multiple phenomena superimposed on geographic features, navigational facilities and other landmarks. The representation of features on such displays is usually in the form of symbols and icons, with numerical values displayed alongside where appropriate.

While the PPI is the most commonly used display format for weather radars, it does not provide a clear picture of the vertical structure of the phenomena being displayed. To display such a picture one makes use of the *range-height indicator* (RHI) mode of display. In this mode the radar antenna is made to scan over a sector in a vertical plane while being held at a constant azimuth angle. Each beam position of the radar is represented by a radial from the point corresponding to the radar position on the display surface, and the elevation angle of the radial line is kept proportional to the actual beam elevation. The returns from the various resolution volumes along each beam position are intensity-modulated or colour-modulated on the radial. The result is a map of the vertical-plane distribution of the parameter being displayed. Examples of RHI display may be found in Colour Plates 6 (bottom frame) and 11.

Another useful display mode is the *velocity-azimuth display* (VAD) which shows the Doppler or radial wind components as the radar beam performs a conical scan about a vertical axis. In a way, the VAD scan is the counterpart of the RHI scan, because the former involves varying the beam azimuth angle at a constant elevation while the latter consists of scanning the beam in elevation at a constant azimuth. A layer of uniform horizontal wind would cause a sinusoidal variation of the radial component of the wind over the corresponding range interval as the beam completes one azimuthal scan. By performing a harmonic analysis of the Doppler variation over the scan cycle corresponding to each range it is possible to obtain the uniform horizontal wind component as a function of height, and continuation of this process in time yields the evolution of this vertical wind profile. The two-dimensional depiction of the height–time variation of horizontal winds is a more common mode of display in the VAD. A sample VAD output is shown later in Colour Plate 4.

The display types discussed in this section pertain to the basic products of Doppler weather radars. These are usually the most universally displayed products and contain the most basic information from which human operators can draw a variety of inferences. However, as already mentioned, modern aviation weather surveillance systems are designed so as to maximise the speed of response by minimising human involvement. Even nonaviation applications today utilise automated data processing to a large extent because of the economic availability of complex computing hardware and sophisticated and reliable software to perform such processing. Modern Doppler weather radars thus generate a host of higher-level data products with specific utility in different fields of application. Such data products have their own formats for efficient display. Some of these higher-level products and their display formats are discussed in the following chapter.

204 Aviation weather surveillance systems

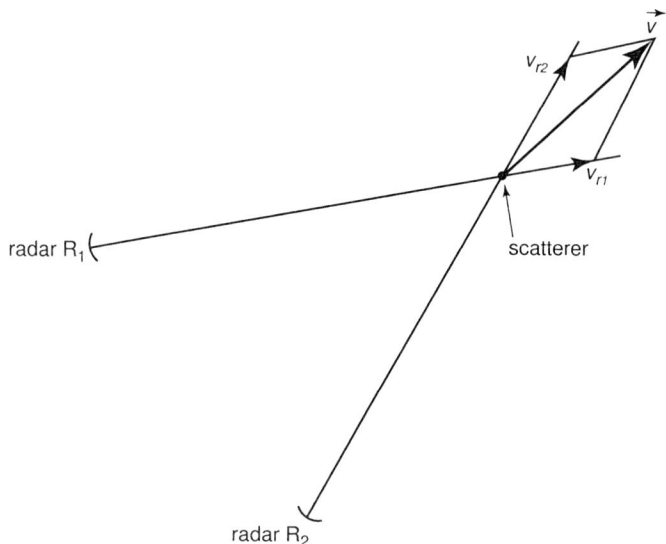

Figure 6.17 Geometry of dual-Doppler-radar observation of a scatterer to determine its velocity vector

6.6.5 Derivation of vector wind fields

A Doppler radar inherently senses only the radial component of wind relative to itself. The component of wind (i.e. scatterer motion) which is perpendicular to the radial from the radar in the direction of the scatterer causes no Doppler shift, and is therefore not sensed by the radar. This is a distinct limitation of the Doppler radar, which presents an incomplete picture of complex wind fields.

In many applications such an incomplete picture is adequate. Trained human operators or specially written software can usually recognise signatures of canonical wind field patterns or phenomena even in a radial velocity field as mapped by a single Doppler radar (see Chapter 10). However, to obtain a more complete picture of wind fields and for better identification and parameter estimation of atmospheric phenomena, it is often necessary to map the vector winds in designated regions of the atmosphere.

The velocity vector of a given scatterer in two dimensions (e.g. on a horizontal plane) can be determined if two noncollinear components of its velocity can be measured independently. This is possible by observing the scatterer simultaneously from two different aspects with two spatially separated Doppler radars, each yielding the velocity component of the scatterer in its own direction. The geometry for such an observation is shown in Fig. 6.17. To reconstruct the velocity vector most accurately, the radials from the scatterer to the two radars (i.e. the lines of sight) should intersect at right angles. However, good results may still be obtained for intersection

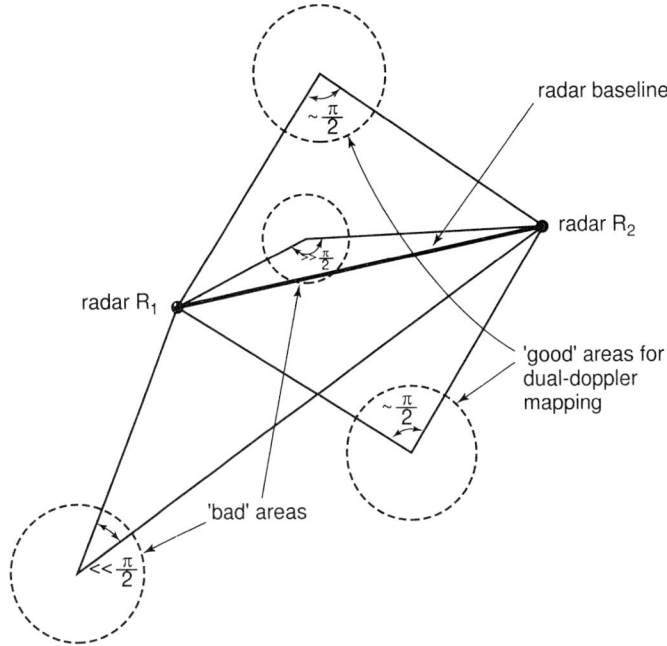

Figure 6.18 Radar-relative location of suitable ('good') areas and examples of unsuitable ('bad') areas for accurate wind velocity mapping by dual-Doppler observation

angles somewhat away from $\pi/2$. Thus the area around the point of normal intersection of the radar radials is well suited for *dual-Doppler mapping* (Fig. 6.18). Areas where the angle between the radials is too acute or too obtuse are ill suited for this purpose, yielding less accurate velocity vectors.

A dual-Doppler configuration yields the velocity vector only in the plane containing the radars and the point of observation. To obtain the three-dimensional wind field, the principle can be extended to observation by three radars, but the catch is that these radars must be noncoplanar with respect to the point of observation. In fact, analogous to the dual-Doppler case, the three lines of sight to the scatterer must be nearly orthogonal mutually. Such a geometry is nearly impossible to set up using earth-based radars.

Owing to the impracticality of triple-Doppler mapping, three-dimensional wind mapping is usually performed by using a dual-Doppler observation to obtain the horizontal wind components, and utilising the continuity property of fluid flows to compute the vertical component. The continuity equation for a compressible fluid like air is somewhat complex, but if the density variation due to the flow itself is neglected (i.e. air is assumed to be locally incompressible), it takes the simple form

$$\nabla \cdot (\rho_a \mathbf{v}) = 0 \qquad (6.36)$$

where $\mathbf{v}$ is the three-dimensional wind vector and ρ_a is the density of the ambient air, which depends essentially on the altitude.

If the two dual-Doppler-observed gradients of wind are fed into the continuity equation, then it would yield the gradient in the third dimension, the vertical. This gradient can be integrated from the ground upwards to obtain the vertical wind component at the required altitude. The boundary condition to start the integration is that the vertical velocity of air immediately above the ground is zero. If the heights involved are large so that the lines of sight from the radars to the point of observation depart significantly from the horizontal, then a cylindrical co-ordinate system may be employed to perform the computations (Doviak and Zrnic', 1993, Section 9.2). Also, for such tilted beams, the Doppler component due to the fall velocity of hydrometeors (see Fig. 6.10) must be removed before processing the data further.

An extended area under dual Doppler coverage would consist of many pixels (resolution volumes). The reconstruction of the velocity vector at each pixel would yield a complete picture of the true velocity field over the entire area. However, this process is not quite so straightforward. This is because the size and orientation of the pixels corresponding to each radar vary with their range and azimuth from the radar. Thus the data grid density corresponding to each radar varies over the mapping area, and at each point in the area it differs between the two radars. To combine the radial velocity data from the two radars to generate a common vector velocity map, it is therefore necessary to reduce the data from each radar to a common grid structure. This involves a stage of data preprocessing consisting of interpolation in two dimensions.

It is not always possible to obtain dual-Doppler mapping of all phenomena of interest because of the restrictions on what constitutes a 'good' area for such mapping (Fig. 6.18). On the other hand, data from single weather radars are available over more extensive areas. There are certain methods available for wind vector mapping based on single-Doppler data. A simple method is to divide the instantaneous data field (e.g. reflectivity or Doppler velocity) over a mapped area into smaller patches and track the movement of each patch over subsequent scans of the radar through a process of correlation (e.g. Mahapatra and Zrnic', 1984). The assumption here is that the data patches being tracked remain internally stable at least over a few successive scans, and are only spatially translated by the local wind vector. Each small data patch therefore essentially acts as a 'tracer' of the local wind. Obviously, since the data patches would comprise many pixels (larger data patches would yield better correlation), the resolution of the velocity field obtained through the correlation process would be much poorer than the basic pixel size of the radar measurements. Note here that since the tracer field can consist of reflectivity data, correlation-based wind retrieval is possible, in principle, even with conventional (non-Doppler) radars. However, as modern Doppler weather radars provide even reflectivity data more accurately and with greater resolution than the conventional radars of the earlier generation, they are the more preferred sensors for such wind retrieval even when their velocity data may not be explicitly used in the process.

More recently, an adjoint method has been used to retrieve two-dimensional wind fields from single Doppler radar data (Qiu and Xu, 1992; Xu et al., 1993). This method also utilises a tracer field and obtains the best estimate of the velocity vector that minimises a defined cost function over a few successive scans (optimally four). Xu et al. (1993) have utilised this technique to realistically map microburst wind fields from single-Doppler data (Fig. 6.19). Another method proposed to derive wind fields from single Doppler radar data is the Tangential Velocity Assumed Display (TVAD) (Takahashi et al., 1991).

It is also possible to get vector wind fields from a single Doppler radar if it can be made to move during the data collection process and 'look' at each resolution volume from two different aspects. Such a 'pseudo-dual-Doppler method' of data collection may be performed with airborne Doppler radars. The dual-aspect viewing of individual resolution volumes can be achieved by flying the radar along paths with near-orthogonal segments, e.g. L-shaped (Fig. 6.20) or U-shaped trajectories (e.g. Ray et al., 1985). The antenna beam is usually scanned in a plane perpendicular to the flight direction. A more sophisticated implementation of the moving-Doppler-radar technique is the Fore/Aft Scanning Technique (FAST) (Jorgensen and DuGranrut, 1991) wherein the airborne radar beam alternately scans along symmetric forward- and backward-looking cones with their common axis oriented along the aircraft heading (Fig. 6.21a). If a limited region is to be covered in a pseudo-dual-Doppler mode, then one may first scan the region with only the forward antenna orientation, and then scan the same region with only the backward antenna orientation as the aircraft recedes. It is also possible to employ two separate radars with their antennas mounted back-to-back such that their beams scan along the two cones simultaneously. As the aircraft carrying the radar flies along a straight line path, each point is 'seen' twice, once by each of the two beams, from two different directions (Fig. 6.21b). This provides the two Doppler projections necessary to retrieve the wind velocity vector at each point.

The success of the pseudo-dual-Doppler method is based on the assumption that the velocity vector at each point being observed does not change significantly during the time the aircraft takes to fly between the positions from which the radar takes the two 'looks' at the point. The shorter this time, the more accurate would be the velocity estimates provided by the method. The FAST technique has the advantage that the pseudo-dual-Doppler data can be obtained from a single straight-line flight, unlike the L- or U-shaped flight patterns necessary for single-beam radars. This cuts down the observation time significantly besides simplifying the flight and navigation procedures, and helping gather data for a much larger swath from a single flight.

Three-dimensional vector wind fields may be obtained by the pseudo-dual-Doppler method by using the continuity equation (eqn. 6.36) in a manner

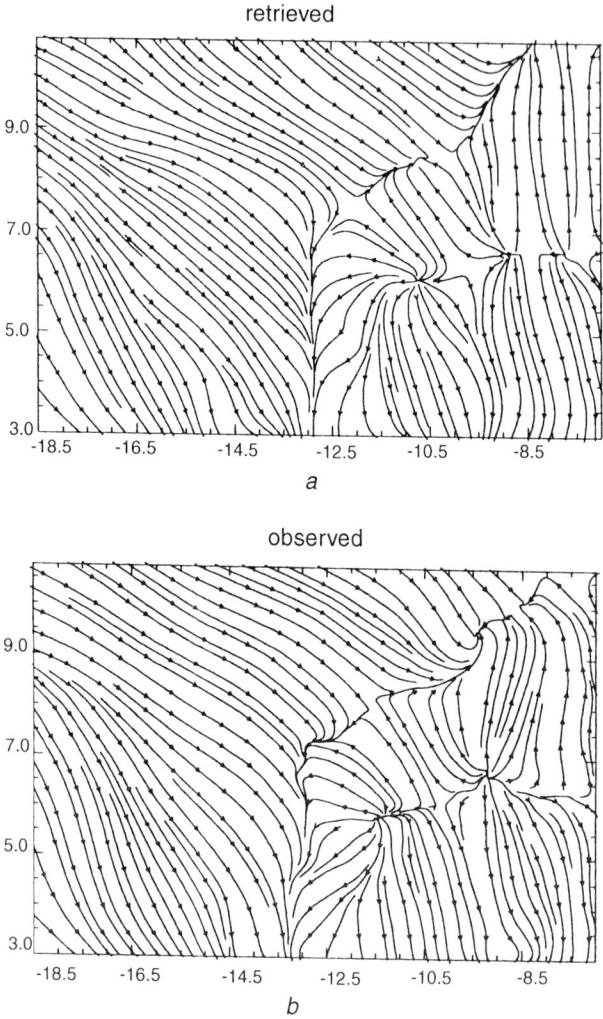

Figure 6.19   Example of wind field in a microburst outflow retrieved by an Adjoint Method (a), which compares well with the true winds as determined by dual-Doppler observation (b) (from Xu et al., 1993, by author's permission)

similar to the true dual-Doppler method. An example of the horizontal and vertical velocity fields obtained by using the technique is shown in Fig. 6.22.

6.7 Summary

In this chapter the modern Doppler weather radar has been introduced as a primary sensor to generate and feed data into the aviation weather

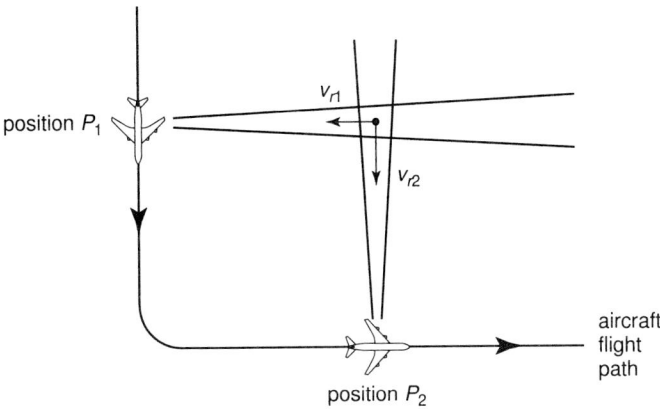

Figure 6.20 Pseudo-dual-Doppler observation of wind field from L-shaped flight path. The Doppler radar determines the two radial velocity components v_{r1} and v_{r2} at a given point from the two positions P_1 and P_2, respectively

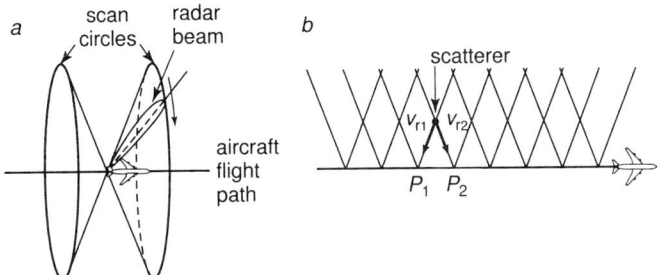

Figure 6.21 The pseudo-dual-Doppler scheme of vector wind retrieval: (a) beam scanning cones, and (b) observation of a given point by 'fore' (bold lines) and 'aft' (thin lines) beams from points P_1 and P_2 along the straight-line flight path

surveillance systems being designed and installed for current and future aviation systems. The modern Doppler weather radar provides high-resolution three-dimensional coverage of a large volume of space, and provides an accurate picture of the distribution of such parameters as rainfall intensity, wind velocity and turbulence within the observation volume. These parameters are intrinsically of high significance from the point of view of aviation safety and efficiency, and the pattern of their distribution helps in the recognition of specific atmospheric features and the estimation of their hazard potential. Modern aviation weather surveillance systems perform many of these functions in an automated manner, for which the primary data

210 *Aviation weather surveillance systems*

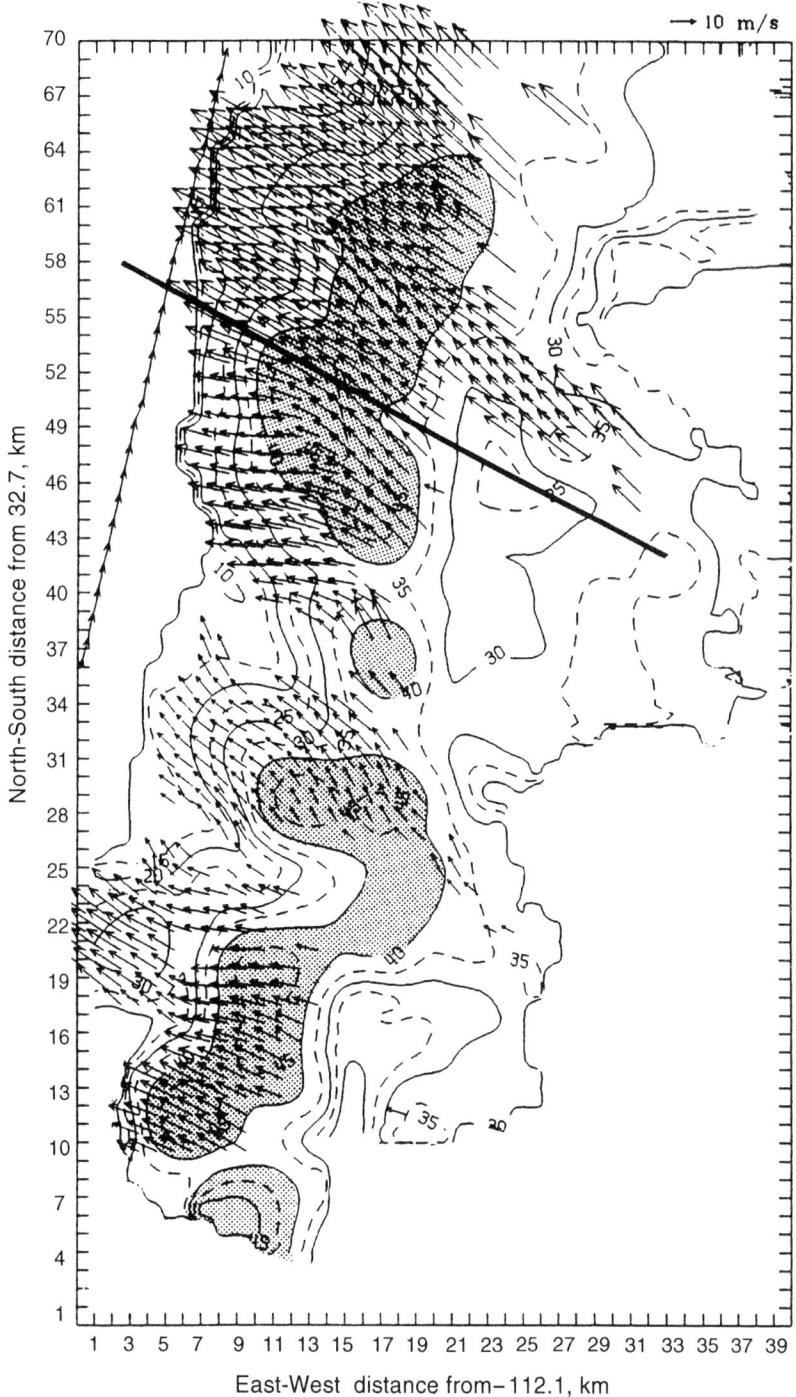

a

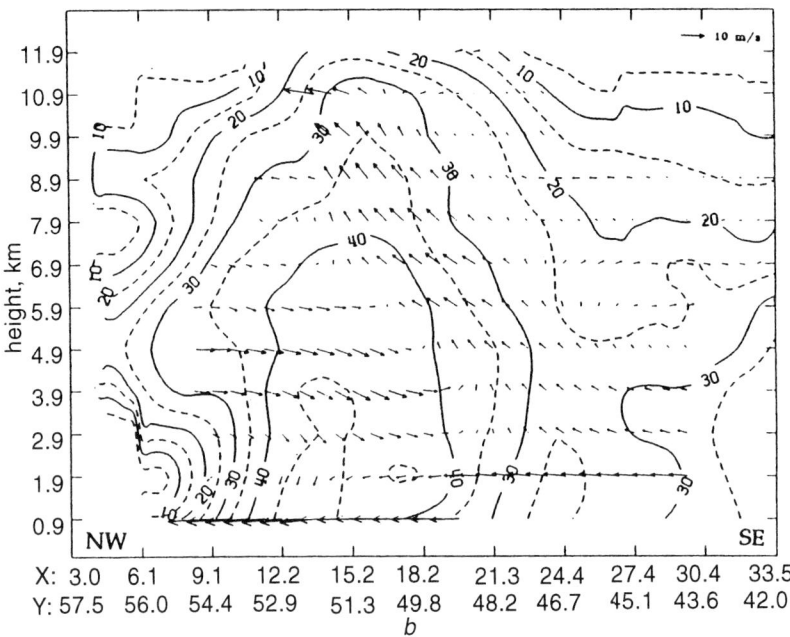

Figure 6.22 Example of a pseudo-dual-Doppler measurement of wind field (in a mesoscale convective system): (a) horizontal winds at a height of 900 m above sea level, and (b) winds in the vertical plane shown as a heavy line in (a). All winds are system-relative, i.e. the translational motion of the storm system as a whole has been subtracted from the wind velocities in these plots. The wind velocity scale for each plot is shown on the upper right. The contours in both the Figures show reflectivity levels as marked (from Bartels et al., 1991, by author's permission)

from the Doppler weather radars serve as the basic input. The nature and features of such automated data processing for the generation of higher-level weather products leading to operational decisions, alerts and warnings are discussed in a subsequent chapter.

In addition to high resolution and wide coverage, a major advantage of the weather radar in the context of aviation weather surveillance lies in a matching of the resolution requirement of the surveillance system with the inherent resolution capability of radars. This is especially true of terminal area surveillance, which constitutes the most demanding segment of the aviation weather surveillance process. The accuracy of weather observation, measured in terms of resolution, is normally required to be the highest at the runway complex and is relaxed progressively at farther distances from the complex. Radars inherently provide higher resolution at closer ranges than at longer distances, making themselves well suited for data collection for aviation applications. Modern Doppler weather radars, while preserving this basic advantage, also yield accurate multiparameter observation of the surveillance volume. This makes them ideal sensors and data sources for aviation weather surveillance system operation.

6.8 References

ATLAS, D. (1964): 'Advances in radar meteorology', *Adv. Geophys.*, **10**, pp. 317–748
ATLAS, D., and KESSLER, E. (1957): 'A model atmosphere for widespread precipitation', *Aeronaut. Eng. Rev.*, **16**, pp. 69–75
ATLAS, D., and ULBRICH, C.W. (1977): 'Path- and area-integrated rainfall measurement by microwave attenuation in the 1–3 cm band', *J. Appl. Meteorol.*, **16**, pp. 1322–1331
BARTELS, D.L., JORGENSEN, D.P., and SMULL, B.F. (1991): 'Airborne Doppler analysis of a mesoscale convective system observed during the SW area monsoon project'. Preprints of 25th International Conference on Radar Meteorology, Paris, France, 24–28 June 1991 (American Meteorological Society, Boston), pp. 486–489
BATTAN, L.J. (1973): 'Radar observation of the atmosphere' (University of Chicago Press, IL)
BLAKE, L.V. (1990): 'Prediction of radar range', in M.I. Skolnik (Ed. in Chief): 'Radar handbook' (McGraw-Hill, New York, 2nd edn.), Chap. 2
BLANCHARD, D.C. (1953): 'Raindrop size distribution in Hawaiian Rains', *J. Meteorol.*, **10**, pp. 457–473
BURROWS, D.R., and ATTWOOD, S.S. (1949): 'Radio wave propagation' (Academic Press, New York)
CAIN, D.E., and SMITH, P. L., Jr. (1976): 'Operational adjustment of radar estimated rainfall with rain gage data: A statistical evaluation'. Preprints, 17th Conference on Radar Meteorology (American Meteorological Society, Boston), pp. 533–538
CHADWICK, R.B., and MORAN, K.P. (1980): 'Long term measurements of C_n^2 in the boundary layer', *Rad. Sci.*, **15**, pp. 355–362
COLLIER, C.G. (1989): 'Applications of weather radar systems' (Ellis Horwood, Chichester)
DOVIAK, R.J., and ZRNIC', D.S. (1988): 'The Doppler weather radar', in E. Brookner, Ed., 'Aspects of Modern Radar', (Artech House , Boston, MA), Chap. 9
DOVIAK, R.J., and ZRNIC', D.S. (1993): 'Doppler radar and weather observations' (Academic Press, San Diego, 2nd edn.)
GOSSARD, E.E., and STRAUCH, R.G. (1983): 'Radar observation of clear air and clouds' (Elsevier, Amsterdam)
GUNN, K.L.S., and MARSHALL, J.S. (1958): 'The distribution with size of aggregate snowflakes', *J. Meteorol.*, **15**, pp. 452–466
JONES, D.M.A. (1955): '3 cm and 10 cm wavelength radiation backscatter from rain'. Proc. 5th Weather Radar Conf. (American Meteorological Society, Boston), pp. 281–285
JOSS, J., and WALDVOGEL, A. (1970): 'A method to improve the accuracy of radar-measured amounts of precipitation'. Preprints, 14th Conference on Radar Meteorology (American Meteorological Society, Boston, MA), pp. 237–238
JORGENSEN, D.P., and DUGRANRUT, J.D. (1991): 'A dual-beam technique for deriving wind fields from airborne Doppler radar'. Preprints of 25th International Conference on Radar Meteorology, Paris, France, 24–28 June 1991 (American Meteorological Society, Boston, MA), pp. 458–461
KNOTT, E.F. (1990): 'Radar cross section', in M.I. Skolnik (Ed. in Chief): 'Radar handbook' (McGraw-Hill, New York, 2nd edn.), Chap. 11
LAWS, J.O., and PARSONS, D.A. (1943): 'The relationship of raindrop size to intensity', *Trans. Am. Geophys. Union*, **24**, pp. 452–460
LEE, R.W. (1978): 'Performance of the poly-pulse-pair Doppler estimator', Lassen Research Memo 78-03
MAHAPATRA, P.R., and ZRNIC', D.S. (1983): 'Practical algorithms for mean velocity estimation in pulse Doppler weather radars using a small number of samples', *IEEE Trans. Geosci. Remote Sens.*, **GE-21**, pp. 491–501

MAHAPATRA, P.R., and ZRNIC', D.S., (1984): 'A physical basis for NEXRAD data update rates,' *J. Aircr.*, **21**, pp. 840–850

MARSHALL, J.S., HITSCHFELD, W., and GUNN, K.L.S. (1955): 'Advances in radar weather', *Adv. Geophys.*, **2**, pp. 1–56

MARSHALL, J.S., and PALMER, W.McK. (1948): 'The distribution of raindrops with size', *J. Meteorol.*, **5**, pp. 165–166

MILLER, K.S., and ROCHWARGER, M.M. (1972): 'A covariance approach to spectral moment estimation', *IEEE Trans. Inf. Theory*, **IT-18**, pp. 588–596

QIU, C.J., and XU, Q. (1992): 'A simple adjoint method of wind analysis for single-Doppler data', *J. Atmos. Ocean. Technol.*, **9**, pp. 588–598

RAY, P.S., JORGENSEN, D.P., and WANG, S.-L. (1985): 'Airborne Doppler radar observations of a convective storm', *J. Clim. Appl. Meteorol.*, **24**, pp. 688–698

RINEHART, R.E. (1991): 'Radar for meteorologists'. Available from the author, PO Box 6124, Grand Forks, ND 58206-6124, USA (2nd edn.)

RUMMLER, W.D. (1968a): 'Introduction of a new estimator for velocity spectral parameters'. Bell Telephone Labs., Whippany, NJ, Tech. Memo. MM-68-4121-5

RUMMLER, W.D. (1968b): 'Two-pulse spectral measurements'. Bell Telephone Labs., Whippany, NJ, Tech. Memo. MM-68-4121-15

RYDE, J.W. (1946): 'The attenuation and radar echoes produced at centimetre wavelengths by various meteorological phenomena' *in* 'Meteorological factors in radio wave propagation', Report of a conference held 8 April 1946 by The Phyical Society and Royal Meteorological Society. (Phyical Society, London)

SAXTON, J.A., and HOPKINS, H.G. (1951): 'Some adverse influences of meteorological factors on marine navigational radar', *Proc. IEE*, **98**, P. III, p. 26

SEKHON, R.S., and SRIVASTAVA, R.C. (1971): 'Doppler radar observations of drop size distributions in a thunderstorm', *J. Atmos. Sci.*, **28**, pp. 983–994

SERAFIN, R.J. (1990): 'Meteorological radar', in M.I. Skolnik (Ed. in Chief): 'Radar handbook' (McGraw-Hill, New York, 2nd edn.), Chap. 23

SHRADER, W.W., and WEIL, T.A. (1987): 'Lens-effect loss for distributed targets', *IEEE Trans. Aerosp. Electron. Syst.*, **AES-23**, pp. 594–595

SIRMANS, D., and BUMGARNER, B. (1975): 'Numerical comparison of five mean frequency estimators', *J. Appl. Meteorol.*, **14**, pp. 991–1003

SKOLNIK, M.I. (Ed. in Chief) (1990a): 'Radar handbook' (McGraw-Hill, New York, 2nd edn.)

SKOLNIK, M.I. (1990b): 'An introduction to radar', in M.I. Skolnik (Ed. in Chief): 'Radar Handbook' (McGraw-Hill, New York, 2nd edn.), Chap. 1

SRIVASTAVA, R.C. (1971): 'Size distribution of raindrops generated by their breakup and coalescence', *J. Atmos. Sci.*, **28**, pp. 410–415

STRAUCH, R.G., KROPFLI, R.A., SWEEZY, W.B., MONINGER, W.R., and LEE, R.W. (1978): 'Improved Doppler velocity estimates by the poly-pulse-pair method'. Preprints, 18th Conference on Radar Meteorology, American Meteorological Society, Boston

TAKAHASHI, N., UYEDA, H., KIKUCHI, K., and OKAZAKI, M. (1991): 'A method to describe the fluctuation and discontinuity of horizontal wind fields by a single Doppler radar'. Preprints of 25th International Conference on Radar Meteorology, Paris, France, 24–28 June 1991 (American Meteorological Society, Boston), pp. 642–645

TAYLOR, J.W., JR. (1990): 'Receivers', in M.I. Skolnik (Ed. in Chief): 'Radar handbook' (McGraw-Hill, New York, 2nd edn.), Chap. 3

VAUGHN, C.R. (1985): 'Birds and insects as radar targets: a review', *Proc. IEEE*, **73**, pp. 205–227

WEIL, T.A. (1973): 'Atmospheric lens effect: Another loss for the radar range equation', *IEEE Trans. Aerosp. Electron. Syst.*, **AES-9**, pp. 51–54

WOODLEY, W.L., OLSEN, A.R., HERNDON, A., and WIGGERT, V. (1975): 'Comparison of gage and radar methods of convective rain measurement', *J. Appl. Meteorol.*, **14**, pp. 909–928

XU, Q., QIU, C.-J., YU, J.-X., GU, H.-D., and WOLFSON, M. (1993): 'Adjoint-method retrievals of microburst winds from TDWR data'. Preprints of 26th International Conference on Radar Meteorology, Norman, OK, 24–28 May 1993 (American Meteorological Society, Boston), pp. 433–434

ZRNIC', D.S. (1977): 'Spectral moment estimates from correlated pulse pairs', *IEEE Trans. Aerosp. Electron. Syst.*', **AES-13**, pp. 344–354

ZRNIC', D.S. (1979): 'Estimation of spectral moments for weather echoes', *IEEE Trans. Geosci. Electron.*, **GE-17**, pp. 113–128

ZRNIC', D.S., and MAHAPATRA, P.R. (1985): 'Two methods of ambiguity resolution in pulse Doppler weather radars', *IEEE Trans. Aerosp. Electron. Syst.*, **AES-21**, pp. 470–483

Aviation weather surveillance systems 215

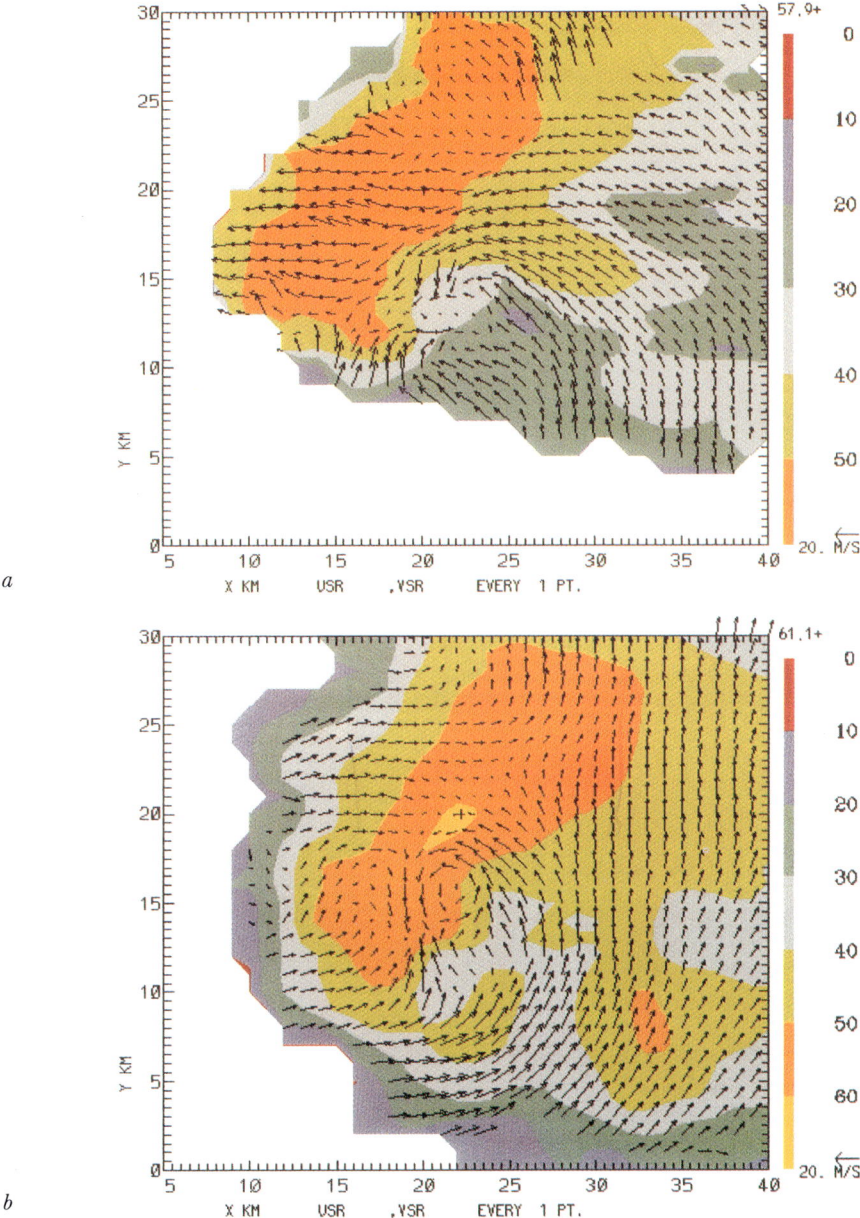

a

b

Colour Plate 1 Cross-sections through a tornado-producing supercell storm showing wind fields and precipitation activity at four different height levels: (a) 0.5 km, (b) 3.5 km, (c) 7.5 km and (d) 13.5 km. The wind pattern is shown by arrows according to the scale shown at the bottom right of each frame, and the precipitation is indicated by its radar reflectivity in dBZ, colour-contoured as per the ribbon scale on the right. Both reflectivity and winds are sensed by a 3 cm airborne Doppler radar flying at an altitude of 5 km. A fore/aft scanning technique (FAST) (see Section 6.6.5) is used

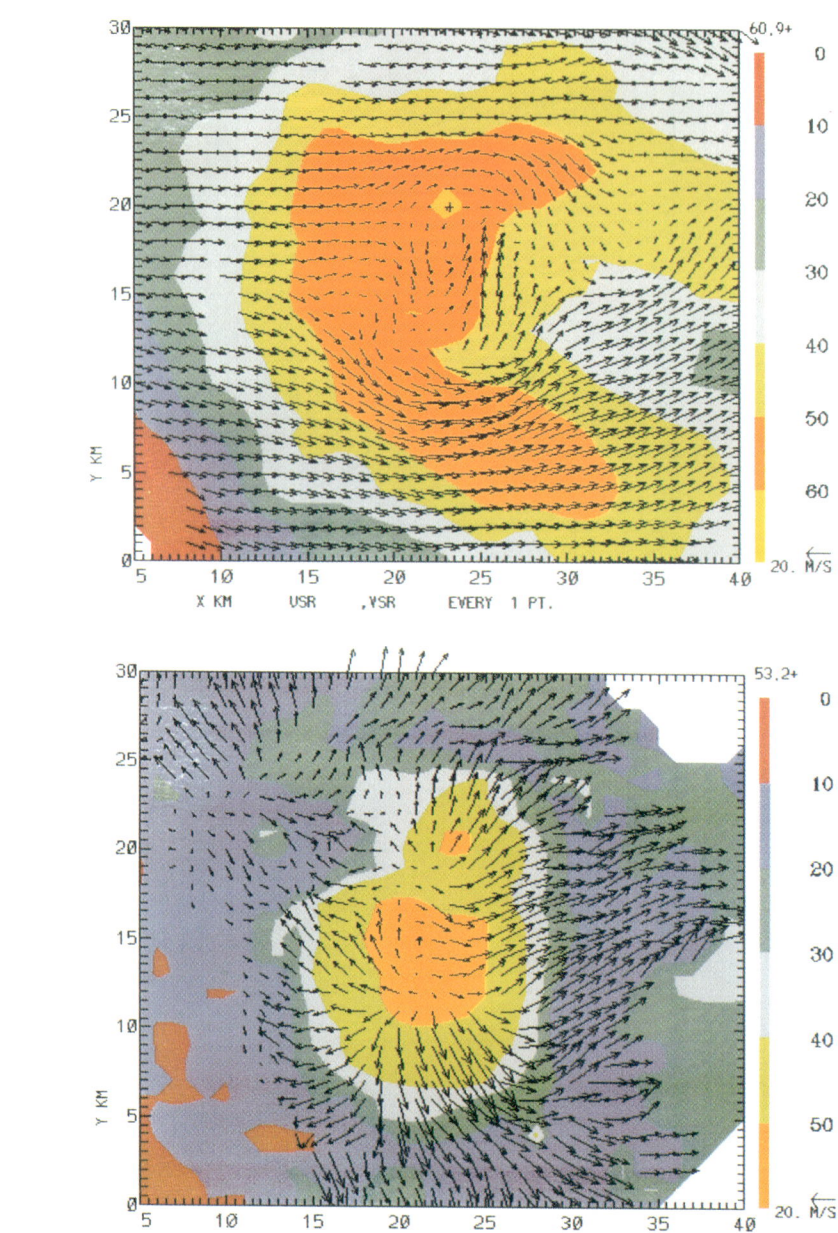

c

d

Colour Plate 1 cont. for data collection, and vector winds are synthesised using pseudo-dual-Doppler processing (see Section 6.6.5). The pronounced rotational motion of the air within the storm is apparent in the three lower levels (a,b,c), and the strong storm-top divergence is clearly marked at the highest level (*d*). (Reference: Dowel, D.C., Bluestein, H.B. and Jorgensen, D.P.: 'Airborne Doppler radar analysis of supercells during COPS-91', *Month. Weather Rev.*, 1997, **125**, pp. 365–383. Courtesy D.C. Dowell, University of Oklahoma)

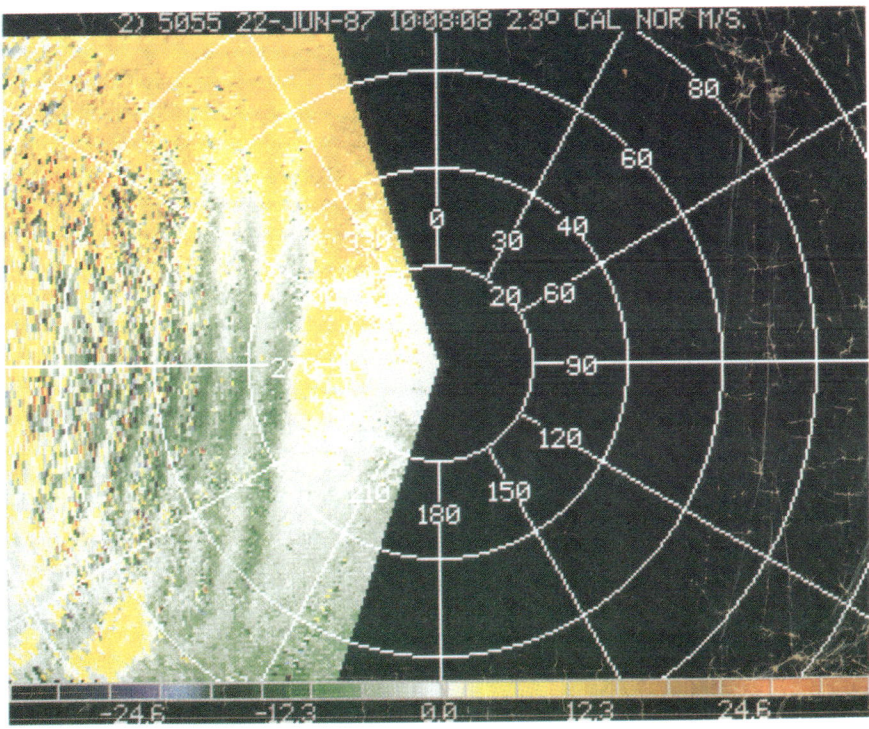

Colour Plate 2 Radial velocity field of an undular bore phenomenon in central Oklahoma, observed with a 10 cm Doppler radar. The elevation angle is 2.3° and the range circles are in kilometres. As many as eight cycles of undulation are visible. (Courtesy R.J. Doviak, NOAA National Severe Storms Laboratory, Norman, Oklahoma)

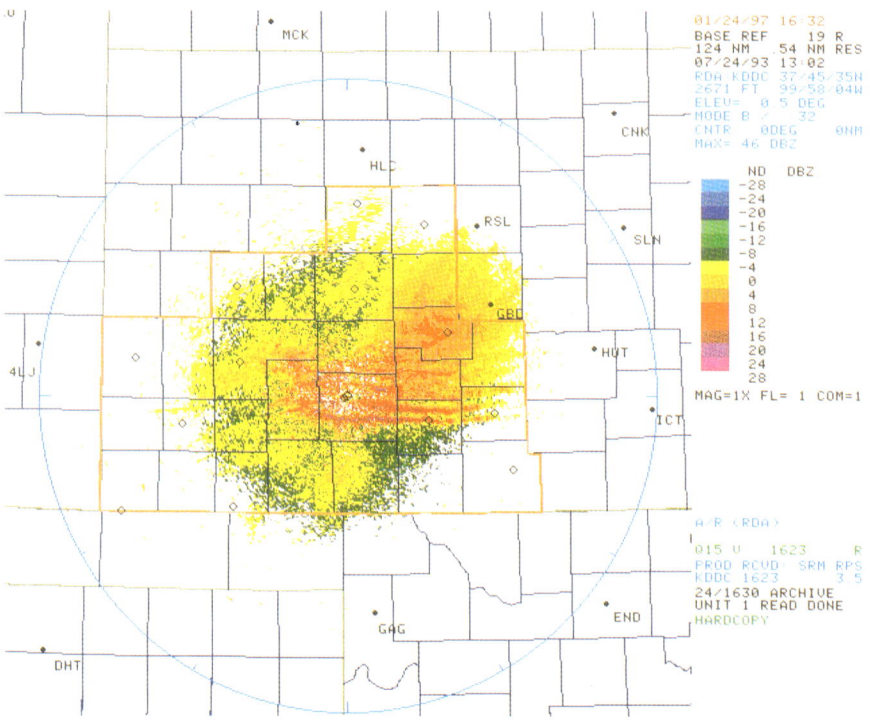

Colour Plate 3 Composite reflectivity map from a WSR-88D radar, showing parallel bands which indicate Kelvin-Helmholtz instabilities. The waves have been caused by a strong storm, but are occurring in a clear-air environment. The wind shear and turbulence associated with such waves (see Colour Plate 4) are a potent source of hazard for aviation. (Reference: Johnson, J.: 'Identification of a clear-air roll turbulence and low level wind shear event by the WSR-88D radar'. Preprints of 6th Conference on Aviation Weather Systems, Dallas, Texas, 15-20 January 1995, pp. 491–496. Courtesy J. Johnson, National Weather Service Office, Dodge City, Kansas)

Aviation weather surveillance systems   219

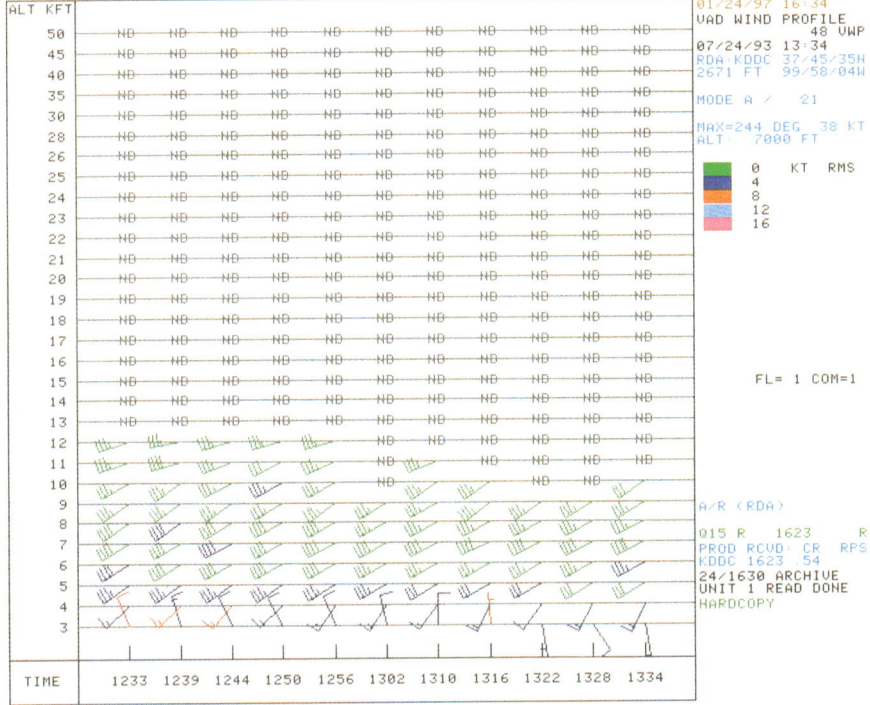

Colour Plate 4 VAD (velocity azimuth display) wind profile of the phenomenon shown in Colour Plate 3, taken with the same WSR-88D radar. The wind barbs show the magnitude and direction of the wind and the colour of the barb indicates the turbulence according to the code shown on the right. Notice the rapid wind change (shear) at the lowest altitudes, and the turbulence, exceeding 8 knots (~4 m s^{-1}) RMS. 'ND' indicates 'no data'. (Reference: Johnson, J.: 'Identification of a clear-air roll turbulence and low level wind shear event by the WSR-88D radar'. Preprints of the 6th Conference on Aviation Weather Systems, Dallas, Texas, 15-20 January, 1995, pp. 491–496. Courtesy J. Johnson, National Weather Service Office, Dodge City, Kansas)

Colour Plate 5 A weather radar antenna and pedestal assembly.
(Courtesy Enterprise Electronics Corporation)

Aviation weather surveillance systems 221

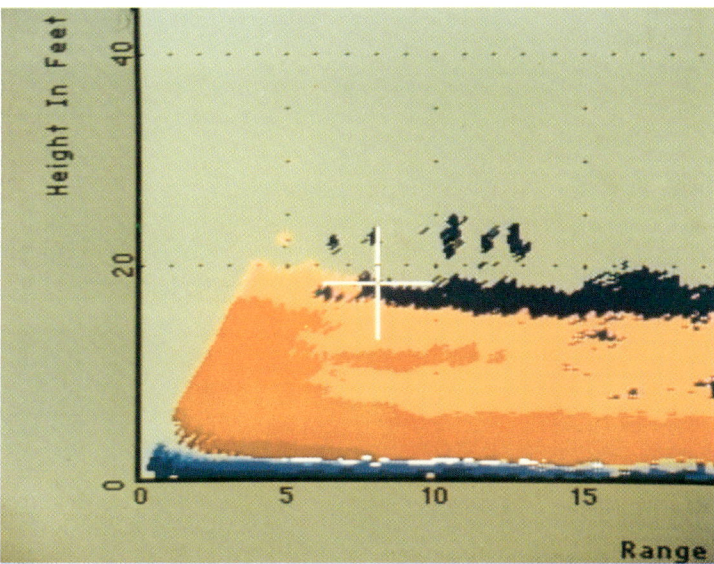

Colour Plate 6 Display of basic ('raw') Doppler velocity data from a C-band (5615 MHz) weather radar on PPI (plan position indicator) (top) and RHI (range height indicator) (bottom) displays. The colour code for the display is shown on the velocity scale on the lower right part of the top frame. Each colour step on the scale represents a velocity interval of 3.54 m s^{-1} (1/8 of the Nyquist interval), except the centre or 'zero' step, which spans the interval of ±0.1 m s^{-1}. The notations A1 ... A4 on the velocity scale signify velocities away from the radar, and T1 ... T4 refer to velocities towards the radar. The T4 colour (the bottom-most colour in the velocity scale) appearing on the top right of the PPI image and the high-altitude portion of the RHI image, immediately after the A4 colour, signify velocity aliasing. For the RHI, the vertical scale is in kilofeet and the horizontal scale is in statute miles.
(Courtesy Enterprise Electronics Corporation)

222 *Aviation weather surveillance systems*

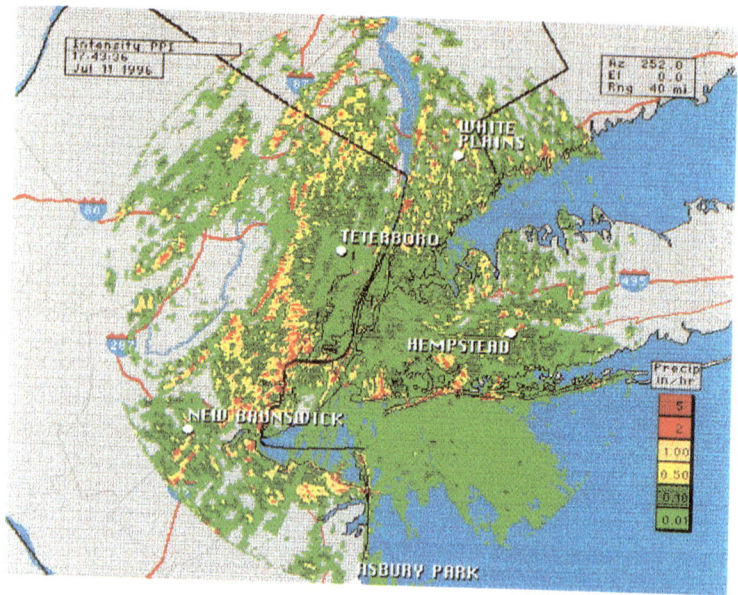

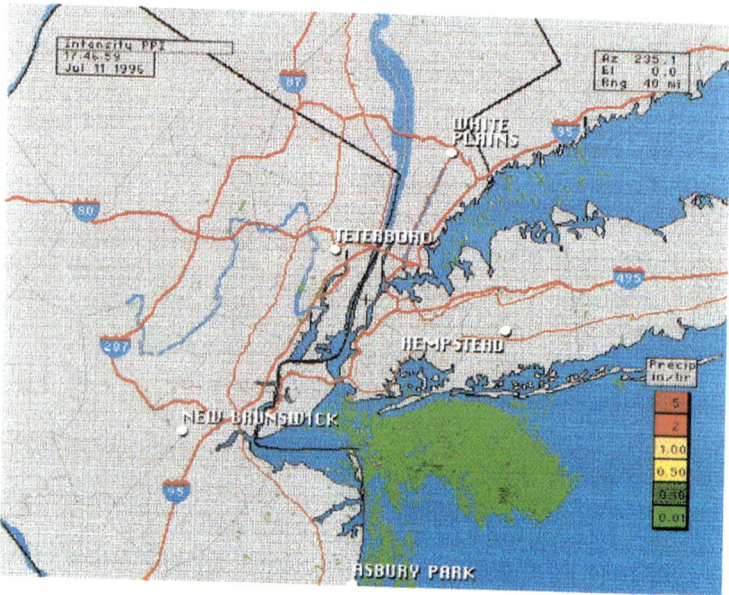

Colour Plate 7 Rain rate map (PPI display) from the DSWR-90C radar system manufactured by Enterprise Electronics Corp. The 5 cm radar, with a 12 ft antenna, is installed on the Rockefeller Center Building in Manhattan, New York. The top field of raw radar returns shows extensive ground clutter contamination which cannot be distinguished from the true rain activity. The bottom image, taken 3.5 min after the top one, is processed through the ground clutter filter. Notice that nearly all of the extraneous green patches have been removed by the filter, showing the patch of light rain activity over the sea. (Courtesy Enterprise Electronics Corporation).

Aviation weather surveillance systems 223

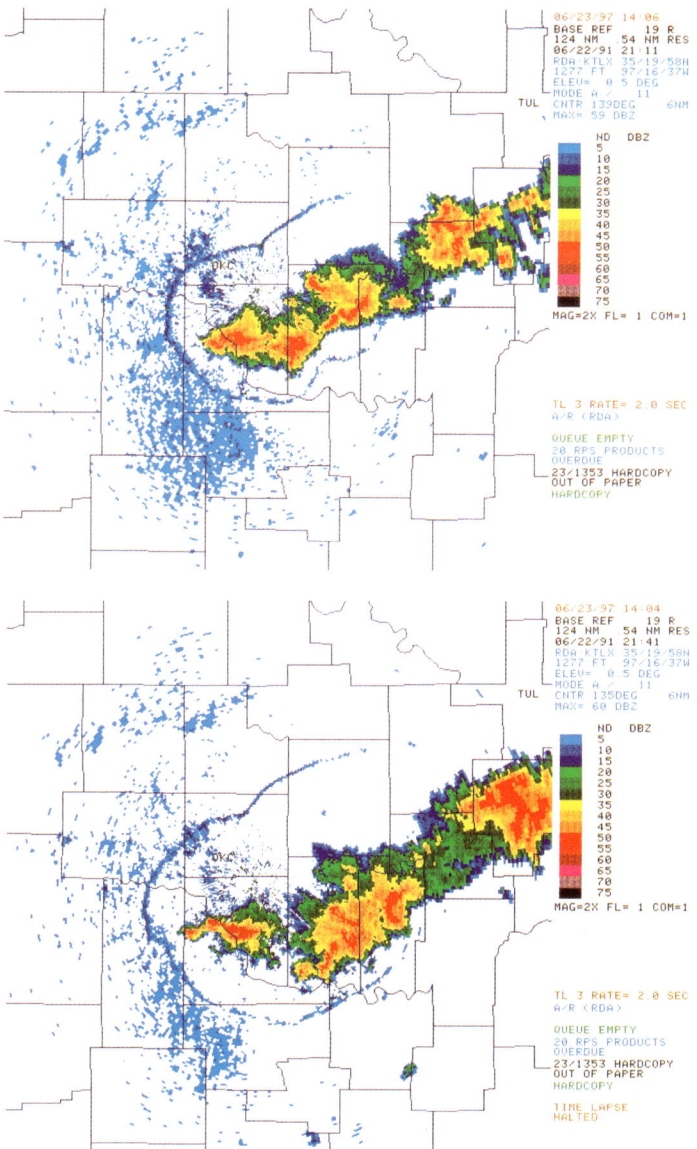

Colour Plate 8 Two reflectivity images of a thunderstorm taken 30 minutes apart by a WSR-88D radar at 0.5° beam elevation, showing storm growth (increase in precipitation area) and progress of the outflow. The maximum reflectivity of the storm is ~60 dBZ. The 'thin line' arc surrounding the left half of the storm is the gust front, which is the frontal boundary of the storm outflow. The significant reflectivity of the gust front, 15 dBZ, is due to the dust and debris raised by the outflow winds, and the clear-air returns arising from the turbulence. Note that in the second frame (bottom) the gust front is as far as 50 km from the reflectivity core (heavy-rain area) of the storm. Notice also the weakening of the gust front with expansion. (Courtesy D.S. Zrnic', NOAA National Severe Storms Laboratory, Norman, Oklahoma)

224 *Aviation weather surveillance systems*

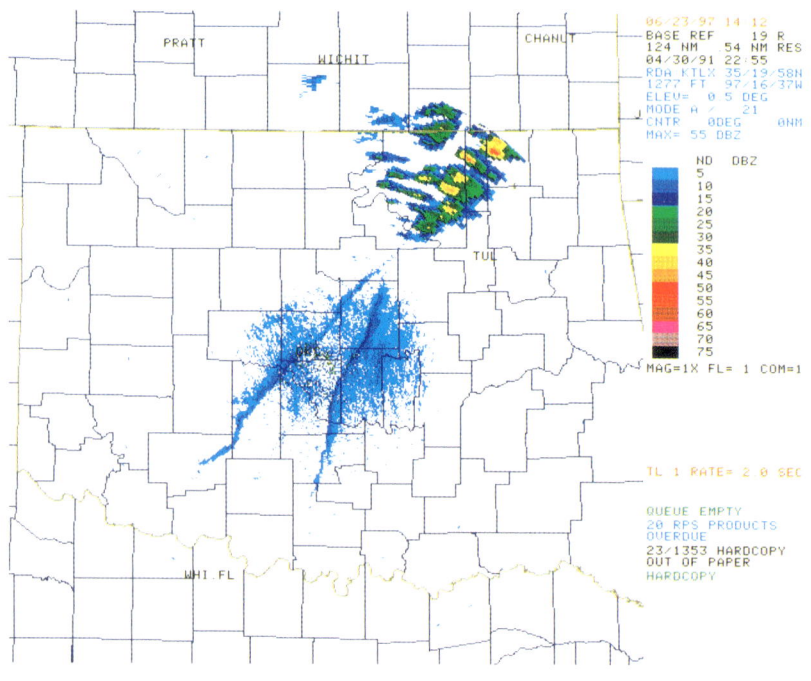

Colour Plate 9 WSR-88D radar reflectivity field showing two linear clear-air features: a cold front (the line on the left) and a dry line (the line on the right). The intersection of the two lines often gives rise to storms. Here a patch of precipitation activity comprising a number of storm cells is found in the zone of intersection. The radar is located near the centre of the picture. The patchy blue shading around the radar position is due to ground clutter. (Courtesy D.S. Zrnic', NOAA National Severe Storms Laboratory, Norman, Oklahoma)

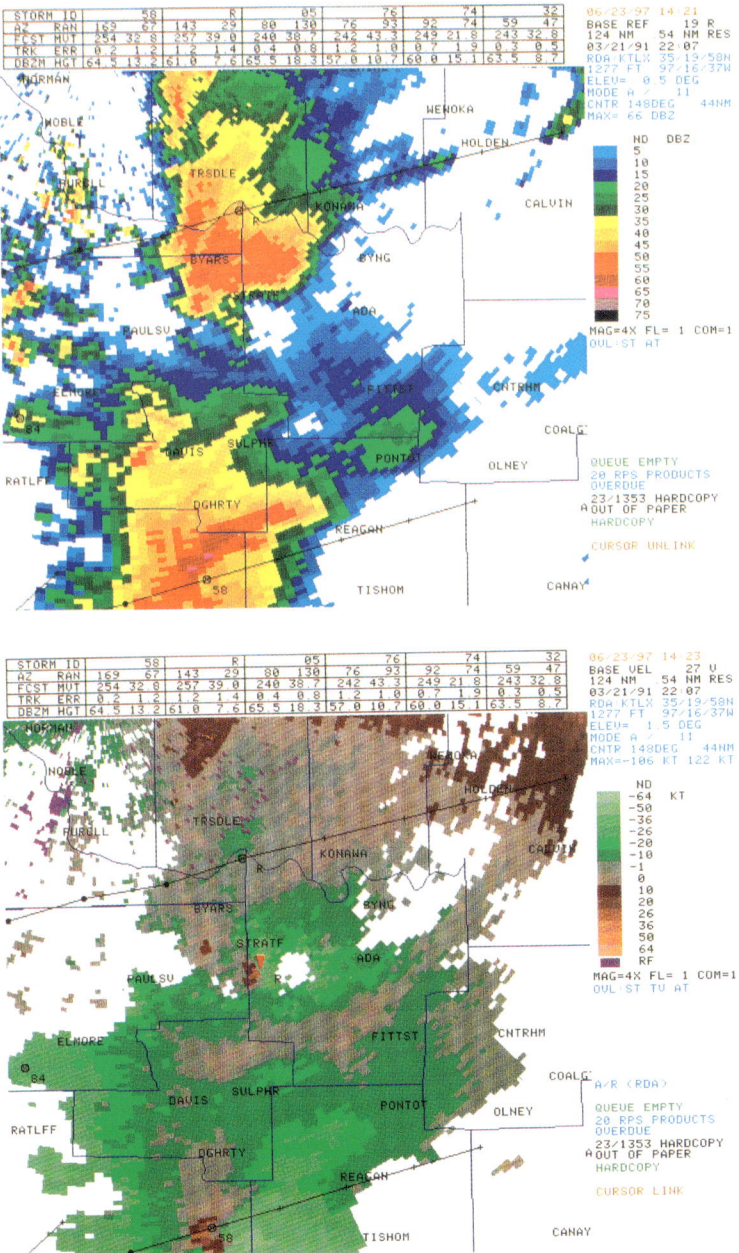

Colour Plate 10 WSR-88D velocity display (bottom) showing a tornado vortex signature (TVS) at the centre (just below the word 'STRATF') with a characteristic Doppler couplet of positive and negative radical velocities. The TVS has been automatically detected by the software, and is flagged with a small red triangle. The reflectivity display (top) of the same scene, however, reveals that the TVS is occurring in a region of weak precipitation activity. (Courtesy D.S. Zrnic, NOAA National Severe Storms Laboratory, Norman, Oklahoma)

226 *Aviation weather surveillance systems*

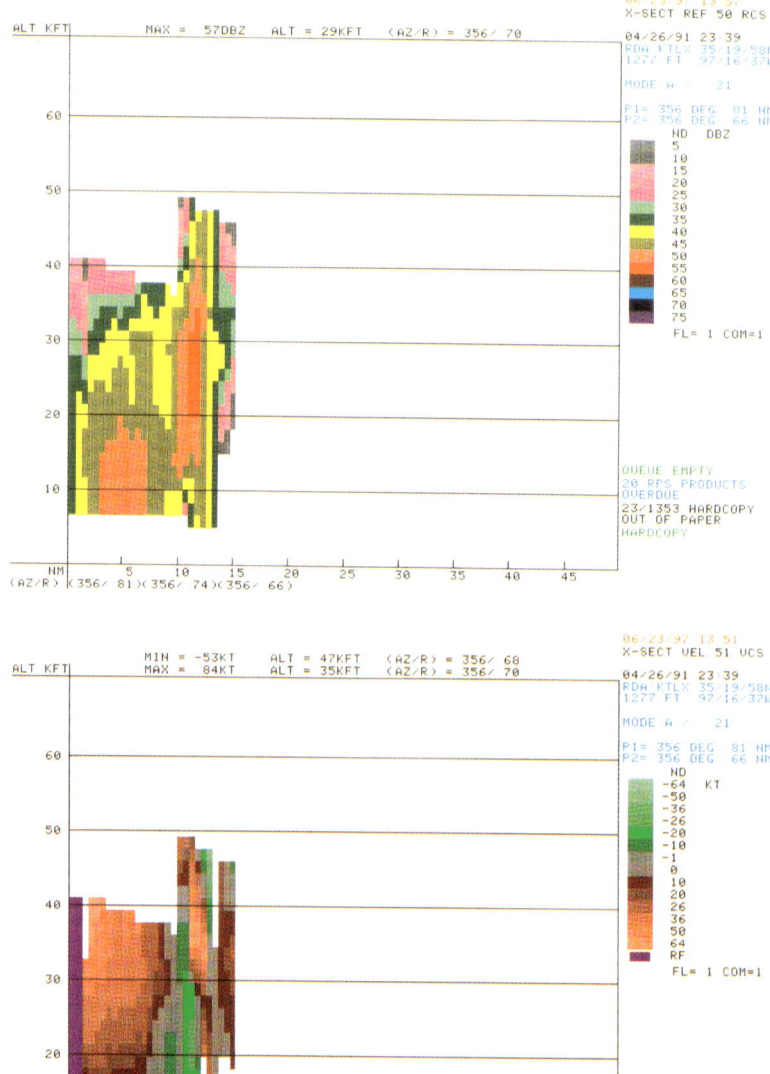

Colour Plate 11 RHI (range height indicator) display of a thunderstorm, synthesised from WSR-88D horizontal scan data at multiple elevation angles, showing the vertical structure of the storm. The reflectivity field (top) depicts the precipitation distribution, and the velocity field (bottom) shows the convergence at low levels which feeds the rising shaft of the thunderstorm. The radar location is to the right of the frames. The green shades indicate velocities towards the radar, and purple/red shades signify velocities away from the radar. (Courtesy D.S. Zrnic', NOAA National Severe Storms Laboratory, Norman, Oklahoma)

Aviation weather surveillance systems 227

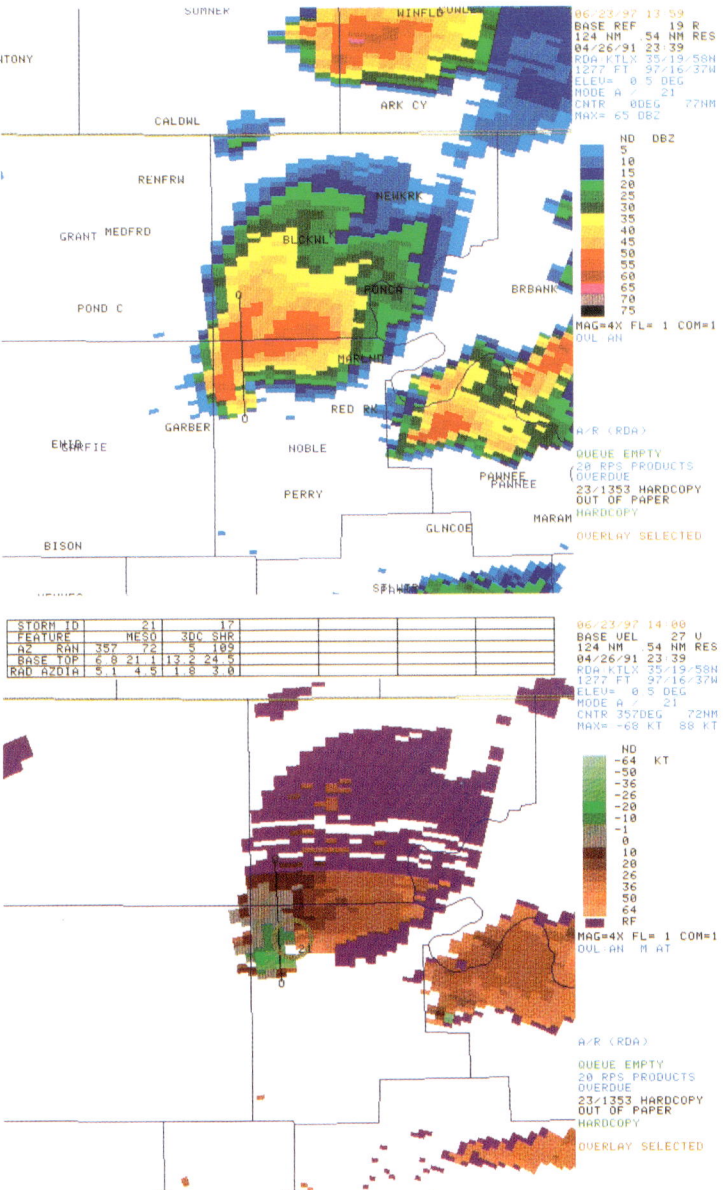

Colour Plate 12 WSR-88D reflectivity (top) and velocity (bottom) pictures of a weather zone. The velocity field clearly shows the signature of a mesocyclone (Doppler doublet of approaching and receding velocities, oriented normal to the radial), which is automatically detected and marked with a circle. The reflectivity field, however, does not show any indication of the existence of the mesocyclone. The radar is located somewhat to the south of the bottom centre of each frame. The vertical line segment with small circles at its ends marks the section of the storm along which the RHI shown in Colour Plate 11 was constructed. (Courtesy D.S. Zrnic', NOAA National Severe Storms Laboratory, Norman, Oklahoma)

228 *Aviation weather surveillance systems*

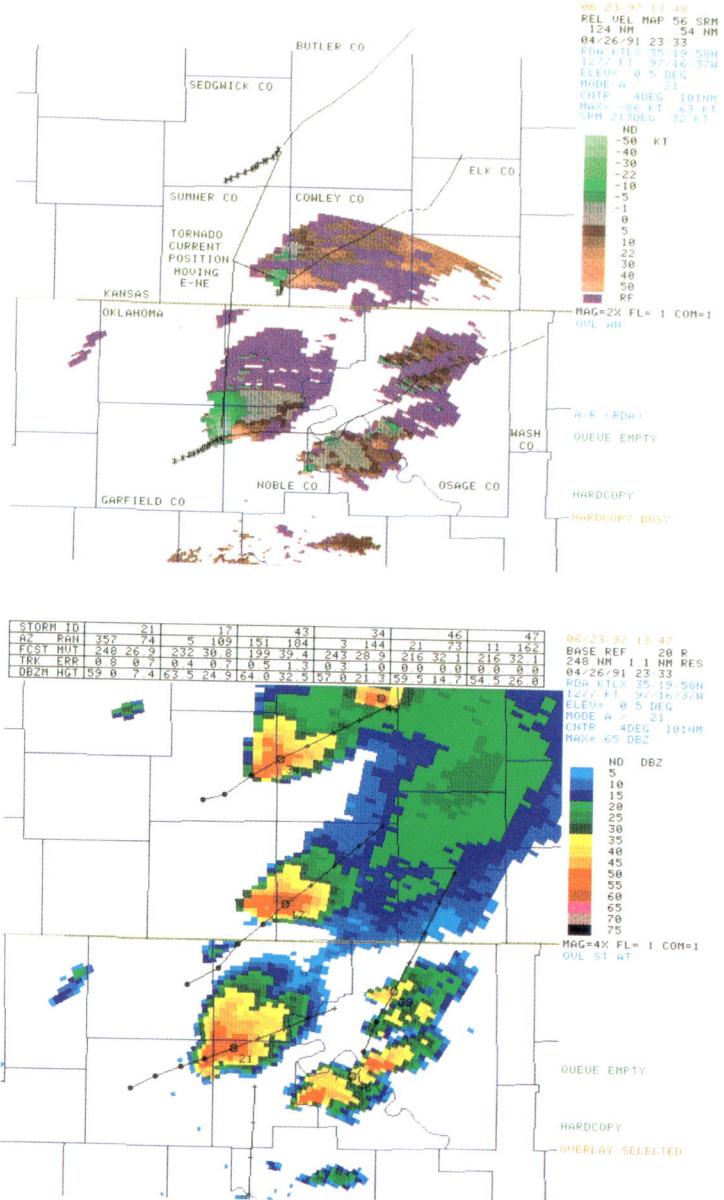

Colour Plate 13 The top frame shows the velocity field of the same scene as Colour Plate 12 (with a few minutes' time gap) to a more compressed scale. The ground tracks of three tornadoes (two tracks long and one short), including one associated with the mesocyclone of Colour Plate 12, are shown as curves with a string of asterisk marks. The bottom frame shows the tracking and prediction of the positions of three storm cells in the reflectivity field of a different scene. The present position of each storm centroid is shown as ⊗, past positions as heavy dots, and the projected future positions as crosses on a line segment. (Courtesy D.S. Zrnic', NOAA National Severe Storms Laboratory, Norman, Oklahoma)

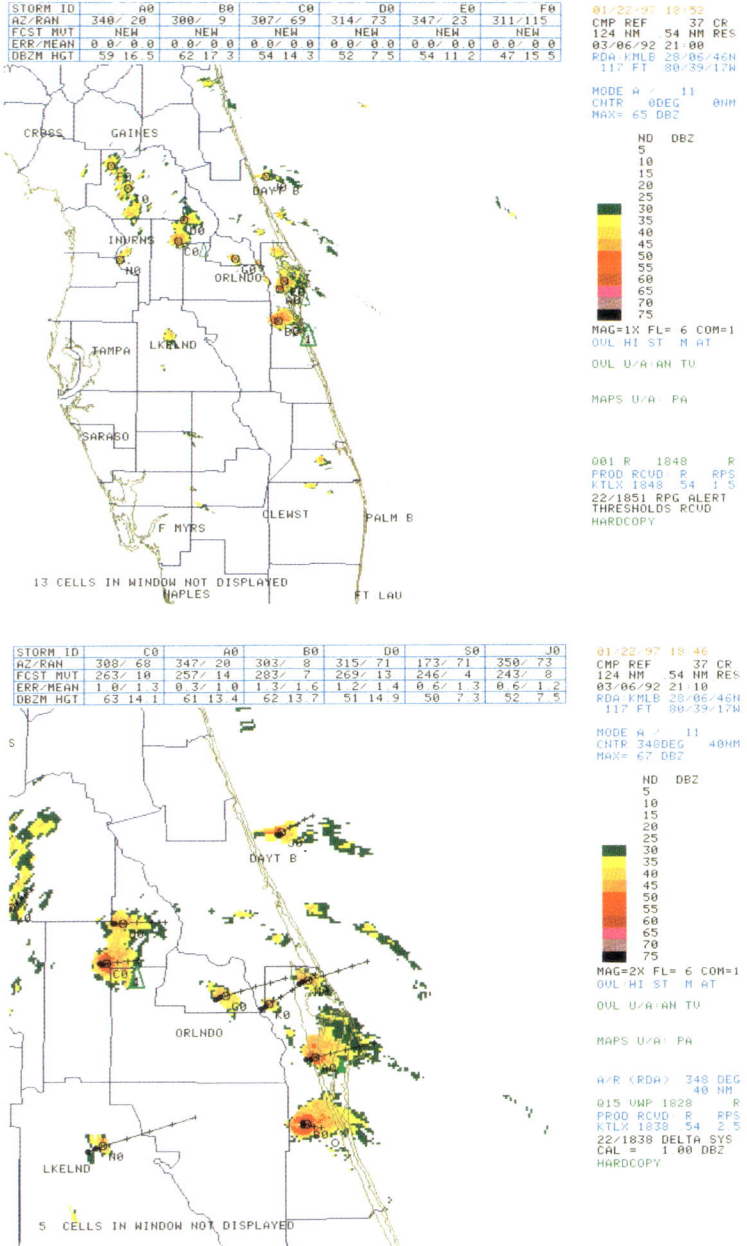

Colour Plate 14 Automatic location, tracking and extrapolation of storm centre positions by Storm Cell Identification and Tracking (SCIT) algorithm. In the top frame, the storm cells have been located and marked with a ⊗ icon. The bottom frame shows a part of the top frame expanded. The past positions of each storm centre are shown as heavy dots (overlapping), and the projected future positions are marked with crosses along a straight line. Both the frames show reflectivity. (Data collected by WSR-88D radar at Melbourne, Florida)

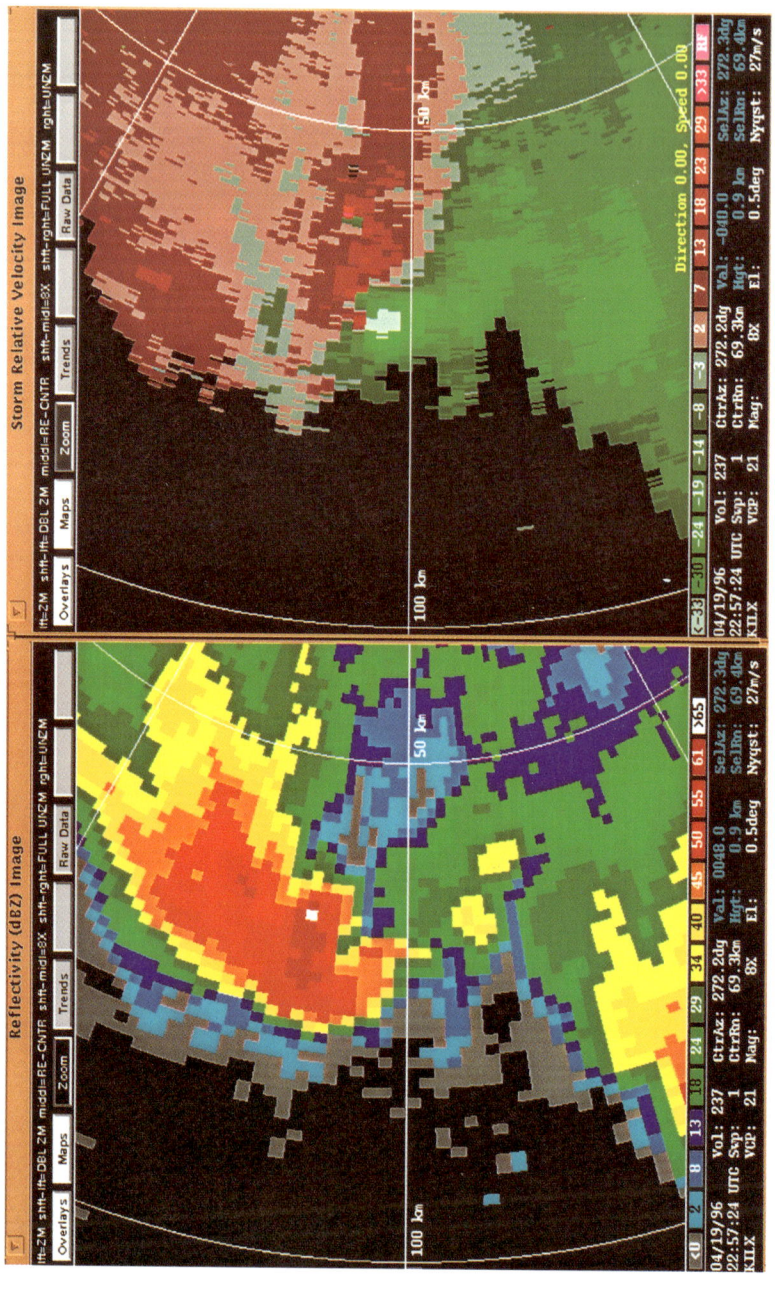

Colour Plate 15 The appearance of a tornado in WSR-88D data. The tell-tale signature with a Doppler couplet normal to the radar line of sight is clearly visible near the centre of the velocity field (right), but a signature is less apparent in the reflectivity field (left). (Courtesy D.S. Zrnic', NOAA National Severe Storms Laboratory, Norman, Oklahoma)

Aviation weather surveillance systems 231

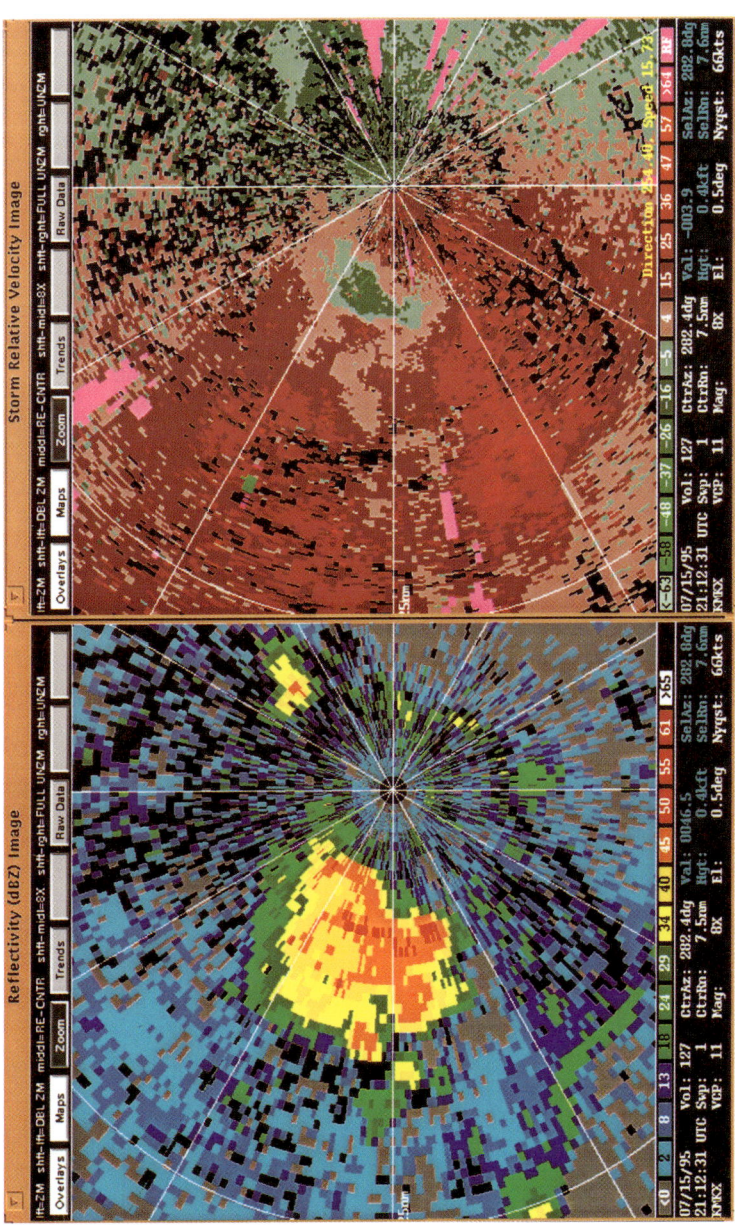

Colour Plate 16 The appearance of a microburst in a WSR-88D data field obtained at 0.5° elevation. In the velocity display (right), the microburst winds appear as a patch of negative velocity (green shades) at the centre of the frame lying within a continuum of positive velocities (red/pink shades). The microburst occurs within the precipitation patch seen in the reflectivity field (left). (Courtesy D.S. Zrnic', NOAA National Severe Storms Laboratory, Norman, Oklahoma)

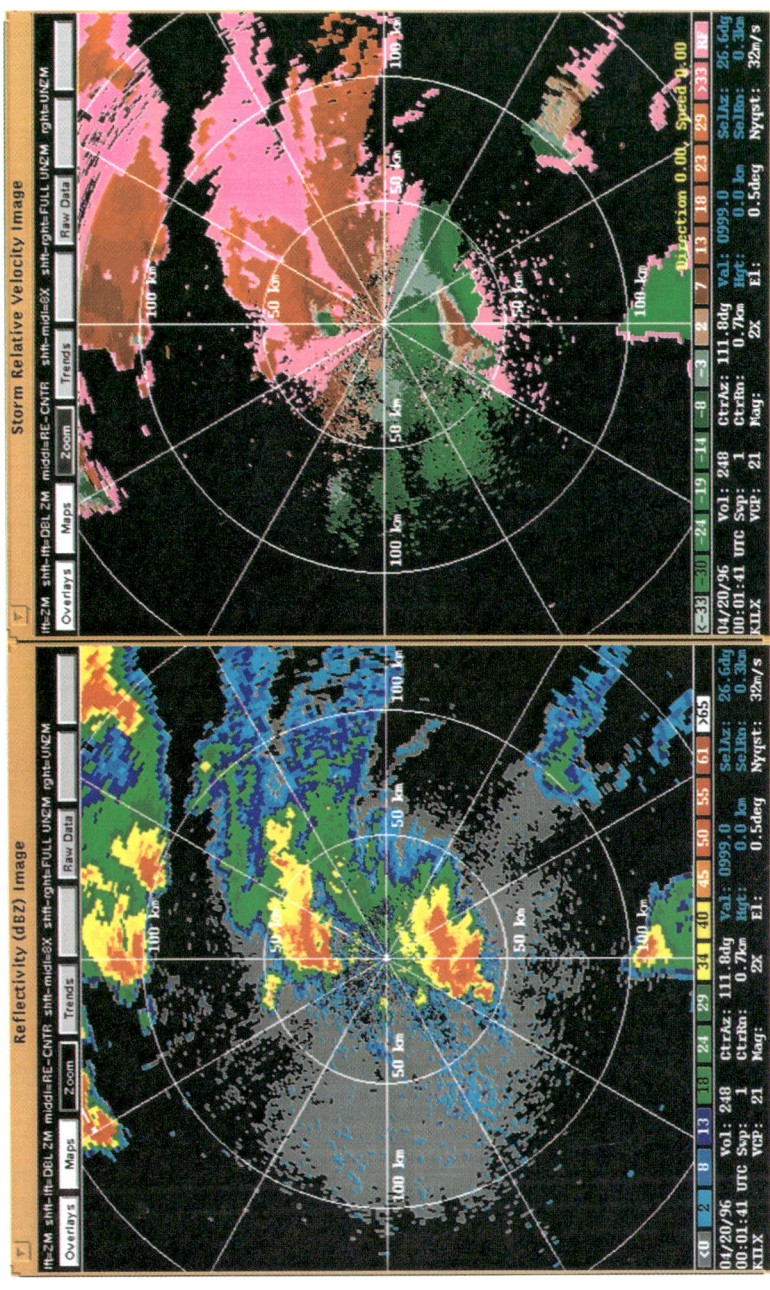

Colour Plate 17 WSR-88D data shown in the reflectivity (left) and velocity (right) fields. The artificial, elongated shapes of the features, apparently wedging into the origin (radar location), are due to range folding or overlay. This effect is particularly distinct in the velocity field. (Courtesy D.S. Zrnic', NOAA National Severe Storms Laboratory, Norman, Oklahoma)

Aviation weather surveillance systems 233

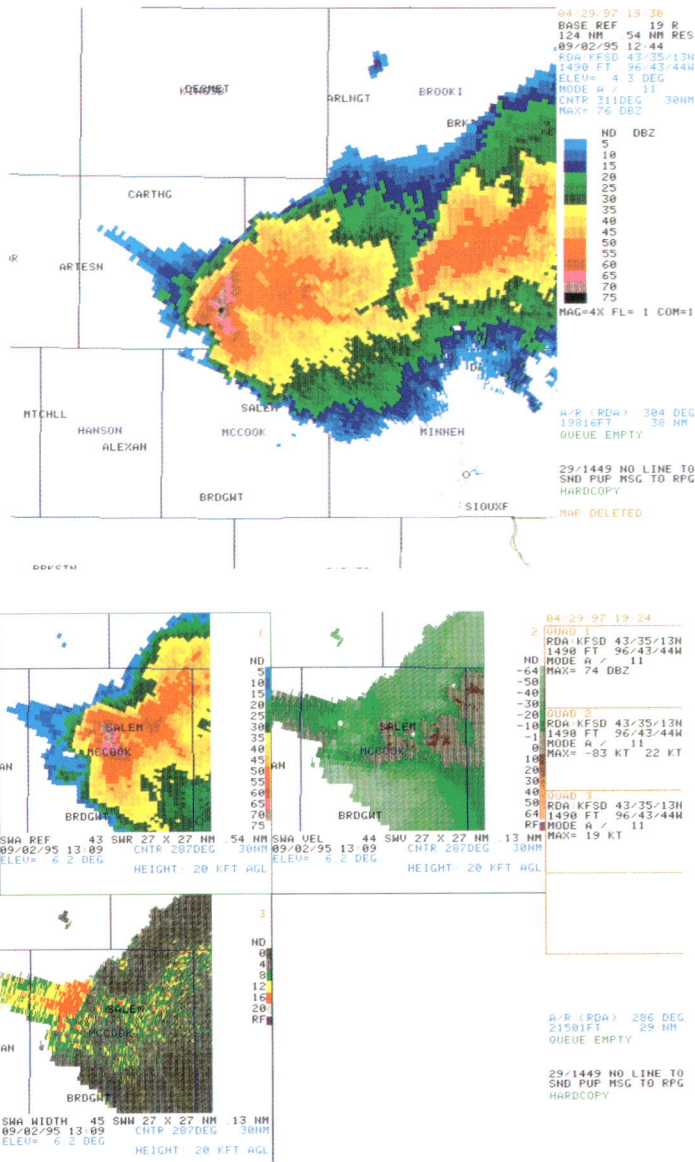

Colour Plate 18 Illustration of an artefact of radar detection of heavy weather. The horn-like projection (to the north west) of the reflectivity map (at 4.3° elevation) behind the very high reflectivity (peak 76 dBZ) region in the top frame is because of the 'three-body effect', i.e. radar echo from precipitation aloft being bounced off the ground and reflected again from precipitation before returning to the radar antenna. The bottom frame shows all the three moments (reflectivity, velocity, spectrum width) of the phenomenon at an elevation angle of 6.2°. The high apparent velocity and turbulence in the artificial elongation is mainly due to the falling speed of precipitation particles. (Courtesy D.S. Zrnic' and L. Lemon, NOAA National Severe Storms Laboratory, Norman, Oklahoma)

234 *Aviation weather surveillance systems*

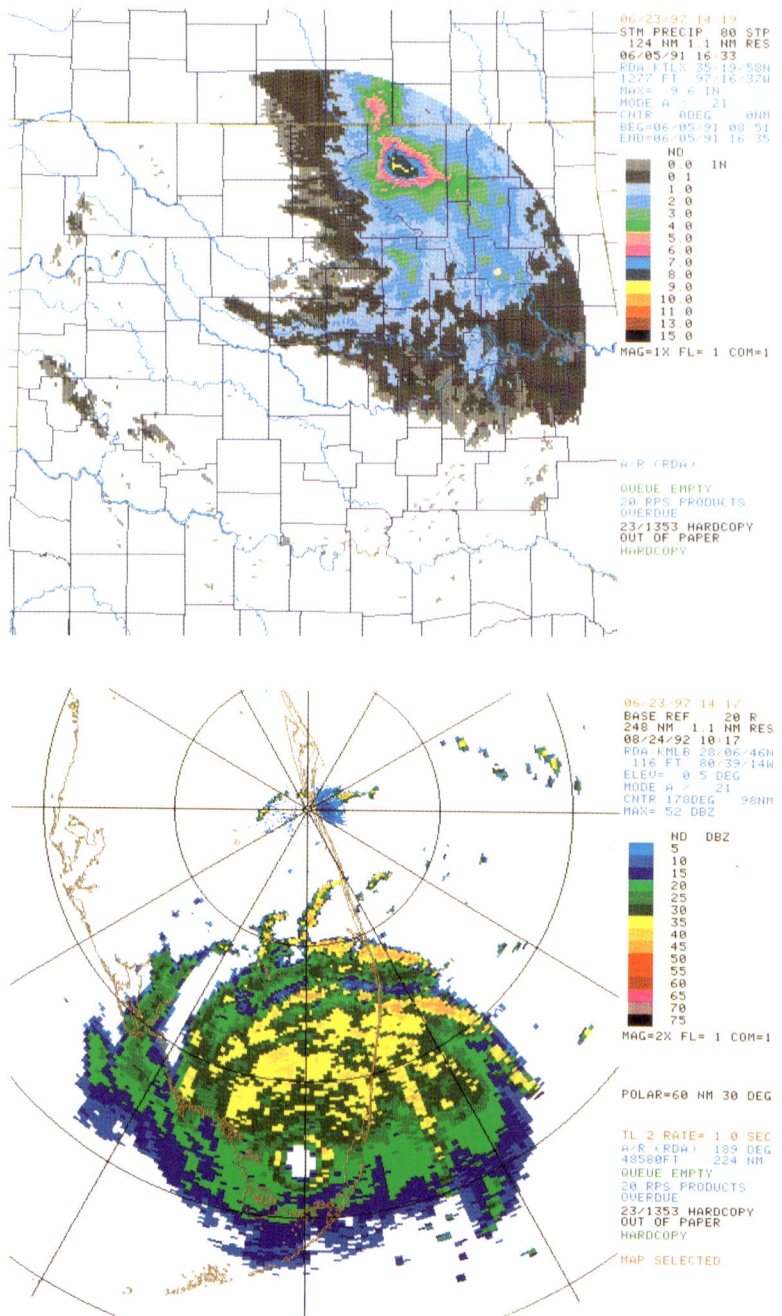

Colour Plate 19 WSR-88D rain accumulation product over a period of about 8 hours (top) and reflectivity map of hurricane Andrew on 24 August 1992 (bottom). (Courtesy D.S. Zrnic', NOAA National Severe Storms Laboratory, Norman, Oklahoma)

Aviation weather surveillance systems 235

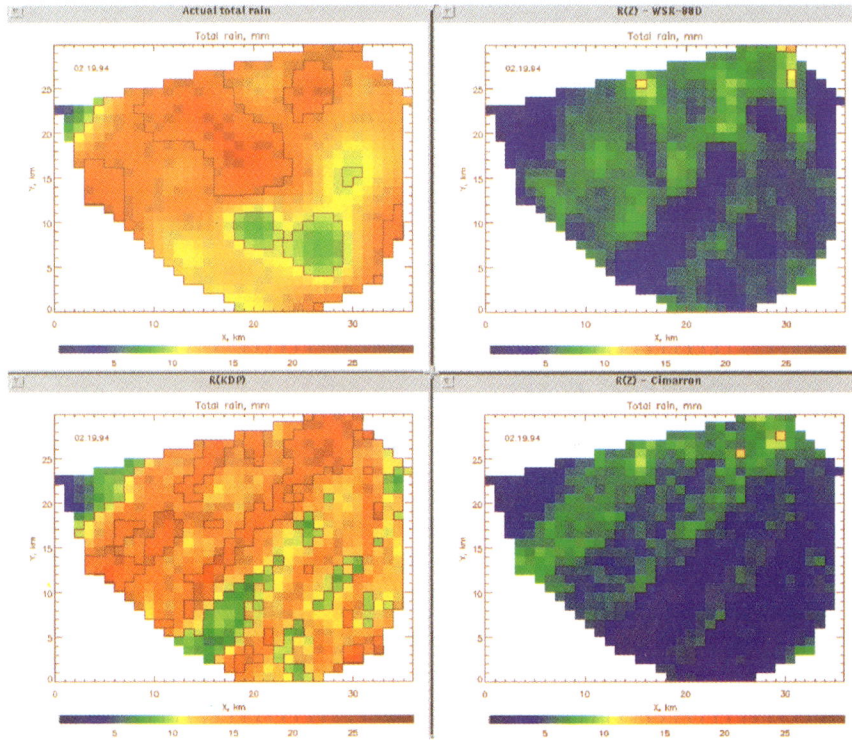

Colour Plate 20 Distribution of rainfall accumulation over a portion of central Oklahoma as derived from four different sources. Left: spatial interpolation from measurements by 42 rain-gauges distributed over the area (top), and estimates from the specific differential phase K_{DP} measured by the polarimetric radar at Cimarron (bottom). Right: estimates from reflectivity only, as measured by a WSR-88D radar (top) and the Cimarron radar (bottom). Note that both radars grossly underestimate the total rainfall based on reflectivity, while a given radar (Cimarron) provides more realistic rain estimates based on K_{DP} than on reflectivity. (Courtesy D.S. Zrnic', NOAA National Severe Storms Laboratory, Norman, Oklahoma)

236 *Aviation weather surveillance systems*

Colour Plate 21 A Geographic Situation Display to provide broad-area weather situational awareness through graphical and alphanumeric displays for use by air traffic controllers and supervisors as an aid to air traffic management.
(Reference: Weber, M.E. and Kays, T.A.: 'An advanced weather surveillance processor for airport surveillance radars'. Preprints of 6th Conference on Aviation Weather Systems', Dallas, Texas, 15–20 January 1995. Reproduced with permission of MIT Lincoln Laboratory, Lexington, Massachusetts)

Colour Plate 22 (opposite) Examples of high-level data products with Terminal Doppler Weather Radar (TDWR) and Integrated Terminal Weather System (ITWS) algorithms. The phenomenon observed is a microburst. Both the frames have a common background consisting of the wind shear map generated by TDWR. Superimposed on this shear map are the microburst outline shapes (red contours) produced by TDWR (top) and ITWS (bottom). Such fully processed meteorological data make them useful even for nonmeteorologists. By comparing with the background shearmap, it is possible to infer that the microburst contours in the lower frame represent the phenomenon more faithfully and precisely. The white line indicates the track of an experimental aircraft flight that penetrated the microburst.
(Reference: Mathews, M.P., and Dasey, T.J.: 'Improving aircraft impact assessment with the Integrated Terminal Weather System Microburst Detection Algorithm'. Preprints of the 5th International Conference on Aviation Weather Systems', Vienna, Virginia, 2–6 August 1993, pp. 45–50. Reproduced with permission of MIT Lincoln Laboratory, Lexington, Massachusetts)

Aviation weather surveillance systems 237

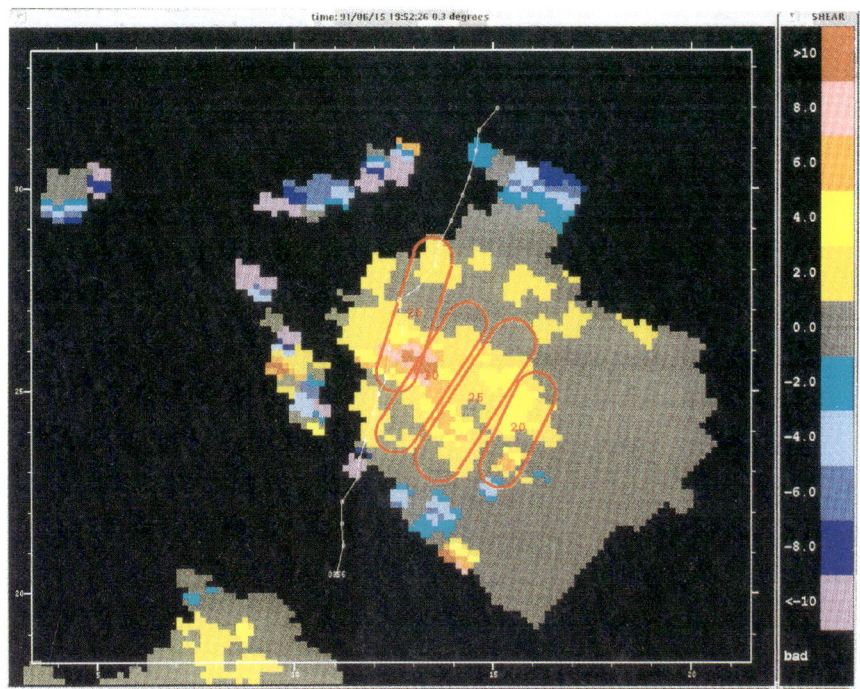

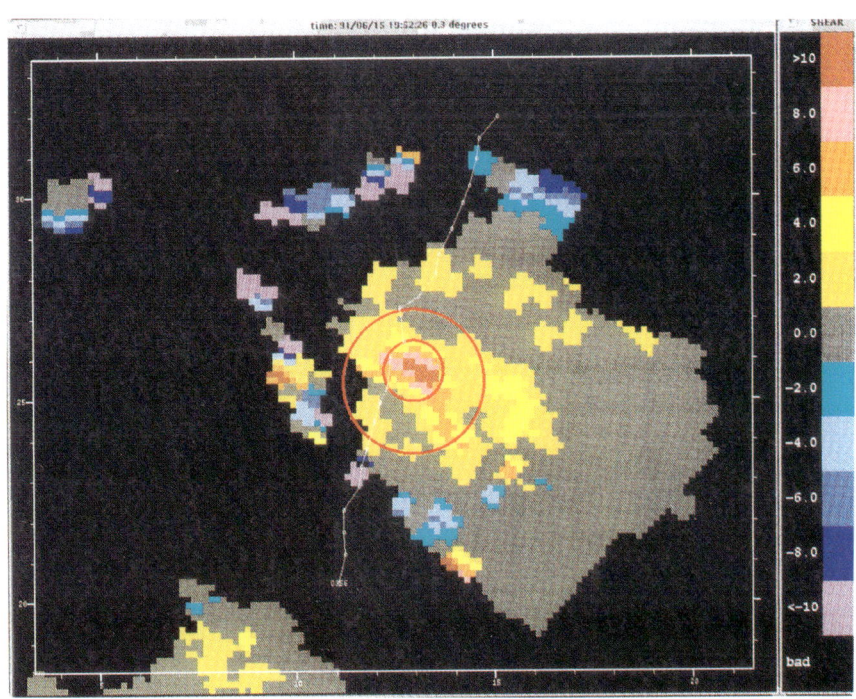

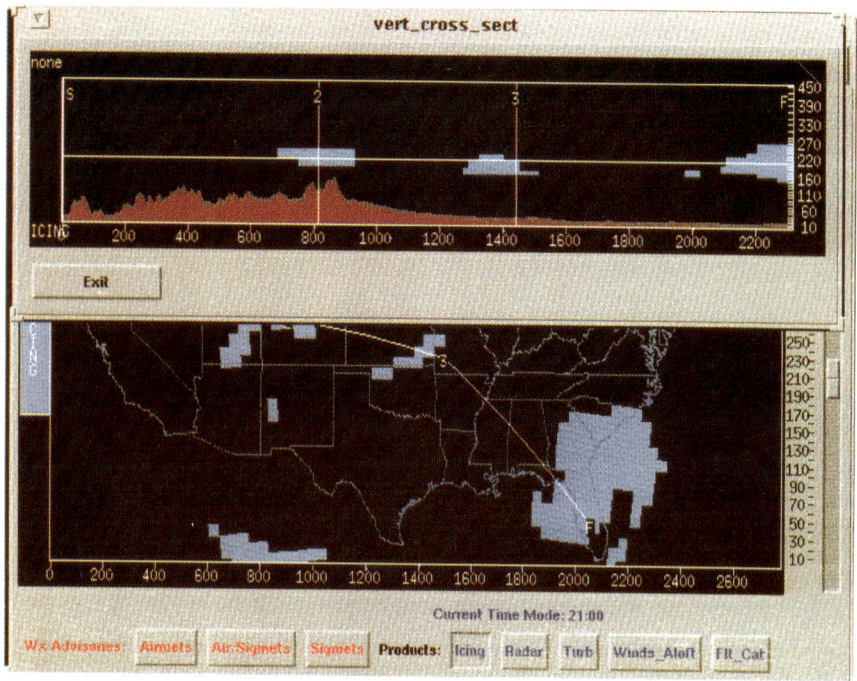

Colour Plate 23 Example of the Icing Potential Product from the Aviation Weather Products Generator (AWPG) system. The bottom frame provides a plan view of the geographical distribution of high icing potential (light blue patches). The top frame shows the same parameter in a vertical section along the flight path indicated by the white line in the bottom frame. The brown area along the bottom edge of the top frame represents the terrain contour along the flight-path.
(Courtesy William P. Mahoney, National Center for Atmospheric Research, Boulder, Colorado. The National Center for Atmospheric Research is sponsored by the National Science Foundation. This research is funded in part by the Federal Aviation Administration (FAA) and the National Weather Service (NWS))

Aviation weather surveillance systems 239

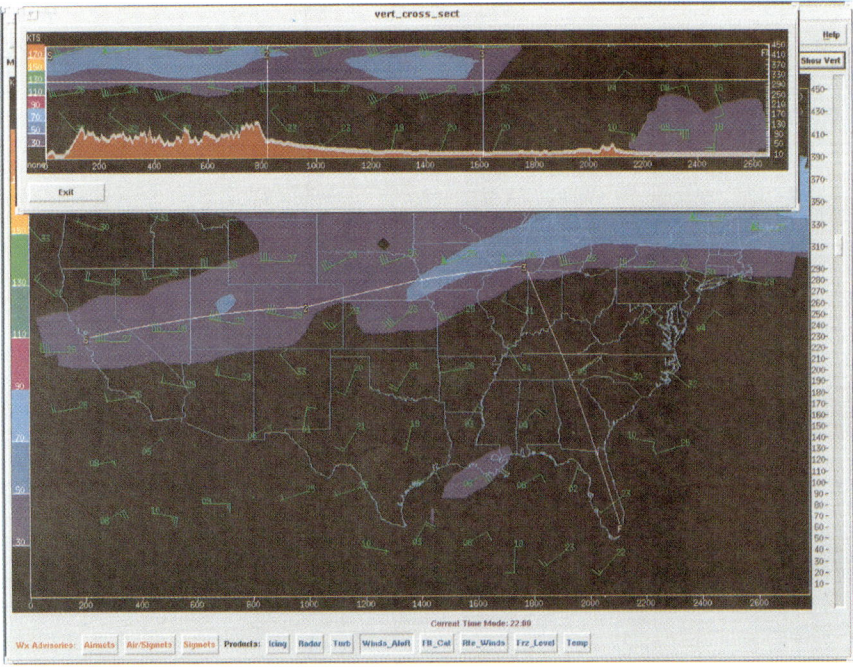

Colour Plate 24 Example of the display of winds aloft from the Aviation Weather Products Generator (AWPG) system. The bottom picture provides a plan view of the geographical distribution of winds at flight altitudes according to the colour code on the left. The top picture shows the winds in a vertical section along the flight path indicated by the white line in the bottom frame. The brown area along the bottom edge of the top picture represents the terrain contour along the flight path.
(Courtesy William P. Mahoney, National Center for Atmospheric Research, Boulder, Colorado. The National Center for Atmospheric Research is sponsored by the National Science Foundation. This research is funded in part by the Federal Aviation Administration (FAA) and the National Weather Service (NWS))

240 *Aviation weather surveillance systems*

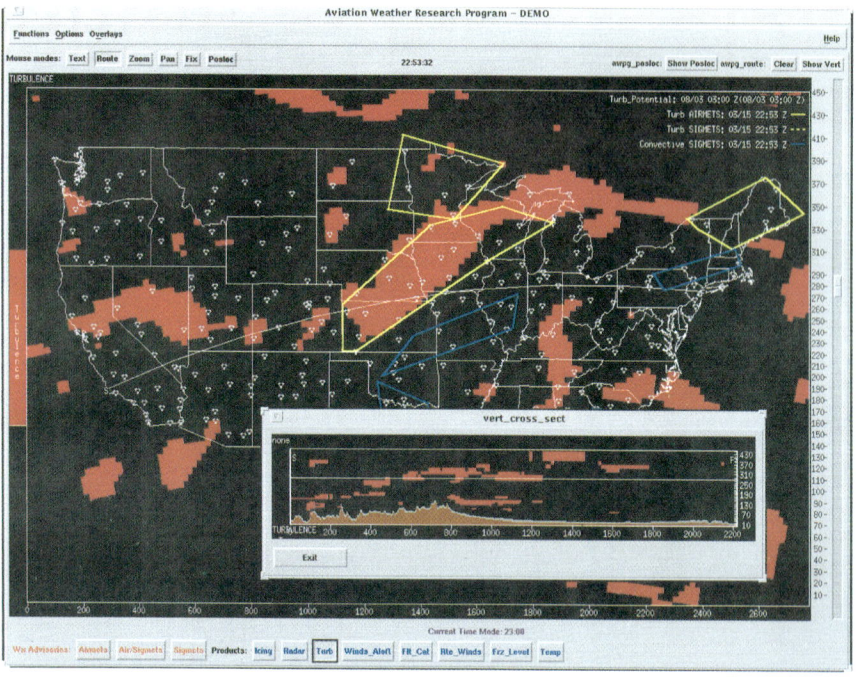

Colour Plate 25 Example of the display of atmospheric turbulence parameter from the Aviation Weather Products Generator (AWPG) system. The main picture provides a plan view of the geographical distribution of turbulence (red patches). The insert shows the occurrence of turbulence in a vertical section along the flight path indicated by the white line in the main picture. The brown area along the bottom edge of the insert represents the terrain contour along the flight path.
(Courtesy William P. Mahoney, National Center for Atmospheric Research, Boulder, Colorado. The National Center for Atmospheric Research is sponsored by the National Science Foundation. This research is funded in part by the Federal Aviation Administration (FAA) and the National Weather Service (NWS))

Aviation weather surveillance systems 241

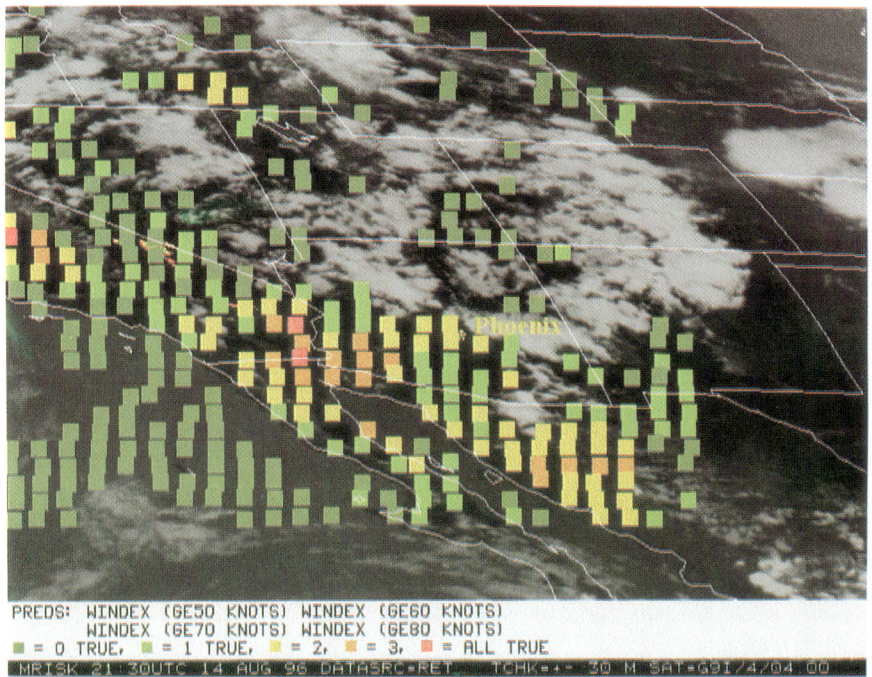

Colour Plate 26 Example of display of Microburst Wind Index (WINDEX) derived from GOES-9 sounder thermodynamic data at 2130 UTC for the severe storms on 14 August 1996 in southern Arizona. Coloured squares show WINDEX values at the appropriate location where GOES-9 soundings were available. Windex is derived from the lapse rate (700–500 mb), height of the melting level, and mixing ratios at middle and low levels. In other areas, the infrared image is displayed. Maximum possible winds of over 60 knots (yellow squares) are shown near Phoenix (marked by asterisk), with winds up to 70–80 knots to the south-west of the city. The city actually experienced winds of over 100 mph. (Courtesy G. P. Ellrod, NOAA National Environmental Satellite Data and Information Services, Washington, DC)

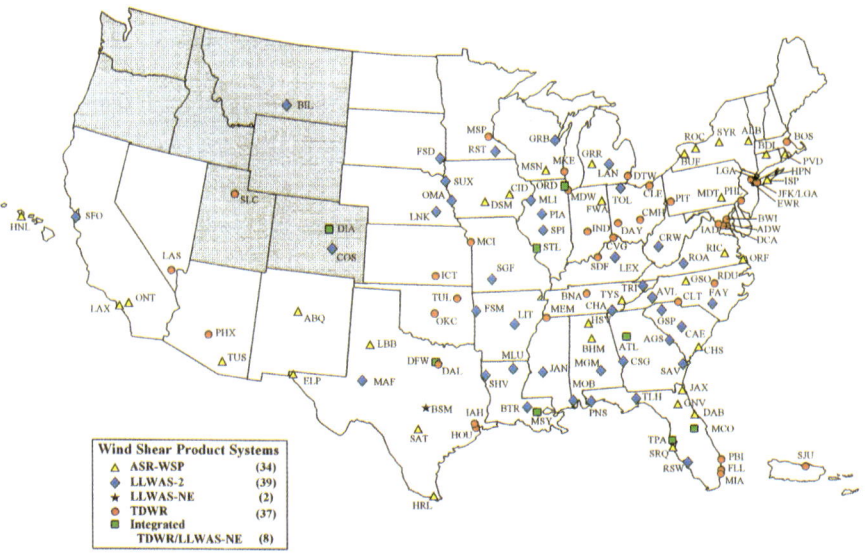

Colour Plate 27 US National Airspace System (NAS) wind shear product systems as of 17 May 1996. For acronyms in the box, please see list of abbreviations. (Courtesy K.M. Starr, TRW Inc.)

Aviation weather surveillance systems 243

Colour Plate 28a A forward scatterometer type visibility sensor at Anchorage International Airport, Alaska.
(Courtesy Teledyne Controls, photographed by Larry Bangert)

Colour plate 28b A close-up view of the scatterometer shown in Colour plate 28a, with its calibration plate mounted (Courtesy Teledyne Controls, photographed by Larry Bangert)

Chapter 7
Modern Doppler weather radars for aviation

7.1 General

The power of the modern Doppler weather radar to provide high-resolution three-dimensional pictures of the weather field in terms of numerous parameters and with a high degree of agility is being harnessed to great advantage in improving the safety and efficiency of aviation systems. Information derived from weather radars has been used for air traffic planning and control for a long time. However, the modern and futuristic aviation weather surveillance systems make use of the weather radar in a much more intimate way. This is done at two levels. First, the radars used for such systems are either optimised specifically with aviation use in mind, or are designed for multiple roles with significant importance given to aviation data product generation. Next, these radars, along with a number of other sensors, are integrated into the aviation weather surveillance system in such a way that the sensor data are available to the system continuously in real time or near-real time for processing, inference generation, communication to air traffic controllers and pilots, and decision-making regarding air traffic management and aviation facility operation. In many cases, the integration of the sensors is intimate enough to allow control of their operational modes and data formats according to the needs of the surveillance situation. In this chapter we concentrate on the first level, i.e. weather radars that are primarily or substantially designed for use as part of modern aviation weather surveillance systems.

The optimisation of weather radar systems for aviation use is carried out in several ways. The first step is the choice of the basic radar parameters such as carrier frequency, pulse repetition frequency, transmitted power level, antenna gain and sidelobe levels, antenna scan rate, receiver characteristics, clutter-rejection characteristics, etc. to optimally fulfil the role envisaged for the radar under the inherent and imposed constraints. Thus weather radars designed to cover terminal areas would be significantly different from those intended for *en route* coverage in terms of their basic parameters. The next step is to design the data formats and communication formats and protocols to integrate the radars into the surveillance system. A third aspect of the optimisation process consists of proper choices of the locational and operating parameters of the radar. As already pointed out, a major advantage of the radar as a sensor for aviation weather is its capability for airport-centric observation, and the consequent matching of resolution requirements with aviation needs. This calls for appropriate placement of the radar relative to the surveillance domain. Among operational factors that require optimisation are the strategies for scanning, data collection and processing, and adaptive choice of variable parameters such as pulse-repetition frequency, pulse staggering, if any, transmitted pulsewidth, and processing and/or display modes for signals and data. Some of these aspects are considered in this chapter in relation to the major weather radar systems developed for aviation weather surveillance support.

7.2 WSR-88D system

The WSR-88D is the largest dedicated weather radar system developed and deployed in the USA during recent years. The name is an abbreviation for <u>W</u>eather <u>S</u>urveillance <u>R</u>adar – <u>88</u> <u>D</u>oppler. The embedded number 88 corresponds to the year in which its developmental phase was assumed to be completed, paving the way for acceptance and deployment. During the developmental phase, the radar project was named Next Generation Radar, or NEXRAD for short. The NEXRAD was an interagency project in the USA, and the resulting radar is designed to serve many sectors including general weather observation, hydrology and aviation. The deployment pattern of the radar envisages covering most of the US land and coastal areas including island territories, as well as certain overseas locations.

The WSR-88D radar system is designed to significantly enhance tornado warnings, improve the detection and measurement of damaging winds, severe turbulence, wind shear and hailstorms, and more accurately delineate areas that are threatened by severe weather. The design philosophy lays emphasis on substantial reduction in the number of false alarms and incorrect forecasts of severe weather, and improvement of the display and dissemination of radar data products.

7.2.1 Architecture

A skeletal schematic diagram of the radar system is shown in Fig. 7.1. The system consists of three major units: the Radar Data Acquisition (RDA) unit, the Radar Product Generator (RPG) and the Principal User Processor (PUP). The first unit performs the basic radar functions such as the transmission, reception and processing of signals, the second performs the computational functions for generating the radar data products, and the last unit maintains interface with users and provides display and control of data at user locations.

Modern Doppler weather radars for aviation 247

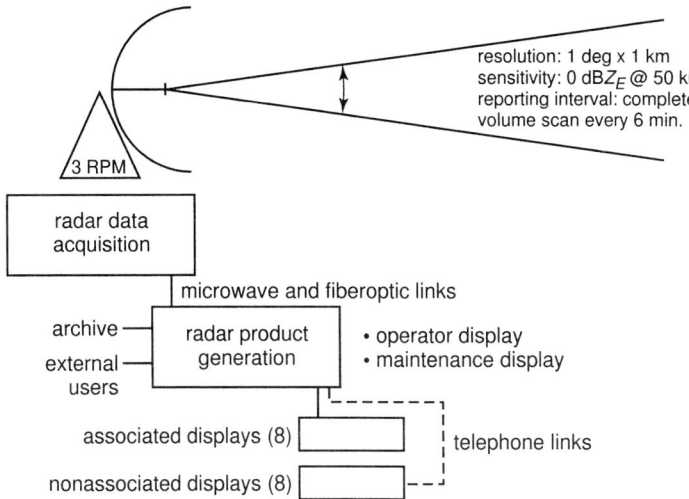

Figure 7.1 The main blocks of the WSR-88D radar system

7.2.2 Parameters

A fully coherent radar operating in the 10 cm wavelength band, the WSR-88D has a parabolic dish antenna of 8.5 m diameter generating a pencil beam of width somewhat less than a degree. The antenna is protected from environmental elements by being housed in a spherical radome of 12 m diameter. Other important basic radar parameters include a transmitted peak power of 750 kW and two possible transmitted pulsewidths of 1.57 and 4.5 μs, each with its associated range of pulse repetition frequencies. Such a two-mode transmission scheme helps in range ambiguity resolution, as discussed below. The parameters and features of the radar system are listed in some detail in Table 7.1.

7.2.3 System features

The WSR-88D radar system has a very high dynamic range of 95 dB, together with a high sensitivity of −8 dBZ at the 50 km range (corresponding to a signal-to-noise ratio of at least 0 dB). These features facilitate weather observation over long ranges, and in both precipitative environments as well as in clear air. The system also has a high clutter-rejection capability which is optimised by utilising five-pole digital infinite-impulse-response filters with software-controlled notch widths (Heiss *et al.*, 1990). A clutter rejection up to 54 dB is achieved for fixed targets.

Special care has been taken in the radar design to achieve a very clean spectrum of the transmitted signal to satisfy radio-interference requirements. A complex self-calibration scheme is built into the system to minimise the

248 *Aviation weather surveillance systems*

Table 7.1 WSR-88D radar system characteristics[a]

Parameter/feature	Value/description
Radar system	
Range of observation:	
Reflectivity	460 km
Velocity	230 km
Angular coverage:	
Azimuth	Full circle or sector
Elevation	Operational limits: $-1°$ to $+20°$
Antenna	
Type	S-band, centre-fed, parabolic dish
Reflector aperture	8.54 m (28 ft) diameter circular
Beamwidth (one-way, 3 dB)	0.96° at 2.7 GHz, 0.88° at 3.0 GHz
Gain	45.8 dB at 2.85 GHz (midband)
Polarisation	Linear horizontal
First sidelobe level	-29 dB
Steerability	360° azimuth, $-1°$ to $+45°$ elevation
Mechanical limits (elevation)	$-1°$ to $+60°$
Rotation rate (maximum)	30° s^{-1} (azimuth and elevation)
Angular acceleration	15° s^{-2} (azimuth and elevation)
Pointing accuracy	$\pm 0.2°$
Radome	
Type	Fibreglass skin foam sandwich
Diameter	11.89 m (39 ft)
RF loss (two-way)	0.3 ± 0.06 dB over 2.7–3.0 GHz band
Transmitter	
Type	Master oscillator power amplifier (MOPA)
Frequency range	2.7 to 3.0 GHz
Peak power output	500 kW into antenna
Pulsewidth (nominal)	1.57 µs (short pulse) and 4.5 µs (long pulse) $\pm 4\%$
RF duty cycle (maximum)	0.002
Pulse repetition frequency:	
Long pulse	322 to 446 Hz $\pm 1.7\%$
Short pulse	322 to 1282 Hz $\pm 1.7\%$
Waveform types	Contiguous and batch

Table 7.1   Continued

Parameter/feature	Value/description
Receiver	
Type	Linear
Tunability (frequency range)	2.7 to 3.0 GHz
Bandwidth (3 dB)	0.63 MHz (short pulse), 0.22 MHz (long pulse)
Phase control	Selectable
Receiver channels	Linear output I/Q, log output
Dynamic range	95 dB max, 93 dB at 1 dB compression
Minimum detectable signal	−113 dBm
Noise temperature	450 K
Intermediate frequency	57.6 MHz
Sampling rate	600 kHz
Signal processor	
Type	Hardwired/programmable
Parameters derived	Reflectivity, mean radial velocity, Doppler spectral width
Algorithms (respective)	Power averaging, pulse-pair, single-lag autocorrelation
Accuracy (standard deviation):	
Reflectivity	<1 dB
Velocity and spectrum width	<1 ms^{-1}
Number of pulses averaged:	
Reflectivity	6–64
Velocity and spectrum width	40–200
Range resolution:	
Reflectivity	1 km
Velocity and spectrum width	0.25 km
Azimuth resolution	1°
Clutter canceller	Digital, infinite impulse response (IIR), 5-pole
Clutter suppression	30–50 dB
Filter notch half-width	0.5–4 ms^{-1} (equivalent radial velocity)
Radar product generator (RPG)[b]	
RPG processor	32 bit general-purpose digital computer
Shared memory	32 MB semiconductor memory, expandable to 96 MB
Wideband communication	1.544 Mbit/s data rate
Narrowband communication	Up to 21 of 9600/4800 bit/s 4-wire
	Up to 26 of 9600/4800 bit/s 2-wire
	(will have 14400/9600/4800 bit/s capability)

250 *Aviation weather surveillance systems*

Table 7.1 Continued

Parameter/feature	Value/description
RPG graphic display processor	
Principal User Processor (PUP)	Fixed point, 32 bit general purpose digital computer
Communications	9600/4800 bit/s 2- and 4-wire (maximum 10 lines) RS449/RS232 converters
Video	Colour, with split-screen and zoom functions
Mass storage	Up to two 600 MB discs

[a] Originally from Mahapatra and Zrnic' (1991), substantially refined/updated with inputs from Mr. Dale Sirmans, formerly with the National Severe Storms Laboratory, Norman, OK
[b] Specifications for computer and communication parameters may undergo continual upgradation

effects of parameter drift and achieve the desired levels of measurement accuracy.

Operationally, the radar system is capable of remote, unattended operation and transmission of raw data products to the RPG. It has two distinct operational modes: the precipitation and clear-air modes. By design, the system is capable of executing as many as eight different types of scan cycles (see Section 7.4.3), though not all of them may be specified or available as options in a given installation.

7.2.4 Data products

The WSR-88D radar system provides full-colour displays of its data products and makes the products available to users in electronic data formats over communication links. Besides the primary outputs of the Doppler spectral moments and VAD wind profile (see Section 6.6.4), the system incorporates software for generating higher-level weather products. The complement of weather products available from a typical WSR-88D installation is listed in Table 7.2. Many of the higher-level products pertain to precipitation parameters (e.g. the rainfall accumulation map shown in the top frame of Colour Plate 19), but some of them refer to the vertical extent of weather phenomena (echo tops) and other parameters of interest to aviation. Parameters relating to wind shear and turbulence levels can be readily derived from the primary Doppler data.

In the aviation context, the WSR-88D reflectivity data are planned to be used initially in the form of composite reflectivity products. These products are generated by vertically combining the data obtained from the various levels of scan into four different altitude layers (OFCMSSR, 1991). The names of these products and the altitude interval over which each product is valid are listed in Table 7.3.

Table 7.2   Typical WSR-88D data products

Reflectivity
Composite reflectivity
Layer composite reflectivity
Mean radial velocity
Echo tops
One-hour rainfall accumulation
Three-hour rainfall accumulation
Storm total rainfall accumulation
Hourly digital rainfall array
Vertically integrated liquid water
Velocity azimuth display wind profile

Levels 1, 2 and 3 referred to in Table 7.3 are also called the low, high and superhigh sectors, respectively. It is clear that layer 1 covers most of the general aviation activity, layer 2 covers much of the airliners and large transport aircraft in cruise, and layer 3 data are intended for high-flying aircraft such as supersonic military and transport aircraft. The layered products are generated with 8 or 16 levels of reflectivity quantisation and 1 or 4 km of spatial resolution, with the 8-level, 4 km product being favoured in the USA for *en route* surveillance (Dunbar and Mittelman, 1993).

7.2.5 Performance

The WSR-88D radar system underwent thorough testing and evaluation prior to its acceptance and deployment, and more data have been accumulated regarding its ability to detect hazardous weather at operational sites. During a period of operational test and evaluation completed in 1989, the system demonstrated a 91% probability of detection and a false alarm rate of 21% for severe thunderstorms. These constituted major improvements over the US averages of 58% and 57% as the corresponding figures for the pre-WSR-88D conventional weather radars (Heiss *et al.*, 1990).

Table 7.3   WSR-88D composite reflectivity products and their altitude bands

Composite reflectivity product	Altitude interval, kft msl*
Composite reflectivity	0–60
Layer 1 composite reflectivity	0–24
Layer 2 composite reflectivity	24–33
Layer 3 composite reflectivity	33–60

* kft msl = kilofeet (thousands of feet) above mean sea level

7.3 Range and velocity ambiguities

At this point, before moving on to describe the next radar system, it is instructive to dwell on a few considerations that are important in the context of using modern Doppler weather radars for facilitating aviation operations. These will be discussed here in relation to the WSR-88D system. However, many of these considerations are valid for other radars as well, and will be referred to in the appropriate context. This section focuses on the ambiguities associated with range and Doppler velocity estimation, and a few other special considerations will be dealt with in the following section.

7.3.1 Nature of problem

Ambiguous mapping of weather fields is a serious problem in Doppler weather radars. Ambiguities may occur in range or velocity or both. *Range ambiguity* or *range folding* results in an uncertain or improper association of the range value to a particular echo signal, and is the result of significant echoes due to a particular transmitted pulse being received after the transmission of the succeeding pulse from the radar. *Velocity ambiguity* or *velocity aliasing* occurs when the Doppler frequency shift due to the scatterers in a given radar resolution volume is higher than can be adequately sampled by the radar's pulse repetition frequency according to the Nyquist criterion.

The effect of range folding is the overlaying of data from different range intervals. This may not only provide misleading range estimates of hazardous phenomena, but also yield erroneous velocity estimates because of the mixing up of Doppler spectra from two or more resolution volumes spaced far apart. Velocity aliasing may lead to grossly lower estimates of the speeds of high-velocity phenomena, resulting in severe underestimation of the aviation hazard potential of such phenomena. Further, ambiguities cause sudden discontinuities in the reflectivity and velocity data fields. While experienced human observers may readily detect such discontinuities, these would adversely affect the automatic hazard detection algorithms which are an important aspect of modern and agile aviation weather surveillance systems.

The origin and effect of the range ambiguity phenomenon in weather radars are shown schematically in Fig. 7.2. If a certain scatterer S lies at a long range r so far away from the radar that its echo return pulse p_A due to a transmitted pulse P_A is received after the transmission of the succeeding pulse P_B, then the radar receiver will tend to associate p_A with P_B rather than with P_A. It will then compute the range of S based on the time difference t' between p_A and P_B while the correct range corresponds to the time interval t between p_A and P_A. This apparent range r' of S differs from the true range r, in this case, by a distance r_u given as

$$r_u = \frac{cT_r}{2} \tag{7.1}$$

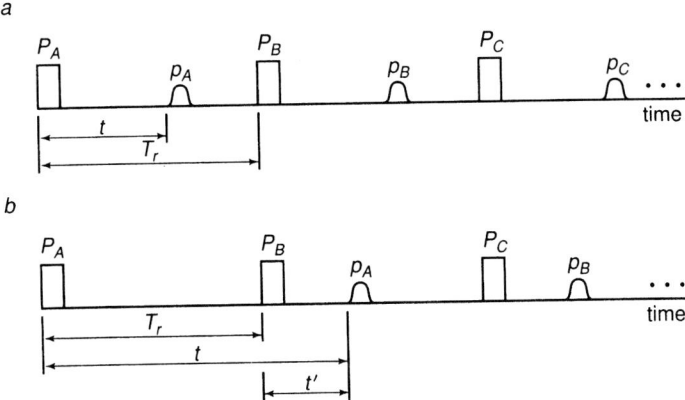

Figure 7.2 Radar pulse timing diagram showing the cause of range ambiguity: (a) looking at targets with round-trip range-delay t less than the pulse repetition time T_r ensures unambiguous ranging, but (b) range ambiguity results when $t > T_r$

where T_r is the pulse repetition time (PRT) of the radar and c is the speed of light. An echo due to a given transmitted pulse received after the succeeding transmitted pulse, as in Fig. 7.2b, is called a *second trip echo*. Depending on the range to S and the temporal spacing between transmitted pulses, it is possible for the echo pulse p_A to arrive after more than one pulse has been transmitted following P_A (i.e. after P_C or P_D, etc.). Such echoes are called *multiple-time-around echoes*, and the difference between the apparent and true ranges of S in such cases will be a corresponding multiple of r_u.

It is clear that if the scatterer S lies within a range equal to r_u then its echo due to a given transmitted pulse will never arrive after the next transmitted pulse, and there will be no possibility of range ambiguity. The distance r_u given by eqn. 7.1 is therefore called the *unambiguous range* of the radar, and is a very important parameter of the radar. The unambiguous range will, of course, change if the PRT of the radar is varied.

In the case of a single scatterer such as S, range ambiguity leads to erroneous range estimation. However, if there is another scatterer S' present at a distance r' from the radar, then the true round-trip propagation delay from S' equals t'. In such a case, the echo pulses from S and S' due to successive transmitted pulses overlap and the radar receiver cannot 'see' them as separate scatterers.

In the case of weather radars the targets are not single points but volumes or continua of scatterers. On the radar scopes (usually the PPI) weather targets appear as patches of variable reflectivity, and range ambiguity causes the patches lying in any of the ambiguous range intervals to fold over and appear erroneously in the unambiguous range interval, as shown in Fig. 7.3. The mapping process involved in the fold-over causes a geometric distortion,

254 *Aviation weather surveillance systems*

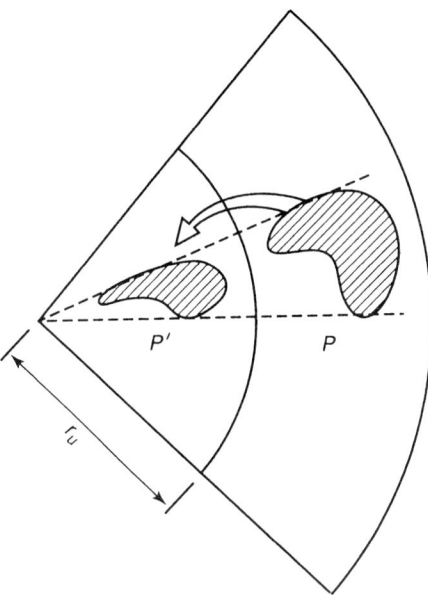

Figure 7.3   Schematic diagram of a sector on a radar plan position indicator (PPI) display showing the effect of range folding. A reflectivity patch P outside the unambiguous range circle maps radially on to the patch P′ inside the circle, undergoing geometric distortion

however. Because of the radial nature of the mapping geometry (wherein each range-ambiguous point is mapped along its radial, and the radial and angular extents of the patches are preserved during mapping) weather patches appear laterally compressed on being folded over to the unambiguous range circle (see Colour Plate 17). The higher the order of the ambiguous range interval from which a patch folds over, the more pronounced would be the compression effect. Because of the known nature of such distortion, experienced weather radar observers familiar with the natural shapes of different types of weather reflectivity patches can usually detect the occurrence of range folding and use their judgment to draw appropriate inferences. However, it is difficult for automated computer-based weather data processing algorithms to detect range folding based on geometric distortions alone.

Range folding also causes weather patches separated approximately by multiples of the unambiguous range (but located in the same direction from the radar) to overlap. Two effects result from such overlap. First, two or more overlapping resolution volumes will appear as one common point in the reflectivity field. Second, the Doppler spectrum of the overlapping echo will be a combination of the spectral components contributed by each of the overlapping resolution volumes, and the spectral moment (mean and width) estimates obtained from such an echo signal will in general be different from those of each of the overlapping phenomena. These effects are a matter of serious concern for Doppler weather radars.

Although two weather patches occurring at different ranges may overlap due to range folding, the actual echo power returned from each of them would still be governed by the radar range equation. Thus if two equally reflecting targets overlap, then the actual echo power from the farther one would be lower than from the closer one, so that the closer target (the one within the unambiguous range) has better 'visibility'. This effect is pronounced in the case of point targets such as aircraft and ships, for which the strength of the return signal decreases as the fourth power of range. However, for weather targets, the return signal power diminishes only as the square of the range, as shown by eqns. 6.14–6.16. Thus the natural advantage of targets within the unambiguous range circle *vis-à-vis* those from the first and further ambiguous range intervals due to range weighting is lower for weather targets than for point targets.

The solution to the range ambiguity problem is relatively straightforward in the case of conventional weather radars. A simple way to eliminate range ambiguity is to choose a large enough value for T_r (i.e. a small enough value for the pulse repetition frequency f_r) such that the corresponding unambiguous range r_u from eqn. 7.1 significantly exceeds the maximum range of observation. Such a solution is adopted in the case of conventional weather radars, which explains the ability of the WSR-57 radar to transmit at pulse repetition frequencies down to 200 Hz (see Table 6.6).

In the case of Doppler radars, however, the ambiguity problem is more complex. Equation 7.1 for unambiguous range is also valid for Doppler radars. In addition, for scatterer radial velocities to be measured unambiguously, they must lie within the limits $\pm v_u$, where

$$v_u = \frac{\lambda}{4T_r} \tag{7.2}$$

is called the *unambiguous velocity* or *Nyquist velocity* and λ is the wavelength of the radar. The Doppler velocity interval lying between $\pm v_u$ is called the *unambiguous velocity interval* or *zeroth aliasing interval* (Fig. 7.4). The nth aliasing interval, where n is a positive or negative integer, is equal in extent to the zeroth aliasing interval, but shifted along the Doppler velocity axis by $2nv_u$. Doppler frequencies lying outside the limits $\pm v_u$ will be undersampled and will appear as an apparent frequency within this band.

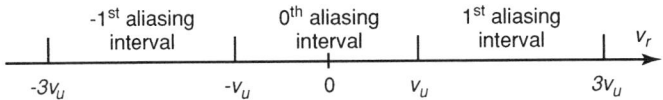

Figure 7.4 The Doppler frequency line showing unambiguous and ambiguous velocity intervals

Combining eqns. 7.1 and 7.2 yields the relation between the unambiguous range and velocity as

$$r_u v_u = \frac{c\lambda}{8} \qquad (7.3)$$

which, in effect, establishes a combined range-Doppler ambiguity envelope. Equation 7.3 shows that in a Doppler radar the unambiguous range and velocity are multiplicatively coupled, so that trying to increase any one of these parameters would result in a reduction of the other. Hence the simple solution of increasing the pulse repetition time T_r to eliminate range ambiguity, as in the case of conventional radars, would severely reduce the unambiguous velocity v_u in a Doppler radar, thus aggravating the velocity ambiguity problem. The choice of the pulse repetition time can therefore be made only through a compromise between the two types of ambiguity.

For a 10 cm wavelength radar such as the WSR-88D, eqn. 7.3 would take the particular form

$$r_u v_u = 3750 \qquad (7.4)$$

where r_u is expressed in km and v_u in m/s. To better appreciate the constraint imposed by eqn. 7.4 a few discrete values of unambiguous range and velocity as given by the equation are shown in Table 7.4. The range values in the table correspond approximately to binary fractions of the maximum range of the WSR-88D system, which is 460 km.

It has been found from experience that an unambiguous velocity of the order of ±30 m/s is necessary for the observation of most severe weather phenomena without undue velocity aliasing. As seen from Table 7.4, keeping the entire WSR-88D observation range of 460 km unambiguous would drive the unambiguous velocity to a low value of ~8 m/s, which would result in severe velocity aliasing and thus render the Doppler data nearly unusable. On the other hand, trying to achieve the desired unambiguous velocity of >30 m/s would reduce the unambiguous range to ~115 km, which is only a quarter of the maximum observation range of the WSR-88D. This would

Table 7.4 Some unambiguous ranges and the corresponding unambiguous velocities for a 10 cm radar

Unambiguous range, km	Unambiguous velocity, m/s
30	125
60	62.5
115	32.6
230	16.3
460	8.2

cause the radar data to be range-folded and overlaid from successive unambiguous range intervals.

It is thus apparent that significant range-Doppler ambiguity is inherent and inescapable in a radar like the WSR-88D, and techniques must be employed to mitigate the effects of such ambiguity if the radar data are to be useful in the widest variety of situations. These methods aim to restore the data to their true range and velocity values by unfolding and dealiasing the data. Some of these techniques are built into the operating scheme of the WSR-88D system, and many others are the subject of research and development.

7.3.2 Minimisation of range overlays

Appropriate mechanisms are provided in the WSR-88D for unambiguous mapping of reflectivities at different elevation angles. The methods are as follows (OFCMSSR, 1992).

7.3.2.1 Low elevation angles

At each of the two lowest elevation angles of scan, usually 0.5 and 1.5°, the radar antenna scans azimuthally twice. The first scan uses a 'continuous surveillance waveform', which is essentially a long-PRT transmission, to generate an echo power array with a reflectivity resolution of 0.5 dB and range resolution of 1 km for each radial spaced 1° apart. These are stored in the RDASC (Radar Data Acquisition Status Control) till the end of the scan. The second scan at the same elevation uses a 'continuous Doppler waveform' which is a low-PRT transmission. The data from this scan are compared and combined with those stored from the previous scan, radial by radial, to generate a single record of Doppler data. The comparison process also yields information on overlays of weather patches caused by range ambiguity.

It was mentioned in the preceding subsection that, to be overlaid, weather patches must be separated radially by a multiple of the unambiguous range of the radar. Since, as suggested by eqn. 7.1, the unambiguous range is a direct function of the operating PRT of the radar, the existence and extent of overlap among weather phenomena occurring in different range ambiguity intervals can be altered by appropriately changing the PRT. The WSR-88D system has an automatic PRT selection algorithm which chooses the best PRT from a set of four discrete values (corresponding to unambiguous ranges lying between 115 and 145 km) so as to achieve the least obscuration due to overlaid echoes over the entire scan circle at the lowest elevation.

7.3.2.2 Middle elevation angles

Here a 'batch mode' is employed. Surveillance (high-PRT) and Doppler (low-PRT) waveforms are interlaced to achieve the required range, while allowing the radar to simultaneously collect base reflectivity, mean radial velocity and spectrum width data. Such an interlaced scheme is used at middle scan levels, but not the lower levels, because the requirement of clutter rejection is more stringent at lower levels, necessitating a uniform PRT.

7.3.2.3 High elevation angles

No special procedure to combat range overlay is considered necessary at high elevation angles. Since storm tops do not exceed 70 000 ft in height, range overlay of velocity data is not expected at high altitudes. The oblique radar beam pointing at high elevation angles tops (i.e. passes over the tops of) storm cells at relatively close ranges (see Fig. 6.4). As a result, all the weather echo is confined to close ranges, and it is possible to employ a relatively high PRF at these elevation angles, which provides a larger unambiguous velocity.

7.3.3 Velocity dealiasing

The WSR-88D system has built-in software for dealiasing velocity data. Although velocity dealiasing is part of the preliminary processing of data, and not a meteorological algorithm, it is performed in the RPG (Radar Product Generator) computer of the WSR-88D, and not in the RDA (Radar Data Acquisition) unit (OFCMSSR, 1991). The algorithm is based on the scheme proposed by Eilts and Smith (1990). The essential basis for the correction of aliases is the continuity of velocities along radials, and between adjoining radials at the same range. When the radar data quality is good (adequate signal-to-noise ratio, no excessive missing data points, and no extreme wind gradients), the algorithm detects sudden Doppler velocity changes (jumps) due to aliasing by examining data along the current radial. For more difficult conditions, neighbouring points from a previous radial are examined. Elaborate error checks are made at every stage of the algorithm to arrest the propagation of any erroneous detection of aliasing. The algorithm is made efficient by making it adaptive with respect to data quality. Thus it uses simpler logic for well behaved data which occur commonly under precipitative severe weather conditions, and reserves more elaborate procedures for data that are difficult to dealias.

7.3.4 Advanced ambiguity resolution methods

The built-in ambiguity resolution methods are useful, but have significant lacunae in performance. Notably, these methods are severely strained, and frequently fail, when patches of good data are spatially separated by intervals of poor or no data which hinders the use of the continuity principle. Such situations occur in the case of storms or storm clusters isolated from one another by intervals of clear air which returns little echo power. More powerful ambiguity resolution methods have been the subject of research and development to enhance the reliability of weather radars under more diverse and complex weather situations (e.g. Zittel and O'Bannon, 1993). Some of the approaches are summarised below.

Ray and Ziegler (1977) have suggested a different one-dimensional (i.e. processing data along a radial) technique for dealiasing that corrects Doppler velocities by multiples of the aliasing interval $2v_u$ such that the corrected velocities have a normal distribution about their mean value. This method can be robust, but suffers from difficulties of implementation in the presence of severe aliasing which causes a large spread of velocities and biases the sample mean. Another continuity-based single-radial method of Doppler velocity dealiasing has been suggested by Bargen and Brown (1980), which compares individual velocity values with an average of the preceding values and minimises their difference to determine the true aliasing interval in which a data point should lie. However, the algorithm requires operator intervention to handle velocity fields with difficult aliasing problems.

The next higher level of algorithms for Doppler velocity dealiasing is performed in two dimensions, i.e. they involve the processing of data from an entire scan. Merritt (1984) proposed a two-dimensional processing approach consisting of three main steps. First, data from a complete azimuthal revolution of the radar antenna at a constant elevation angle is segmented into regions, each of which contains velocities that are close (within some percentage of the unambiguous velocity) to each other. In the second step, the aliasing interval of each region is adjusted so as to minimise shear (differential Doppler velocities) along the borders. The third step uses an estimate of the ambient wind field, derived from a wind model, to determine the proper aliasing interval of the large areas already rendered internally consistent in terms of velocity continuity.

Merritt's algorithm suffers from the possibility of propagation of possible errors in the wind field model. To overcome this problem, Boren *et al.* (1986) added to the algorithm a wind field model monitor which uses a set of rules to determine whether the wind field model is correct. If the model is found to be incorrect at any stage, it is abandoned and replaced with a preceding model that was considered valid, thus arresting the propagation of the model error. The Merritt technique has been further enhanced by Bergen and Albers (1988) through the addition of a noise filter, a mechanism to deal with ground clutter, and the use of a sounding rather than a wind field model to estimate the aliasing present in distant echo fields which do not have continuous velocity measurements out to their range.

Merritt's technique, and the family of algorithms it has spawned, have the common drawback of being highly computation-intensive. The basic methods themselves require high amounts of computation because of the need to manipulate large two-dimensional data arrays (920 range bins × 360 azimuthal positions in the case of WSR-88D, as evident from Table 7.1). In addition, the presence of strong wind shears causes problems in proper dealiasing, the resolution of which may require a second pass through the data field (as in the case of the Bergen and Albers algorithm, which is the most developed member of the family), adding further to the computational load (Eilts and Smith, 1990). In spite of the great strides made in computer-related technologies, two-dimensional velocity dealiasing algorithms need to be made more efficient before they can be robustly implemented in real time at affordable costs.

7.3.5 Potential and futuristic methods

The ambiguity problem in Doppler weather radars is serious enough to attract considerable research attention, and a variety of methods has been suggested which hold the potential for being developed into efficient techniques and algorithms in the near and distant future (Mahapatra et al., 1993).

In general, the range and Doppler ambiguities in coherent radars can be resolved at either the signal processing level or the data processing level. In the *signal processing* approach one treats the radar echoes in such a way as to determine the true (unambiguous) Doppler velocity *before* a numerical value is associated with the velocity and stored as a data point. As a rule, signal processing aims to resolve the ambiguity at each individual resolution volume independently, without reference to other data points within the data field. In the *data processing* approach, on the other hand, the Doppler value of each resolution volume is first recorded as it is sensed, which may or may not be ambiguous. Then an attempt is made to detect and resolve possible ambiguities at each data point by examining the point in relation to other points in the neighbourhood, or even the entire data field. It is possible to adopt a hybrid approach, combining the signal processing and data processing routes, to fight the ambiguity problem. In such an approach, the signal processing route is used to resolve as much ambiguity as possible before data points are recorded, and any residual ambiguity may be resolved by examining each data point in relation to others in the field.

Both the above generic approaches have their respective advantages. The signal processing route potentially holds possibilities of higher speed, robustness and resource saving, while the data processing route to ambiguity resolution offers a wide diversity of algorithms from the very simple to those that are quite complex. Further, since the data processing stage follows signal processing, any deficiencies of the former in resolving ambiguities can be made up by the latter. It is worth pointing out here that some of the methods based on either of the two approaches may require additional capabilities on the part of the radar system, or at least a control over certain operating modes or parameters of the radar. A few potentially powerful methods of ambiguity minimisation and resolution based on the two approaches are briefly outlined below.

7.3.5.1 Spectral decomposition

The spectra of Doppler velocities within single resolution volumes are generally of a Gaussian shape. Overlaid resolution volumes, because of their separation by large distances (one or more unambiguous range intervals), would in general have significantly different spectral mean and width values. Thus, when their echoes are superimposed (e.g. due to range folding), a Fourier analysis of the composite signal would generally show bimodal or multimodal spectra, somewhat like the one shown in Fig. 6.15b. By attempting a Gaussian fit on each of these modes, it should be possible to separate the individual spectra. This would not only yield the Doppler velocities within each of the overlapping resolution volumes, but even the power levels of the returns from each volume (Waldteufel, 1976). Of course this technique would not be successful when the contributing Doppler spectral peaks are so close that the modes cannot be separated, but such cases would statistically constitute only a small fraction of the overlaid pixels. It is to be expected that the method would be easier to implement when only two pixels overlap (i.e. there is a single folding), and would be progressively more difficult and less reliable as the number of overlapping pixels increases.

The spectral decomposition method represents a signal processing approach to ambiguity resolution. A major drawback of the method is its relative complexity. Time series data from each range must be processed through a fast Fourier transform (FFT) algorithm and then subjected to mode separation, which is a fairly complex process especially if more than two pixels overlap. Further, when the spectra of returns from one or more of the overlapping resolution volumes depart significantly from the Gaussian shape, their separation becomes even more difficult. In extreme cases, such as in tornadic storms, the individual spectra may themselves be bimodal, which the mode separation algorithm would find extremely difficult or impossible to handle.

7.3.5.2 Triple-PRF radar observation

This is a data-processing-based method that imposes some special requirements on the radar scan cycle and the corresponding PRF variation scheme. Doppler velocity aliasing is a result of undersampling of the relatively high Doppler frequencies by the radar pulses. Since the true Doppler velocity and the apparent or aliased velocity are related by the unambiguous velocity of the radar system, the observation of a particular Doppler velocity at two (or more) different PRFs (i.e. sampling rates) would yield information that can be used to resolve the velocity ambiguity.

One scheme of varying the PRF is to alternate it between two values in quick succession, even from pulse to pulse. However, there are difficulties with this scheme, especially from the point of view of ground clutter rejection (see discussion of the method below). A simpler scheme for Doppler observation at two different PRFs, while also performing range unfolding, is to scan at a given elevation three times, each time with a different PRF. The first one, with a long PRT (e.g. the 'continuous surveillance waveform' of WSR-88D with an unambiguous range of 460 km) can be used to collect reflectivity data without range ambiguity, as explained in Section 7.3.2.1. The other two are range-ambiguous PRFs or 'continuous Doppler waveforms' that are different in value and thus provide dual sampling of the same Doppler velocities, from which the true or unambiguous Doppler velocity can be derived. Compared with the standard scan cycles of the WSR-88D radar which involve double-scanning at the lowest two levels, the triple-PRF scan would entail a small percentage of increase in the total scan time.

7.3.5.3 Staggered PRT scheme

In the staggered PRT method (Zrnic' and Mahapatra, 1985) the PRT is varied sequentially from pulse to pulse to effectively obtain a large unambiguous velocity. In the simplest and most commonly used technique of this type, a dual PRT is used, with two constant values of PRT alternating between successive pulses. The unambiguous velocity for such a scheme is inversely proportional to the difference between the two PRTs, and can be increased by choosing the two PRTs closer to each other. However, there is a limit to the proximity of these two values, because the velocity estimates also become less accurate (i.e. the variance of the velocity estimates increases) as the difference between the two PRTs decreases. The rule for deciding the values of the two PRTs is to have them as different from each other as possible while providing the required unambiguous velocity, and to check whether the necessary velocity estimation accuracy is obtained with these values. In the case of the WSR-88D radar, it can be shown that alternating PRT values of 0.56 and 0.97 ms can provide an unambiguous range of 230 km (as specified in Table 7.1) and an unambiguous velocity of 60 m/s simultaneously. However, the specified velocity estimation accuracy requirement of 1 m/s can be met only at a rather high signal-to-noise ratio of 20 dB instead of a more desirable 5 dB.

Apart from the difficulty about estimation accuracy, the staggered PRT scheme has certain other major drawbacks. The first is that the measurement of reflectivity and the assignment of correct ranges to overlaid reflectivity values follows a somewhat complex logic (Zrnic' and Mahapatra, 1985). The second difficulty with the staggered PRT scheme is that the realisation of clutter filtering is difficult, and in general larger clutter residues are left over than in the case of uniform PRTs. This is a serious deficiency in the case of sensitive weather observation as in the aviation context, where significant aviation hazard can arise from clear-air phenomena which may get obscured by the enhanced clutter residues. Where ground clutter is not a serious problem, such as for antenna elevation angles above a couple of degrees, the staggered PRT scheme provides a viable solution to the ambiguity problem. Such a scheme was indeed developed for the Interim Terminal Doppler Weather Radar (ITWR), a derivative of the developmental model of the current WSR-88D which was devised to test certain concepts relating to the Terminal Doppler Weather Radar (TDWR), which is described in Section 7.5.

7.3.5.4 Random phase transmission

Echoes due to different transmitted pulses can be 'recognised' by coding the transmitted pulses in distinctive ways. One of the simplest methods of achieving this is to provide a random but known phase shift from pulse to pulse. This would cause each echo signal to be coherent only with respect to the transmitted pulse that caused it, but incoherent with other transmitted pulses (Zrnic' and Mahapatra, 1985). By storing the known phases of the individual transmitted pulses digitally, it is possible to recohere (i.e. render coherent again) separately the echoes from each unambiguous range interval. The overlaid echoes from other range intervals would then appear as white noise (Laird, 1981). The recohered echoes can be adaptively filtered, and a subsequent recohering of the residues will enhance the ratio between the in-trip (i.e. range-unambiguous) signal and the overlaid signal (Siggia, 1983).

7.3.5.5 Systematic discrete phase coding

Sachidananda and Zrnic' (1986) proposed a method for recovering spectral moments from overlaid weather echoes in a Doppler weather radar. The method essentially imparts to the successive transmitted pulses a periodic sequence of discrete phase shifts, some possible sequences being $(0, \pi/4, 0, \pi/4, \ldots)$, $(0, 0, \pi/2, \pi/2, 0, 0, \pi/2, \pi/2, \ldots)$, $(0, 0, 0, \pi, 0, 0, 0, \pi, \ldots)$, etc. This has the effect of splitting the spectrum of second-trip echoes and possible multiple-time-around echoes such that the mean velocity estimates of the first trip echoes (i.e. echoes from within the unambiguous range) are not biased. By repeating the same procedure for the second trip, the biasing effect of the first trip echo on the second can be eliminated. It was shown through analysis and simulation that in a situation of single range folding, both the overlapping signals can be independently recovered even when their power ratio is as high as 15 dB. Like the random phase method discussed above, the systematic phase coding method can, in effect, enhance the unambiguous range readily by a factor of two without sacrificing the unambiguous velocity.

The systematic phase coding method also has some drawbacks, important among which are the following:

1. The accuracy of some of the spectral moment estimates, especially the spectrum width, is highly sensitive to the relative location of the overlaid spectra. For overlapping or closely spaced spectral peaks, the estimate is highly erroneous. The spectrum width estimate is also strongly dependent on the power ratio of the overlaid spectra.
2. The method is realisable quite neatly for resolution of echoes from two trips, but becomes much more complex and error-prone in the presence of echoes from third and higher trips.

3. For comparable power levels of the two overlapping echoes, the estimation of the mean velocity of each has a high variance. To keep this within acceptable limits, a large number of samples would have to be used for autocovariance estimation (see Section 6.6.3.2), which would necessitate a slow scan rate of the antenna.
4. To be able to widen the range of power ratios of overlapping echoes over which spectral moments can be estimated with acceptable accuracy, a combination of spectral processing and autocovariance processing (Section 6.6.3.2) must be used, which is highly computation-intensive and cannot be readily implemented in the case of multiple range folding.

7.3.5.6 Single-pulse Doppler estimation

Since the primary reason for velocity aliasing in Doppler radars is the sampled nature of the radar signals, the problem would not arise if the Doppler velocities were estimated from nonsampled signals. An example is the laser Doppler radar in which velocities are usually estimated from echoes of single transmitted pulses. Since Doppler shifts are high at laser frequencies, a significant number of cycles of the Doppler waveform would be contained within one transmitted pulse, permitting straightforward measurement of the Doppler frequency. Such a principle of measurement is not adopted in microwave radars, as the time window represented by the width of the transmitted pulse is normally a small fraction of a Doppler signal cycle in such radars.

Although a final and accurate estimate of the Doppler velocity is not possible from a single pulse in microwave radars, a very crude mean velocity estimation with an accuracy of the order of the aliasing interval would be very useful in resolving the velocity ambiguity. If even this is not possible, at least an indication of the sign of the Doppler shift would serve to reduce the number of aliasing intervals over which the search for the true velocity is to be conducted. It can be shown that, in the case of a radar like the WSR-88D, Doppler signal phase changes of the order of 0.2° within the transmitted pulsewidth of 1.57 μs would be necessary to resolve Doppler ambiguities when the unambiguous velocity is 30 m/s.

7.4 Other special considerations

Besides paying attention to the range-velocity ambiguity problem, several other factors must be addressed in the context of designing and operating aviation weather surveillance systems based on modern Doppler weather radars. Three such important considerations are discussed in the following subsections.

7.4.1 Coverage

In the aviation context, the primary role of the WSR-88D is as an *en route* weather surveillance device. With a long observation range, the system is well suited for this role. As Table 7.1 shows, the radar is capable of providing the reflectivity picture (and data) up to a range of 460 km and velocity data up to 230 km. With such long observational ranges, a large geographical area can be covered with a relatively small number of radars. Thus, the deployment plan of these radars is such as to cover almost the entire geographical area of the USA, including the surrounding coastal zones and island territories, with the help of ~160 radar systems. The location of these systems is shown in Fig. 7.5. A significant part of the territory has multiple radar coverage to ensure reliable data supply.

The data from these radars can be accessed individually, and the weather picture and data can be combined in a mosaic fashion to get a wide-area view of aviation-significant weather parameters in fine detail. This provides a rich and reliable source of data for aviation weather for *en route* flight operations.

Although the WSR-88D system is ideal for providing *en route* weather data, because of its good resolution and close-range performance, it can also be a valuable source of weather information for terminal area aviation operations if it is suitably located with respect to airports. This aspect is discussed below.

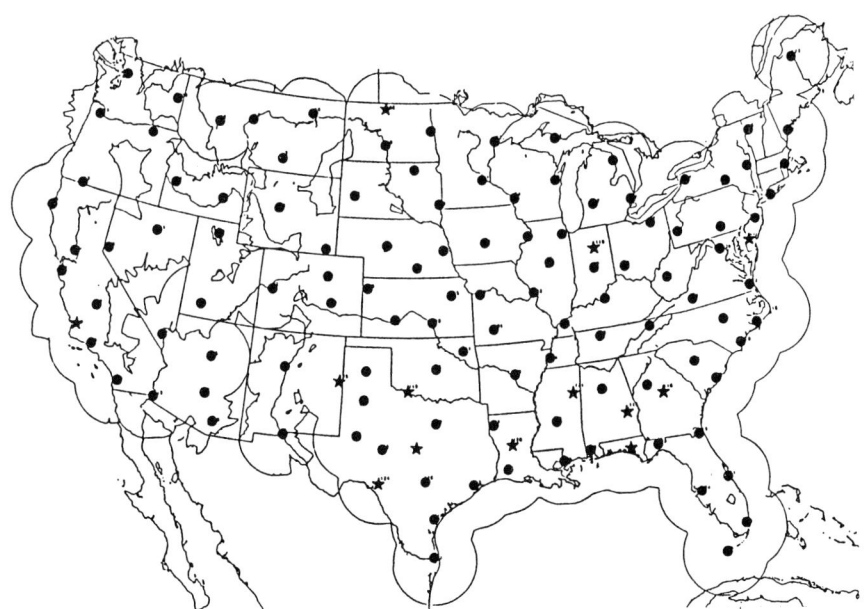

Figure 7.5 Planned coverage of the conterminous USA by a network of WSR-88D radars

7.4.2 Siting for terminal area surveillance

To best satisfy the requirements of terminal area weather surveillance, the location of a weather radar must be chosen carefully relative to the airport. Optimal siting of radars for aviation weather surveillance has been a topic of significant study (e.g. Mahapatra and Zrnic', 1983).

Several factors must be considered carefully to evaluate and optimise radar sites for terminal area weather surveillance. Some of the important ones are as follows.

7.4.2.1 Resolution

As already discussed (in Section 5.5), terminal area weather surveillance requires much finer spatial resolution than *en route* surveillance, the requirement being finest in the central zone of the terminal area. A typical weather radar with a 1° beamwidth would not be able to meet such stringent resolution requirement if it is not located sufficiently close to the centre of the runway complex. If ρ_{aa} is the linear resolution requirement over the airport area of radius r_{aa} (see Fig. 5.3), then the maximum permitted distance $r_{max(aa)}$ of a radar of beamwidth θ_b from the centre of the terminal area, from airport area resolution consideration, would be

$$r_{max(aa)} = \frac{\rho_{aa}}{\theta_b} - r_{aa} \qquad (7.5)$$

Similarly, if ρ_{ta} is the linear resolution requirement over the terminal area (outside the airport area) of radius r_{ta}, then the maximum permitted distance $r_{max(ta)}$ of the radar from the centre of the terminal area, from terminal area resolution consideration, would be

$$r_{max(ta)} = \frac{\rho_{ta}}{\theta_b} - r_{ta} \qquad (7.6)$$

Then the maximum permitted distance $r_{max(\rho)}$ of the radar to meet the resolution requirement everywhere within the terminal area would be the lower of the two limits, i.e.

$$r_{max(\rho)} = \min \{r_{max(aa)}, r_{max(ta)}\} \qquad (7.7)$$

7.4.2.2 Range coverage

Among the phenomena that affect terminal area flight, the weakest from the point of view of radar detection is low-altitude wind shear without precipitation. The farthest distance of radar location is decided by the criterion of detecting such phenomena. The WSR-88D, with its high sensitivity and dynamic range to detect clear-air phenomena, offers considerable flexibility in location relative to the terminal area.

Also of importance in weather radar siting is the minimum observation range of the radar, which is determined by the following three major factors:

1. the recovery time of the receiver (and/or duplexer) following the transmitted pulse
2. the distance to which ground clutter returns received via the main lobe and sidelobes of the antenna pattern saturate the receiver
3. the distance to which the part of the transmitter phase noise reflected by the ground clutter is strong enough to interfere with the weather signal.

For modern radars such as the WSR-88D with very low recovery times, the minimum range is usually decided by the second factor for strongly scattering weather phenomena (e.g. those accompanied by precipitation), and by the third factor for weak ones (e.g. clear-air phenomena).

7.4.2.3 Low-altitude coverage

This is an important criterion as the terminal area, including the airport area, handles aviation activity down to the ground level, and significant wind shear can occur at very low altitudes. Of particular concern are microburst outflows which, as discussed in Section 4.7.2, usually produce their highest wind shear well below 100 m of height from the ground. This explains the lower limit of 200 ft (61 m) of the altitude coverage slab shown in Fig. 5.3. The minimum observable height is primarily determined by the radar horizon, surface obstructions and ground clutter considerations. All these factors are dependent on the location of the radar relative to the centre of the surveillance zone.

7.4.2.4 Zone of blindness

Depending on its scanning scheme, a radar system may be blind to a volume of space directly above itself. If e_{max} is the highest elevation angle of the radar scan cycle and H_{max} the maximum height of the slab of space under surveillance, then the maximum radius of the blind zone r_b (Fig. 7.6) is given as

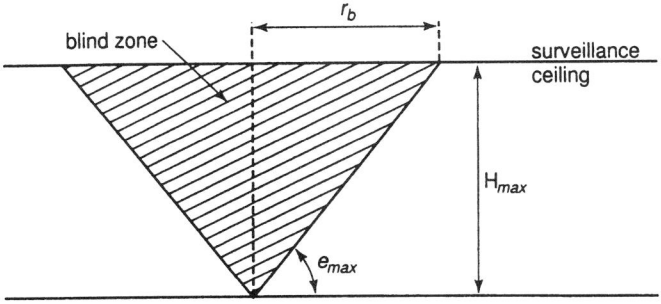

Figure 7.6 Geometry of radar blind zone

$$r_b = H_{max} \cot e_{max} \tag{7.8}$$

The siting of the weather radar must be such that the blind zone stays clear of the important parts of the surveillance space.

It is apparent from eqn. 7.8 that the cone of blindness can be narrowed by increasing the elevation limit e_{max}, and eliminated altogether if the antenna scans all the way up to the vertical (i.e. $e_{max} = 90°$). However, for a given density of coverage (i.e. closeness of successive scan levels) the scan cycle time would increase almost in proportion with e_{max}, correspondingly reducing the data update rate (Mahapatra and Zrnic', 1981). In the dynamic aviation context, such slowing down of data rates is not desirable, and hence the tradeoff normally leaves a significant cone of blindness. The WSR-88D has a mechanical limit of 60° for the elevation angle, though in normal operation its elevation coverage is confined to 20° (see Table 7.1). These limits are, however, not determined from aviation weather surveillance considerations alone, since the system is designed for multiagency use.

7.4.2.5 Range ambiguity and overlaid echoes

These phenomena, which were discussed in detail in Section 7.3, are of significance in the choice of weather radar sites with respect to the zone of surveillance. *Prima facie*, it would appear that range ambiguity should not pose any problem for terminal area weather surveillance since the area is small (see Fig. 5.3) and can be readily made to lie within the unambiguous range of typical weather radars. However, the interfering effect of range ambiguity and consequent echo foldovers and overlays does not disappear by merely ensuring this condition. Because of the large dynamic range of the atmospheric phenomena of interest in aviation management, and the inverse-square dependence of weather echo strength on range (unlike the inverse-fourth-power dependence for point targets) foldover of weather patches from out-of-range locations into the terminal area can pose significant problems. For example, the WSR-88D radar has a nominal unambiguous range of 115 km which is almost twice as large as the terminal area radius. Yet, if intense precipitation of reflectivity 50 dBZ occurs at a range of 160 km from the radar, it would fold over into the unambiguous range and appear at a range of 45 km as a patch of strength 39 dBZ. This is still a high level of apparent reflectivity, and can obscure many types of aviation-hazardous phenomena that may be occurring in the same location. While range folding cannot be eliminated, the radar siting can be planned in such a way that the unambiguous range circle clears the boundaries of the surveillance zone by as large a margin as possible.

7.4.2.6 Airport configuration

As a Doppler radar measures only the radial component of the wind speed relative to the radar, it should be located in line with the runway to be able to sense the true headwind or tailwind along the approach path and the runway. If an airport has multiple runways, the most preferred site for the radar would be the common point of intersection of all the runways. However, such a point may either not exist or not be available for the radar installation. In such a case, the location should be as close to the ideal spot as possible, considering constraints such as height limitations on structures in different parts of the airport, possible blockages in the radar field of view, etc. Such a location will permit the radar to measure a significant component of the headwind or tailwind. The wind along the runway may then be estimated from the radial wind component and the angle between the runway and the radar line-of-sight. If a gust front is detected, the winds in its vicinity may alternatively be derived by assuming the wind direction to be normal to the local tangent to the gust front.

7.4.2.7 Comparison of siting alternatives

Locating a weather surveillance radar within the airport area has the advantages of higher sensitivity to weakly reflecting phenomena such as gust fronts and dry microbursts, less severe problems due to overlaid echoes, ability to observe wind shear close to the ground level over the runways, high-resolution observation in the area of runways and glideslope, less possibility of beam blockage, and smaller blind zones. The disadvantages of such a location include less flexible siting options, more critical location of the blind zone, and serious clutter interference while observing the runway complex, which demands a stringent clutter rejection performance.

In contrast, if the radar is sited outside the airport area, but within the terminal area, then ground clutter interference would cease to be a serious problem while observing the runway complex as well as the critical parts of the glideslope. Further, greater flexibility in radar siting would be available, the airport area can be kept free of the blind zone, and reasonably good resolution can be obtained over the airport area. Such a radar location, however, suffers from the disadvantages of a larger blind zone, a significant main-beam clutter problem, and insufficient resolution over parts of the terminal area.

Finally, locating the radar outside the terminal area provides the possibility of sector scan to cover the terminal area (reducing the scan time), greatest flexibility in finding a site, and clutter-free observation of the entire terminal area. The drawbacks of such radar sites include a high radar horizon which restricts the minimum height of observation over the airport area, insufficient resolution, insensitivity to detection of clear-air phenomena, and stronger interference due to overlaid echoes.

Considering all factors, Mahapatra and Zrnic' (1983) recommended siting at a distance between 10 and 12 km from the centre of the runway complex for a radar such as the WSR-88D. However, most of the considerations leading to this conclusion are generic, and such a distance criterion has been adopted

270 *Aviation weather surveillance systems*

in the case of many TDWR installations, which are primarily intended for terminal area observation.

7.4.3 Scanning strategies and modes

As weather surveillance radars have pencil beams, the beams must scan spatially to collect data from the zone of observation. The scanning is usually carried out in a cyclic fashion, with the beam passing through successive positions in a definite and repetitive time sequence. The *scan cycle* or *scan strategy* is not unique, but must be tailored for each particular surveillance application. In general, optimising a scan cycle would involve a tradeoff between conflicting requirements. Some of the major considerations of a good scanning strategy in the aviation context are discussed below.

Unlike conventional weather radars which usually performed a single-level scan in a nearly horizontal plane, modern weather radars are capable of observation of volumes of space. Such three-dimensional observation gives a more complete picture of the weather phenomena than planar scanning, which essentially 'cuts' a single slice through the weather field. Volume scanning is of special value for aviation because aviation itself is a three-dimensional operation, and detailed weather information at the flight altitude greatly enhances flight safety and facilitates flight management. Further, the weather phenomena of utmost interest to aviation, which are severe local phenomena such as thunderstorms, are faithfully reconstructed and their wind fields accurately estimated and predicted if they are observed in three dimensions. For these reasons, it is imperative for modern aviation weather surveillance radars to perform volume scans.

Scanning at multiple levels, of course, requires correspondingly more time than planar scanning. The beam thus takes several times more time before it returns to a given position, proportionately slowing down the rate of data refreshment or updating. This is a serious drawback in the aviation context, where longer data update intervals may mean failure to observe fast-developing hazardous phenomena. A primary condition on scan strategies for aviation weather surveillance is therefore the minimisation of scan cycle time while providing as complete a volume coverage as possible.

A dense coverage of the scan volume would require contiguous or stacked-beam scanning, i.e. the incremental elevation angle between successive scans should be equal to the beamwidth itself (see Fig. 7.7). Such a scan strategy ensures that no part of the surveillance volume escapes observation. However, stacked-beam scanning is not practical for modern high-resolution radars, which have beamwidths of the order of a degree and would require far too many azimuthal scans to cover a reasonable elevation interval, say 20 or 30 degrees, resulting in unacceptable scan cycle times. Practical scan cycles usually have only their lowest two or three elevation levels contiguously scanned, while the higher scan levels are spaced at convenient angular intervals. The lowest beam positions are contiguous because these elevations

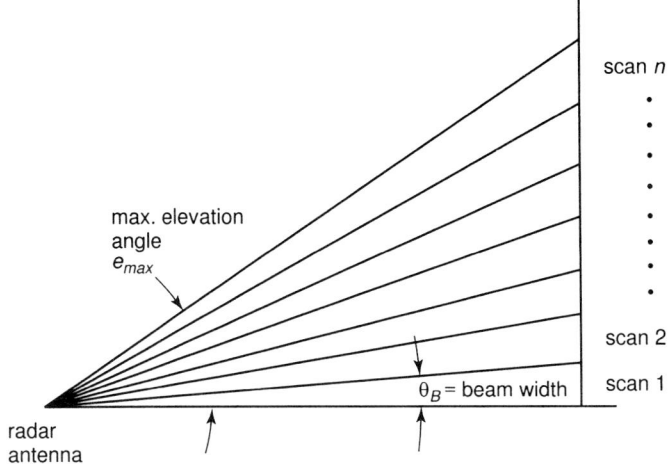

Figure 7.7   Schematic diagram of a stacked-beam scanning scheme

are used for long-range observation. Large gaps at these elevations would result in very coarse vertical sampling of faraway phenomena, and at extreme ranges all but the lowest beam position may miss the phenomena altogether. For radars observing terminal areas alone, such long ranges are not of interest, but still dense scanning is necessary at the lower elevations since the wind fields at the lowest altitudes are of utmost importance for aircraft landing or taking off, and hence must be accurately mapped.

An important parameter of the scan strategy is the angular speed of scanning or the *scan rate*. A rapid scan rate of the radar antenna requires powerful and heavy drive mechanisms which are expensive and power-consuming. The maximum scan rate capability of weather radars is therefore normally kept as low as possible while meeting the operational requirements. However, a slower scan rate makes the radar take a longer time to complete a given scan cycle, which is undesirable for aviation weather surveillance. For this reason, modern weather radars providing aviation support have fairly high scan rates. In the case of the WSR-88D the maximum scan rate capability is 36°/s in both azimuth and elevation (see Table 7.1). This is equivalent to a rotational speed of 6 rev/min in azimuth; in the elevation dimension such a measurement does not carry much meaning under normal scanning since elevation scanning is carried out only over limited angles or sectors.

A weather radar does not, however, always perform scanning at its maximum capability merely to minimise the scan cycle time. It is often necessary to scan well below the maximum limit. The chief reason for slowing down the scanning is to collect a larger number of echo samples during the time that the beam dwells on any given resolution volume. Within certain limits, a higher number of pulses used for signal processing can yield better spectral parameter estimates and provide improved clutter rejection. Slow scanning is usually necessary for the lower beam elevations (where ground clutter interference is higher) and for clear-air observation (for which the signal-to-noise ratio is poor).

Scan cycles of modern aviation weather surveillance radars often involve repeated azimuthal scanning of certain elevation levels, usually the ones close to the ground. One reason for repeat scanning is related to range ambiguity resolution, as discussed in Section 7.3.2. Another reason is to meet simultaneously the conflicting requirements of dense volume coverage (which requires a correspondingly long scan time) and fast data update (which calls for a short scan time). A scheme of sequential scanning through monotonically increasing or decreasing elevation levels means that each level is visited only once during the scan cycle. A compromise may be achieved by breaking the monotonicity of the ascent or descent of the beam to revisit the lowest one or two elevation levels at about the middle of the scan cycle. These extra scans add only marginally or modestly to the total scan cycle time, but double the frequency of updating data from the lowest levels of the atmosphere which harbour the phenomena of strongest hazard potential for aviation, including microburst-induced divergence.

A typical volume scan of the WSR-88D radar in the precipitation mode for severe weather (Volume Coverage Pattern 11) has 14 scan levels from 0.5 to 19.5° in elevation (the angles denote elevation of the beam centreline or boresight). As shown in Fig. 7.8, the lowest seven beam positions, up to a beam boresight elevation angle of 6.2°, are stacked or contiguous, and the higher beam positions have gaps among them. The lowest two beam positions (i.e. 0.5 and 1.45°) are scanned twice, for reasons explained in Section 7.3.2.1, giving a total of 16 cycles of antenna rotation to cover one volume scan. The angular scan rate of the antenna varies between 2.69 and 4.45 rev/min among the scan levels, the higher rates being applied to the top seven scan levels where clutter interference is low. The time taken for this volume scan is ~5 min.

A second typical scan mode for precipitation (Volume Coverage Pattern 21) has nine scan levels, also between 0.5 and 19.5° elevation, with the lowest five levels in common with Pattern 11 above. Here again the lowest two levels are scanned twice, giving a total of 11 circular scans within the volume scan. The angular scan rate of the antenna is slower in this scheme, between 1.86 and 2.40 rev/min. In spite of the lower number of scan levels, this scan cycle takes a longer time of ~6 min.

WSR-88D scan modes for clear air take even longer for a volume scan. For example, the Volume Coverage Pattern 31 (clear air mode/long pulse) scans between elevation angles of 0.5 and 4.5°, with 1 deg. increments, and the two lowest levels are scanned twice each. At a uniform scan rate of only 0.84 rev/min, such a cycle takes ~10 min to complete.

Modern Doppler weather radars for aviation 273

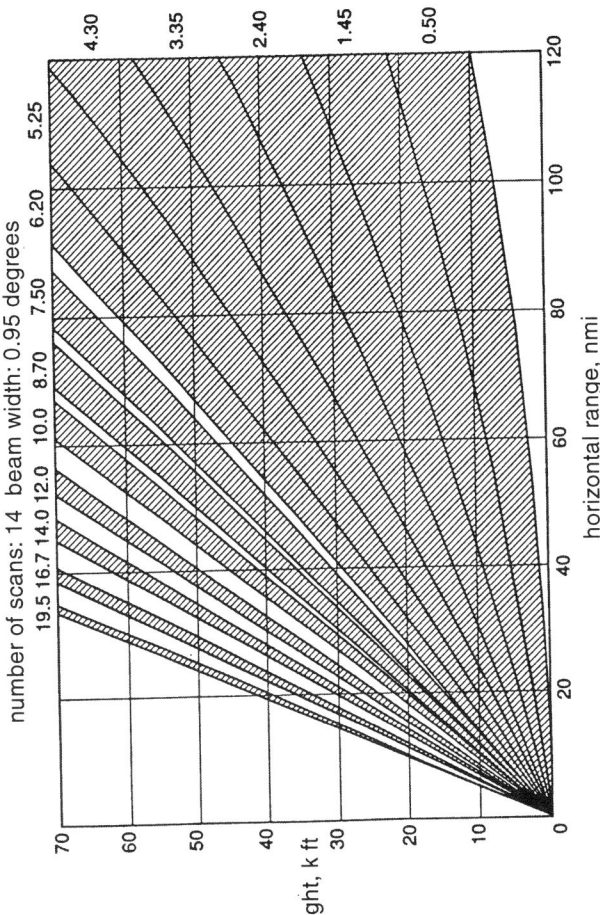

Figure 7.8 Vertical section through a typical WSR-88D scan cycle. The higher beams appear thinner as the horizontal scale of distances is more compressed than the vertical scale. The beam edges are slightly curved because of the combined effect of the earth's curvature (straightening the curved surface of the earth bends lines of constant elevation angle upwards in a range–height plot) and the atmospheric refraction (from Dunbar and Mittelman, 1993)

7.4.4 Data lag

The relatively long time taken for completing a volume scan is cause for some concern in the aviation context. As seen above, a typical volume scan cycle of the WSR-88D radar in the precipitation mode takes ~5–6 min. The radar processor collects and stores data from all the scan levels or 'tilts' in a volume scan before combining them together to generate the composite products needed for aviation use. Thus, data from the first scan level, usually the lowest, is already 5 or 6 min old by the time that the data begins to be processed. Adding the processing time, the time for data handling within the air space management system, and the wait till the next update, some part of the data may be as old as 12 min before being available for aviation use. This phenomenon is called *data aging* or *data lag*, and is undesirable from the aviation point of view. More discussion about the effects of data lag follow in the subsection below.

7.4.5 Comparison with air route surveillance radar

As already mentioned, the WSR-88D would not fully meet the stringent weather observation requirements in the terminal areas, and its primary use in aviation is as a weather sensor for *en route* air navigation. It is instructive to compare its effectiveness in this role with an existing radar that performs the same task. The reference here is to the Air Route Surveillance Radar (ARSR), which is currently used by *en route* air traffic control centres to track aircraft as well as to observe weather.

The ARSR is primarily an aircraft surveillance radar, but it also possesses weather observing capability. It employs a vertical fan beam scanning horizontally at 5 rev/min, and has a nominal range of ~280 km. Its weather channel categorises weather (rainfall) intensity into just two levels: 'moderate' and 'heavy'. The moderate rain here has a reflectivity lying in the band 30–41 dBZ, which corresponds to level 2 of the US National Weather Service (see Table 6.2). All weather patches with reflectivity higher than 41 dBZ are classified and displayed as heavy weather. ARSRs usually combine reflectivity data from 12 consecutive scans. Since each scan takes 12 s, the data update interval is usually ~144 s (Federal Aviation Administration, 1989). Aircraft positions are also depicted in the same display as small red squares, each with a pair of accompanying numbers showing the beacon identification and flight altitude level (altitude in hundreds of feet) of the aircraft.

The WSR-88D, being a more modern and dedicated weather radar, is designed to provide weather data of better quality than the ARSR. It also has altitude discrimination, i.e. the capability to provide the weather picture at different altitudes, while the ARSR does not have this capability; it provides only a single reflectivity value for the entire depth of its beam at each range location. Further, WSR-88D can provide many levels of reflectivity discrimination (typically 8 or 16 levels for aviation use, more for other purposes) compared to only two levels for ARSR. Yet, the ARSR retains the advantage of a faster data update, and hence a less serious problem of data lag, compared to the WSR-88D. This and other points of difference have prompted comparative studies between the two systems as weather sensors for *en route* aviation (Dixon, 1992; Dunbar and Mittelman, 1993). The conclusions of the latter cited study are summarised below:

1. The ARSR makes a coarse presentation of weather cell boundaries, while the WSR-88D presentation is precise.
2. Reflectivity level thresholding is rather variable. Thus even two or more ARSRs observing the same weather patch often provide significantly different boundaries for the patch. The ARSR also tends to overestimate the area covered by heavy weather. In contrast, the detection of reflectivity level by WSR-88D is accurate and consistent.
3. The ARSR has relatively high levels of missed detection and false alarm for weather, i.e. it may miss significant weather over certain areas, while it may show nonexistent patches of weather elsewhere. The WSR-88D is more consistent in its depiction of weather fields.
4. The depiction of weather by the ARSR has less correspondence with the pilot's view of weather than does the depiction by WSR-88D. Study of aircraft flying in active weather environments showed that many more aircraft flew through areas depicted as having heavy weather by the ARSR than through the corresponding areas marked by WSR-88D. Since aircraft have a more direct view of weather ahead through their own airborne radars and/or visual observation, comparisons showed weather delineation by the WSR-88D to be far more realistic than that by the ARSR.
5. The effect of data lag in WSR-880 is clearly visible in the case of fast-moving weather cells. However, the problem appears to have little operational impact.
6. The overall conclusion is that the weather product output of the ARSR class of radars is not well suited for tactical air traffic control, while the WSR-88D is better suited for this purpose. The drawback of data lag in the case of the latter system is overwhelmingly compensated by its much better capabilities and much higher accuracy and reliability in weather field presentation.

The operational implication of this observation on the aviation system is that the ARSR can mislead air traffic controllers while guiding aircraft in bad weather environments, and maintaining a higher margin of weather avoidance to account for observational uncertainties would reduce the handling capacity of the air traffic control system. By obviating these difficulties, the WSR-88D can enhance flight safety and system capacity, and reduce pilot-to-controller communication loads.

7.5 Terminal Doppler weather radar (TDWR)

During the development of the WSR-88D radar system, it became evident that such a multipurpose radar would not be able to observe the air space in and around terminal areas with the degree of focus, reliability, agility and versatility required for supporting modern high-density and dynamic aviation operations. This is because of the compromises necessitated by the multiagency role of the radar, as well as the task of wide-area surveillance.

The origin of the Terminal Doppler Weather Radar (TDWR) is in response to a need felt among aviation planners for closer and more dedicated surveillance of aviation-significant weather, especially wind shear, in terminal areas. In particular, the development of the TDWR was strongly motivated by the requirement of timely and reliable detection of microbursts (Turnbull *et al.*, 1989; Turnbull, 1995). Its other major functions are the detection of gust fronts and the prediction of wind shifts due to such fronts, as well as the estimation and graphic display of precipitation over terminal areas. Yet other uses or potential uses of the TDWR are in the prediction of storm movements, estimation of turbulence in areas of precipitation, tornado detection, and the prediction of the surface impact of microbursts and the initiation of convection leading to the formation of thunderstorm cells. The TDWR, being located within the airport area for dedicated terminal area surveillance, can effectively observe microburst outflows which have depths varying from 300 to 1200 m, with the strongest winds in the lowest 100 m (see Section 4.7.2). To facilitate the measurement of wind shear at such low altitudes, efficient ground clutter rejection is a primary design feature of the TDWR (Michelson *et al.*, 1990). Further, in its scan cycles spanning ~5 min, the lowest elevation level is visited once every minute to conform to the microburst detection criterion established in Section 5.4.

Table 7.5 lists some of the basic parameters of the TDWR system. Unlike the WSR-88D, which has a transmitted wavelength in the vicinity of 10 cm (S-band), the TDWR operates in the 5 cm wavelength band (C-band). A major reason for the choice of such a wavelength was the lack of spectrum allocation in the S-band (Michelson, *et al.*, 1990; Evans and Turnbull, 1989) which is a sensitive band in the vicinity of airports. The 5 cm wavelength has the advantages, relative to the 10 cm band, of lower radio-frequency interference and better signal-to-clutter ratio from equivalent precipitation (Michelson *et al.*, 1990). However, it has the disadvantage of higher attenuation rates while propagating through heavy rain. For example, at a temperature of 18°C, rain with a reflectivity factor of 50 dBZ attenuates 5 cm radar waves by as much as 0.2 dB/km (Kessler, 1990). The effect of such attenuation is to cause a significant underestimation of reflectivity, and consequently precipitation intensity, at locations which are screened from the radar by a thick layer of heavy precipitation (Hogg, 1969; Lin, 1975; Allen *et al.*, 1981).

It is possible to correct for the rain attenuation effects indirectly. Such correction is based on the fact that the attenuation rate is a function of the local rainfall rate, which in turn is expressible in terms of the reflectivity factor. By using the relationship between the local reflectivity factor and the attenuation rate along the radar beam, the true rainfall rates may be estimated through an iterative procedure (see e.g. Hildebrand, 1978) which

Table 7.5   Basic parameters of terminal Doppler weather radar

Parameter	Value
Radar system	
Range of observation:	
Radial velocity	89 km (±1 m/s accuracy)
Reflectivity	460 km (±1 dB accuracy)
Antenna	
Beamwidth	<0.55° (pencil beam)
Gain	50 dB
Sidelobes:	
Near-in	−27 dB
Far-out	−40 dB
Transmitter	
Frequency	5.60–5.65 GHz
Power:	
Peak	250 kW
Average	550 W
Pulsewidth	1.1 μs (165 m range equivalent)
Pulse repetition frequency (PRF)	2000 Hz (maximum)
Receiver	
Linearity	61 dB
Noise figure	2.3 dB
Dynamic range	129 dB
Sensitivity-time control (STC)	26 dB
Automatic gain control (AGC)	42 dB
System clutter suppression	55 dB

is usually valid within certain altitude limits. However, it must be remembered that the reflectivity–attenuation relationship has appreciable uncertainties because of its dependence on temperature and the hydrometeor distribution in the precipitation which are not sensed by the radar. This leads to significant residual errors in the reflectivity estimation by the iterative methods (Kessler, 1990). The iterative correction procedures are known to be particularly sensitive to radar reflectivity calibration errors, even those as low as 1 or 2 dBZ (Hildebrand, 1978; Hitschfeld and Bordan, 1954), which is the order of accuracy specified for modern weather radars. This effect is pronounced for intense precipitation, corresponding to reflectivity factors of the order of 60 dBZ or higher, which is precisely the class of precipitation most hazardous for aviation. In fact, for such high precipitation rates, iterative correction schemes often become intractable because of a tendency to diverge (Hildebrand, 1978). More about the problem of rain rate underestimation due to attenuation, and advanced methods for its mitigation, follow in Chapter 13 on polarisation diversity in weather radars.

A 5 cm wavelength also increases the susceptibility to Doppler ambiguities relative to 10 cm. This is evident from eqn. 7.2, which shows that the unambiguous velocity limit for a 5 cm radar would be only half that for a 10 cm radar if the PRT is kept constant. Hence Doppler velocities that are not aliased when sensed by a 10 cm radar may get aliased by a 5 cm radar, and velocities aliased by the former would undergo aliasing of a higher order by the latter. Attempts to restore the unambiguous velocity by increasing the PRF (i.e. decreasing the PRT) results in a proportionate reduction of the unambiguous range as per eqn. 7.1. The tradeoff between the unambiguous range and unambiguous velocity for a 5 cm radar is given by the combined range-Doppler ambiguity envelope

$$r_u v_u = 1875 \tag{7.9}$$

which may be compared with the envelope (eqn. 7.4) for 10 cm wavelength.

The functioning of the TDWR is, however, not unduly impaired by its relatively restricted ambiguity envelope because it is required to perform surveillance only over a limited zone. Table 7.5 shows that the range of Doppler velocity observations of the TDWR is 89 km, which is enough to cover the terminal area and an appreciable distance beyond. Compare this with the corresponding requirement of 230 km for the WSR-88D system.

The limited surveillance zone does not make the TDWR free from the ambiguity problem, however. As discussed in Section 7.4.2 in the context of range ambiguity and overlaid echoes for terminal area surveillance, overlay of distant storms (i.e. those occurring outside the zone of interest) can cause serious problems due to the obscuration of weather features within the surveillance zone. The TDWR is capable of selecting its operating PRF to mitigate the problem of overlay over specified regions. In each scan cycle it performs a low-PRF surveillance scan with an unambiguous range of 460 km (Table 7.5). Data from this scan can be used to predict the amount of obscuration as a function of PRF, and a value of PRF can then be chosen adaptively and automatically that minimises the obscuration effect over the selected area (Michelson *et al.*, 1990; Crocker, 1988).

It is interesting to note that both TDWR and WSR-88D have the same range value of 460 km as the outer limit for the observation of reflectivity. This limit is set by geometrical factors related to the earth's curvature, atmospheric refraction and the maximum expected storm height, and not by the inherent capabilities of the radar systems such as transmitted power and receiver sensitivity.

In addition to microburst detection, the monitoring of gust fronts and the resulting wind shifts in the airport area are important functional requirements of the TDWR. The passage of gust fronts often results in sustained change in the prevailing winds. This has serious implications for airport operations such as the decision to change the active runway(s) and the direction of aircraft landing and takeoff, with consequent changes in the approach and departure corridors. To facilitate orderly change, air traffic control supervisors must be advised of the wind changes ~20 min before their actual occurrence (Hansen, 1989). The monitoring of wind shifts requires an estimation of the wind velocity vector (i.e. wind speed and direction) behind the moving gust front, where the reflectivities are typically low (Merritt *et al.*, 1989). The capability of the TDWR to observe over tens of kilometres around the runway complex enables it to detect gust fronts well before they move into the runway area, and thus provide significant warning time. In addition to improving aircraft safety during landing and takeoff and the operational efficiency of the airport, optimal terminal area flight management based on accurate wind shift data can save large amounts of fuel and money, as discussed in Section 3.2.5.

A major advanced feature of the TDWR system is its design for automated operation for the generation and dissemination of high-level weather products. This is possible through the use of powerful software to help the system detect and estimate the parameters of aviation-significant phenomena within its surveillance zone, and present the data, rapidly updated, to users in simple formats. The output data are presented to supervisors in the control tower and the Terminal Radar Control (TRACON) in the form of a 'geographic situation display' in colour. The display, such as the example shown in Colour Plate 21, overlays the location of microbursts and their intensities (wind speed differential), location and speed of gust fronts, and the location and intensity of precipitation on a plan of the runways and the adjacent areas. TDWR output data are also available to air traffic controllers in alphanumeric formats to facilitate their control functions and also transmission to pilots. Wanke and Hansman (1992) have made an experimental comparison of certain microburst alert displays.

Much of the primary software for TDWR has been subjected to rigorous field testing at the Stapleton International Airport (serving the city of Denver, CO) and the Kansas City International Airport in the USA. Extensive tests, during the developmental phase, on detection algorithms using data from various sources showed (Turnbull *et al.*,1989) that microbursts with differential velocities below 20 m/s were detected with ~90% probability, while the stronger ones with differential velocities exceeding 20 m/s were detected with probabilities >98%. The false alarms were in the region of 4 to 5%. Based on data for the Denver area, the detection of strong gust fronts, defined as those with a velocity differential >15 m/s, had a probability of 91%, and the false alarm probability associated with gust front detection was as low as 2%. In 1988 the US Federal Aviation Administration awarded a contract for 47 TDWR systems for installation at various airports within the country (Fig. 7.9 shows their geographical distribution), with the option to procure 55 more systems at a later stage (US Department of Transportation, 1989; Michelson *et*

280 *Aviation weather surveillance systems*

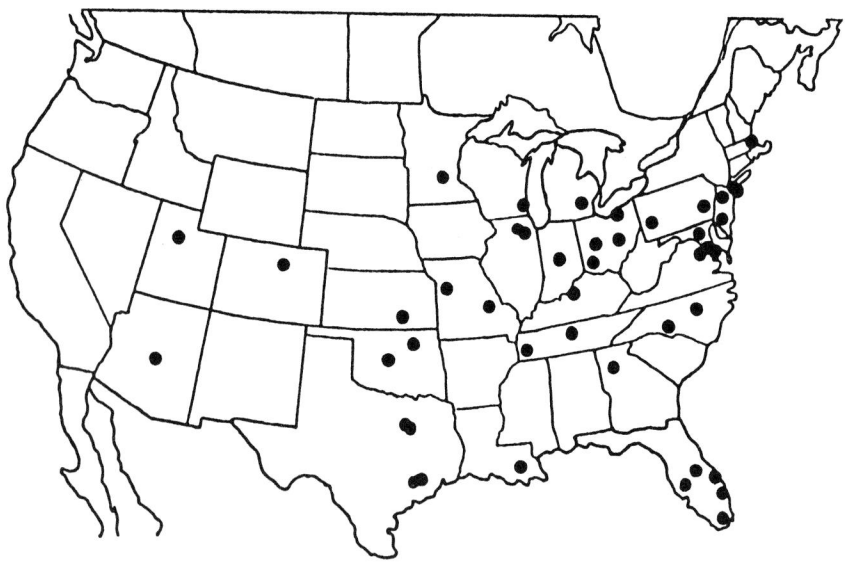

Figure 7.9 Location of 47 TDWR systems in the USA during the first phase of the procurement and installation programme

al., 1990). The allocation of TDWR systems to different airports depends on the importance of the airports in terms of traffic density, and the probability of occurrence of microbursts and thunderstorms in their vicinity.

As discussed in Section 5.2.1, a major advantage of radar surveillance of weather over *in situ* sensing is that radar observes the atmosphere at several elevation levels, providing a detailed volumetric picture of the weather field. This makes it possible to predict the occurrence of phenomena that have damaging effects at the ground level by observing precursors that may occur at higher levels. Indeed, precursor observation may be necessary to meet certain weather information requirements. One example is the stipulation by the US Federal Aviation Administration that a wind shear warning must be issued at least 1 min prior to the time that an airport actually encounters hazardous wind shear (Hansen, 1989). The possibility of microburst precursor observation has been practically verified in the case of the TDWR (Campbell, 1989; Campbell and Isaminger, 1989). In one particular instance, on the basis of the detection of a reflectivity core and rotation aloft, a microburst precursor signature was detected 9 min in advance of the surface outflow, and the microburst alarm itself was advanced in time by 1 min (Merritt *et al.*, 1989). Even this order of prediction time lead for hazard warning is considered significant in the highly dynamic environment of the terminal area aviation operations.

7.6 Airport surveillance radar with weather channel

Weather radars are not the only radars employed in the service of aviation. As discussed in Section 2.6, radars perform a number of functions in aircraft navigation and air traffic control. The use of dedicated terminal area weather surveillance radars such as the TDWR would add one more radar to the already complex and costly aviation management system. While this may be justifiable in the case of the larger and busier airports, such an option would not be viable for all airports. If the essential capabilities of the modern weather radar can be incorporated into another existing radar of the aviation system, considerable savings in overall system complexity can be achieved and the benefits of weather radar coverage can be afforded at a larger number of airports. Additionally, combining the functions of two radars in one would reduce the demand for suitable radar sites within the airport area, the need for additional frequency allocation, and electromagnetic interference. One such radar is the airport surveillance radar (ASR), which is normally located in the area of the runway complex and is therefore well placed to perform weather surveillance of the terminal area if the requisite hardware and software are built into it. This is the logic behind the design of the ASR-9 radar system with a dedicated weather channel.

Although a single radar combining airport surveillance and weather roles is functionally desirable, there are a number of major structural and operational differences between the optimal configurations of ASRs and weather radars. The most important among them pertain to the respective antenna patterns and scan cycles. An ASR has a fan beam scanning at a uniform elevation angle. In contrast, a typical Doppler weather radar has a narrow pencil beam, scanning at several levels. This difference strongly influences the vertical resolution, clutter rejection capability, and signal processing features of the radars. Further, while the ASR is usually located in the vicinity of the runways, the optimum site for a weather radar for terminal area weather surveillance is ~10–12 km away from the runway complex (see the last paragraph of Section 7.4.2.7). A major achievement of the ASR-9 is to obtain an acceptable fusion of the ASR and weather functions in spite of such basic differences.

The earlier version of the ASR in the same series, the ASR-8, which is operational at many airports, does have a limited weather observation capability. It can provide a single level outline of storms within its observation range, but the echoes are uncalibrated and nonquantised (Puzzo *et al.*, 1989). Because of the single threshold level employed in mapping precipitation areas, no detail or structure is discernible within the radar display of storm areas. In contrast, the ASR-9 weather channel is designed to provide air traffic control personnel with accurate, quantised and clutter-free representation of the precipitation field. In Section 6.2.3 (Table 6.2) we had been familiar with the 6-level US National Weather Service scale of reflectivity corresponding to different intensities of precipitation. The ASR-9 is capable of quantising the

282 *Aviation weather surveillance systems*

reflectivity field into these six levels (or seven levels, considering the no-rain level with reflectivity <18 dBZ), and the air traffic control personnel may select and display any two of these levels.

An important improvement of the ASR-9 system configuration compared with its predecessors is the provision of separate channels for aircraft detection and weather reflectivity measurement (Puzzo *et al.*, 1989; Taylor and Brunins, 1985; Troxel, 1989). The radar utilises both linear and circular polarisation. In normal operation a linear (vertical) polarisation is used for both transmission and reception. In this mode the weather channel receives the same signal as the target aircraft channel. However, when weather reflectivity >41 dB (corresponding to National Weather Service Level 3 in Table 6.2) is observed over large areas, the radar switches over to a circular polarisation which reduces interference due to rain echoes. In this mode, a very large portion of the weather echo has a polarisation sense (i.e. right-hand or left-hand circular) opposite to that of the transmitted signal, while the target echo is divided almost evenly between the two senses of circular polarisation. The target channel receives a circularly polarised signal of the same sense as the transmitted signal, but the weather channel is connected to the orthogonal port of the polariser to receive the signal of the opposite polarisation. Such an arrangement reduces the polarisation-induced loss of the weather signal inherent in single-channel receivers optimised for aircraft detection, as in the case of the earlier ASRs, and ensures optimal detection of both target and weather signals.

The ASR-9 achieves a high degree of ground-clutter rejection by incorporating in the weather signal processor a filter bank controlled by a clutter map of the area surrounding the installation. Four filter choices are possible, each with different clutter-rejecting notches. Included among these is a zero-notch-width filter (or 'allpass' filter) for areas without significant ground clutter.

The basic parameters of the ASR-9 radar system related to weather observation are listed in Table 7.6 (Puzzo *et al.*, 1989). The radar is capable of unattended operation, and has facilities for remote performance monitoring, fault isolation, and control. The first ASR-9 radar was commissioned in the USA in May 1989, and over a hundred such radars may be installed in the foreseeable future.

The ASR-9 radar employs a dual-beam antenna producing two fan-shaped beams separated in elevation, as depicted in Fig. 7.10. The antenna has a rapid scan rate of 12.5 rev/min, which provides raw data updates at intervals of ~5 s. However, data from several successive scans are normally smoothed to reduce statistical fluctuations, resulting in an operational update rate of the order of half a minute. Rapid weather data update is a strong advantage of the ASR-9 system in the aviation context.

An additional weather data processing channel has also been developed for the ASR-9 to automatically detect regions of low-altitude wind shear. An ASR-9 system with such a facility has been shown to be able to perform adequate clutter suppression, estimation of radial wind speed near the ground, and

Table 7.6 Basic parameters of the ASR-9 radar system

Parameter	Value/description
Radar system	
Maximum range	110 km (60 nmi)
Reflectivity map	Six levels, of which any two may be selected for display
Antenna	
Beam shape	Fan beam
No. of fan beams	2, vertically stacked
Beamwidth (-3 dB):	
Elevation	6° (min. 4.8°)
Azimuth	1.4°
Beam separation	3.5° in vertical plane
Polarisation	Vertical or circular
Gain	34 dB
Scan rate	12.5 rev/min
Transmitter	
Frequency	2.7–2.9 GHz
Transmitter type	Klystron
Peak power	1.10 MW
Pulsewidth	1 μs
Transmitted signal format (typical)	Block stagger mode: 8 pulses at 940 Hz, 10 pulses at 1200 Hz
Receiver	
Noise figure	4.1 dB (max)
Sensitivity	-108 dBm
A/D word size	12 bits
Range sampling	115.8 m
Clutter rejection	45 dB
No. of clutter filters	4
Clutter filter type	Variable-notch-width (clutter-map controlled)

automatic wind shear hazard detection in the case of wet microbursts (which have a higher reflectivity than the dry microbursts). The fan beam of the radar is not ideal for detecting shallow phenomena such as microburst outflows, since the beam covers much greater heights than these phenomena, and hence upper level winds are also 'seen' by the beam. However, low-altitude winds can be separated from the winds aloft by comparing the signals received through the high and low beams of ASR-9. Using the difference

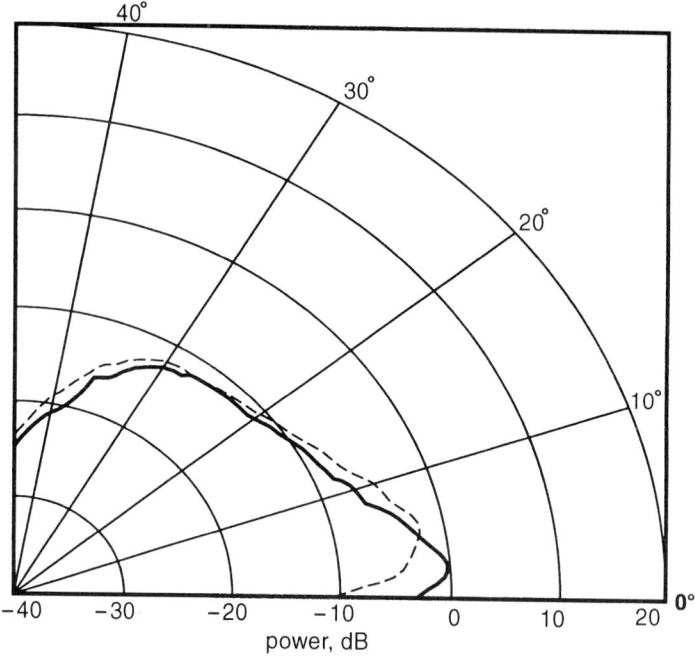

Figure 7.10 Elevation beam patterns of ASR-9. The low and high beams are plotted with solid and dashed lines, respectively (from Weber and Kays, 1995, reprinted with permission of MIT Lincoln Laboratory, Lexington, MA, USA)

between the spectra observed through the two beams of the WSR-9 radar, detection probabilities >90% and false alarm probabilities <4% have been achieved for microbursts occurring within a range of 12 km and having a velocity differential >10 m/s in their divergent outflows. The performance is even better for more intense microbursts (Weber and Noyes, 1989). With such a weather processor, the ASR-9 can be used in a stand-alone mode for wind shear warnings at airports which do not have a more dedicated and advanced sensor such as the TDWR.

7.7 Summary

In view of the overwhelming advantages of Doppler weather radars in the aviation context, a number of such radars with dedicated or important application to aviation have been developed recently and are in various stages of proving, procurement and installation. A few such selected systems have been discussed in this chapter. The WSR-88D, which is a long-range radar with multiple applications, is discussed in detail, including the problems and special considerations obtaining in the context of aviation. One particular problem in simultaneous range-velocity measurement, common to all Doppler weather radars, is that of range and velocity ambiguities. Their implications and methods of alleviation have been presented.

Other considerations of Doppler weather radars in aviation applications include coverage, location relative to aviation activity areas such as airports, appropriate scanning schemes and data currency. These are interrelated, with each aspect affecting the others. For example, coverage up to higher elevation angles with a given density (beam spacing) increases the scan time, which impairs the data currency. Some of the tradeoffs and compromises have been discussed in the chapter.

The Terminal Doppler Weather Radar and the Airport Surveillance Radar with weather channel are two other radars presented. These radars are optimised for supporting the dense and low-altitude flight environment of terminal areas. In particular, they are capable of detecting such low-altitude aviation-hazardous phenomena as microbursts and gust fronts. Potent software capabilities are being evolved for the generation of high-level data products that can be used directly by personnel without specialised meteorological training.

While the TDWR and ASR-9 are expected to perform the terminal area weather surveillance role in the foreseeable future, new ideas are already being generated for this function with additional facilities and capabilities. One such concept is the Terminal Area Surveillance System (TASS) (Rogers, 1995), which is an integrated aircraft and weather surveillance system intended to replace the TDWR and ASR systems, and also allow the decommissioning of the Low Level Wind Shear Alert Systems (see Section 8.4). Using a phased array antenna with a pencil beam, the TASS is expected to overcome many of the limitations of the TDWR and ASR systems arising from siting, scan strategy, blind zone and beam shape.

7.8 References

ALLEN, R.H., BURGESS, D.W., and DONALDSON, R.J. (1981): 'Attenuation problems associated with a 5-cm radar', *Bull. Am. Meteorol. Soc.*, **62**, pp. 807–810

BARGEN, D.W., and BROWN, R.C. (1980): 'Interactive radar velocity unfolding'. Proceedings of 19th Conference on Radar Meteorology, Miami, FL (American Meteorological Society, Boston), pp. 278–283

BERGEN, W.R., and ALBERS, S.C. (1988): 'Two- and three-dimensional dealiasing of Doppler radar velocities', *J. Atmos. Ocean. Technol.*, **5**, pp. 305–319

BOREN, T.A., CRUZ, J.R., and ZRNIC', D.S. (1986): 'An artificial intelligence approach to Doppler weather radar velocity dealiasing'. Proceedings of 23rd Conference on Radar Meteorology, Snowmass, CO (American Meteorological Society, Boston), pp. 107–110

CAMPBELL, S.D. (1989): 'Use of features aloft in the TDWR microburst recognition algorithm'. Proceedings of 24th Conference on Radar Meteorology, Tallahassee, FL (American Meteorological Society, Boston), pp. 167–168

CAMPBELL, S.D., and ISAMINGER, M.A. (1989): 'Using features aloft to improve timeliness of TDWR hazard warning'. Proceedings of Third International Conference on Aviation Weather System', Anaheim, CA (American Meteorological Society, Boston), pp. 184–189

CROCKER, S.C. (1988): 'TDWR PRF selection criteria'. Report DOT/FAA/PM-87-25, Prepared for the Federal Aviation Administration by MIT Lincoln Laboratory, 15 March 1988

DIXON, M. (1992): 'ARSR/NEXRAD Comparison Study Phase One (Third Draft), National Center for Atmospheric Research, Boulder, CO, 15 December 1992

DUNBAR, B., and MITTELMAN, J. (1993): 'The Next Generation Weather Radar (NEXRAD)/Air Route Surveillance Radar (ARSR) operational comparison'. US Department of Transportation, Federal Aviation Administration Report DOT/FAA/SE-93/4, July 1993

EILTS, M.D., and SMITH, S.D. (1990): 'Efficient dealiasing of Doppler velocities using local and environmental constraints', *J. Atmos. Ocean. Technol.*, **7**, pp. 118–128

EVANS, J., and TURNBULL, D. (1989): 'Development of an automated wind shear detection system using Doppler weather radar', *Proc. IEEE*, **77**, pp. 1661–1673

FEDERAL AVIATION ADMINISTRATION (1989): 'Multiple radar processing', National Airspace System Configuration Management Document NAS-MD-320, Federal Aviation Administration Technical Center, Atlantic City International Airport, New Jersey, 4 August 1989

HANSEN, A.L. (1989): 'Ground based weather radar for aviation'. Proceedings of Third International Conference on Aviation Weather System, Anaheim, CA (American Meteorological Society, Boston), pp. 420–421

HEISS, W.H., McGREW, D.L., and SIRMANS, D.S. (1990): 'Nexrad: Next generation weather radar (WSR-88D)', *Microw. J.*, **33**, pp. 79–98

HILDEBRAND, P.H. (1978): 'Iterative correction for attenuation of 5 cm radar in rain', *J. Appl. Meteorol.*, **17**, pp. 508–514

HITSCHFELD, W., and BORDAN, J. (1954): 'Errors inherent in the radar measurement of rainfall at attenuating wavelengths', *J. Meteorol.*, **11**, pp. 58–67

HOGG, D.C. (1969): 'Statistics on attenuation of microwaves by intense rain', *Bell Syst. Tech. J.*, **48**, pp. 2949–2962

KESSLER, E. (1990): 'On low-level windshear alert systems (LLWAS) and Doppler radar in aircraft terminal operations', *J. Aircr.*, **27**, pp. 423–428

LAIRD, B.G. (1981): 'On ambiguity resolution by random phase processing'. Preprints of 20th Conference on Radar Meteorology (American Meteorological Society, Boston), pp. 327–331

LIN, S.H. (1975): 'A method for calculating rain attenuation distribution on microwave paths', *Bell Syst. Tech. J.*, **54**, pp. 1051–1086

MAHAPATRA, P.R., and Zrnic', D.S. (1981): 'Scanning strategies for air traffic control radars'. Proceedings of 37th Annual Meeting of the Institute of Navigation, Annapolis, MD, 9–11 June 1981, pp. 100–106

MAHAPATRA, P.R., and ZRNIC', D.S. (1983): 'Optimum siting of NEXRAD to detect hazardous weather at airports', *J. Aircr.*, **20**, pp. 363–371

MAHAPATRA, P.R., and ZRNIC', D.S. (1991): 'Sensors and systems to enhance aviation safety against weather hazards', *Proc. IEEE*, **79**, pp. 1234–1267

MAHAPATRA, P.R., ZRNIC', D.S., and EILTS, M.D. (1993): 'Strategies for mitigating range and Doppler ambiguities in the WSR-88D'. National Oceanic and Atmospheric Administration, National Severe Storms Laboratory, Norman, OK, Internal Report, August 1993

MERRITT, M.W. (1984): 'Automatic velocity dealiasing for real-time applications'. Proceedings of 22nd Conference on Radar Meteorology, Zurich, Switzerland (American Meteorological Society, Boston), pp. 528–533

MERRITT, M.W., KLINGLE-WILSON, D., and CAMPBELL, S.D. (1989): 'Wind shear detection with pencil-beam radars', *Lincoln Lab. J.*, **2**, pp. 483–510

MICHELSON, M., SHRADER, W.W., and WIELER, J.G. (1990): 'Terminal Doppler weather radar', *Microw. J.*, **33**, pp. 139–148

OFFICE OF THE FEDERAL COORDINATOR FOR METEOROLOGICAL SERVICES AND SUPPORTING RESEARCH (OFCMSSR) (1991): 'Federal Meteorological Handbook No. 11: Doppler radar meteorological observations, Part C – WSR-88D products and algorithms'. FCM-H11C-1991 (Interim Version One), Washington, DC, April 1991

OFFICE OF THE FEDERAL COORDINATOR FOR METEOROLOGICAL SERVICES AND SUPPORTING RESEARCH (OFCMSSR) (1992): 'Federal Meteorological Handbook No. 11: Doppler radar meteorological observations, Part D – WSR-88D unit description and operational applications'. FCM-H11D-1992 (Interim Version One), Washington, DC, April 1992

PUZZO, D.C., TROXEL, S.W., MEISTER, M.A., WEBER, M.E., and PIERONEK, J.V. (1989): 'ASR-9 weather channel test report'. Report DOT/FAA/PS-89/3, Prepared for the Federal Aviation Administration by MIT Lincoln Laboratory, 3 May 1989

RAY, P., and ZIEGLER, C. (1977): 'Dealiasing first moment Doppler estimates', *J. Appl. Meteorol.*, **16**, pp. 563–564

ROGERS, J.W. (1995): 'Terminal area surveillance system'. Proceedings of IEEE International Radar Conferece, Alexandria, VA, 8–11 May 1995, pp. 9–10

SACHIDANANDA, M., and ZRNIC', D.S. (1986): 'Recovery of spectral moments from overlaid echoes in a Doppler weather radar', *IEEE Trans. Geosci. Remote Sens.*, **GE-24**, pp. 751–764

SIGGIA, A. (1983): 'Processing phase-coded radar signals with adaptive digital filters'. Preprints of 21st Conference on Radar Meteorology (American Meteorological Society, Boston), pp. 167–172

TAYLOR, J.W., JR., and BRUNINS, G. (1985): 'Design of a new airport surveillance radar (ASR-9)', *Proc. IEEE*, **73**, pp. 284–289

TROXEL, S.W. (1989): 'ASR-9 weather channel test report, Executive summary'. Report DOT/FAA/PS-89/6, Prepared for the Federal Aviation Administration by MIT Lincoln Laboratory, 3 May 1989

TURNBULL, D., MCCARTHY, J., EVANS, J., and ZRNIC', D.S. (1989): 'The FAA terminal Doppler weather radar (TDWR) program'. Proceedings, Third International Conference on Aviation Weather System, Anaheim, CA, pp. 414–419

TURNBULL, D.H. (1995): 'Aviation weather radar'. Proceedings of the IEEE International Radar Conference, Alexandria, VA, 8–11 May 1995, pp. 748–751

US DEPARTMENT OF TRANSPORTATION (1989): 'The Federal Aviation Administration plan for research, engineering and development, Volume II: Project descriptions'. Federal Aviation Administration, January 1989

WALDTEUFEL, P. (1976): 'An analysis of weather spectra variance in a tornadic storm'. US National Oceanic and Atmospheric Administration Technical Memo. ERL NSSL-76

WANKE, C., and HANSMAN, R.J. (1992): 'Experimental evaluation of candidate graphical microburst alert displays'. 30th Aerospace Sciences Meeting and Exhibit, 6–9 January 1992 (American Institute of Aeronautics and Astronautics, New York), Paper AIAA-92-0292

WEBER, M.E., and KAYS, T.A. (1995): 'An advanced weather surveillance processor for Airport Surveillance Radars'. Preprints of 6th Conference on Aviation Weather Systems, Dallas, TX, 15–20 January 1995, pp. 396–401

WEBER, M.E., and NOYES, T.A. (1989): 'Wind shear detection with airport surveillance radars', *Lincoln Lab. J.*, **2**, pp. 511–526

ZITTEL, W.D., and O'BANNON, T. (1993): 'On the performance of three velocity dealiasing techniques over a range of Nyquist velocities: preliminary findings'. Preprints, 28th Conference on Radar Meteorology (American Meteorological Society, Boston), pp. 53–55

ZRNIC', D.S., and MAHAPATRA, P.R. (1985): 'Two methods of ambiguity resolution in pulse-Doppler radars', *IEEE Trans. Aerosp. Electron. Syst.*, **AES-21**, pp. 470–483

Chapter 8
Other sensors and systems for aviation weather

8.1 General

Doppler weather radars are beginning to play the central role in the sensing of weather phenomena in the evolving and futuristic scenarios of aviation management and safety enhancement. However, it must be remembered that Doppler weather radars are elaborate and expensive equipment which cannot be installed everywhere, and there will always be a considerable amount of aviation activity that will not have the benefit of these radars to the required degree. In terms of gross area, of course, weather radar coverage may be available over most parts of many countries in the foreseeable future, but the data from such radar networks will not meet the stringent requirements of aviation support, especially in terminal areas, over the entire zone of coverage. In particular, the data adequacy and quality with respect to low-altitude coverage (because of radar horizon limitations) and resolution, which are dependent on the distance from the radar, will be highly nonuniform over the coverage zone. Further, the multipurpose weather radars used for network coverage of large areas will not, in general, provide the data update rates and the high-level and special-purpose software capabilities required for supporting modern and future aviation. This fact, together with the limitations on the number of dedicated radars with sophisticated aviation weather surveillance capability (such as the TDWR and ASR-9), means that significant aviation activity will remain outside the type of weather radar coverage that provides important information such as low-altitude divergence and wind shear.

There is thus need for additional instrumentation which would be more widely distributed and be cost-effective for providing the basic aviation-significant weather information, if necessary over limited areas. These can supplement Doppler radar data where their effective areas overlap with the radars, and can form an independent source of weather information about wind shear and shifts where Doppler radar support is not available. A number of such sensing instrument systems have been developed, and some of the important ones will be covered in this chapter.

290 *Aviation weather surveillance systems*

8.2 Wind profilers

It may be recalled from the discussion in Section 3.3.1 that a significant form of wind shear from the point of view of aviation safety is the vertical wind shear. This type of shear refers to the vertical rate of variation of horizontal wind speed and/or direction. A schematic diagram of this form of wind shear and its effect on aircraft was presented in Fig. 3.3. Sensing such wind shear is of great benefit for aviation. A plot of the horizontal wind velocity with respect to height from the ground is called a *wind profile*, and an instrument or instrument system that can measure these wind variations is called a *wind profiler*.

8.2.1 Conventional wind profiling

The wind profiles in the lowest few tens or even hundreds of metres are usually obtained by anemometers mounted at various heights on instrumented towers (Alexander and Camp, 1985). Such towers yield wind profiles continuously in time, but at discrete heights from the ground.

A conventional method of wind profiling to higher altitudes consists of releasing a balloon and tracking its movement with a precision radar. The tracking is usually aided by a transponder carried as a part of the balloon payload. The transponder return signal also contains information derived from other sensors included in the payload, such as temperature and pressure. The radar performs three-dimensional tracking, providing both height and wind velocity information which is utilised for plotting the wind profile.

The balloon method of profiling is very convenient, useful and popular, and is routinely carried out around the globe by meteorological agencies. Balloon data have extensive meteorological applications including those in support of aerospace activities. An example is the use of the balloon method by the US National Aeronautics and Space Administration (NASA) to obtain wind profiles that help decide about spacecraft launches from the Kennedy Space Center in Florida. These profiles can be obtained up to 18 km altitude with 1 m/s accuracy and 50 m of height resolution (Chadwick *et al.*, 1984).

From the point of view of aviation management, however, wind profiling by balloon tracking has some serious drawbacks. These are as follows:

1. The method is very slow. Since the balloon drifts in the atmosphere only by natural air currents and buoyancy forces, it rises slowly to its operational ceiling, requiring a time interval of the order of one or two hours for a profiling operation.
2. The wind profiles are obtained only along the flight path of the balloon, on which the operating personnel have no control. Depending on the local winds, the balloon may drift large distances away from the vertical at the launch point, and there is a great deal of uncertainty in ensuring that

profiling is done over a small area of interest such as an airport area or a given flight corridor. Indeed, balloon profiling presents a paradox. The balloon will rise nearly straight upward and thus provide a vertical profile only when the atmosphere is calm, i.e. has very little wind shear. On the other hand, when there is significant shear, and hence profiling would be of interest, the balloon would drift and yield a profile far away from the vertical

3. Each balloon launch is a discrete operation in time. Since a single profiling may take an hour or more, balloon launches are separated by even longer intervals. Routine meteorological balloon launches are normally carried out twice a day from any given location. The launches may be made somewhat more frequent for dedicated aviation support such as from airport sites, but still the launch frequency would not be commensurate with the needs of modern aviation operations.

8.2.2 Radar wind profilers

Electromagnetic wind profilers provide a more modern method of obtaining the vertical profile of wind velocities. These profilers are essentially radars that are configured and optimised to receive echo returns from layers of air above their location and process the signal to derive their velocity parameters as functions of height.

It may be recalled from the discussions in Section 6.6.4 that the general purpose Doppler weather radar is capable of providing wind profiles when operated in the velocity-azimuth display (VAD) mode. However, since such general weather radars are designed and operated to perform a variety of functions, they are operated in the VAD mode only occasionally. It is, of course, possible to dedicate a Doppler weather radar for wind profiling alone, but such use would amount to only a partial utilisation of the radar's capabilities, and represent a costly solution to the profiling problem. The modern special-purpose electromagnetic wind profilers are also Doppler radars, but these are far less versatile, and consequently simpler, less costly and often smaller in size than the multifunction Doppler weather radars discussed in the preceding chapter.

Since profiling is required to be done in all weather conditions, electromagnetic profilers are designed to operate satisfactorily with clear-air echoes from the atmospheric layers. As discussed in Section 6.2.4, clear-air radar echoes are primarily caused by small-scale fluctuations in the atmospheric refractive index, which in turn arise from inhomogeneities or irregularities in the local temperature and water vapour concentration. Further, the spatial scale of such fluctuations which contributes most to the echo power is one that corresponds to a half of the radar wavelength. Electromagnetic profilers typically utilise frequencies in the VHF and UHF bands in preference to microwave frequencies since the longer wavelengths provide stronger echo returns from the refractive index fluctuations

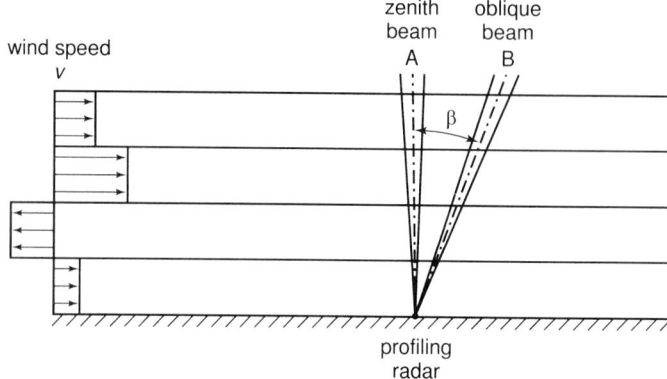

Figure 8.1 An ideal horizontally layered atmosphere. The vertical beam A would not sense any Doppler, while the oblique beam B would 'see' a radial component of the horizontal wind speed in each layer

associated with relatively large eddies that are in the inertial subrange (see Section 11.4.1), and hence do not dissipate rapidly. Centimetre-scale eddies, which are efficient scatterers at microwave frequencies, decay rapidly after their creation due to viscous dissipation of energy.

Hydrometeoric scatterers such as raindrops are not considered to be good atmospheric tracers for profiling purposes even though they return strong echoes and hence can greatly reduce the transmitted power and/or sensitivity requirements of profiling radars. This is because hydrometeors are not present at all altitudes at all times. Further, since radar profilers 'look' essentially in a vertically upward direction, the considerable fall velocity of hydrometeors introduces additional Doppler shift which affects the accuracy of wind estimates (Wuertz et al., 1988). The use of the longer wavelengths in the VHF/UHF bands significantly reduces echoes from hydrometeors, but the velocity estimates may still be perceptibly affected by strong precipitation.

In an idealised profiling scenario, depicted in Fig. 8.1, the atmosphere may be assumed to be horizontally layered, with each layer moving at a different velocity in general. In such a case, a profiling radar looking vertically upwards would not sense any mean Doppler shift since the beam axis is normal to the layer velocities. To be able to sense the horizontal velocities in the atmospheric layers through a Doppler shift, the beam should be tilted away from the vertical by a finite angle. A tilt of the order of 10 or 15 degrees is considered adequate for radar profiling. A larger tilt would, of course, provide a higher Doppler shift for a given horizontal wind speed and thus result in improved sensitivity for the measurement of low speeds. However, a large tilt angle would take the beam far away from the vertical, especially at high altitudes, and the wind velocities measured along the beam may not accurately represent the profile along the vertical. Further, for a given highest altitude of profiling, the slant range from the radar increases with the tilt angle, which would demand larger transmitted power on the part of the profiler.

If the wind velocity vector is horizontal and lies in the vertical plane through the profiler beam axis in its tilted position, the radial component v_r of the velocity along the beam axis is

$$v_r = v \sin \beta \quad (8.1)$$

where v is the wind speed and β is the beam tilt angle. Then, from eqn. 6.23, the Doppler shift of the echo relative to the transmitted signal, as a function of height, is given by the relation

$$f_d(h) = -\left(\frac{2}{\lambda}\right) v(h) \sin \beta \quad (8.2)$$

where λ is the wavelength of the radar. The velocity profile $v(h)$ can be readily deduced from the measured values of the Doppler frequency at each height (i.e. at the corresponding slant ranges).

In general the horizontal velocity vector will not lie in the vertical plane through the inclined beam. The profiler will then measure only the projection of the velocity vector on the vertical plane, which is the wind velocity component along the azimuth direction of the oblique beam. To obtain the vector velocity itself, its components along two beam positions must be measured such that the two beams do not lie on a common vertical plane. Preferred beam positions are along orthogonal directions in azimuth (Fig. 8.2a) because they would measure the orthogonal components of the horizontal wind velocity vector.

Although a two-oblique-beam configuration is adequate for wind vector profiling under the assumption of horizontal wind velocity at all altitudes, such an assumption is not necessarily true in nature, with local winds at all but the layer immediately in contact with the ground often having significant vertical components. In such cases, eqns. 8.1 and 8.2 would not be valid (the angle of the velocity vector with respect to the horizontal plane must be algebraically added to the tilt angle β), and the two-beam profiler can yield significantly erroneous wind profiles. To estimate or eliminate the third (i.e. vertical) component, a third beam position would be necessary. The so-called *tripod configuration* may have three beams arranged along the vertices of an equilateral triangle (Fig. 8.2b). Alternatively, one of the beams may be vertical, with the other two tilting at equal angles along orthogonal directions in azimuth (Fig. 8.2c).

Wind profiling with the minimum configuration consisting of three beam positions is not very accurate. Since the different beam positions observe wind components at spatially separated points, the separation being as much as a few kilometres at high altitudes, nonuniformities in wind fields can cause

294 Aviation weather surveillance systems

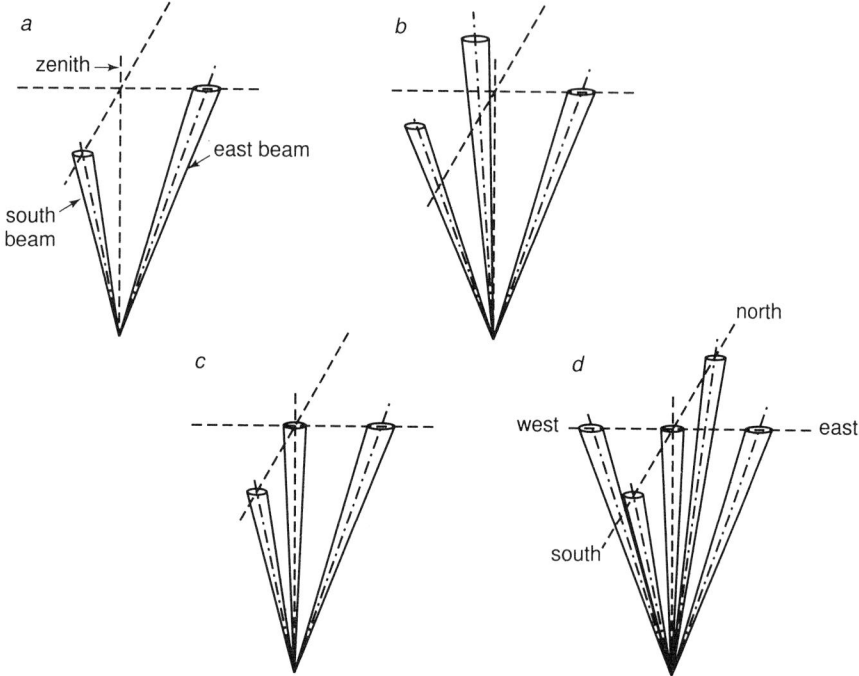

Figure 8.2 Simple profiler beam configurations: (a) two beams in orthogonal vertical planes, (b) three beams along vertices of an equilateral triangle, (c) three beams in orthogonal vertical planes, and (d) five beams in orthogonal vertical planes

significant errors in the estimation of wind velocities. Such errors can be compensated to a certain extent by using more beam positions than the minimum of three necessary to estimate the wind components under ideal conditions. Proper signal processing and data averaging schemes can minimise the effects of noise and random wind fluctuations, as well as wind field nonuniformities, on profiling accuracy (May and Strauch, 1989).

A common beam configuration for radar wind profiling employs five beams as shown in Fig. 8.2d. Here, a vertical beam position is surrounded by four tilted beams in the manner of a cross. Such a configuration provides a degree of redundancy of information which permits more accurate profiling by averaging the common velocity components sensed by each pair of opposite beams. In each orthogonal plane there are three beams, which may be considered as two pairs with a common beam position. Each pair of beams provides an independent estimate of the wind velocity component in the plane, and averaging the two values would serve to reduce the statistical uncertainty associated with the estimates. The difference of the winds sensed by the two beam pairs in each plane would indicate the variation of the corresponding wind component across the baseline of the beam configuration, but such differential wind estimates have a high noise component.

The use of a few fixed beam positions for profiling obviates the need for a fully steerable antenna. This is a great advantage which substantially simplifies the profiler system, leading to significant cost saving. The advantage is particularly great in the case of large-sized profilers operating at relatively low frequencies. In particular, if array antennas are used, the generation of a few fixed beam positions greatly simplifies the phase shifter network and its control apparatus.

The diverse and often conflicting requirements of electromagnetic profilers in terms of operating frequency, antenna size, altitude coverage, etc. have led to the development of different types of profilers, many of which can provide useful data of direct use for aviation (Frisch *et al.*, 1986; Strauch *et al.*, 1989a). A versatile profiler is the *tropospheric wind profiler*, which is a clear-air radar operating at 404.37 MHz, and is conveniently referred to as a 405 MHz profiler. This device can observe wind profiles up to a height of ~17 km, covering the entire troposphere and the lower reaches of the stratosphere. It has a phased-array antenna producing a 6° conical beam, and has five discrete beam positions comprising one vertical beam and four north–south–east–west beams tilted at 15° with respect to the vertical as shown in Fig. 8.3. It is capable of automatic and continuous operation, and can generate wind profiles every hour with a height resolution of ~1500 ft (500 m) and velocity measurement accuracy of 1 m/s for horizontal wind components. A prototype of this profiler was set up at Platteville, CO, USA, in September 1988, and a number of these units have subsequently been installed in the form of a network in the central zone of the USA (Strauch *et al.*, 1984). The locations of the profilers in the network are shown in Fig. 8.4. Sample outputs from members of this network are presented later in this section. In addition to helping in improving general weather forecasting, the data from such networks of profilers is of great value in air traffic routing, especially of commercial airliners, to either avoid or take advantage of jet streams depending on the direction and wind profile of the jet streams relative to the flight heading and altitude.

While profiler networks can improve air traffic routing and *en route* flight efficiency by providing wide-area profile patterns, individual profilers can greatly aid local aviation operations such as in terminal areas. Very useful information about vertical wind shear and wind shifts in support of aircraft approach, landing and takeoff operations in terminal areas may be obtained from a single wind profiler suitably located within the airport area. A profiler for such an application does not have to cover as much altitude as a tropospheric profiler, but needs to 'see' only up to a maximum height of the order of 4–6 km, which is the ceiling of most terminal area operations. A profiler observing up to such heights is called a *lower tropospheric wind profiler*. As explained in Section 6.2.4, clear-air reflectivity is generally much higher in the lower troposphere than at higher altitudes. Hence lower tropospheric wind profilers require significantly lower transmitted power and/or sensitivity than tropospheric profilers, and can operate at higher frequencies. Also,

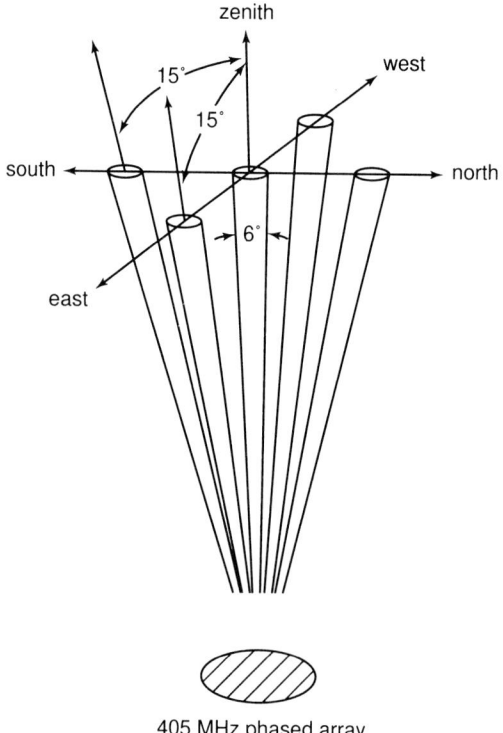

Figure 8.3 Example of a 405 MHz radar wind profiler with five discrete beam positions

because of stronger echo power, there is less need for signal integration, leading to a finer height resolution of the order of 450 ft (150 m) and faster profile updates at 10 min intervals. Profilers for such an application have been built at nominal operating frequencies of 405 and 915 MHz. A profiler of the latter type was installed at Stapleton airport in Denver, CO, USA, as early as 1983 (Strauch *et al.*, 1989a). The relatively high operating frequency and low power and sensitivity requirements of lower tropospheric wind profilers make them simple, small and affordable for installation at numerous airports, and the compact configuration makes the system transportable (Moran *et al.*, 1989).

Profilers can be made even simpler, smaller and cheaper if the maximum height of observation is reduced further relative to the lower tropospheric profiler. Profilers which provide wind data up to ~2 or 3 km of altitude are called *boundary layer profilers*. Even such a height coverage is quite adequate for a large part of the aviation operations in terminal areas. As the earth's boundary layer has well developed eddy structures with significant refractive index variations due to the mixing of moisture arising from the ground, the

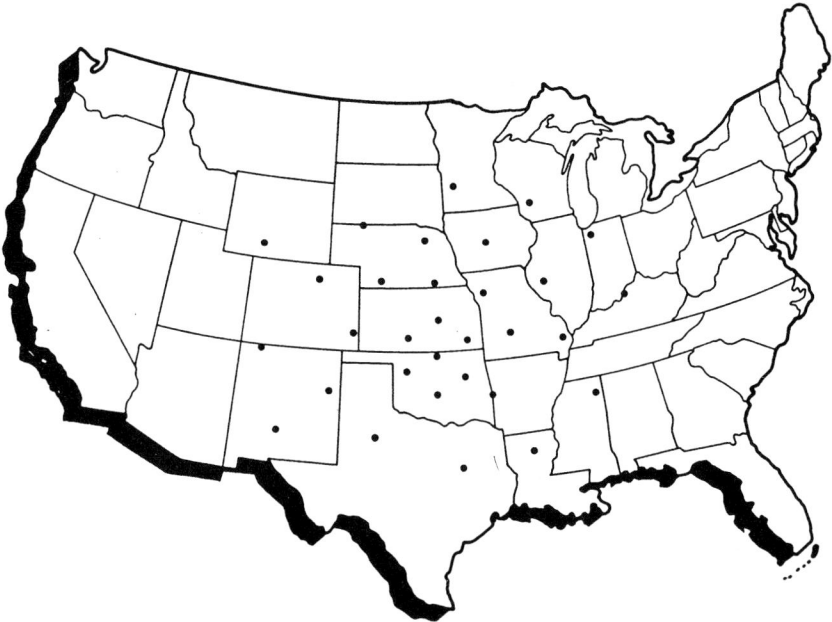

Figure 8.4   The US 405 MHz tropospheric wind profiler network

clear-air reflectivity of this layer is relatively high, thus greatly reducing the power and sensitivity needs of the profiler. Ecklund *et al.* (1988) present a boundary layer profiler operating at 915 MHz which employs a small mechanically steered antenna, providing a much greater degree of flexibility in profiling than fixed-beam profilers. The profiler provides a good height resolution of 300 ft (100 m) and fast profile updates at 10 min intervals. Its relatively high operating frequency and bandwidth permit the measurement of winds much closer to ground, down to ~300 ft (100 m), than other profilers. Profiling down to such low levels is particularly advantageous in the context of terminal area aviation because of the extreme hazard potential of low-altitude wind shear for aircraft during landing and takeoff.

As already mentioned, reducing the altitude limit of profiling greatly reduces the cost of profilers. Strauch *et al.* (1989a) mentioned cost figures, in 1987 US Dollars, in the range of 40 000 to 60 000 for boundary layer profilers, 120 000 to 150 000 for lower tropospheric wind profilers, and 350 000 to 450 000 for tropospheric profilers. It is observed that the cost of profilers rises almost directly with altitude capability. These figures, in turn, may be compared with the costs of the order of 1 to 3 million dollars or more for fully fledged modern Doppler weather radars. The low cost of boundary layer profilers makes them highly affordable even for installation at small airports.

To complete the discussion on profilers, mention may be made of a class of much larger profilers called *stratospheric-tropospheric* (ST) *profilers* with height coverage up to ~20 km, good height resolution (450 ft or 150 m) and short data update interval (10 min). These profilers necessarily operate at a lower frequency, of the order of 50 MHz, are expensive (cost $3–5 million), but have been justified in the context of supporting space vehicle launch operations (Strauch *et al.*, 1989a; Chadwick *et al.*, 1984), in addition to scientific research. An even larger system is the *mesospheric-stratospheric-tropospheric* (MST) *radar*, also typically operating in the 50 MHz band. These are sophisticated systems sensitive enough to receive echoes from the ionosphere, but can readily estimate winds in the tropospheric and stratospheric altitudes. These radars are too large and expensive to be deployed operationally, and there are only a few of these systems globally. However, they can provide useful wind profile information in the vicinity of their location.

A sample wind profile obtained from an MST radar is shown in Fig. 8.5. The Figure plots the Doppler spectrum of the wind, as seen by a beam inclined 20° west of zenith, at successive altitudes. The displacement of the spectrum mean from the zero-Doppler line at any given altitude indicates the horizontal wind speed. The 50 MHz radar, located at Tirupati in southern India, has a basic velocity measurement constant of 2.83 m/s/Hz, which yields ~8.6 m/s/Hz (i.e. 2.83/sin 20°) for the 20° beam. The narrow well defined spectral peaks indicate quiet and streamlined wind flows. In Fig. 8.6 the radar beam points straight up, and the Doppler spectrum plots at various heights correspond to the vertical winds. The mean of the spectrum still indicates the average vertical wind, but notice the very large width of the spectra here compared with those in Fig. 8.5. Larger spectrum widths signify higher levels of turbulence. The noisy or rippled appearance of the baselines in the upper parts of Figs. 8.5 and 8.6 is due to the weak echoes from the upper reaches of the atmosphere, providing poor signal-to-noise ratios.

For all their advantages, radar wind profilers do have certain important limitations. These arise from (i) the spatial separation of their beam positions, (ii) the need for assumptions regarding the uniformity of wind fields, (iii) the existence of upper and lower limits of altitude coverage (the latter being critical for terminal area operations), (iv) the ability to profile only the atmospheric layers vertically above the installation, and (v) susceptibility to errors due to precipitation. In particular, profiler performance may be appreciably degraded in stormy conditions which usually involve both precipitation and turbulent wind fields. In contrast, microwave Doppler radars operating at centimetre wavelengths appear to be able to perform profiling under such conditions (Zrnic' *et al.*, 1986).

In spite of such limitations, radar profilers can be very effective in providing wind shear and wind shift information of direct significance to aviation. The normal output of a radar profiler is a plot (and/or electronic data file) of horizontal wind velocity with respect to height from ground. This is usually given in the form of two graphs showing the wind speed and direction, respectively. Alternative forms of the plot are possible. For example, the two orthogonal components of horizontal wind may be plotted separately. It is

Other sensors and systems for aviation weather   299

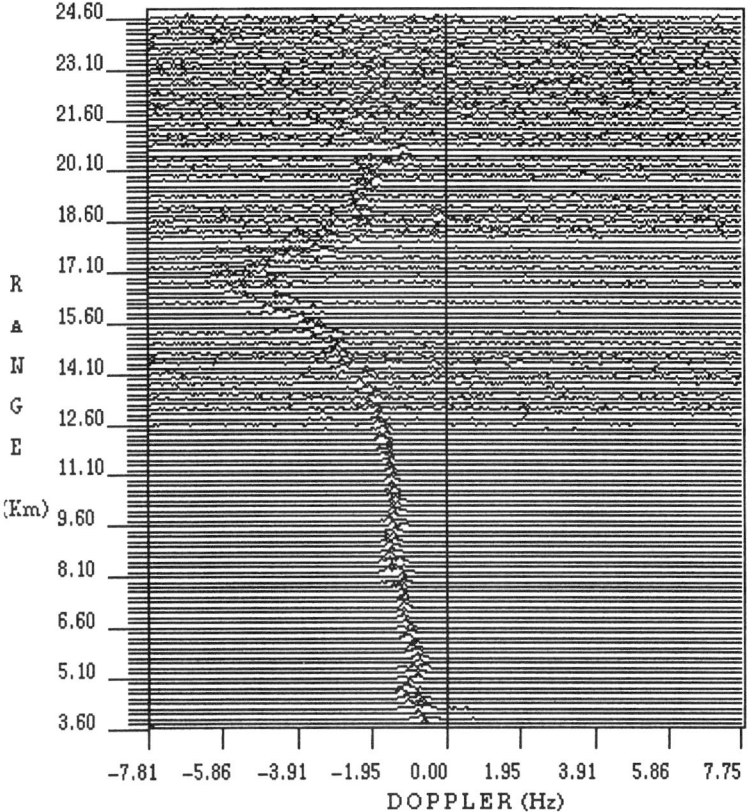

DATE : 19 /7 /93   Time : 11:6:2
FFT Points: 256 Rangebins: 141 Coherent integ.: 64 Incoh.Integ.: 4
Interpulse Period: 1000 microsec   Pulse Width: 16 microsec
CODE: CODED, BEAM: WEST, Data : SPECTRUM, Comments :

Figure 8.5   Example of MST radar wind profiling. The compressed individual plots at various altitudes show Doppler spectra of winds as received by an oblique beam. (Courtesy P.B. Rao, former Director, Indian National MST Radar Facility)

also possible to show the total wind velocity with respect to altitude in the form of 'wind barbs' placed at discrete locations along the height. A mathematical differentiation or differencing operation of the wind vector samples with respect to height would yield the wind shear along the vertical from the profiler. The existence of sharp discontinuities or strong gradients at certain altitudes in the wind profile is indicative of the presence of strong wind shear at those altitudes.

300 Aviation weather surveillance systems

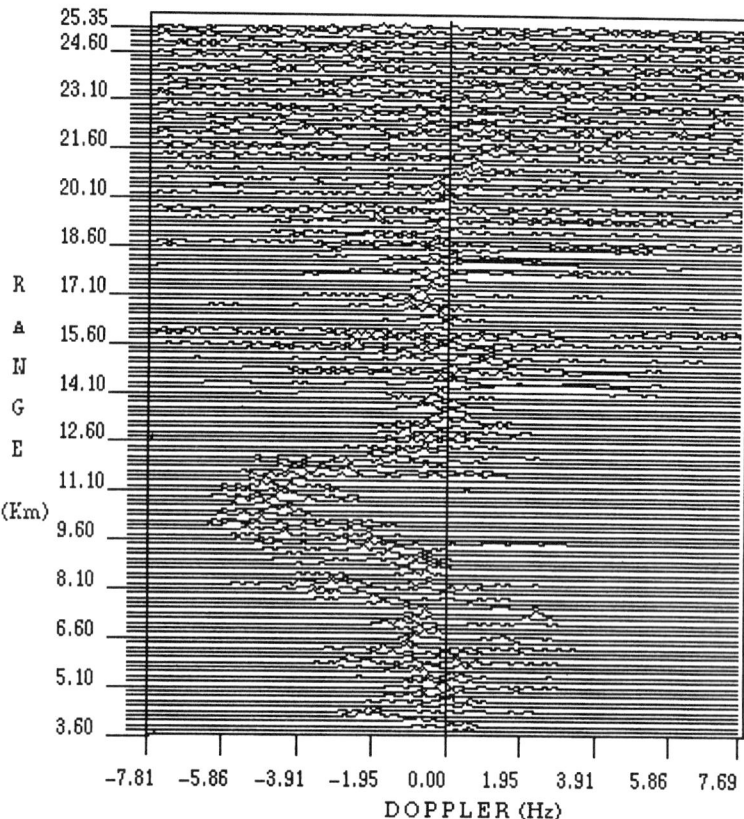

DATE : 29/9/93   Time : 19:42:42

FFT Points: 128 Rangebins: 201 Coherent integ.: 64 Incoh.Integ.: 1

Interpulse Period: 1000 microsec   Pulse Width: 16 microsec

CODE: CODED, BEAM: ZENITH_X, Data : SPECTRUM, Comments :

Figure 8.6   Doppler spectra observed by the vertical beam of MST radar. (Courtesy P.B. Rao, former Director, Indian National MST Radar Facility)

A very useful output of a radar profiler is a *time–height plot*. This is generated by obtaining a series of wind-barb plots of instantaneous wind profiles, preferably at regular intervals, over a period of time and placing them in a seqence displaced proportionately along the time axis. Such a plot provides a clear visual picture of the evolution of the wind field, and in particular the shear zones, over the profiler location. If a loose assumption is made that the wind field is frozen and drifts as such over the profiler site, then the time–height plot would provide a pseudo-spatial picture of the wind field over a vertical section of the atmosphere passing through the profiler location. Some interesting time–height plots obtained from actual radar profilers are

Other sensors and systems for aviation weather 301

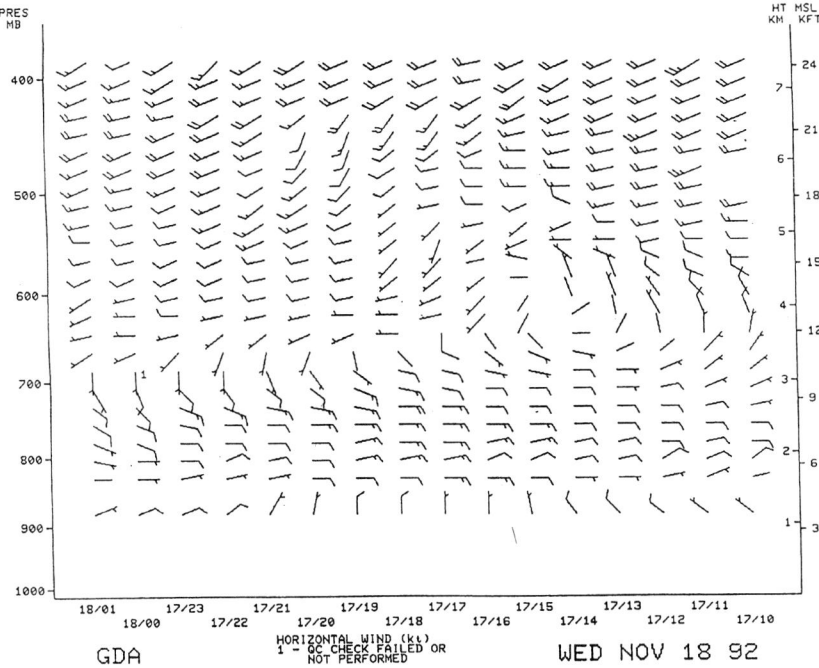

Figure 8.7 Time-height plot of horizontal winds on 17–18 November 1992 over the 405 MHz profiler at Granada, Colorado, which is a member of the network shown in Fig. 8.4. Wind profiles are shown over a period of 15 h at 1 h intervals starting from 1000 h on 17 November. The plot shows a clear and narrow wind shear zone in a height band of ~10–12 kft or 3–3.6 km (pressure height ~700 mbar). (Courtesy S.K. Beckman, National Severe Storms Forecast Center, Kansas City, MO, USA)

provided in Figs. 8.7–8.10. It is easy to see that such a detailed picture of the vertical structure of the wind field even in the absence of precipitation would be very useful in flight management from the point of view of safety and economy.

The utility of profilers is particularly high in situations where weather radar coverage of adequate quality is not available. Even in areas covered by weather radars, profilers can usefully supplement radar observations. This is because profilers are optimised for clear-air observation, can observe down to very low altitudes without significant clutter interference, and provide wind information in a vertical direction which is within the blind zone (see Fig. 7.6) of Doppler weather radars in their normal operating modes. The high value of radar profilers in terminal area wind surveillance has motivated nations around the world to consider installing this device to support their aviation systems (e.g. Korhonen, 1989).

302 *Aviation weather surveillance systems*

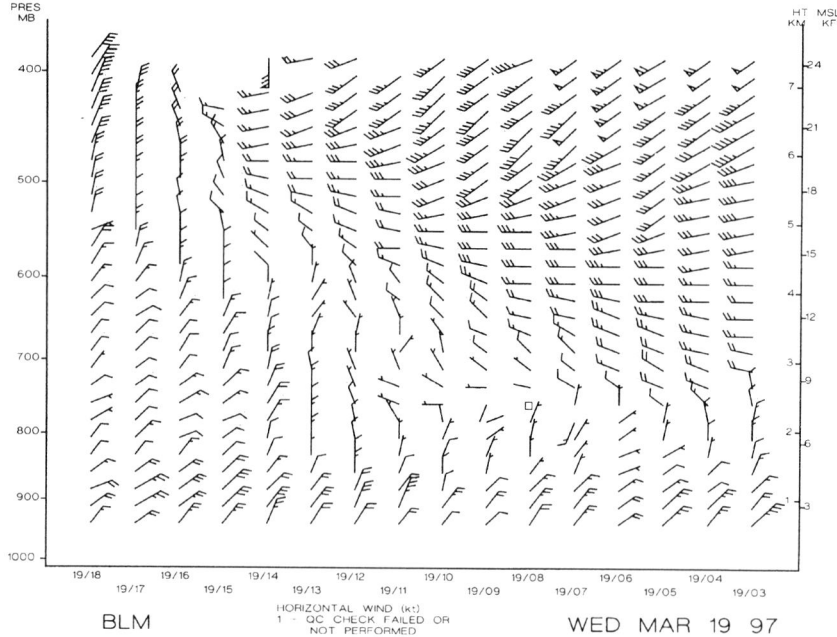

Figure 8.8   Plot similar to Fig. 8.7 over the profiler at Bloomfield, MD, USA, on 19 March 1997. The wind profiles, over a period of 15 h beginning from 0300 h, show a wind shear zone starting at ~6 kft (2 km) altitude and rising and broadening with time to a height band of ~18–26 kft (5–8 km). The random orientation of wind barbs indicate the occurrence of turbulence. (Courtesy R.J. Williams, National Aviation Weather Advisory Unit, Kansas City, MO, USA)

8.3 Radio-acoustic sounding systems (RASS)

Radars, including electromagnetic profilers, basically sense the intensity and Doppler shift of the echoes returned by the atmosphere. They have no direct way of sensing local atmospheric parameters such as temperature and pressure, since the reflectivity and wind fields are only indirectly and weakly influenced by these parameters. These are, however, very important atmospheric parameters, and it would be very useful to be able to remotely sense them. In particular, local ambient temperature is a strong factor in determining the conditions conducive to ice accretion on aircraft (see Section 3.5).

8.3.1 Basic system

An instrument called the radioacoustic sounding system (RASS) is a useful augmentation of the radar wind profiler which can measure temperature profiles (i.e. the variation of the atmospheric temperature with height) in the lower troposphere (May *et al.*, 1988; Strauch *et al.*, 1989b). The RASS is a

Other sensors and systems for aviation weather 303

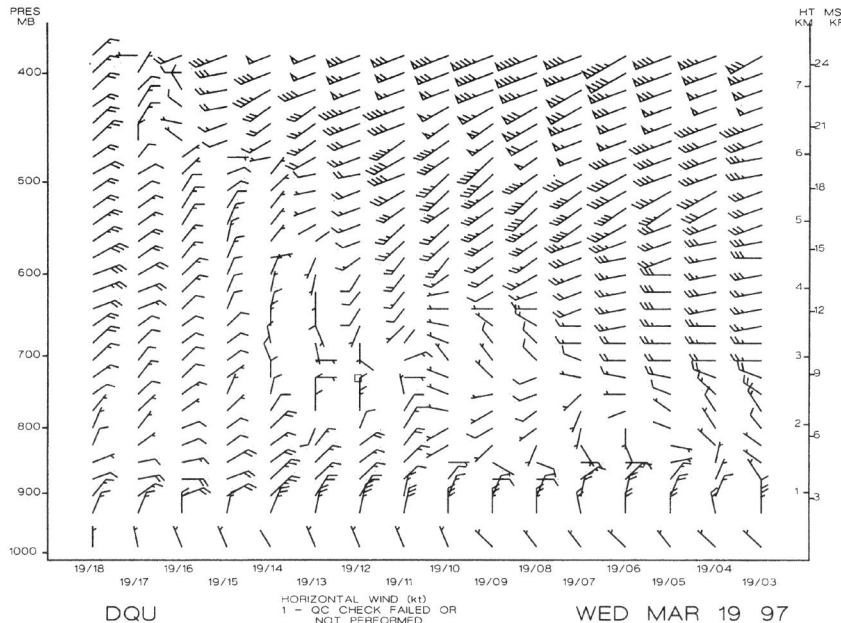

Figure 8.9 Plot in the same time frame as Fig. 8.8 obtained from the profiler at Dequeen, AR, USA. (Courtesy R.J. Williams, National Aviation Weather Advisory Unit, Kansas City, MO, USA)

combination of a radar profiler and an acoustic radiator which beams powerful sound waves into the same atmospheric zone where the profiler beams lie. The sound waves cause cyclic density change of air, effectively forming a grating which scatters the electromagnetic radiation of the radar profiler through the Bragg effect. A schematic diagram of the system is shown in Fig. 8.11. The frequency of the acoustic radiator is swept over an interval, and the acoustic frequency corresponding to the maximum electromagnetic echo return from each location along the beam is used to deduce the temperature profile.

As already mentioned in Section 3.5 hazardous conditions from the point of view of ice formation on aircraft are most often found at temperatures between -10 and $0°C$, and the RASS, with temperature profiling accuracies better than $0.7°C$, can help in delineating such zones (Westwater and Kropfli, 1989).

Profiling radars operating at 50, 405 and 915 MHz have been used in RASSs. The ones using the lowest frequency can perform temperature profiling up to altitudes of ~10 km. The altitude ceiling decreases for higher operating frequencies. This is because the frequency of acoustic radiation in a RASS must increase in proportion to the electromagnetic frequency to enable the Bragg effect to be manifest, and the sound waves of higher frequencies suffer more rapid attenuation through the atmosphere.

304 *Aviation weather surveillance systems*

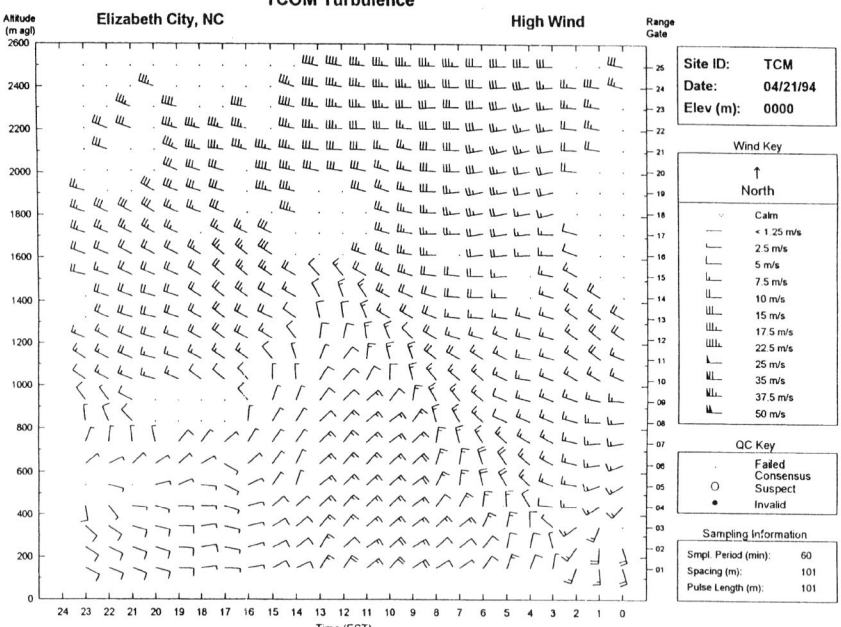

Figure 8.10   *24 h time–height cross-section of low-altitude winds (up to 2.6 km) obtained on 21 April 1994 from the 915 MHz profiler at Elizabeth City, NC, USA. Winds above ~1.5 km altitude are nearly steady throughout the day, but below this height there is strong and variable wind shear. The turbulence, measured by the standard deviation of the mean winds, is significant, reaching values up to 1.8 m/s early in the day* (from Winston et al., 1995, by authors' permission)

8.3.2 *RASS augmentation for sensing aircraft icing conditions*

Although the basic RASS can obtain the temperature profile of the atmosphere, accurate and reliable prediction of icing potential of given regions of the atmosphere cannot be made based on temperature data alone. Referring again to Section 3.5, prediction of aircraft icing conditions requires knowledge of the distribution of supercooled liquid water in the atmosphere, which in turn depends on the vertical profiles of both temperature and liquid water concentration. The latter, as well as the water vapour content, can be obtained by using dual-wavelength microwave radiometers operating at 21 and 32 GHz (Guiraud *et al.*, 1979; Hogg *et al.*, 1980). A combination of a RASS and a dual-channel microwave radiometer thus provides a viable method of monitoring aircraft icing conditions.

Such a sensor combination may be further augmented by one or more of the following sensing instruments for more accurate and reliable nowcasting of atmospheric conditions conducive to aircraft icing (Westwater and Kropfli, 1989; Stankov and Bedard, 1990):

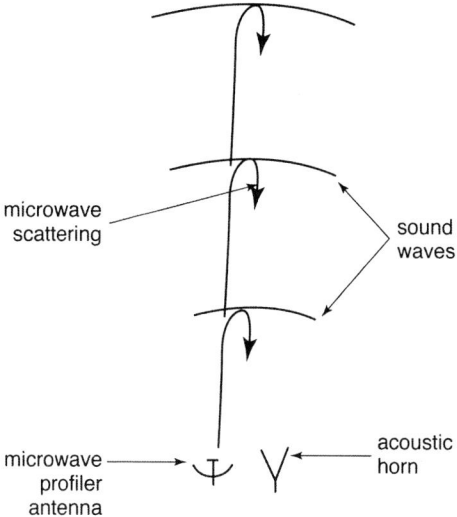

Figure 8.11   Schematic diagram showing the principle of the RASS

1. airborne radiometer which helps in determining the horizontal distribution of supercooled liquid water
2. Doppler radars for mapping supercooled liquid water distribution and water vapour flux
3. lidar, a radar-like sensor operating at laser wavelengths, which is useful for remote sensing of winds, temperature, water vapour concentration and cloud-base height
4. GOES satellite data for estimating cloud top temperature (see Section 8.8).

Exploring the use of such multisensor data to detect the existence of aircraft icing conditions and to forecast icing possibilities up to 4 h, possibly even up to 12 h, has been the goal of the Icing Research Program of the US Federal Aviation Administration (US Department of Transportation, 1989a).

8.4 Low-level wind shear alert system (LLWAS)

It has been repeatedly pointed out hitherto in this book that wind shear in the lowest layers of the atmosphere constitutes perhaps the most severe and frequent source of hazard for aviation operations. Thus, even in the absence of a comprehensive surveillance mechanism to monitor the entire weather picture affecting the aviation system, a simpler observation system dedicated to the detection of low-altitude wind shear should greatly enhance aviation safety by minimising aircraft encounters with such shear during landing and takeoff. This was the motivation for the development of the low-level wind shear alert system (LLWAS) before the modern generation of weather radars became available.

The LLWAS was conceived as a specific technological response to the weather-induced air disasters of the 1970s such as the highly visible and well investigated Eastern Airlines accident (National Transportation Safety Board, 1976) at John F. Kennedy Airport, New York, in 1975. Many of these accidents were attributed to low-altitude wind shear, and had attracted the particular attention of the US National Academy of Sciences (National Academy of Sciences, 1983).

8.4.1 Concept and basic configuration

The concept and design of the LLWAS system are based on the premise that phenomena which generate wind shear in the lowest layers of the atmosphere must manifest themselves through detectable ground-level signatures. Thus proper sensing, processing and interpretation of such signatures should lead to conclusions regarding the existence or otherwise of hazardous levels of wind shear at low altitudes.

The LLWAS is an instrument system based on *anemometers* which are *in situ* wind sensors. Anemometers are usually either of the vane type or cup type. In the former, a wind vane with its horizontal axis aligned along the wind direction is driven by the local wind and rotates at a rate which is a monotonic function of the wind speed. A reading of the rotational speed of the vane therefore serves to indicate the local wind speed. The latter type of anemometer has a set of cups, usually three in number, mounted cyclically with their individual axes horizontal, on a rotor with a vertical axis. Because of such an arrangement, the rotor is insensitive to the horizontal wind direction and turns at a rate dependent on the wind speed. With either type, the capability to sense wind direction can be incorporated by adding a weathercock mechanism. Modern anemometers have built-in transducers to convert their readings into electrical impulses which can be transmitted over wires or wireless links for remote monitoring.

Each weathercock–anemometer combination measures the horizontal wind vector only at one point. To sense the existence of wind shear over a linear path and estimate its intensity, a minimum of two anemometers may be used, one at each end of the path segment, and their difference taken. For more reliable and detailed information regarding the wind distribution along the path, additional sensors may be placed along the path. For sensing wind shear over an area, the area may be appropriately circumscribed with a number of vector anemometers; a comparison of their readings would serve to indicate the strength of wind shear across the area. A denser coverage of the area with anemometers, including some located in its interior, would enhance the reliability of the wind shear information and provide a more detailed picture of the wind field.

The LLWAS is based on this principle, and is optimised for aviation support. It utilises an anemometer array to cover the most sensitive parts of airport areas from the point of view of low-altitude wind shear. These include the runway complex and the adjoining areas over which flights occur at very low altitudes. In the original version of the LLWAS (Goff, 1980), the array consists of one anemometer positioned close to the centre of the runway complex and five more at outlying spots located, where possible, close to the approach paths of individual runways. Each anemometer is mounted on a mast at a height ranging between 4 and 20 m from the ground depending on the local air flow quality (velocity perturbations, turbulence, etc.) and obstruction considerations. A central processor monitors the individual anemometer readings received through a radio link at 10 s intervals. The data from the centre-field anemometer are averaged over a 2 min period on a moving-window basis. The instantaneous readings of each outfield anemometer are compared with this averaged centre-field wind, and a wind shear advisory is issued if any of the outlying anemometers shows a difference >15 knots relative to the centre-field wind.

The LLWAS was originally intended to be an interim solution to the low-altitude wind shear warning problem, to be eventually replaced by more elaborate systems based on Doppler weather radars. Specifically, the initial design of the LLWAS was tailored to satisfy the requirements of detecting the wind shear associated with gust fronts and other frontal phenomena not originating from thunderstorms (US Department of Transportation, 1989b). The installation of LLWAS units started in 1977–1978, and over a hundred airports in the US have been equipped with the system. Owing to the success of the system, the LLWAS is no longer regarded as a temporary solution, but has been incorporated as an element of the 1987 Integrated Windshear Program Plan of the US Federal Aviation Administration (US Department of Transportation, 1987). Because of the simplicity and cost-effectiveness of the system, other countries have experimented with, designed and/or considered installation of the system or its variants.

8.4.2 Enhanced system

Although the original LLWAS does provide useful indication and warning of low-altitude wind shear, this basic configuration has its own limitations (Kessler, 1990; National Transportation Safety Board, 1983). These limitations arise essentially from the small number of sensors in the basic system as well as their location. Indeed the effects of the two factors are interrelated. Because of the small number of sensors in the original LLWAS configuration, only a limited area can be covered. Thus, the outfield anemometers are located so as to cover only a short terminal segment of the aircraft glideslope. Wind shear zones outside this coverage are not sensed by the LLWAS array. This situation can be remedied to an extent by enlarging the array coverage by locating the outfield anemometers farther out along the aircraft approach

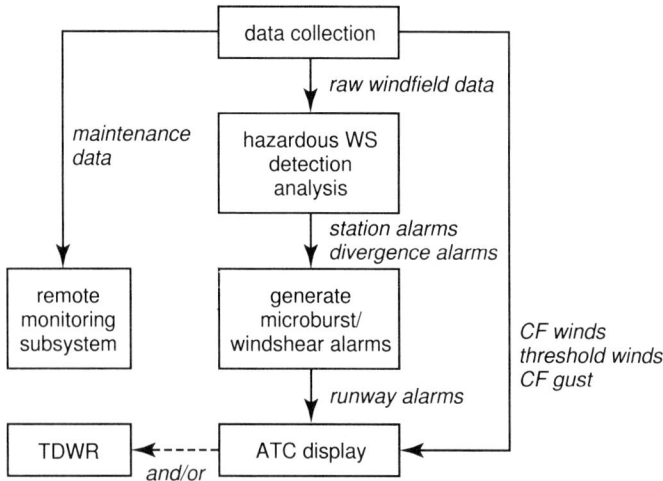

Figure 8.12 LLWAS functional data flow diagram (from Nilsen and Starr, 1995, courtesy K.M. Starr, TRW Inc.)

paths. But it must be remembered that intense low-altitude wind shear such as that due to microbursts and gust fronts can be quite localised. Thus, spreading out a limited number of sensors over an expanded boundary could lead to spatial undersampling of the wind field, resulting in missing or underestimating localised strong shear occurring within the nominal area of coverage.

To overcome such limitations, the original LLWAS has been enhanced in many ways. These include (US Department of Transportation, 1989b; Goff and Gramzow, 1989; Jaffe, 1989):

1. network expansion, i.e. increasing the number of wind sensors beyond the original six
2. network and software design enhancements to identify microbursts and gust fronts and to provide runway-oriented wind shear information
3. elimination of site effects on wind sensors.

The functional data flow in an LLWAS-NE (i.e. LLWAS with network expansion) is shown in Fig. 8.12. Two enhanced 'test bed' systems were installed at New Orleans and Denver airports in the USA in 1984 and 1985, respectively. An enhanced array of 12 wind sensors tested at Denver, along with a wind shear and microburst detection algorithm, was found to perform appreciably better than the conventional method of merely comparing the outer sensor readings with the averaged centre-field sensor reading (Smythe, 1989).

With these improvements, the scope of performance of the LLWAS has been greatly enhanced. Against its original expectation of providing only general indication of the existence of low-altitude wind shear in the runway area and its close environs, essentially due to gust fronts, the enhanced versions of LLWAS can perform detection and recognition of major shear-producing phenomena such as microbursts and gust fronts, and provide more accurate quantitative estimation of the wind shear along the runway direction, which is the main parameter of interest for decisions regarding landing and takeoffs. With such improved performance, the LLWAS has emerged as a viable stand-alone system of great value in aviation safety and efficiency augmentation at airports not provided with more expensive, elaborate and versatile aviation weather surveillance systems, and a very useful supplement to such systems where they exist. The system has been credited with helping avert major accidents in real life (e.g. Hughes, 1990).

When an aircraft encounters a wind shear field and senses gain or loss of altitude and/or speed, it must immediately initiate a recovery procedure to escape the damaging effect of the shear. Such recovery procedures are a standard part of many crew training procedures now. However, the proper in-flight recovery techniques for a head-on encounter with a microburst and with nonconvective low-level wind shear are opposite (Jackson, 1991). The wrong choice of recovery procedure can make the difference between a successful escape and a mishap. Thus recognition of the type of wind shear phenomenon ahead of an aircraft is of paramount importance for aviation safety against low-altitude wind shear. The ability of the enhanced LLWAS to recognise the nature of wind shear fields is therefore one of its greatest advantages in promoting aviation safety.

Irrespective of the level of its sophistication, the LLWAS does, however, suffer from certain drawbacks which are inherent to ground-based *in situ* sensors. The chief one among them is that surface winds detected by the LLWAS sensors may be significantly different from those along the aircraft flight paths during landing and takeoff. For example, the results of one study (Eilts, 1987) showed that winds at a few hundred feet altitude, even as low as a hundred feet (30 m) above the ground, were on average ~60% higher than those at heights corresponding to the location of the LLWAS sensors. The difference between the true and the surface-sensed winds in terms of effects on flight is even higher since wind effects on aircraft vary as the square of the wind speed (Section 2.3).

Such limitations notwithstanding, the LLWAS is a useful system for enhancing aviation safety. In particular, it can perform a very useful role in conjunction with terminal area Doppler weather radars such as the TDWR. In such a combination, the LLWAS would provide wind data very close to the ground level, for which heights radar data are often either unavailable or unreliable due to radar horizon limitations and ground clutter contamination.

8.5 Airborne wind shear detection

All the systems for wind shear detection discussed hitherto in this work are ground-based systems. Irrespective of whether they are based on *in situ* sensors or remote sensors, a common feature of all these systems is that information on wind shear is obtained at a location on the ground. If the information is to be used by the aircraft pilot for being aware of the weather scene and/or taking decisions regarding the flight path and operations, it must be transmitted to the aircraft in raw or processed form. Alternatively, reports, advisories or warnings based on the ground-sensed weather data may be transmitted to the pilot.

A measure of autonomy can be achieved by the pilot if wind shear could be sensed on board the aircraft itself, minimising the need for extensive ground support and communication links. Such an arrangement would have important advantages. First, the time involved in collecting, processing and possibly interpreting and confirming the observations, and in multiple transmissions (e.g. from the sensing system to the air traffic control centre, and then on to the pilot) would be eliminated. This may add precious moments to the time available for the pilot to be aware of any impending danger due to wind shear and take evasive action or other crucial flight decisions. A second major advantage is that aircraft so equipped could have a high degree of protection from wind shear hazards even in areas and air spaces not provided with adequate ground-based shear-warning systems. The quality of weather coverage will be nonuniform over large areas (e.g. continents) for a considerable period in the future, and may never be uniform globally. Aircraft with on-board capability to sense wind shear would be less dependent on ground support, and can minimise wind shear hazards even while conducting flight operations in the less developed parts of the world.

8.5.1 In situ sensing

Wind shear may be detected on board aircraft either *in situ* or by remote sensing. In the former, air motion is sensed by measuring its effect on the aircraft trajectory and dynamics (McLean, 1988). Perturbations in aircraft flight paths and body attitudes are readily detected with the help of inertial sensors such as accelerometers and gyroscopes, which are now part of the standard navigational instrumentation on all but the simplest and smallest aircraft.

The F-factor was discussed as a measure of wind shear in Section 3.3.1. This factor can be derived in flight by monitoring certain flight parameters such as speed, climb/descent rate, thrust level and angle of attack. The aircraft may be declared to have entered a strong wind shear zone when the sensed F-factor exceeds a certain threshold. It may be recalled from the earlier discussion that most commercial passenger and transport aircraft, which constitute perhaps the most significant segment of aviation activity, would inevitably lose height and/or speed if subjected to wind shear with F-factor values somewhere in the range from 0.11 to 0.17. Hence a threshold of 0.10

to 0.15 may be applied, depending on the type of aircraft, to determine entry into a strong wind shear field.

Measuring the local F-factor on board and applying thresholds to judge its severity, however, is not without difficulties. Strong wind shear in nature does not appear as a neat streamlined flow field with spatial gradients, but is usually accompanied by strong turbulence or gustiness (see Section 3.3.2). The effects of these gusts generally cancel out among themselves and do not affect the aircraft trajectory significantly. However, they can induce sharp local spikes in the variation of the F-factor, giving it a noise-like characteristic. Straightforward thresholding of the raw instantaneous F-factor data may then give rise to false alarms and multiple threshold crossings. The high-frequency signals caused by gustiness must be filtered out before the *in situ* measured F-factor can be meaningfully used for wind shear detection and evasion. Such a smoothing filter, called *gust filter* or *turbulence filter*, would introduce its own time delay into the data chain, which would correspondingly delay the shear detection process. Thus, airborne *in situ* wind shear sensing not only does not provide any advance warning of approaching shear, but actually delays the data compared to real time. This delay is undesirable, and would have an effect opposite to advance warning, i.e. the aircraft would be deeper into the shear field before the pilot even realises the fact.

Experiments in *in situ* wind shear detection have been conducted by the US National Aeronautics and Space Administration using an instrumented Boeing 737 research aircraft to penetrate microbursts. The details of the experiment and the results are discussed by Lewis *et al.* (1992) and Oseguera (1992). A stand-alone *in situ* wind shear warning system comprising a warning computer and associated cockpit displays is described by Aeronautical Radio Inc (ARINC, 1988). The system uses data pertaining to aircraft movement with respect to air to detect and annunciate a wind shear condition. Optionally it may also provide instrument guidance to the crew indicating the optimum pitch to endure the wind shear encounter.

8.5.2 Forward-looking remote sensing

In situ detection of wind shear by aircraft in flight is quite straightforward and can provide useful warning to the pilot before maximum levels of shear are encountered. It can also facilitate the incorporation of automatic compensation mechanisms to mitigate the effects of shear. However, it has the drawback that the aircraft would notice the shear only after it actually enters the shear zone. It would be preferable to sense wind disturbances along the flight path in advance of their encounter with the aircraft to provide some warning time to the pilot to take remedial measures. The advantages of such advance detection of wind shear over reactive mitigation of wind shear effects have been discussed by Hinton (1990).

Even very short warning times can have dramatic effects on the ability of aircraft to survive adverse wind fields. In a study related to the Integrated

Wind-Shear Program undertaken jointly by the US National Aeronautics and Space Administration, Federal Aviation Administration and industry, it was found that the factor which most strongly affects the ability of aircraft to recover from microburst-induced wind shear is the time at which the recovery is initiated. Improving the alert time by just 5 s generally provided greater increase in recovery performance than could be achieved by changing the recovery strategy. Forward-look alerts given 10 s prior to entry into a microburst permitted recoveries to be made with negligible altitude loss (Hinton, 1992). At an approach speed of 150 knots (~ 75 m/s), a 5 s warning time corresponds to a distance of ~ 375 m, and 10 s to ~ 750 m. In another study, Hinton and Oseguera (1993) have noted that all classes of commercial transport aircraft (i.e. those with two, three or four engines) can escape the 'worst-case' microburst located straight ahead if given 20 s of warning time, or ~ 1.5 km of warning distance. This has been proven in practice by the hazard-free escape of the aircraft that followed in the landing queue the Lockheed L-1011, which crashed after encountering a strong microburst at the Dallas–Fort Worth airport on 2 August 1985 (National Transportation Safety Board, 1986; Fujita, 1986). Four other aircraft successfully passed through an equally strong microburst in Denver in 1988 (Schlickenmaier, 1989), and one more is known to have performed the same feat in the subsequent year (Hughes, 1990).

Warning times of such orders can be obtained readily and autonomously through remote sensing of the wind field ahead by employing forward-looking instruments. Airborne remote sensing of wind shear is an area of research. It is a complex topic that involves not only hazard definition and sensor selection and optimisation, but also system integration and flight management (US Department of Transportation, 1989a). Disciplines and studies contributing to a better understanding and solution of the problem include aircraft simulation, mesoscale atmospheric modelling and analysis, and instrumented flight tests.

Besides providing the much-needed advance warning time, wind shear detection ahead of aircraft using forward-looking airborne sensors has another advantage. This pertains to gust filtering. The necessity of filtering out gust signals corrupting the local F-factor values was discussed above in connection with *in situ* wind shear detection where it was mentioned that such smoothing results in an unwanted time delay. This was caused because the *in situ* sensor has to actually fly over the distance corresponding to the integration time before it collects the data used for smoothing. The delay problem does not exist with forward-looking remote sensing, since the wind field ahead is 'seen' instantaneously and hence filtering can be performed in real time. Thus no part of the lead time provided by remote sensing is eaten up by the smoothing process.

It is necessary to comment on the spatial scales involved in the smoothing of the F-factor. For this purpose, it is useful to define an *equivalent average F-factor* $\bar{F}$ as

$$\bar{F}(s_0, L) = \frac{1}{L} \int_{s_0}^{s_0+L} F \, ds \qquad (8.3)$$

where F is the local F-factor, and the averaging is carried out over a length L starting at the position s_0. In terms of the aircraft parameters the equivalent average F-factor may be written as

$$\bar{F}(s_0, L) = \frac{1}{L} \int_{s_0}^{s_0+L} \left(\frac{T-D}{w} \right) ds - \frac{\Delta(V_a^2)}{2gL} - \frac{\Delta h}{L} \qquad (8.4)$$

where

T = engine thrust of the aircraft
D = drag acting on the aircraft
w = weight of the aircraft
Δh = change in the altitude of the aircraft over the interval L
$\Delta(V_a^2)$ = change in the squared air speed of the aircraft over the interval L
g = acceleration due to gravity.

Lewis et al. (1994) have made a comparison of the performance limit curves for transport aircraft, with $\bar{F}$ values estimated from a variety of real wind shear events, to conclude that an averaging interval of the order of 1 km is optimum for discerning hazardous shear. Lower averaging intervals will be dominated by turbulence, which is not a performance threat (in the sense of loss of flight altitude/speed, though turbulence by itself is a serious hazard in a different way; see Section 3.3.2), and significantly higher averaging intervals may result in smoothing out hazardous events of small spatial extent, making them look like having low $\bar{F}$ values. In each of the cases they studied, the hazardous wind shear event crossed the appropriate aircraft performance limit curve at scale lengths of ~1 km. This order of scale length provides a basis of filtering or averaging the wind field detected by forward-looking wind shear sensors. The threshold for shear warning and activation of evasive manoeuvres would depend on the aircraft performance characteristics in the takeoff and landing configurations and the initial energy state of the aircraft at entry into the shear field. The US Federal Aviation Administration has specified an alert threshold boundary for reactive wind shear detection systems (US Department of Transportation, 1990).

Airborne Doppler radar, lidar and infra-red sensors are candidate devices for remote sensing of wind shear ahead of an aircraft from the aircraft itself. Each of these sensors has its own special advantages, but the Doppler radar scores higher on the important point of all-weather wind shear detection. A small low-power Doppler radar may have a maximum detection range of the order of 2 km ahead of the aircraft, but a more powerful device may 'see' up to 10 km or more, providing a longer warning time. Doppler radars also have a minimum range or *blind range*, which may be as much as a kilometre, within which they would not be able to observe the wind shear.

The basic principle of wind determination by airborne Doppler radars is the same as that for ground-based Doppler radars, but there are significant differences in terms of details. An obvious difference is that the airborne radar, unlike its ground-based counterparts, itself moves along with the aircraft relative to the surrounding air, and this motion causes a Doppler shift which is superimposed on the Doppler signal due to the absolute motion (i.e. relative to the earth) of air. This component must be removed to retrieve the true atmospheric Doppler signals which provide a picture of the wind fields ahead. The radar-motion-induced Doppler shift is a function of the aircraft velocity, which is available on board from the aircraft's navigation system, and the pointing angle of the antenna relative to the aircraft velocity vector, which can be read from the antenna pointing mechanism. The cancellation of the Doppler component due to radar motion is therefore a straightforward procedure that can be carried out in real time, but its completeness depends on the accuracy of the velocity and pointing angle data. Incomplete cancellation would leave a residual component that contaminates the atmospheric velocity estimates. The cancellation process may also be aided by comparing the atmospheric Doppler signals with the returns from stationary ground clutter, if the latter is available in sufficient strength.

Sensing wind shear ahead of aircraft using airborne Doppler radar has a natural advantage compared to performing the same task with ground-based radars. As the airborne radar looks ahead along the flight direction, the component of wind variation along the path, which is of most concern for the flight, appears to the radar as a radial component. The Doppler radar, which can sense only the radial component of wind, can therefore realistically estimate the severity of the wind shear which the aircraft is likely to encounter ahead of the current position. This is not necessarily true in the case of ground-based radars which would, in general, observe the path ahead of the aircraft from an oblique aspect, and therefore sense a different component of the wind shear than the one along the flight path. Such a situation is depicted in Fig. 8.13. The ground-based radar data may, of course, be used in an indirect way by recognising the nature of the shear field (e.g. that due to microburst or gust front), modelling the field based on its observation, and then computing the wind variation along the expected flight path.

An important limitation of airborne weather radars relative to those based on the ground is with regard to the choice of operating frequency. From size and weight considerations, the antennas of airborne radars are of limited aperture, and therefore utilise higher operating frequencies to achieve a narrow beamwidth. Frequencies in the X-band (wavelengths of the order of

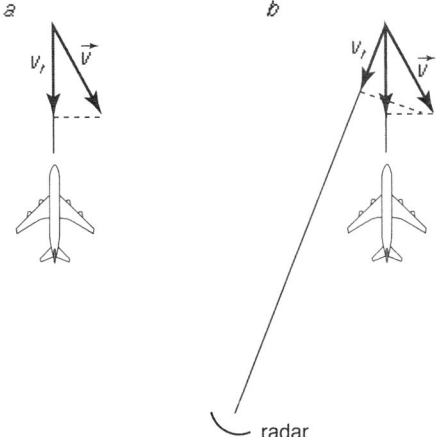

Figure 8.13 Difference in the geometry of wind (shear) detection by airborne (a) and ground-based (b) radars. Each measures the component of wind in its direction, and the components are not the same in general

3 cm) are commonly used, with those in the C-band (wavelengths of the order of 5 cm) a possibility. These higher frequencies have their attendant problems of enhanced rain attenuation (see the relevant discussion in Sections 7.5 and 13.4) and lower sensitivity for clear-air observation.

While operating at higher altitudes, such as during *en-route* flight, airborne radars have the advantage of negligible ground clutter interference. However, at the higher altitudes, where the refractive index fluctuations due to turbulent mixing are feeble because of lower water vapour concentration (see the relevant discussion in Sections 6.2.4 and 8.2), the detection of wind shear and turbulence in clear air becomes difficult. On the other hand, these phenomena can be very well detected in the presence of precipitation.

At altitudes close to the ground level the reflectivity of clear-air phenomena is much better, but the problem of ground-clutter contamination becomes significant. For such operation, Doppler lidars have an advantage in clear air (as lidars do not have the ground-clutter problem), but are ineffective in seeing through precipitation and fog. Airborne Doppler radars would perform quite well in light rain, but heavy rain would impair their performance due to absorption of their shorter wavelengths.

Sensor choice and operation as well as system configuration for airborne remote sensing of wind shear under a wide variety of operating conditions is thus a complex task, and no easy solutions exist. The airborne Doppler weather radar has the potential to provide useful wind data in a variety of situations of interest, and may find wide usage in this role in the future when the technology is perfected.

8.6 Airborne turbulence measurement

Like wind shear, turbulence can also be sensed and estimated from observations made with the help of instruments mounted on aircraft. At present this is not a routine and formalised procedure, and the main source of turbulence information consists of pilot reports which are subjective and provide only an integrated or average sense of the turbulence as being light, moderate, etc. Airborne measurement, in contrast, has the potential to provide continuous and quantitative records.

As discussed in Section 3.3.2 atmospheric turbulence causes random displacements of the aircraft centre of gravity as well as random changes in the aircraft attitude. These are readily measurable *in situ* using inertial instruments such as accelerometers and gyroscopes. These sensors are part of the inertial navigation systems carried on nearly all modern commercial and other large and high-performance aircraft. The most important components of turbulence-induced aircraft motion are the vertical movement of the centre of gravity and the pitching motion about this point. This is so because the large lift force which balances the weight of the aircraft is essentially vertical, and its variation due to the wind fluctuations caused by turbulence are also in a vertical plane. The variations in the lift magnitude cause vertical acceleration of the aircraft, and changes in the lift magnitude as well as position give rise to pitching moments.

The motion induced in aircraft by turbulence may be studied by using linear systems theory. The air-speed variations due to turbulence, and the resulting fluctuations in the forces acting on the aircraft, act as a forcing function on the aircraft, and the response function of the aircraft acts as a filter whose output would correspond to the resultant incremental aircraft motion. Thus by measuring the aircraft response to turbulence (with the help of inertial instruments), and knowing the transfer function of the aircraft, the strength of the ambient turbulence can be deduced.

Cornman *et al.* (1993) have used this method to develop a practically usable method of *in situ* turbulence measurement. They have used a two-degree-of-freedom model of aircraft motion, consisting of pitch and vertical displacement, and have assumed the aircraft to be a rigid body, i.e. neglected the flexibility of the aircraft structure. The model is basically developed in the frequency domain, incorporating autopilot effects and simple corrections for Mach number, aspect ratio, wing sweepback effects, unsteady aerodynamic effects, and the quasi-steady downwash effect of the wing on the tail. The power spectral density of turbulence as a forcing function is assumed to vary as the inverse 5/3rd power of frequency, and its strength depends on the 2/3rd powers of the *eddy dissipation rate* and the aircraft's air speed The concepts involved in the description of atmospheric turbulence have been dealt with by Panofsky and Dutton (1984), and will be discussed in some detail in Chapter 11. The simple but realistic method of modelling the problem permits real-time evaluation of the eddy dissipation factor (Chapter 11) as the flight progresses, while providing good accuracy of the estimates.

8.7 Automated weather observing systems

While high-resolution three-dimensional weather data such as those provided by radar systems and networks are of great value for modern aviation support and management, accurate and current monitoring of basic atmospheric parameters is still a matter of high utility for aviation. These parameters usually include such fundamental quantities as temperature, pressure, dew point, humidity, wind speed and direction, rainfall rate and accumulation, etc. In the specific context of aviation certain other parameters such as airport local visibility, visibility ceiling, sky condition, runway wetness or ice covering, etc. are also of great importance. These basic weather parameters are normally measured and disseminated irrespective of the availability of more detailed weather data such as those derived from Doppler radars, profilers, LLWAS, etc. At many small airfields essentially supporting general aviation activities, a few of the basic weather parameters constitute the only weather information available.

The basic weather-related parameters are usually observed and measured *in situ* by using instruments that sense the parameters in their immediate vicinity. The process is traditionally human-centred, i.e. the data reading, recording, interpretation and dissemination require direct and routine human involvement. Such operation makes the weather observation system expensive, and slow in the aviation context.

The Automated Weather Observation System (AWOS) is a system intended to improve the accuracy, currency and availability of basic weather data. It performs measurement, processing and direct dissemination of weather observations to pilots without routine human involvement (Kraus and Mayou, 1989).

The primary focus of the AWOS design is on providing accurate and reliable weather information to pilots in real time to aid them in landing and takeoff operations. To serve this purpose best, the AWOS sensors are recommended to be located in the close vicinity of the touchdown area of the runway in order to accurately sense the conditions to be encountered by aircraft. The system processor updates the data continuously (once every minute), providing highly current data to pilots directly through computer-generated voice accessible through NDBs, VORs (see Section 2.4.2) or dedicated frequencies, or even through telephone for preflight planning. This eliminates the need for an interactive query-response mode of information dissemination, reducing time delays and pilot workload during the critical phases of takeoff and landing operations. It also reduces demands on and for personnel manning the airports for responding to such queries.

318   *Aviation weather surveillance systems*

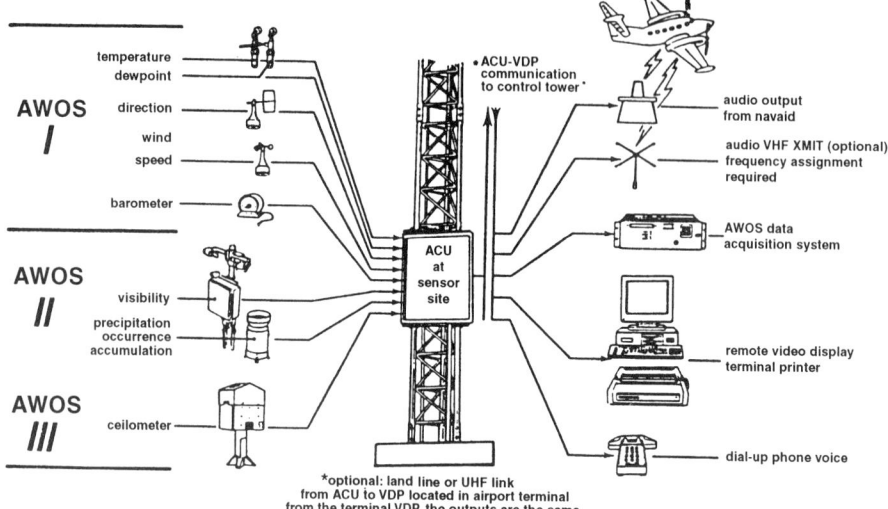

Figure 8.14   Schematic diagram showing the three levels of AWOS configuration (from Hodges and Starr, 1995, courtesy K.M. Starr, TRW Inc.)

AWOS data may also be displayed at critical locations in the airport, and are archived for later retrieval.

The Federal Aviation Regulation (FAR) of the US Federal Aviation Administration stipulates three levels of AWOS implementation ((Kraus and Mayou, 1989). The AWOS-1 reports altimeter setting, wind, temperature, dew point and density altitude. The AWOS-2 provides visibility data in addition to AWOS-1 observations. Further addition of visibility ceiling or cloud base height data comprises the capability of the AWOS-3 level. The three levels of AWOS configuration are shown schematically in Fig. 8.14. Definite rules have been stipulated regarding the approval of these systems for usage at different classes of airports. The systems are designed with a modular architecture such that additional sensors can be integrated readily to upgrade the level of a given AWOS. The ADAS (AWOS Data Acquisition System) is intended to collect, process, archive, and disseminate aviation surface weather observations including data from a maximum of 137 AWOSs (Hodges and Starr, 1995).

Another automated system for weather monitoring is the Automated Surface Observing System (ASOS), which is the product of an interagency programme of the US National Weather Service, Federal Aviation Administration and Department of Defense. The ASOS is functionally quite similar to the AWOS, but may use somewhat different hardware.

Because of the simplicity, affordability and high utility of the automated weather monitoring and information systems, such systems have a high rate of acceptance and installation. Hundreds have been installed in the US alone

Other sensors and systems for aviation weather 319

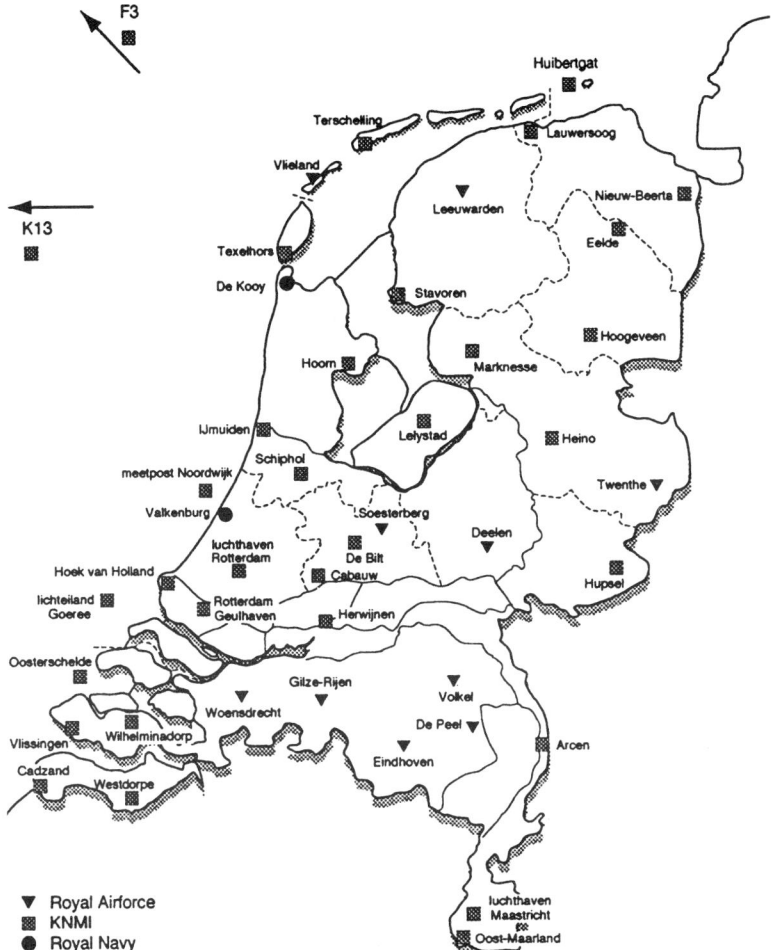

Figure 8.15a Network of automatic meteorological observing stations in the Netherlands (from van Dijk, 1993, courtesy W.C.M. van Dijk, Royal Netherlands Meteorological Institute, De Bilt)

within the first few years of their availability (during the late 1980s and early 1990s), and other nations have either established networks of automated weather stations or have plans for doing so in support of their aviation programmes. Hellroth and Olsson (1989) described the Swedish effort in this area. Figure 8.15a shows the network of automatic meteorological observing stations in the Netherlands, and Fig. 8.15b depicts a sample screen of the graphical presentation of meteorological data at Schiphol Airport, showing wind, visibility and cloud base data.

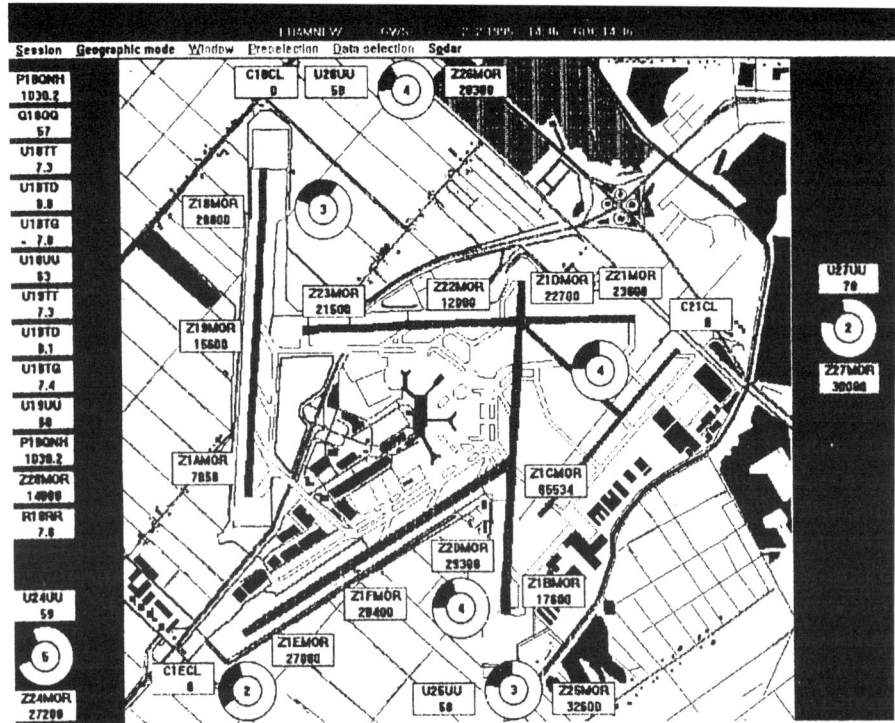

Figure 8.15b   A sample screen of the graphical presentation of meteorological data at Schiphol Airport, the Netherlands. MOR refers to Meteorological Optical Range (from van Dijk, 1993, courtesy W.C.M. van Dijk, Royal Netherlands Meteorological Institute, De Bilt)

The limitations of automated weather observing and reporting systems are inherent to their basic nature, arising from their dependence on *in situ* sensors and their nonuse of human perception, intelligence and subjective judgment. Because of the latter, messages from unmanned automated weather reporting stations do not contain remarks about qualitative and visual observations such as prevailing overall visibility, quadrant-wise visibility differences, present weather type and sky conditions, cloud types, etc. Incorporation of such capabilities will require further developments in sensors and induction of artificial intelligence techniques into the systems.

8.8 Radiometric satellite observation

We now move on to weather observation on a different scale. While individual long-range weather radars provide detailed observations over areas of the order of 1000 km in diameter, terminal area surveillance radars over tens or hundreds of kilometres, and *in situ* observation systems such as LLWAS and

AWOS/ASOS indicate specific conditions over localised areas, satellites are capable of providing pictorial views of weather patterns over subcontinental, continental and global scales. Since these scales match the geographical scales of operation of aviation entities such as airline companies and aviation regulating agencies, satellite data are of good value for aviation. Satellite weather monitoring has the added advantage of detailed observation such that on the lower end of the spatial scale the resolution may match the sizes of the local phenomena to which aviation is highly sensitive.

Weather data from satellites usually comes in the form of multispectral imagery and sounding data. The multispectral imagery consists of photographic images taken at multiple optical wavelengths falling within the visible and infra-red bands. Visible observation is naturally confined to local daylight hours, while infra-red sensing can be performed round the clock. The infra-red data are processed with the help of computers to generate quantitative pictures of cloud top temperatures. Remote sounding of atmospheric profiles of temperature and water vapour are derived from multichannel radiometric measurements. In recent decades these data have proved to be very valuable inputs for monitoring, modelling and forecasting of weather on a global as well as regional basis.

The wide coverage of the fields of view of satellites enables them to track the evolution of large-scale features such as tropical cyclones, movement of weather systems, cold/warm fronts, etc. However, in the context of nowcasting of local weather for aviation support, the spatial and temporal resolution of weather satellites have been considered inadequate. For example, the spatial resolution of the highly useful Geostationary Operational Environmental Satellites (GOES) is nominally 1 km in the visible band imaging, 8 km in infra-red imaging and 14 km in sounding. The latter two are inadequate for the detection of small-scale phenomena such as microburst-producing cells which are of concern to aviation, though they are very useful for observing, for example, mesoscale convective complexes which are the cause of much of the weather activity over large parts of the USA (Maddox, 1980; 1981). In contrast, the visible band, with its much better spatial resolution, does often show gust fronts or thunderstorm outflow boundaries in the form of arc clouds (Purdom, 1989), and also the buoyancy or 'gravity' waves generated by these outflows (Erickson and Whitney, 1973). It can also show the parent storms or storm systems that create strong wind shear or turbulence, e.g. the storm that created the hazardous weather conditions leading to the 1985 Dallas-Fort Worth accident (Fujita, 1986), but the details are usually not sufficient to be used by themselves for specific and localised aviation hazard warning.

Temporally, satellite data, which are usually updated at intervals of 30 min or 1 h, are too slow for the dynamic aviation environment in terminal areas. Satellite data, including soundings, are, however, useful in determining the environmental conditions of storms (Ellrod, 1989). Colour Plate 26 shows the spatial distribution of a microburst wind index derived from GOES-9 satellite data. The use of satellite data for local forecasting (Birkenheuer, 1989) can help in the enhancement of the efficacy of dedicated aviation weather instrumentation by alerting the operating personnel to increase the level of readiness of these systems and adjust their operational parameters during times of anticipated weather activity.

Satellite data are indeed of great use in many aspects of modern weather monitoring and forecasting, but of special concern to aviation activity is the ability of satellites to observe thunderstorms and their associated phenomena. Scofield and Oliver (1977) reported a scheme for quantitative estimation of convective rainfall at mid-latitudes. The technique involves enhancing satellite infra-red images in a particular way so as to emphasise cold cloud top temperatures below $-30°C$ and provide strong constrasts between selected temperature intervals between -30 and $-70°$ C. Certain patterns and their changes in this enhanced temperature pattern as well as in visible images indicated the presence of heavy local convective rainstorms.

In the context of thunderstorm observation, an important capability of satellites pertains to the observation of the intensity and progress of gust fronts, which are of particular hazard value for aviation. However, satellites, unlike ground-based Doppler radars, cannot directly sense the horizontal wind discontinuities that characterise gust fronts. Satellites therefore detect gust fronts through observation of their associated visual patterns. For example, the presence of strong outflows may be inferred from the shapes of individual storms as monitored by satellites or radars (Fujita, 1978). A more definitive indicator of gust fronts is the cloud arc that often forms along the outflow boundary or gust front. Gurka (1976) found correlation between satellite-observed characteristics of such cloud arcs and the corresponding wind observations on the ground. He postulated the following rules to infer the location of adverse wind fields (which are of concern to aviation) in a gust front using satellite imagery:

1. Wind shear and turbulence occur at (or very near) the leading edge of the cloud arc.
2. The regions of vigorous convection coincide with high gradients of the brightness observed on enhanced infra-red images.
3. The strongest winds occur beneath the part of the cloud arc that is closest to the most vigorous convection.
4. Filled cloud arcs in a cloudy area with active convection signify stronger gusts than thin arcs with clear skies in their interior.
5. Rapidly progressing cloud arcs are generally associated with strong winds close to the ground.

The zones of intersection of arc-shaped lines of convective clouds have a high potential for development of fresh intense convective clouds and severe local storms (Purdom, 1973; 1974). An intersection of two fronts, as detected by a weather radar, is shown in Colour Plate 9. The ability to precisely locate such boundaries in modern satellite imagery and to detect or anticipate their

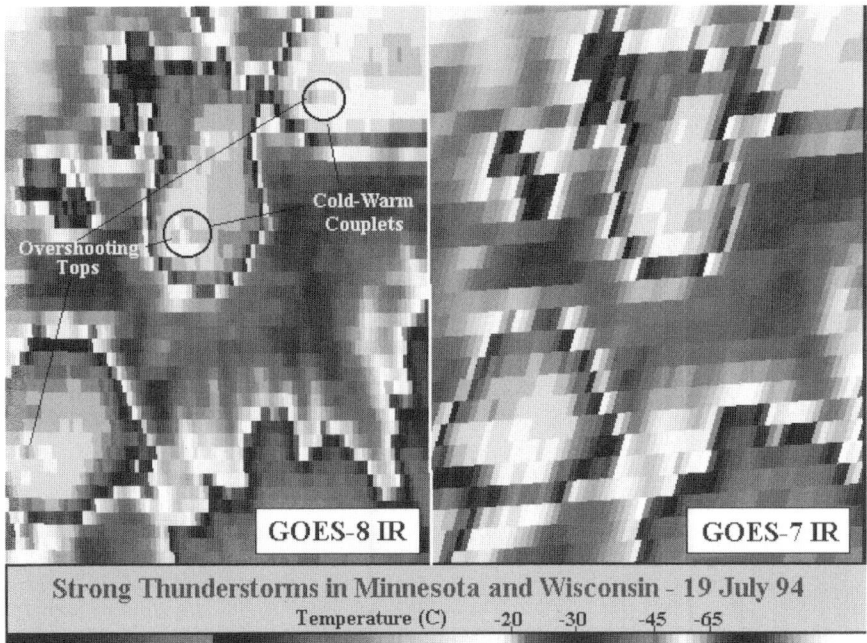

Figure 8.16   Infra-red images at a wavelength of 11 μm from GOES-8 (a) and GOES-7 (b) satellites showing the enhanced tops of two strong thunderstorms seen by both. The former shows two cold–warm couplets which are much less clearly visible in the latter (Courtesy G.P. Ellrod, NOAA National Environmental Satellite Data and Information Services, Washington, DC)

intersection provides a basis for short-term forecasting of intense convective storms of such origin.

The improvement of the spatial and temporal resolution parameters of weather satellites has brought them closer to the needs of aviation weather surveillance (Ellrod and Nelson, 1995). The qualitative enhancement in the ability of satellites to observe local phenomena hazardous to aviation with improvement in resolution is illustrated in Figs. 8.16 and 8.17, which compare the infra-red imagery from GOES-7 and GOES-8 satellites. It is easily seen that the latter, with a spatial resolution of ~7 km against ~14 km for the former, is much better able to depict phenomena such as cold-warm couplets in cloud tops, and mountain waves.

A better potential for detailed surveillance is offered by the GOES I-M series of satellites, which are improved versions of the original GOES satellites. The former are three-axis-stabilised satellites which have almost three times the weight and 30 times the electrical power output compared to the latter. They also have a significantly higher number of imaging and sounding channels. Importantly, the GOES I-M series has much better

324 Aviation weather surveillance systems

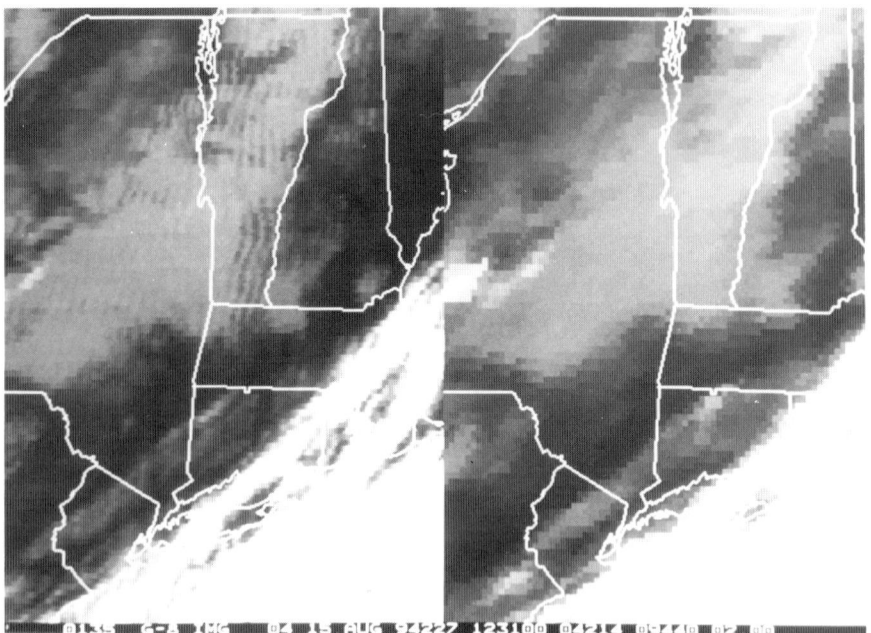

Figure 8.17 Infra-red images from GOES-8 (a) shows mountain wave clouds that are almost invisible in the GOES-7 image (b). (Courtesy G.P. Ellrod, NOAA National Environmental Satellite Data and Information Services, Washington, DC)

resolution compared to the GOES satellites, as shown in Table 8.1 (Mosher, 1989).

Another important advantage of the GOES I-M satellites is their flexible scanning scheme, which permits greatly improved data update rates in times of necessity. Their normal imaging frequency is once every 30 min, but the interval may be reduced to 15 min every few hours before the occurrence of severe weather, and further to a rapid scan rate of once every 5 min just before and during the occurrence of severe weather. For the imager, it is even possible to perform a super-rapid scan of state-sized areas at ~30 s intervals

Table 8.1 Resolution parameters of GOES I-M and GOES satellites

Function or parameter	GOES I-M resolution	GOES resolution
Infra-red imaging	4 km	8 km
Sounding	8 km	16 km
Grey scale	1024 shades	64 shades

(Purdom, 1989; Savides and Reseck, 1989). This would enable the observation of the evolution of fast-growing hazardous phenomena that are of the greatest significance to aviation. With these advanced capabilities the utility of satellite data in local severe weather nowcasting is significantly enhanced (Mosher, 1989).

Rapid data updates also permit accurate tracking of specific visual and infra-red features in satellite imagery. Tracking of clouds enables estimation of the winds that drive their motion. By tracking clouds at different levels, it is possible to derive the wind profile across the depth of the atmosphere. This profile may not be as accurate or detailed as those obtained from ground-based wind profilers (see Section 8.2), but has the advantage of being available over wide areas, including those not provided with such ground equipment. Vertical motion such as ascent rates of cloud tops may also be estimated from infra-red satellite data available at 5 min intervals or faster (Adler and Fenn, 1979). The ascent speed of the tops of thunderstorms may act as an indicator of their intensity, with severe storms rising faster than nonsevere ones.

Remote sounding from weather satellites using multichannel radiation data can yield vertical profiles of the atmospheric temperature and water vapour concentration. Again, these may not be as accurate as the RASS (see Section 8.3), but would be more widely available. These profiles can help in identifying regions of air space that harbour conditions conducive to aircraft icing. Examples of icing products obtained from satellite data are shown in Fig. 8.18. The grey areas in Fig. 8.18*a* indicating regions with high icing potential have been determined according to the simple criteria of relative humidity exceeding 50% and temperature lying between 0 and $-20°C$. However, such a gross product showing extensive areas of perceived hazard potential is not very useful since it generally overestimates the hazard zones and imposes undue restrictions on aviation activities. A more refined product for the same situation is shown in Fig. 8-18*b*. This product is derived from the one in Fig. 8.18*a* by removing clear (cloudless) areas and areas with relatively warm cloud tops. The refinement not only shrinks the potential hazard area, but also reveals mesoscale details of icing fields. A more sophisticated icing product based on a decision tree utilising imager data from a maximum of three GOES-8 visible and infra-red channels is shown in Fig. 8.19.

8.9 Airport visibility measurement

The effects of poor visibility on flights were briefly discussed in Section 3.6. Poor visibility due to the turbidity of the atmosphere, as caused by several types of weather phenomena, is to be differentiated from that due to the lack of sufficient ambient light such as during twilight, night and overcast conditions. Most airports of significant size are generally provided with elaborate lighting systems to help pilots discern runway and airport

326 Aviation weather surveillance systems

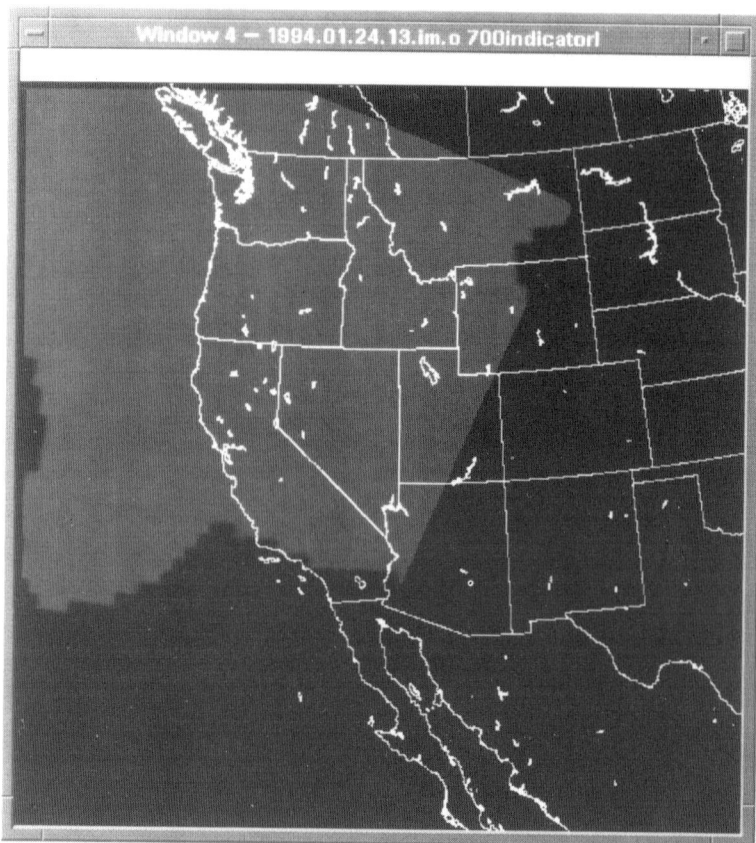

Figure 8.18a   Example of icing products derived from infra-red data obtained from Advanced Very High Resolution Radiometer (AVHRR) aboard NOAA polar orbiting satellites. Grey areas show grossly delineated icing potential regions (from Lee and Clark (1995) Courtesy T.F. Lee, Naval Research Laboratory, Monterey, CA, USA)

boundaries, taxiways, etc. during conditions of poor light. However, the effectiveness of these visual aids, as well as natural-light visibility, is greatly reduced due to phenomena which render the atmosphere relatively opaque.

Since adequate visibility is crucial to landing operations, both in operational and legal senses, the level of visibility obtaining at airports must be observed and recorded in a quantitative way and made available to pilots and air traffic controllers as necessary. A quantity called *runway visual range* (RVR) is an estimate of how far a pilot can see along a runway. Methods of observing and reporting RVR have been systematised and standardised by national and international bodies (International Civil Aviation Organisation, 1981).

Visibility levels may be inferred or measured either by human or automated observations. Human observation is straightforward and does not require

Other sensors and systems for aviation weather 327

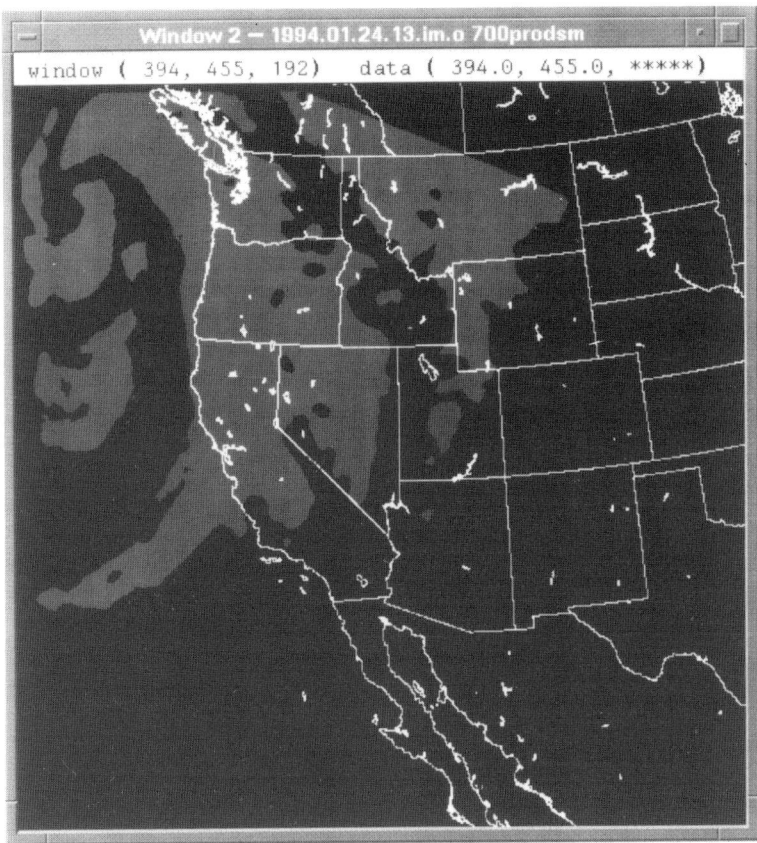

Figure 8.18b A more sophisticated icing product than the one shown in Fig. 8.18a obtained by utilising cloud and other data (from Lee and Clark (1995) courtesy T.F. Lee, Naval Research Laboratory, Monterey, CA, USA)

expensive equipment. This method of visibility estimation was used commonly in the early years of aviation, and is still used at small airports. At larger airports, the method is useful in the case of failure of specialised visibility-measuring equipment. However, human estimation of visibility is subjective, and is usually based on the observation of a limited number of objects or structures which are not standardised or optimum for the purpose of quantitative deduction of visibility.

Because of the overwhelming advantages of automated monitoring of atmospheric visibility for aviation purposes, most significant airports today are equipped with instruments for this purpose. RVR estimation is based on the measurement of light scattered and absorbed over a standard path length in the atmosphere. Since these two components of light are difficult to measure

328 Aviation weather surveillance systems

Figure 8.19 Example of an icing product (for 20 September 1996, 1215 UTC) based on a multistep decision logic utilising GOES-8 imager data from a maximum of three channels consisting of visible and two infra-red (10.7 and 3.9 µm) wavelengths. Dark areas show potential for icing. Pilot reports of icing intensity within 1 h of the image time are shown by letters L (light) and M (moderate). (From Ellrod and Nelson, 1997. Courtesy G. P. Ellrod, NOAA National Environmental Satellite Data and Information Services, Washington, DC)

directly, in practice the amount of light transmitted over the path length is measured, and the sum of the scattered and absorbed components inferred by comparing the measured light intensity with the intensity expected through a clear air path.

The transmitted light is measured by a *transmissometer*. A collimated light beam of known intensity is generated by a *transmissometer projector* and incident on a *transmissometer detector* separated by ~75 or 150 m. A data converter compares the detector output with a standard value corresponding to clear air conditions, and generates a visibility value after incorporating other visibility-affecting factors such as the time of the day and the absolute runway light intensity. The visibility value represents an estimate of the distance a pilot can see with acceptable clarity during an approach to the runway. This value is made available to the air traffic controllers through a remote digital display.

Factors such as fog, rain, snow, dust and smoke responsible for low visibility are often in a highly dynamic state, being wind-driven and undergoing turbulent mixing with surrounding air masses. This causes a strong spatial and temporal variability in the local transmission factor of light. Thus care must be taken that the runway visibility as seen by the pilot, which is an integrated effect over a relatively long air path and over the timescale of the approach and landing operations, is correctly inferred from the transmissometer readings.

The effect of temporal variability of local transmissivity is reduced through averaging. To take into account the spatial variability to a reasonable extent, each runway at important airports is normally provided with a dedicated RVR located by its side near its midpoint. In the case of high-traffic airports, two or even three RVRs may be located to cover different parts of individual runways.

An important quantity derived from sensor measurements is the atmospheric *extinction coefficient* which, along with other parameters, yields the visibility estimates through standard equations. There are separate equations representing the human visual range for a specific type of ideal targets, usually light or black objects, and particular ambient conditions, e.g. day or night (International Civil Aviation Organisation, 1981; Pawlak and Burnham, 1997; West and Burnham, 1997).

A new generation of RVR systems under deployment by the US Federal Aviation Administration determines the atmospheric extinction coefficient by measuring the forward scattering of light (Miles *et al.*, 1995). A photometer-based ambient light sensor to measure the background luminance, which is required in RVR estimation, is also part of the RVR system. The *forward scatterometer* uses a hooded transmitter and a receiver with intersecting beams as shown schematically in Fig. 8.20. The receiver collects and measures the light scattered in its direction by the particles (e.g. fog) in the common volume between the two beams, called the sample volume, which are illuminated by the transmitter. The scattered light intensity is proportional to the extinction coefficient. The forward scatterometer method of measurement of extinction coefficient has the advantages, over the transmissometer, of being compact, all-weather and less sensitive to effects of dirty windows, and having a much greater dynamic range. However, unlike the latter, it is not self-calibrating and depends on the latter for its absolute calibration. Calibration can be transferred from calibrated to uncalibrated scatterometers by using a calibration plate mounted along the plane of symmetry of the device. A new generation forward scatterometer is shown in Colour Plate 28, which also shows a close-up view with the calibrator plate mounted.

Automated visibility estimates are also available from the AWOS and ASOS systems discussed in Section 8.7. These systems also use forward scatterer sensors. The *airport visibility* (AV) obtained from these systems is related to the RVR obtained from RVR systems, as shown in Fig. 8.21 for night and day conditions. Under identical atmospheric conditions, RVR values are generally greater than the airport visibility.

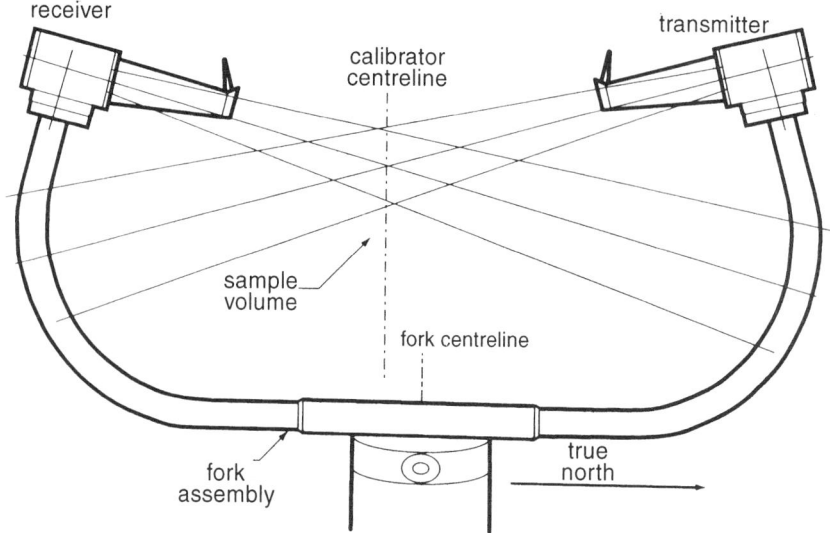

Figure 8.20 Outline diagram of the forward scatterometer of a new generation runway visual range system. (Courtesy Teledyne Controls)

8.10 Summary

A variety of atmospheric sensors can usefully supplement radar data in performing surveillance, nowcasting and forecasting for aviation support. The structure, working principle and capabilities of some of the important ones among them have been discussed in this chapter. The microwave wind profiler is a special-purpose Doppler radar optimised for receiving clear-air echoes from the layers of air above, and computing the winds as a function of height in a cost-effective way. In aviation applications, it can directly provide wind shear and turbulence information of certain scales, and a network of profilers can be a source of useful data for modelling and forecasting. Augmented with an acoustic sounding facility, the profiler is capable of yielding temperature profiles of the atmosphere.

The Low Level Windshear Alert System, based on an anemometer array and related data processing, provides a simple yet effective way of sensing and warning of close-to-ground wind shear of strength hazardous to aviation. The original six-sensor array can detect wind shear in a compact area around the landing zone. Recent enhancements in the number of sensors as well as the sophistication of processing enable the system to cover a larger area and recognise significant wind shear sources such as the microburst. In addition to being a cost-effective safety-enhancing aid by itself, especially at relatively small airports, the system can usefully augment the low-altitude observation and feature recognition capability of more sophisticated systems such as the Terminal Doppler Weather Radar. At higher altitudes, instrumentation

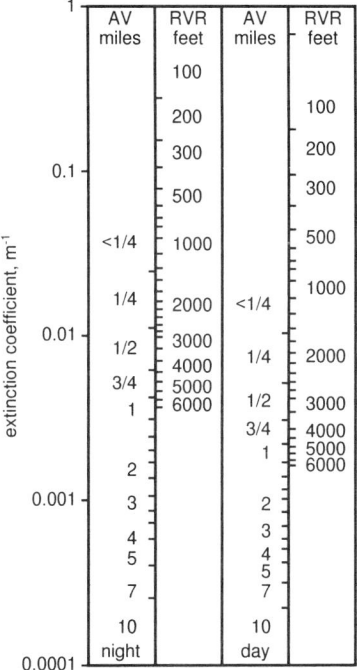

Figure 8.21 Comparison of Runway Visual Range (RVR) with Airport Visibility (AV) under day and night conditions, using standard US reporting units for each parameter. (Courtesy R. Pawlak, John A. Volpe National Transportation Systems Center, Cambridge, MA. USA)

mounted on aircraft themselves can help detect both wind shear and turbulence in the environs or even ahead of the aircraft in flight.

While aviation stands to gain greatly from the detailed high-quality multiparameter data from modern radars and other sensors, the importance of basic atmospheric data such as pressure, temperature and visibility to aviation remains. Sensing and reporting systems to automate and systematise the availability of such data have been designed and are being installed on a large scale to facilitate aviation activity in general, and the operation of small airports in particular. The need for verbal query–response and operating personnel is greatly reduced, and data availability is enhanced by multimode data broadcast, particularly computer-synthesised voice reporting.

Satellite surveillance of weather is normally considered useful in monitoring large-scale weather patterns, but recent and planned improvements in the resolution, accuracy and spectral diversity of satellite sounding and imaging has improved the prospect of monitoring local phenomena such as thunderstorms, mountain waves and gust fronts, which are of direct significance to aviation.

Visibility through the atmosphere is of crucial importance in the vicinity of airports, and there is great merit in monitoring atmospheric visibility parameters by automatic means. Runway visual range measuring systems employing transmissometers are in use in a large number of airports, but a new generation of visibility sensors based on forward scattering measurement are promising to provide numerous advantages when certain aspects such as calibration are perfected.

8.11 References

ADLER, R.F., and FENN, D.D. (1979): 'Thunderstorm vertical velocities estimated from satellite data', *J. Atmos. Sci.*, **36**, pp. 1747–1754

ALEXANDER, M.B., and CAMP, D.W. (1985), 'Analysis of low-altitude wind speed and direction shears', *J. Aircr.*, **22**, pp. 705–712

ARINC (1988): 'Windshear warning and guidance system'. ARINC Characteristic 742, January 1988.

BIRKENHEUER, D.L. (1989): Use of GOES data for local forecasting'. Program, GOES I-M Operational Satellite Conference, Arlington, VA, pp. 70–75

CHADWICK, R.B., FRISCH, A.S., and STRAUCH, R.G. (1984): 'A feasibility study on the use of wind profilers to support space shuttle launches'. NASA Contractor Report 3861, December 1984

CORNMAN, L.B., MORSE, C.S., and CUNNING, G. (1993): 'Estimation of atmospheric turbulence severity from in-situ aircraft measurements'. Preprints of 5th International Conference on Aviation Weather Systems, Vienna, VA (American Meteorological Society, Boston), pp. 152–156

ECKLUND, W.L., CARTER, D.A., and BALSLEY, B.B. (1988): 'A UHF wind profiler for the boundary layer: Brief description and initial results', *J. Atmos. Ocean. Technol.*, **5**, pp. 432–441

EILTS, M.D. (1987): 'Low altitude wind shear detection with Doppler radar', *J. Clim. Appl. Meteorol.*, **26**, pp. 96–106

ELLROD, G. (1989): 'Dallas microburst storm environmental conditions determined from satellite soundings. Proceedings of 3rd International Conference on Aviation Weather System', Anaheim, CA (American Meteorological Society, Boston), pp. 15–20

ELLROD, G.P., and NELSON, J.P., III (1995): 'Benefits of the advanced GOES satellites to aviation users'. Preprints of 6th Conference on Aviation Weather Systems, Dallas, TX, 15–20 January 1995 (American Meteorological Society, Boston), pp. 144–148

ELLROD, G.P., and NELSON, J.P. (1997): 'An experimental GOES image product to identify conditions favorable for aircraft icing'. Preprints of 7th Conference on Aviation, Range, and Aerospace Meteorology, Long Beach, CA, 2–7 February 1997 (American Meteorological Society, Boston), pp. 112–115

ERICKSON, C.O., and WHITNEY, L.F. (1973): 'Gravity waves following severe thunderstorms', *Mon. Weather Rev.*, **101**, pp. 708–711

FRISCH, A.S., WEBER, B.L., STRAUCH, R.G., MERRITT, D.A., and MORAN, K.P. (1986): 'The altitude coverage of the Colorado wind profilers at 50, 405 and 915 MHz', *J. Atmos. Ocean. Technol.*, **3**, pp. 680–692

FUJITA, T.T. (1978): 'Manual for downburst identification for Project NIMROD'. SMRP Paper No. 156 (University of Chicago, Chicago, IL), 104 pp

FUJITA, T.T. (1986): *DFW Microburst*. The University of Chicago, Chicago, IL

GOFF, R.C. (1980): 'The low-level wind shear alert system (LLWSAS)'. Federal Aviation Administration Report DOT/FAA/RD-80/45

GOFF, R.C., and GRAMZOW, R.H. (1989): 'The Federal Aviation Administration's Low Level Windshear Alert System: A project management perspective'. Proceedings of 3rd International Conference on Aviation Weather System, Anaheim, CA (American Meteorological Society, Boston), pp. 408–413

GUIRAUD, F.O., HOWARD, J., and HOGG, D.C. (1979): 'A dual-channel microwave radiometer for the measurement of precipitable water vapor and liquid', *IEEE Trans. Geosci. Electron.*, **GE-17**, pp. 129–136

GURKA, J.J. (1976): 'Satellite and surface observations of strong wind zones accompanying thunderstorms', *Mon. Weather Rev.*, **104**, pp. 1484–1493

HELLROTH, B., and OLSSON, E. (1989): 'MET 90, a project for the development of the future Swedish aviation weather system'. Proceedings of 3rd International Conference on Aviation Weather System, Anaheim, CA (American Meteorological Society, Boston), pp. 228–232

HINTON, D.A. (1990): 'Relative merits of reactive and forward-look detection of wind shear encounters during landing approach for various microburst strategies'. NASA TM-4158, DOT/FAA/DS-89/35, February 1990

HINTON, D.A. (1992): 'Forward-look wind-shear detection for microburst recovery', *J. Aircr.*, **29**, pp. 63–66

HINTON, D.A., and OSEGUERA, R.M. (1993): 'Microburst avoidance crew procedures for forward-look sensor equipped aircraft'. AIAA Aircraft Design, Systems and Operations Meeting, Monterey, CA, 11–13 August 1993 (American Institute of Aeronautics and Astronautics, Washington, DC), Paper AIAA 93-3942

HODGES, S.M., and STARR, K.M. (1995): 'A status and update of weather sensors projects in the Federal Aviation Administration (FAA). Preprints of 6th Conference on Aviation Weather Systems, Dallas, TX, 15–20 January 1995 (American Meteorological Society, Boston), pp. 35–39

HOGG, D.C., GUIRAUD, F.O., and BURTON, E.B. (1980): 'Simultaneous observations of cool cloud liquid by ground-based microwave radiometry and icing of aircraft', *J. Appl. Meteorol.*, **19**, pp. 893–895

HUGHES, D. (1990): 'LLWAS credited with helping 737 survive major microburst', *Aviat. Week Space Technol.*, 16 July 1990

INTERNATIONAL CIVIL AVIATION ORGANISATION (1981): 'Manual of runway visual range observing and reporting practices'. Document 9328-AN/908, 1st edn.

JACKSON, R.L. (1991): 'Low-level wind shear terminology'. Preprints of 4th International Conference on Aviation Weather Systems, Paris, France, June 24–28, 1991 (American Meteorological Society, Boston), pp. 13–15

JAFFE, K.D. (1989): 'Improvement of the performance of sensors in the Low-Level Wind Shear Alert System (LLWAS)'. Proceedings of 3rd International Conference on Aviation Weather System', Anaheim, CA (American Meteorological Society, Boston), pp. 356–361

KESSLER, E. (1990): 'On Low-Level Windshear Alert Systems (LLWAS) and Doppler radar in aircraft terminal operations', *J. Aircr.*, **27**, pp. 423–428

KORHONEN, O.P. (1989): 'Profiling winds for air traffic – an application test at Helsinki airport'. Proceedings of 3rd International Conference on Aviation Weather System', Anaheim, CA (American Meteorological Society, Boston), pp. 146–147

KRAUS, K., and MAYOU, L. (1989): 'Impact of automated weather observing system on aviation'. Proceedings of 3rd International Conference on Aviation Weather System', Anaheim, CA (American Meteorological Society, Boston), pp. 108–111

LEE, T.F., and CLARK, J.R. (1995): 'Aircraft icing products from satellite infra-red data and model output'. Preprints of 6th Conference on Aviation Weather Systems, Dallas, TX, 15–20 January 1995 (American Meteorological Society, Boston), pp. 234–236

LEWIS, M.S., ROBINSON, P.A., HINTON, D.A., and BOWLES, R.L. (1994): 'The relationship of an integral wind shear hazard to aircraft performance limitations'. NASA Technical Memorandum 109080, February 1994

LEWIS, M.S., YENNI, K.R., VERSTYNEN, H.A., and PERSON, L.H. (1992): 'Design and conduct of a flight experiment to detect and penetrate microburst windshears in a transport category aircraft'. AIAA 6th Biennial Flight Test Conference, Hilton Head, South Carolina, 24–26 August 1992 (American Institute of Aeronautics and Astronautics, New York), Paper AIAA-92-4092

MADDOX, R.A. (1980): 'Mesoscale convective complexes', *Bull. Am. Meteorol. Soc.*, **61**, pp. 1374–1387

MADDOX, R.A. (1981): 'Satellite depiction of the life cycle of a mesoscale convective complex', *Mon. Weather Rev.*, **109**, pp. 1583–1586

MAY, P.T., and STRAUCH, R.G. (1989): 'An examination of wind profiler signal processing algorithms', *J. Atmos. Ocean. Technol.*, **6**, pp. 731–735

MAY, P.T., STRAUCH, R.G., and MORAN, K.P. (1988): 'The altitude coverage of temperature measurement using RASS with wind profiling radars', *Geophys. Res. Lett.*, **15**, pp. 1381–1384

MCLEAN, D. (1988): 'Airborne detection of wind shear'. Proceedings of 2nd International Symposium on Aviation Safety, Toulouse, France, November 1986 (Cepad, Toulouse), pp. 227–244

MILES, C.S., BURNHAM, D.C., and KARIMI, G. (1995): 'New generation runway visual range system'. Preprints of 6th Conference on Aviation Weather Systems, Dallas, TX, 15–20 January 1995 (American Meteorological Society, Boston), pp. 347–350

MORAN, K.P., STRAUCH, R.G., EARNSHAW, K.B., MERRITT, D.A., WEBER, B.L., and WUERTZ, D.B. (1989): 'Lower tropospheric wind profiler'. Proceedings of 24th Conference on Radar Meteorology, Tallahassee, FL, (American Meteorological Society, Boston), pp. 728–731

MOSHER, F.R. (1989): 'Application of geostationary satellite data, Uses of geostationary satellite data: Past, present and future, Severe local storms'. Program, GOES I-M Operational Satellite Conference, Arlington, VA, pp. 225–228

NATIONAL ACADEMY OF SCIENCES (1983): 'Low altitude wind shear and its hazard to aviation' (National Academy Press, Washington, DC)

NATIONAL TRANSPORTATION SAFETY BOARD (1976), 'Eastern Airlines, Inc., Boeing 727-225, N8845E, John F. Kennedy International Airport, Jamaica, New York, June 24, 1975', Report NTSB AAR-76-8

NATIONAL TRANSPORTATION SAFETY BOARD (1983): 'Pan American World Airways Clipper 759, N4737, Boeing 727-235, New Orleans International Airport, Kenner, Louisiana, July 9, 1982'. Report NTSB AAR-83-02

NATIONAL TRANSPORTATION SAFETY BOARD (1986): 'Aircraft accident report: Delta Airlines, Inc., Lockheed L-1011-385-1, N726DA, Dallas/Fort Worth International Airport, Texas, August 2, 1985'. Report NTSB/AAR-86/05, August 1986

NILSEN, J.D., and STARR, K.M. (1995): 'Low Level Windshear Alert System (LLWAS) project status' Preprints of 6th Conference on Aviation Weather Systems, Dallas, TX, 15–20 January 1995 (American Meteorological Society, Boston), pp. 45–47

OSEGUERA, R.M. (1992): 'NASA wind shear flight test in situ results'. Proceedings of Fourth Combined Manufacturers' and Technologists' Conference, Sponsored by NASA and FAA, Williamsburg, VA, 14–16 April 1992, NASA CP 10105 Part 1, DOT/FAA/RD-92/19-I, September 1992, pp. 45–58

PANOFSKY, H.A., and DUTTON, J.A. (1984): '*Atmospheric turbulence*' (Wiley, New York)

PAWLAK, R., and BURNHAM, D. (1997): 'Visibility products from automated systems'. Preprints of 1st Symposium on Integrated Observing Systems, Long Beach, CA, 2–7 February 1997 (American Meteorological Society, Boston), pp. J28–J32

PURDOM, J.F.W. (1973): 'Meso-highs and satellite imagery', *Mon. Weather Rev.*', **101**, pp. 180–181

PURDOM, J.F.W. (1974): 'Satellite imagery applied to the mesoscale surface analysis and forecast'. Preprints of 5th Conference on Weather Forecasting and Analysis, St. Louis, MO (American Meteorological Society, Boston), pp. 63–68

PURDOM, J.F.W. (1989): 'Satellite observations of convection and severe storms'. Program, GOES I-M Operational Satellite Conference, Arlington, VA, pp. 256–273

SAVIDES, J., and RESECK, K.G. (1989): 'GOES I-M system characteristics'. Program, GOES I-M Operational Satellite Conference, Arlington, VA, pp. 285–314

SCHLICKENMAIER, H.W. (1989): 'Windshear case study: Denver, Colorado, July 11, 1988'. Report DOT/FAA/DS-89/19, November 1989

SCOFIELD, R.A., and OLIVER, V.J. (1977): 'A scheme for estimating convective rainfall from satellite imagery', NOAA Environmental Laboratories, Boulder, CO, Tech. Memo. NESS 86

SMYTHE, G.R. (1989): 'Evaluation of the 12-station enhanced Low-Level Windshear Alert System (LLWAS) at Denver Stapleton International Airport'. Proceedings of 3rd International Conference on Aviation Weather System', Anaheim, CA (American Meteorological Society, Boston), pp. 41–46

STANKOV, B., and BEDARD, A. (1990): 'Atmospheric conditions producing aircraft icing on 24-25 January 1989: A case study utilizing combinations of surface and remote sensors'. Proceedings of AIAA 24th Aerospace Sciences Meeting, Reno, NV, (American Institute of Aeronautics and Astronautics, New York), Paper AIAA-90-0197

STRAUCH, R.G., MERRITT, D.A., MORAN, K.P., EARNSHAW, K.B., and VAN DE KAMP, D. (1984): 'The Colorado wind profiling network', *J. Atmos. Ocean. Technol.*, **1**, pp. 38–49

STRAUCH, R.G., MERRITT, D.A., MORAN, K.P., MAY, P.T., WEBER, B.L., and WUERTZ, D.B. (1989a): 'Doppler radar wind profilers for support of flight operations', *J. Aircr.*, **26**, pp.1009-1015

STRAUCH, R.G., MORAN, K.P., MAY, P.T., BEDARD, A.J., and ECKLUND, W.L. (1989b): 'RASS temperature soundings with wind profiler radars'. Proceedings of 24th Conference on Radar Meteorology, Tallahassee, FL (American Meteorological Society, Boston), pp. 741–745

US DEPARTMENT OF TRANSPORTATION (1987): 'Integrated FAA windshear program plan: Development and logistics, aviation standard and air traffic', Federal Aviation Administration

US DEPARTMENT OF TRANSPORTATION (1989a): 'The Federal Aviation Administration plan for research, engineering and development, Volume II: Project descriptions', Federal Aviation Administration, January 1989

US DEPARTMENT OF TRANSPORTATION (1989b): 'Project implementation plan for the Low Level Windshear Alert System (LLWAS)'. Federal Aviation Administration Directive 6560.24

US DEPARTMENT OF TRANSPORTATION (1990): 'Airborne windshear warning and escape guidance systems for transport airplanes'. Federal Aviation Administration Technical Standard Order C117, 24 July 1990

VAN DIJK, W.C.M. (1993): 'Automation of observations in the Netherlands'. Preprints of 5th International Conference on Aviation Weather Systems, Vienna, VA (American Meteorological Society, Boston), pp. 179–182

WEST, M.D., and BURNHAM, D.C. (1997): 'Analysis of runway visual range errors'. Preprints of 1st Symposium on Integrated Observing Systems, Long Beach, CA, 2–7 February 1997 (American Meteorological Society, Boston), pp. J33–J38

WESTWATER, E.R., and KROPFLI, R.A. (1989): 'Remote sensing techniques of the Wave Propagation Laboratory for the measurement of supercooled liquid water: Application to aircraft icing'. NOAA Technical Memorandum ERL WPL-163, May 1989

WINSTON, H., RIESE, C., and BADESHA, S. (1995): 'A comparison of turbulence derived from a 915 MHz wind profiling radar and an aerostat-mounted wind anemometer'. Preprints of 6th Conference on Aviation Weather Systems, Dallas, TX, 15–20 January 1995 (American Meteorological Society, Boston), pp. 153–154

WUERTZ, D.B., WEBER, B.L., STRAUCH, R.G., FRISCH, A.S., LITTLE, C.G., MERRITT, D.A., MORAN, K.P., and WELSH, D.C. (1988): 'Effects of precipitation on UHF wind profiler measurements', *J. Atmos. Ocean. Technol.*, **5**, pp. 450–465

ZRNIC', D.S., SMITH, S.D., WITT, A., RABIN, R.M., and SACHIDANANDA, M. (1986): 'Wind profiling of stormy and quiescent atmospheres with microwave radars'. NOAA Technical Memorandum ERL NSSL-98

Chapter 9
Integrated system approaches

9.1 General

In the preceding few chapters we have discussed individual instruments and instrument systems that can sense the atmospheric parameters in their respective fields of observation and process the data in a fairly sophisticated manner to generate a host of meteorological data products. However, aviation, as a spatially distributed activity, extends over more area than is covered by any one sensor. It also requires more kinds of data products than can be generated by any one instrument. Combining data from as many diverse sources as possible would not only enhance spatial coverage but also provide a more comprehensive product set for aviation support. This has led to a systems approach to the aviation weather surveillance problem. Such systems gather and fuse data from many sources and provide high-level products specifically tailored for aiding aviation activities. With parallel development of sensors, aviation weather data are available from a number of sources with different characteristics and wide spatial distribution. Colour Plate 27 shows the diversity and distribution of sensors for generating wind shear products.

In the modern information age with powerful hardware and software available for voluminous data handling in many fields of activity, designing systems for collecting and processing weather data from many sources would seem to be a straightforward affair, but the problem is rendered complex by the widely different types of information sources. First, there are a host of generically diverse sensing instruments with differing coverage, types of observed parameters, update rates, and data formats. These include a variety of radars, *in situ* sensors and sensor clusters, weather satellites, and others, with their primary data in digital, graphical or even photographic forms. Then there is the problem of multiple and overlapping coverage by similar instruments in certain areas, while other areas may not be covered at all. Further, these devices operate asynchronously, without any centralised or master control to co-ordinate their operation. Finally, and most importantly, there is a large amount of very valuable information available from noninstrument sources such as pilot reports, meteorologists' impressions and interpretations, and model-based computations. Combining data from all these sources to provide reliable data products involves a high level of ingenuity.

338 *Aviation weather surveillance systems*

Further complexity is added to comprehensive aviation weather data systems by their ambitious specifications driven by the demands of aviation support. To facilitate the utilisation of their output data by operating personnel of diverse backgrounds and even by automated systems, the modern aviation weather data systems are expected to perform highly sophisticated processing which mimics or replaces human capabilities in some ways, especially in cognitive and inferential aspects. Thus the multi-source data are not only fused and collated, but are used to recognise the nature of the hazardous phenomena, locate their spatial boundaries, estimate their intensity, and evaluate their hazard potential. In the end the data products are to be delivered in a final form that can be used directly and immediately for aviation-related decisions. In addition to these sophisticated processing functions, the high-level weather data systems also interface automatically with other air traffic management systems to share the data and make them available in a timely fashion at locations where they are required.

Yet another requirement of modern aviation weather surveillance systems is predictive ability. It is not enough to know accurately just the current weather scenario. To operate the aviation system smoothly and to minimise the disrupting effect of sudden decisions on individual flights as well as support facilities, it is necessary to know of hazardous weather developments in advance. Even a few or several minutes of warning in the case of fast-evolving phenomena such as microbursts and gust-frontal wind shifts is beneficial. Longer lead times are both desirable and possible in predicting changes in less agile parameters such as visibility, snowfall intensity and environmental winds. Predictions may be made from computational forecast models run by specialised agencies, by extrapolating observed phenomena and parameters, and/or by sensing and recognising precursor phenomena. The inclusion of all these possibilities imparts a great degree of sophistication to modern high-level aviation weather data processing systems, and makes their design a challenging task. It is only in the 1990s that such comprehensive data systems are being attempted.

9.2 Integrated terminal weather system

A prime example of a high-level weather data processing system for aviation support is the evolving Integrated Terminal Weather System (ITWS). Because of the special needs and great importance of weather surveillance in terminal areas, the ITWS has been conceptualised to focus on that particular function and thus help in improving terminal planning, traffic capacity and safety.

The ITWS is being developed by the Lincoln Laboratory of Massachusetts Institute of Technology as a major project of the US Federal Aviation Administration in the area of aviation weather, and is expected to attain a mature stage by the year 2000. Unlike other systems for sensing, processing and disseminating weather information, such as the LLWAS and the AWOS/ASOS which have their own dedicated sensors of raw weather data, the ITWS is a complex and comprehensive system that is designed to provide high-level weather information by combining and processing data and products from sensors and systems operated by the Federal Aviation Administration and National Weather Service. The data sources consist of sensors (both ground-based as well as airborne) located in or near the terminal area as well as local and regional forecast information. The ITWS has several advantageous and user-friendly features. Important among them are the following.

9.2.1 Data integration

Vigorous research and development activity of recent decades has led to the evolution of a number of sensors for aviation weather surveillance, many of which have been discussed earlier in this work. Because of their differing capabilities and strengths, a number of different sensors would be required to be located in or near a given terminal area to provide comprehensive surveillance of all types of aviation-significant weather phenomena. The result is multiplicity and parallelism of information generation and display. As an example of the above situation, wind shear information can be provided independently by the TDWR, the LLWAS, the ASR-9 with a Doppler weather channel, and the vertical wind profiler, and their data may also be combined in various ways. Further, not all of these sensors or systems may be present at a given airport. Because of their unique features already discussed (LLWAS senses ground-level wind shear in the vicinity of the runway complex, TDWR provides high-resolution data and high-level weather products over a wider area, ASR-9 additionally performs aircraft detection, and Wind Profiler senses only vertical wind shear), none of these systems can replace any of the others.

Similarly, wind speed data can be obtained from both LLWAS and AWOS/ASOS systems. All the three major radar systems discussed in Chapter 7, i.e. WSR-88D, TDWR and ASR-9, can provide accurate storm reflectivities, and the TDWR system, augmented with data from the LLWAS, can locate gust fronts, extrapolate their future positions, and predict the wind shifts they can cause. Information on turbulence may be available from the three weather radars besides wind profilers as well as pilot reports. Surface observations such as temperature, pressure, humidity, runway visibility and cloud ceiling are also measurable by a host of independent instruments.

Such multiple, parallel and overlapping data sources, though providing valuable redundancy and completeness of information, have their own drawbacks. These are: (i) too many display devices showing data from independent weather sensors and systems would put additional demands on the scarce spaces in front of control personnel and pilots in their respective locations (these spaces are already crowded by the number of air traffic control displays); (ii) a large number of additional displays for weather information would divert the attention of controllers from air traffic control displays and functions; (iii) there may be differences between two or more devices depicting the same phenomenon or parameter, requiring the operating personnel (e.g. meteorologist, controller or pilot) to resolve the conflict; and (iv) the operating personnel would have to digest the information from the diverse displays and mentally integrate the information to obtain a total weather picture of the terminal area and its environs. The net effect of these factors is to overload the air traffic management personnel and interfere with their primary function of ensuring aircraft separation. Requirements of multiple observation and interpretation also introduce time delays into weather-related decision processes.

The ITWS helps overcome these problems by performing optimum integration of data from multiple contributing sources including dedicated aviation weather instrumentation as well as weather forecast centres. Data integration permits the presentation of aviation-significant weather information in a compact manner to enhance aviation safety and capacity in the terminal area, while minimising controller overload.

9.2.2 Automated operation and fully processed output

The ITWS is conceptualised as a fully automated system with stringent specifications for weather data processing and product transmission. To start with, the displays in the control tower are alphanumeric, with the hazard warnings transmitted to aircraft over the voice link. The display in the Terminal Radar Control (TRACON) is graphical, showing the location of hazardous weather zones of different types (hail, turbulence, etc.) and localised phenomena such as microbursts, tornadoes and gust fronts. Geographic situation displays in colour, similar to those of the TDWRs, are available to supervisors in the control tower as well as TRACON (Evans, 1991). The ITWS displays will be integrated into an Advanced Automation System later.

A major advantage of the ITWS is that its data products are in a fully processed and final form that does not require meteorological interpretation. Thus the system output is directly usable by personnel who are not meteorologists by training such as air traffic controllers and pilots, and entails no loss of time on account of interpretation before being utilised for operational planning and decisions. Such product simplification is especially useful for air traffic controllers because their additional work load to digest and disseminate weather information is minimised, leaving them more time and mental resources to concentrate on their primary responsibility of separating aircraft.

The ITWS will also communicate with air traffic automation systems to provide data regarding the present and future location of hazardous weather zones to enable automated trajectory planning for avoiding such zones in the terminal area (Sankey, 1993).

9.2.3 Performance enhancement, versatility and adaptability

Integration of data from several sources, besides simplifying presentation and minimising controller overload, may actually provide a higher level of performance in terms of product generation compared to each of the contributing sources. It is also possible for the ITWS to derive more parameters of aviation-significant phenomena than may be the case with individual sensors. For example, each of the three Doppler radar systems described in Chapter 7, i.e. WSR-88D, TDWR and ASR-9 with a Doppler weather channel, is capable of detecting and estimating the parameters of microbursts, but assumes them to be symmetric, while in practice microbursts may exhibit considerable asymmetry (see Section 4.7.3). Processing data from multiple sensors may permit addressing the asymmetry directly in generating microburst warnings (Evans, 1991). Similarly, the storm cell boundaries normally depicted in the TRACON, based on ASR-9 weather channel reflectivity data, can be made much more accurate by adding three-dimensional reflectivity and Doppler data from WSR-88D and TDWR, possibly with further augmentation from temperature soundings and lightning sensors. Yet another important function of the ITWS is to provide a three-dimensional wind field in the terminal area to the new TATCA system.

In addition to delineating areas of aviation-hazardous weather, the ITWS has the capability to provide warnings of spot phenomena in the terminal area, such as tornadoes, microbursts and hail storms, generated by weather forecast centres.

Another very useful capability of the ITWS is to provide its range of weather products tailored to the needs of individual users such as pilots, air traffic controllers, terminal area traffic managers, and terminal automation systems (Ducot, 1993). The last category of systems includes the Terminal Air Traffic Control Automation (TATCA) and wake vortex advisory systems. Such adaptability in terms of focused data generation for specific user categories simplifies data assimilation on the part of the recipient, and minimises data redundancy and inessential communication load on the system.

9.2.4 Predictive capability

Predicting the occurrence and strength of aviation-significant weather phenomena even by a short lead time of the order of 5 to 10 min helps in vectoring approaching and departing aircraft so as to minimise hazards and flight disruptions. The ITWS is designed to provide data products that may contain short-term predictions, up to 30 min (Sankey, 1993). Some success in predicting microbursts has been achieved by using TDWR data, with mean warning times of ~5 min but significant probabilities of missed predictions (Campbell, 1991). However, improved prediction, especially of the outflow strength, is possible by combining temperature profile data available from sources such as RASS (see Section 8.3) and the ARINC (Aeronautical Radio Inc.) Communication and Retrieval System (ACARS). Humidity profiles and information on lightning activity are also of value in aiding microburst prediction. Such data integration is possible with the ITWS, potentially permitting better microburst prediction.

9.3 Aviation gridded forecast system

The need for weather forecast is not confined to the terminal area. *En route* flight can also benefit immensely from prior knowledge of the weather developments in the air space as a whole. To generate such information, the US Federal Aviation Administration is sponsoring the development of the Aviation Gridded Forecast System (AGFS) at the Forecast Systems Laboratory of the National Oceanic and Atmospheric Administration.

The meteorological condition of the atmosphere can be described in terms of the *state-of-the-atmosphere variables* (SAVs) such as temperature, humidity and wind vector. A Mesoscale Analysis and Prediction System (MAPS) uses data from all sources including automated temperature and winds reported through the ACARS, and provides the current and predicted (up to 6 h) SAVs on a national scale with a spatial resolution of 60 km, updated every 3 h. A more elaborate Eta model updates the SAVs at intervals of 6 h. With the availability of larger and faster computers, especially with massively parallel processing architectures, the spatial resolution may be improved to the level of 30 or even 15 km and the updating interval brought down to an hour.

The next step is to derive from the SAVs the *aviation impact variables* (AIVs), which include icing, turbulence, cloud ceiling, visibility and wind speed and direction (note that wind speed and direction are listed among both SAVs and AIVs). The Aviation Gridded Forecast System (AGFS) is a computer program that performs this task. The system is so named because it yields the AIVs at points arranged in a lattice-like grid in four dimensions (3-D space and time). A user can derive numerical or graphical data along any desired flight path or plane from such gridded data. Examples of such data utilisation will be given in the following section.

A local AGFS has also been developed which produces data over a 600 km square, but provides finer spatial and temporal resolution. The system is based on the Local Analysis and Prediction System (LAPS) and Regional Atmospheric Modelling System (RAMS) (Kraus, 1993). The LAPS generates hourly three-dimensional analyses of temperature, winds, specific humidity, clouds, cloud water and cloud ice on a grid with 61×61 horizontal points spaced 10 km apart and 21 levels separated by 50 mbar of pressure difference. The RAMS utilises the same horizontal grid, but the vertical spacing varies from 250 m to 1 km as a function of height above ground (Cram *et al.*, 1993). It utilises observations from additional sources, including high-resolution data from the TDWR and WSR-88D radars. A more compact version of LAPS, called T-LAPS, has been developed for use in terminal areas as a part of the ITWS discussed in the preceding section (Cole and Wilson, 1993). It produces data at 5 min intervals with a horizontal resolution of 2 km.

To cope with the complexities of the future aviation management scenario, the US Transportation Systems Center is developing an Advanced Traffic Management System (ATMS) with emphasis on a high level of automation in the conduct of flight operations (Jesuroga et al., 1991; Jesuroga, 1993). The ATMS operates by integrating the air traffic management information with current and forecast weather data in real time. The relevant primary weather products for integration consist of wind information (including jet stream location, altitude and strength) and radar reflectivity and lightning strike location data at 5 min intervals. Such integration is also the responsibility of the Forecast Systems Laboratory (Kraus, 1993). The required weather information would come from the AGFS in the form of AIVs, to be combined with air traffic data on interactive workstations called Aircraft Situation Displays (ASDs). Such integration is also the responsibility of the Forecast Systems Laboratory (Kraus, 1993).

The AIVs are not direct products of the numerical models, and hence great care is required to estimate and predict them with accuracy and reliability. Prediction of the severity of turbulence and icing poses a particular challenge. Since the effect of these hazards depends on the aircraft involved (besides their intensities), they may be expressed in terms of appropriate impact indices related to aircraft types (Sankey, 1993).

An example of the use of gridded data is shown in Fig. 9.1, which is a microburst risk image derived from the estimation of three individual risk factors. The darkest shade represents the coincidence of all the three predictor factors, the medium dark shade signifies the overlap of two factors, and only one factor is true in the light areas. Gridded AIVs may also be edited with inputs from observational data to improve their reliability. Figure 9.2 illustrates a case of editing gridded icing forecast data using pilot reports.

9.4 Aviation weather products generator

The Aviation Weather Products Generator (AWPG) is yet another important project of the US Federal Aviation Administration in the aviation weather area. This system, being developed at the National Center for Atmospheric Research (NCAR), utilises the AIV data available from the AGFS discussed in the preceding section and generates final weather products that can be used directly by the aviation end users. For example, a pilot may depend on the AWPG to obtain displays of the weather picture of his or her interest along the vertical section of the atmosphere through a particular proposed flight. The data sources and display system components of a prototype AWPG system are shown in Fig. 9.3. Table 9.1 lists the products of the system at various stages. Examples of AWPG products are shown in Colour Plates 23–25.

The AWPG, by providing improved and comprehensive weather products in a user-tailored manner, is expected to greatly assist the flight planning

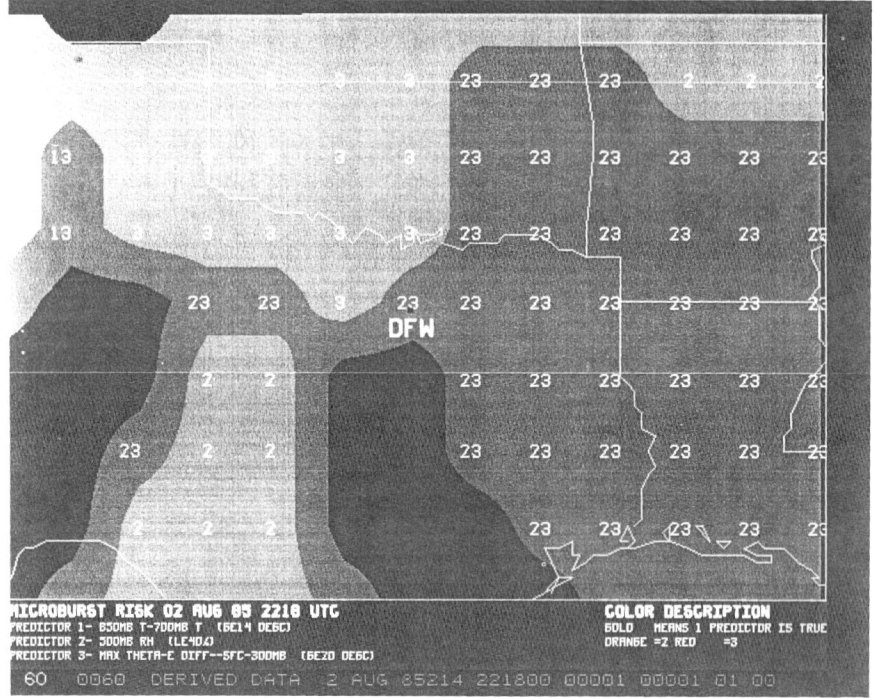

Figure 9.1 Example of a microburst risk image based on three predictor factors derived from gridded data (from Nelson and Ellrod, 1995, courtesy G.P. Ellrod, NOAA National Environmental Satellite Data and Information Services, Washington, DC)

process as well as air traffic management. It would also assist pilots, controllers and other service providers in performing their tasks more efficiently by reducing their workload.

An important goal of the system planning in relation to the AWPG development is to make its data products available to the pilot in the cockpit via the Mode S transponder link (see Section 2.7) or other data links that may be commercially available. This would enable pilots to take considered decisions regarding weather avoidance rather than merely follow controller instructions. Indeed, with a common database of detailed and final (i.e. without requiring further meteorological interpretation) weather products available to the pilot, controller and any intermediary data handlers simultaneously, the possibility of misunderstanding will be minimised and co-operative and/or autonomous decision-making facilitated. Such an approach is consistent with the evolving free-flight aviation scenario, which aims to make individual flights more pilot-centred and less dependent on centralised controller intervention.

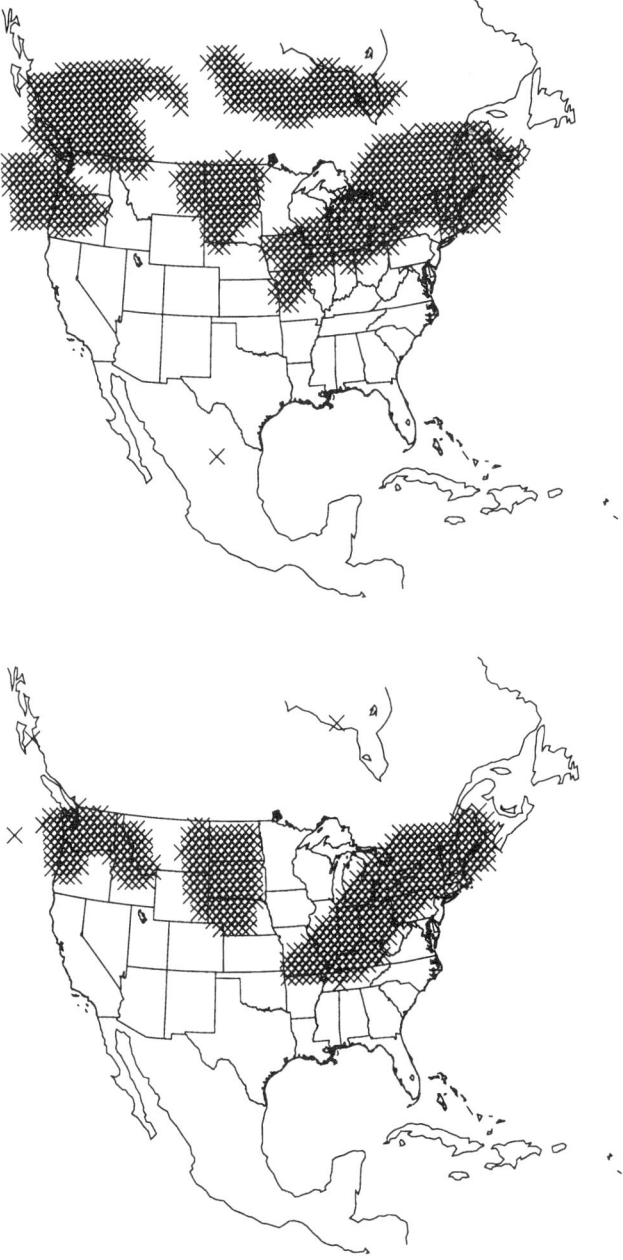

Figure 9.2 Example of editing gridded AIV data: (a) raw map of icing forecast (cross-hatched areas); (b) edited map showing addition and deletion of areas with icing potential. The data grid spacing is 80 km horizontally, and has 39 vertical levels at 1000 ft intervals covering nearly the entire aviation altitude limit (from Mahoney et al., 1997, courtesy J.L. Mahoney, NOAA Forecast Systems Laboratory, Boulder, CO, USA)

346 Aviation weather surveillance systems

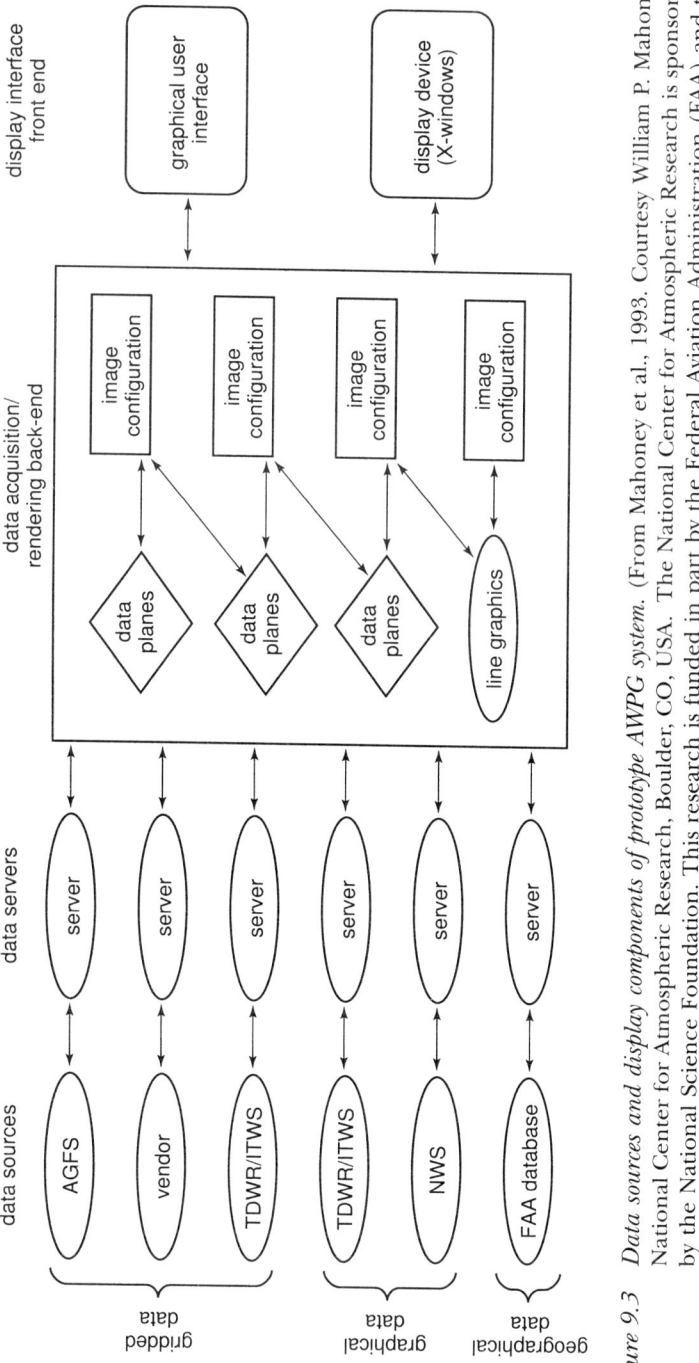

Figure 9.3 Data sources and display components of prototype AWPG system. (From Mahoney et al., 1993. Courtesy William P. Mahoney, National Center for Atmospheric Research, Boulder, CO, USA. The National Center for Atmospheric Research is sponsored by the National Science Foundation. This research is funded in part by the Federal Aviation Administration (FAA) and the National Weather Service (NWS))

Integrated system approaches 347

Table 9.1   Summary of AWPG products[a]

1996 operation	1998–2000 operation	Post-2000 operation
Current ceiling/visibility	Terminal approach category	Ceiling/visibility forecast
Icing diagnosis/forecast	Radar composite	Volcanic ash current/forecast
Turbulence current/forecast	Cloud base/tops current/forecast	Tropopause height current/forecast
Temperatures current/forecast	Storm location-forecast	Weather impacted air space current/forecast
Wind current/forecast	Lightning	Airport weather impact notice
Current weather impacted air space	Wind shear current/forecast	
Current terminal weather depiction	PIREPS	
Freezing level current/forecast	Ground icing diagnosis/forecast	

[a] From Carmichael (1993). Courtesy Bruce Carmichael, National Center for Atmospheric Research, Boulder, CO. The National Center for Atmospheric Research is sponsored by the National Science Foundation. This Research is funded in part by the Federal Aviation Administration (FAA) and the National Weather Service (NWS)

Like the AGFS, the AWPG has a facility for generating its data products at national as well as regional levels. Accordingly, it has two components (Carmichael, 1993): (i) a National Aviation Weather Products Generator (NAWPG) with focus on national-scale products used primarily for strategic planning, including both traffic management and flight planning; and (ii) Regional Aviation Weather Products Generator (RAWPG) to provide high-resolution regional-scale data products for use primarily in a tactical environment (e.g. by *en route* controllers, pilots in the cockpit and flight service specialists).

The development of the AWPG system is planned in four phases stretching to the year 2008. Table 9.2 shows the availability of products in the various phases. The Real-Time Weather Processor (RWP) mentioned in the Table will initially perform ingest, mosaicing and delivery of WSR-88D data to the air traffic control system in real time. In the later stages it will satisfy a broader range of user requirements.

When operational, the AWPG will represent a distinct enhancement in the quality of weather data support for aviation. The improvement over existing aviation weather data systems will be in the following ways:

Table 9.2 Products/services available in the four developmental phases of AWPG[a]

Phase	Products/services available	Year[b]
I	Products for commercial services including: • Meteorologist's Weather Processor (MWP) • Regional Automated Flight Service Station Graphic Weather Display System (GWDS) • Direct User Access Terminal Services (DUAT) • Enhanced Traffic Management System (ETMS)	1996
II	Real-Time Weather Processor (RWP) Enhancements	1998
III	Stand-alone National Aviation Weather Products Generator (NAWPG)	2002
IV	Stand-alone distributed Regional Aviation Weather Products Generator (RAWPG)	2008

[a] Adapted from Carmichael (1993)
[b] Projected year of attaining operational capability

1. finer spatial and temporal resolution
2. rich four-dimensional database covering three-dimensional space and time
3. integration of more types of data from more diversified sources
4. user-tailored products that match aircraft flight plans
5. generation of products in final form which require little or no further meteorological interpretation
6. extensive use of graphics for data presentation
7. user-queried presentation of flight-path-specific views of the weather scene and its evolution with time
8. preflight products suitable for distribution via computer networks, and to television companies, flight service stations and commercial weather data distributors
9. availability of final weather products to the pilot in the cockpit via Mode S or commercial links
10. facility for possible interface with flight management systems
11. simultaneous availability on controllers' and traffic managers' workstations and at flight watch and airline dispatch positions.

9.5 Summary

The thrust towards modernisation of aviation weather surveillance initially concentrated on the design and development of sophisticated and potent individual sensing instruments. A more recent emphasis is on systems that gather data from a wide variety of sources and integrate them in an intelligent way to provide a complete picture of the weather ambience of aviation and minimise the workload on operating personnel. The final goal of this operation is to facilitate strategic planning of aviation operations so as to enhance the aviation system capacity while maximising safety and passenger comfort, minimising delays and promoting operating economy.

System and data integration is carried out at two levels. In the first level, the aim is to provide as complete a weather picture as possible, in a timely, user-friendly and user-tailored manner. The data must not only be complete, but the products must be in a final form that can be used directly by end users such as pilots and controllers without the help of professional meteorologists. This calls for the generation of aviation impact variables, which are of direct concern to aviation personnel, from the basic meteorological variables associated with the atmosphere. Further, each user must be able to extract the data from the system in a way that suits him or her the best. To achieve this, modern integrated weather data systems organise their data in a four-dimensional space–time gridded format from which, with the aid of suitable software and interactive display systems, it is possible for a user to obtain graphic presentation of the weather field along a proposed flight path or over a given controlled air space. The database not only has the latest weather information, but even short-term forecasts (from a fraction of an hour to a few or several hours) to facilitate predictive weather avoidance and strategic route planning. Weather data integration may be on a national or regional scale, or on a more local scale such as over terminal areas. The spatial and temporal data resolution generally becomes finer with contraction in the scale of the overall data field.

The second level of integration occurs between weather data and flight management data, the latter comprising current and projected aircraft positions and other air traffic flow information. Such integration provides the controllers and pilots with a more complete picture of the air space than viewing the two types of data separately. Synthesis of air traffic and weather data, both current and predicted, permits direct viewing of potential hazard situations and facilitates optimal strategic planning of aviation operations.

In all these sophisticated systems of data generation, prediction and integration, a heavy emphasis is laid on automation. Complex systems such as those planned for future aviation, which involve many stages of massive data collection, processing and manipulation, would be greatly slowed down if they require human intervention on a routine basis. The process would also become expensive and error-prone. Evolving automated systems that are robust and reliable while being versatile and user-friendly involves challenges which are being overcome through painstaking and careful developmental strategies, experimentation and validation.

9.6 References

CAMPBELL, S.D. (1991): 'Performance results and potential operational uses for the prototype TDWR microburst prediction product'. Preprints of 4th International Conference on Aviation Weather Systems, Paris, France, 24–28 June 1991 (American Meteorological Society, Boston), pp. J33–J36

CARMICHAEL, B. (1993): 'The Aviation Weather Products Generator'. Preprints of 5th International Conference on Aviation Weather Systems, Vienna, VA (American Meteorological Society, Boston), pp. 8–12

COLE, R.E., and WILSON, F.W., JR. (1993): 'ITWS Gridded Analysis'. Preprints of 5th International Conference on Aviation Weather Systems, Vienna, VA (American Meteorological Society, Boston), pp. 56–60

CRAM, J.M., SNOOK, J.S., ALBERS, S.C., and McGINLEY, J.A. (1993): 'A brief description of, and preliminary results from, the LAPS modeling/4DDA system'. Preprints of 5th International Conference on Aviation Weather Systems, Vienna, VA (American Meteorological Society, Boston), pp. J10–J14

DUCOT, E.R. (1993): 'On designing and engineering the Integrated Terminal Weather System'. Preprints of 5th International Conference on Aviation Weather Systems, Vienna, VA (American Meteorological Society, Boston), p. 31

EVANS, J.E. (1991): 'Integrated Terminal Weather System (ITWS)'. Preprints of 4th International Conference on Aviation Weather Systems, Paris, France, 24–28 June 1991 (American Meteorological Society, Boston), pp. 118–123

JESUROGA, R.T. (1993): 'Using ATMS weather products for air traffic strategic planning'. Preprints of 5th International Conference on Aviation Weather Systems, Vienna, VA (American Meteorological Society, Boston), pp. 395–398

JESUROGA, R.T., RAMER, J., ALBERS, S., and WRIGHT, R.D. (1991): 'Validation of aviation weather products for the Advanced Traffic Management System'. Preprints of 4th International Conference on Aviation Weather Systems, Paris, France, 24–28 June 1991 (American Meteorological Society, Boston), pp. 287–290

KRAUS, M.J. (1993): 'The Forecast Systems Laboratory's role in the FAA's aviation weather development program'. Preprints of 5th International Conference on Aviation Weather Systems, Vienna, VA (American Meteorological Society, Boston), pp. 5–7

MAHONEY, J.L., HENDERSON, J.K., MILLER, P.A., SHERRETZ, L.A., RODGERS, D.M., MOORE, M., and SIMS, D. (1997): 'Using the FSL prototype AIV editor to edit icing grids: A preliminary evaluation using the RTVS'. National Oceanic and Atmospheric Administration, NOAA Technical Memorandum ERL FSL-20, February 1997

MAHONEY, W.P. III, CARON, J., HAGE, F., DELP, S., ROACH, D., BLACKBURN, G., and BITER, C. (1993): 'The prototype Aviation Weather Products Generator: A vehicle to assess user needs'. Preprints of 5th International Conference on Aviation Weather Systems, Vienna, VA (American Meteorological Society, Boston), pp. 388–391

NELSON, J.P., and ELLROD, G.P. (1995): 'Development of a microburst risk image product derived from satellite sounder data'. Preprints of 6th Conference on Aviation Weather Systems, Dallas, TX, 15–20 January 1995 (American Meteorological Society, Boston), pp. 89–94

SANKEY, D.A. (1993): 'An overview of FAA-sponsored aviation weather research and development'. Preprints of 5th International Conference on Aviation Weather Systems, Vienna, VA (American Meteorological Society, Boston), pp. 1–4

Chapter 10
Automatic detection and tracking of hazardous weather features

10.1 General

We ended the preceding chapter with some thoughts on automation in the context of modern aviation weather surveillance systems. We continue with that theme in a concrete way in this chapter. Modern technological advancement has provided us with not only sensitive instruments for observing nature, but also powerful data handling and computing devices and matching software to mimic or substitute many of the intelligent functions hitherto performed only by human beings. Aviation weather surveillance, by its very nature (of being observation-, data- and interpretation-intensive, and quasi-real-time), is an area that can benefit immensely from all these strengths of modern technology. In particular, the high levels of physical understanding of weather phenomena currently available, coupled with algorithmic and software expertise, make it possible to automate many intelligence-requiring functions such as detection, recognition, tracking, prediction, and impact or hazard potential assessment of aviation-significant weather phenomena.

Traditionally, sensing instruments have been used to generate atmospheric data which, if necessary after a certain amount of preprocessing, are presented to meteorologists for interpretation. The meteorologist, after viewing the weather scene, e.g. a radar display of reflectivity, can recognise tell-tale patterns in the field that correspond to specific types of weather phenomena which may be of concern to aviation. Although computers have not yet caught up with the trained human being in terms of versatility and overall cognitive abilities, it is possible to impart to machines the ability to recognise a limited set of phenomena in a field of suitably prepared data. If performed with acceptable reliability and accuracy, such a function can greatly reduce the human-dependence of aviation weather surveillance systems, leading to faster data product generation and minimisation of errors due to human subjectivity and fatigue.

This chapter deals with the basic features of the process of automating the detection and recognition of some of the weather phenomena that are of direct concern to aviation. However, before starting such a discussion it must be pointed out that this is an area of immense breadth, scope and diversity for the following reasons: (i) diverse nature of weather phenomena; (ii) multiplicity of types and specifications of sensors; (iii) options on different levels of automation desired; (iv) diverse nature and format of the data products; and (v) varying capacities and sophistication of the host computers used for implementing the automation. Further variations in the automation approach are provided by the background and algorithmic preferences of the developers. Finally, numerous attempts have been made independently by many research and industry groups and individuals around the world to develop automation algorithms. The result is an exceptionally wide array of computer programs for automatic detection and recognition of aviation-significant weather features and phenomena, with considerable parallelism, duplication and overlap among some of the programs intended for similar applications. Many of these automation programs were or are only concept-proving attempts, but a number of programs have matured, been field-tested and implemented in real systems.

The situation on this front is fast evolving, with newer algorithms being developed for different sensors and sensor combinations, as well as for various end uses. The algorithms are being continuously improved, fine-tuned and validated to enhance reliability and accuracy using field data which include independent redundant observations. It would not be possible to capture all these nuances here. We confine ourselves in this chapter to a discussion of the basic principles involved in the automatic detection of some of the important weather phenomena of concern in aviation. The aim is to provide a flavour of the approaches used in some of the successful programs for automatic weather feature detection.

10.2 Basis of automated weather feature detection

Each type of weather phenomenon produces its own characteristic pattern or signature of parameter variation within a data field. Machines can detect and recognise the phenomena in basically the same way as humans do – by looking for these signatures. The difference is that while the recognition in human observers is based on some relatively unknown processes associated with experience and training, the logic of recognition must be formalised and quantified for use by machines. Such logic has been evolved and fine-tuned through concerted research at many centres, and has reached operational status for many common phenomena of significance to aviation.

A basic prerequisite of data fields used for recognising embedded weather features is that the data must be of appropriately high resolution. Details in the data field facilitate recognition of tell-tale signatures of the different types of phenomena. One good source of data for weather feature recognition is the radar, which provides detailed 'pictures' of the weather field in its zone of observation. Data from radar systems with overlapping coverage can be combined or 'mosaiced' to produce larger data fields to detect and track features over greater areas.

Automated feature recognition based on single data fields, e.g. radar reflectivity alone, is more straightforward than using multiple data fields for the same function. Examples of multiple data fields are provided by combinations of radar reflectivity and velocity, radar and satellite data, etc. However, quite often the signatures of phenomena in a single field may not be distinctive enough to provide reliable and unambiguous detection. Further, data from different fields may be necessary to estimate the intensity or hazard potential of a given phenomenon. Thus, in spite of the attendant difficulties of program development and data organisation, multiple data fields are utilised in some programs for automated weather feature detection and parameter estimation.

A major difficulty encountered in developing programs for automatic weather feature recognition arises from the fact that given generic features may have greatly variable shapes, sizes and strengths or intensities. Thus, a microburst may have a range of diameters, depths and strengths of outflows, may or may not be symmetric, and may or may not be accompanied by precipitation. Similarly, a gust front may be straight or curved, have unknown and variable wind shifts across itself, may or may not be embedded in an environment of precipitation, and may have a random orientation with respect to the radar. The feature recognition algorithm should therefore be based on the generic *pattern* of parameter variation corresponding to the phenomenon rather than the *values* of the parameters. Parameter values may, however, be used as thresholds to filter out the weaker or less probable candidates from consideration. The bases of automatically detecting some specific types of weather features are discussed in the following sections.

10.3 Thunderstorm cells

Thunderstorm cells are most readily recognised in the reflectivity field of radar data. As the radar beam scans horizontally, it provides the reflectivity map of a cross-section of the thunderstorm, the height of the section depending on the elevation angle of the radar beam and the distance of the storm cell from the radar. The cross-section of the precipitation shaft of a single thunderstorm cell typically has the maximum intensity at an interior point, with the intensity decreasing outwards. Two possible ways of idealising such a reflectivity contour are shown in Fig. 10.1. A distribution of this nature forms a guide for the automatic identification of single thunderstorm cells.

The storm recognition program delineates within a radar reflectivity data field the patches of high reflectivity (exceeding a set value), falling within specified size limits, as potential storm cells. It then determines the maximum reflectivity within the patch and fits a plausible basis function having such a maximum value. If the storm is assumed to be symmetric, then a rotated Gaussian or half-cosine function can serve as the basis function. More

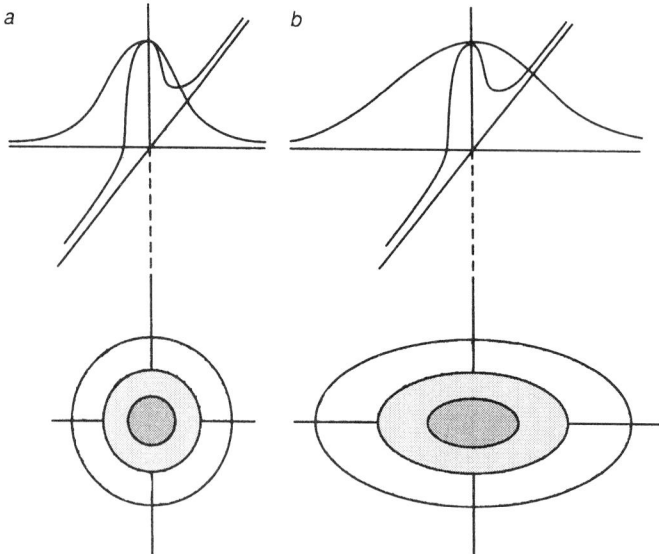

Figure 10.1 Simple models for idealising reflectivity patterns of the cross-sections of individual storm shafts: (a) circular contouring; (b) elliptical contouring. The darker shades represent higher reflectivity. The top Figures show a Gaussian intensity variation corresponding to each contour

complex programs may utilise two-dimensional basis functions with unequal width parameters along orthogonal directions to model asymmetric storm cells. A good fit, indicating the presence of a storm cell, may be declared if the root-mean-square (RMS) departure of the reflectivity distribution across the storm shaft stays within specified bounds.

In practice, the distribution of reflectivity within a thunderstorm may depart strongly from the ideal. The reflectivity maximum may be significantly displaced from the centre of the rain shaft, and the distribution of reflectivity may even be multimodal, i.e. display more maxima than one. In such cases there is a distinct probability of thunderstorms being missed by the recognition program. The detection probability can be enhanced by examining multiple cross-sections of the thunderstorm. Fortunately, as modern weather radars perform volume scan at several elevation levels, they can usually observe many cross-sections of the thunderstorms, and the reflectivity profiles of the storm at some of these cross-sections may be closer to the idealised model than at others. Thus, application of the recognition program to reflectivity data at multiple levels ensures that the storm is detected with a higher probability through these well behaved cross-sections.

An important requirement of weather sensors and systems for serving the aviation sector is that they must not only sense and recognise existing phenomena but, as far as possible, detect the phenomena in their incipient stages. The earlier that a phenomenon is detected and recognised in its development stage, the more would be the precious lead time available for pilots and controllers to take precautionary, corrective or evasive action. In the case of thunderstorms the precipitation activity, which is the primary indicator of the phenomenon to a radar sensor, starts in a small core high aloft in the cumulonimbus cloud, and the precipitation shaft gradually stretches downward to reach the ground as rain. Thus examining the high-elevation cross-sections of thunderstorms provides the best chance of detecting the storms in their early stages.

The storm cell information algorithm adopted for the Integrated Terminal Weather System (ITWS), which was discussed in Section 9.2, provides an example of automatic processing of storm cell data (Dasey *et al.*, 1995). The cell contouring algorithm embedded in this overall scheme analyses each precipitation image to generate contours corresponding to levels 3 and above given in Table 6.2 (level 2 and above are processed by the text product generator and provided directly to pilots). The contours are generated by searching in two dimensions for runs of adjacent pixels crossing a given threshold, and grouping such pixels to form cells. Once a contour at a certain level is formed, the search for pixels of the next higher level is made only within this contour. This ensures that contours of different levels do not cross each other. Cells that have locally the highest reflectivity are called 'hotspot' cells. The storm cell information is assigned to these hotspots, which are the foci of interest for pilots and air traffic controllers. The boundaries of hotspots are shown as dark lines for emphasis. Figure 10.2 shows an ITWS precipitation map with the hotspot contours found by the Storm Cell Information algorithm.

The broader structure of the ITWS Storm Cell Information algorithm, of which the cell contouring algorithm forms a part, is shown in Fig. 10.3. This broader system performs other functions such as hail gridding and generation of lightning flash rate maps, examples of which are shown in Figs. 10.4 and 10.5, respectively.

A useful and commonly performed step following the automated detection of storm cells is *tracking*. Tracking involves following the path taken by a moving phenomenon for a certain period of time. Since the shape and size of a thunderstorm may change during this period, tracking usually involves tracing the path of an important point within the phenomenon, e.g. its centroid. The centroid of a thunderstorm cross-section may be defined as the reflectivity-weighted centre of the part of the section that exceeds a certain reflectivity threshold. Alternatively, the location of the peak of the unimodal basis function best fitting the reflectivity distribution may be declared as the centroid. If a single location of a thunderstorm is to be specified based on observations at multiple levels, the centre of the centroids of the cross-sections at the various levels may be taken as the storm position.

Once the centroid of a thunderstorm is determined, the process can be repeated at intervals to update it as time progresses. A data file containing the successive positions of the thunderstorm centroid over a time segment is

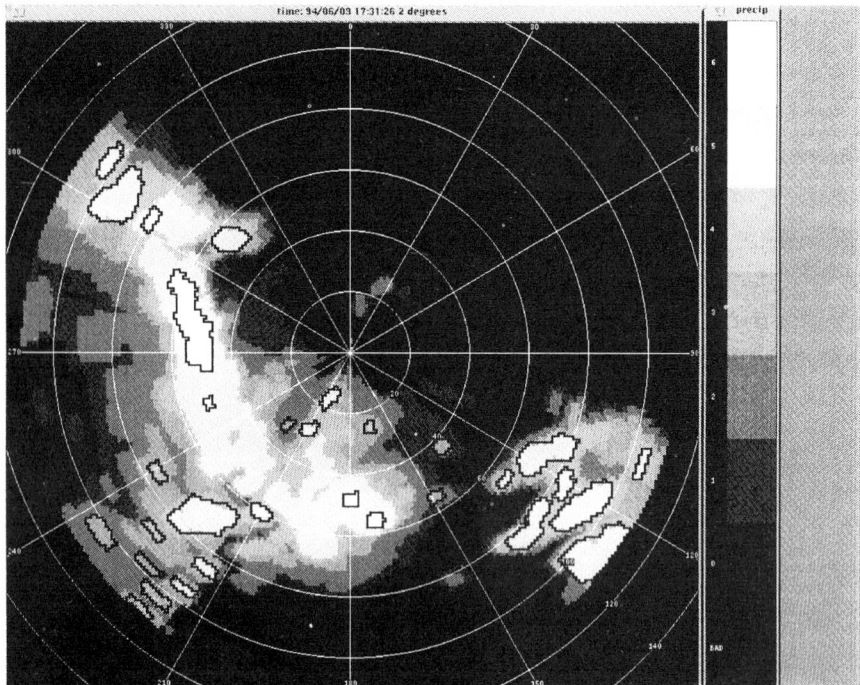

Figure 10.2   ITWS precipitation map with the 'hotspot' contours found by the Storm Cell Information (SCI) algorithm (from Dasey et al., 1995, reprinted with permission of MIT Lincoln Laboratory, Lexington, MA, USA)

called its *track*. The track, after suitable smoothing if necessary, may be shown as a line segment on a plan-position display to depict the path taken by the storm in coming to its current position. However, a more important use of the track is to *predict* the possible positions of the storm in the near future. Short-term storm position prediction, performed by extrapolating the track, can be of great use for air traffic planning, route charting, planning and optimising airport operations, and piloting decisions to enhance safety, comfort and economy. The bottom frames of Colour Plates 13 and 14 show the tracking and prediction of storm positions.

10.4 Mesocyclones

Recognition of phenomena involving rotating masses of air requires very different types of algorithmic approach than the thunderstorm detection. The data field itself is different for the detection of the two types of phenomena. As rotation refers to a particular pattern of velocity distribution,

Automatic detection and tracking of hazardous weather features 357

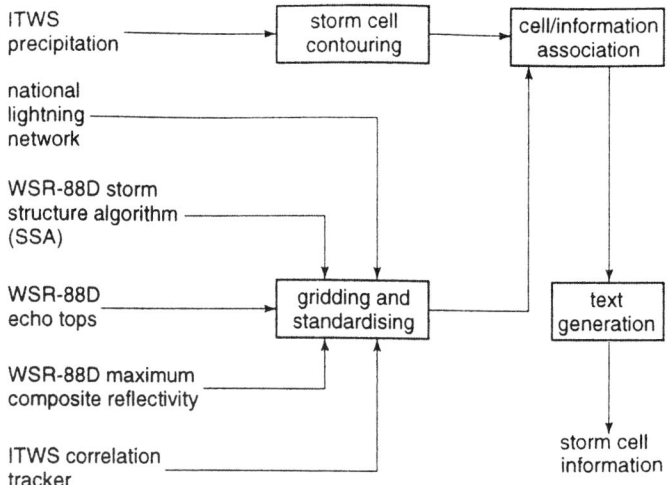

Figure 10.3 Data flow diagram and the functional components of the ITWS Storm Cell Information algorithm (from Dasey et al., 1995, reprinted with permission of MIT Lincoln Laboratory, Lexington, MA, USA)

the Doppler velocity field is the proper data field in which one looks for the existence of phenomena such as mesocyclones.

A cross-section of a mesocyclone, as obtained by a constant-elevation radar scan, would appear as a vortex. This can be recognised by comparison of its Doppler signature with that of an idealised vortex model. The most common model used for this purpose is the *Rankine vortex* model. In this model, shown schematically in Fig. 10.6, a cylindrical core shaft of air is assumed to rotate like a solid (i.e. have constant vorticity), so that tangential velocities of air particles in the shaft increase linearly with distance from the rotation axis. Outside this shaft, the tangential velocity falls in proportion with the increase in distance from the axis.

It is instructive to look at the Doppler behaviour of the Rankine vortex. Such a behaviour is most readily seen in terms of *isodops* which are contours of constant Doppler velocities. Inside the core shaft of uniform vorticity, the isodops are parallel line segments oriented along the direction towards the radar. The ends of each line segment are closed by arcs which represent the isodops outside the core. If the centre of the Rankine vortex is stationary relative to the radar then the isodop pattern would be symmetric, with the zero-Doppler isodop passing through the vortex centre. This is obvious by observing that the air particles on the radial passing through the vortex centre move in a tangential direction relative to the radar, producing no Doppler shift.

The 'Doppler doublet' resulting from the vortex is a tell-tale signature of the mesocyclone (see Colour Plate 12). The doublet consists of two lobes of opposite polarity (i.e. sign of Doppler velocity) existing side by side and

358 *Aviation weather surveillance systems*

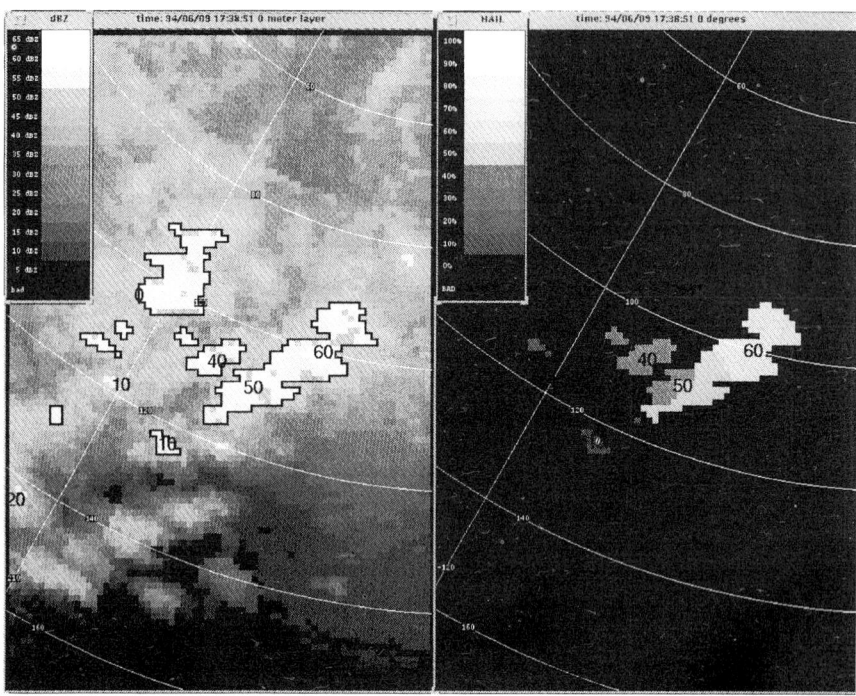

Figure 10.4   *WSR-88D maximum composite reflectivity (left), with 55 dBZ contours and hail point detections used by the hail gridding process, and the resulting hail grid (right). The numbers are the POSH (Probability of Severe Hail) values* (from Dasey et al., 1995, reprinted with permission of MIT Lincoln Laboratory, Lexington, MA, USA)

aligned perpendicular to the radar radial. Such a clue is used routinely by human operators for visual identification of a mesocyclone, and algorithms for their automatic recognition are also usually based on this property. The use of expected values of physical dimensions and velocities for thresholding and comparison enhances the probability of detection. For example, Burgess (1976) estimated that a mesocyclone associated with rotating updrafts in the Oklahoma area of the USA have peak tangential wind speeds lying between 20 and 25 m/s, and diameters of 5 to 6 km at their points of maximum tangential speed.

The idealised Doppler pattern of the Rankine vortex is affected in different ways in the presence of other forms of air motion. Translation of the vortex in a radial direction relative to the radar shifts the zero-Doppler isodop away from the vortex centre, rendering the isodop pattern (and hence the Doppler couplet) asymmetric, and translation at constant range from the radar causes the apparent vorticity of the cyclonic phenomenon to be reinforced or weakened depending on the relative directions of rotation of the vortex

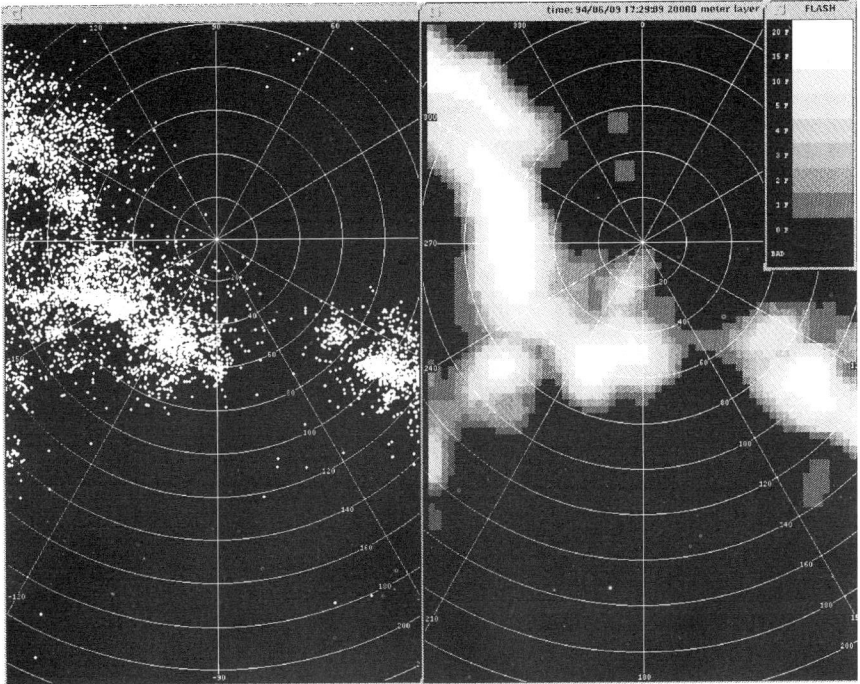

Figure 10.5 Lightning detections over a 2 min period (left), and the resulting lightning flash rate grid (right) (from Dasey et al., 1995, reprinted with permission of MIT Lincoln Laboratory, Lexington, MA, USA)

(about its centre) and its centre (about the radar). At low altitudes cyclonic motion often coexists with storm outflows or inflows which produce superimposed divergent or convergent flow patterns. Such flows cause the isodop pattern to be rotated with respect to that of the ideal vortex by an angle dependent on the relative strengths of the rotation rate and divergence rate. The direction of rotation is governed by the senses of the vortex rotation (cyclonic or anticyclonic) and divergence (divergence or convergence). The rotation of the isodop pattern causes the Doppler couplet to be correspondingly rotated from its ideal tangential orientation with respect to the radar. Thus, a robust software for automatic detection of mesocyclones must be able to take into account nonideal effects on the Doppler couplet such as asymmetry and rotation.

It is also possible to detect mesocyclonic patterns by tracing the trajectories of maximum radial and tangential gradients of the radial velocity in a manner similar to that used for gust fronts discussed in the following section (Uyeda and Zrnic', 1986). Haneda and Uyeda (1997) have applied this technique to actual Doppler data and found that the method can yield perceptible difference in the size and location of a given mesocyclone compared to a

360 Aviation weather surveillance systems

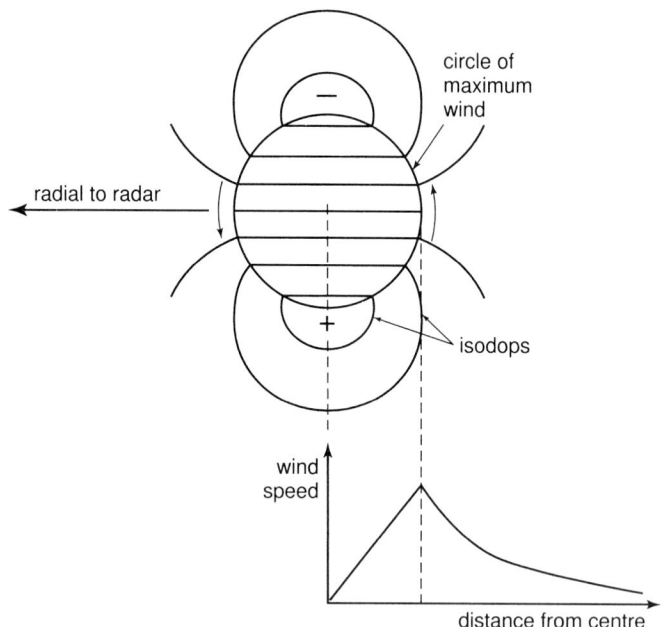

Figure 10.6 Rankine vortex model of a cyclonic phenomenon located far from the radar and with its centre stationary relative to the radar. The circle represents the line of maximum wind velocities, which is an indicator of the 'size' of the phenomenon. The other lines are the isodop contours. The two halves of the vortex produce Doppler shifts of opposite sign. Shown below the vortex is the variation of the wind speed as a function of the distance from the vortex centre

Rankine vortex model. A mesocyclone detected by the Doppler velocity gradient method is shown in Fig. 10.7.

Tornadoes, which are rotating shafts of air with much smaller spatial dimensions than mesocyclones (see Section 4.5.4), can also be modelled as Rankine vortices. However, their small dimension (cross-section) relative to radar resolution parameters makes their recognition different from that of mesocyclones. A tornado may typically occupy only a few to several radar resolution volumes (see Colour Plate 15), from which it is difficult to reconstruct isodops. Since a Doppler radar produces only one set of moment estimates for each resolution volume, each moment represents an average of its value over the volume. In the case of a tornado, this averaging effect is very pronounced because of the rapid variation of its velocity field within the scale of the radar resolution volume, which has typical dimensions of the order of 100 m or more. The averaging results in underestimation of the peak circular wind speeds, but the strong wind variation within the resolution volume serves to widen the Doppler spectrum, yielding a high value for the second moment estimate.

Automatic detection and tracking of hazardous weather features 361

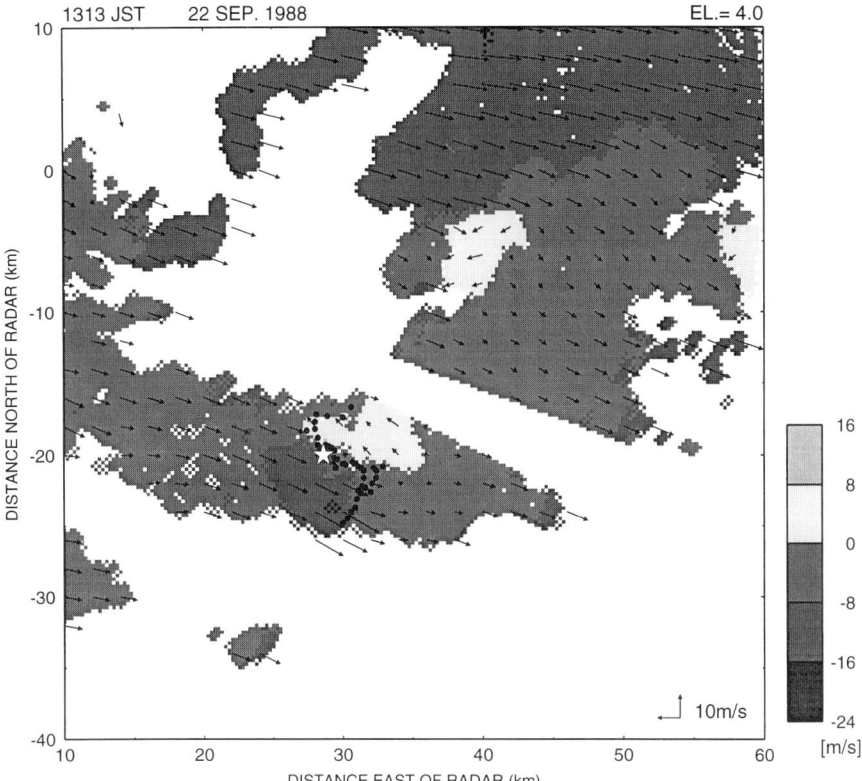

Figure 10.7 A mesocyclonic pattern (heavy dots in a rough 's' shape on either side of the star) detected by the wind gradient method. The arrows show the wind field obtained by the tangential velocity assumed display (TVAD) method (the wind speeds are also grey-shade contoured), and the heavy dots represent points of maximum gradient of the radial wind in radial and tangential directions. Note the characteristic wind reversal associated with the vortex, and the 'Doppler couplet' (with dark and light lobes) oriented approximately perpendicular to the radial from the radar. The radar is located at co-ordinates (0,0). The star marks the ground location of a tornado embedded in the mesocyclone. The clear strip through the grey area is caused by loss of radar data over several radials. (Courtesy H. Uyeda, Hokkaido University, Sapporo, Japan)

These observations can be used to derive the tornado vortex signature (TVS). Colour Plate 10 shows the automatic detection of a TVS in a WSR-88D system. Because of the compact size and high circulation speed of tornadoes, their Doppler couplets are small, closely spaced and intense, besides being tangentially or azimuthally aligned with respect to the observing radar (Fig. 10.8*a*). Thus, sharp changes in Doppler velocity between adjacent radar

362 Aviation weather surveillance systems

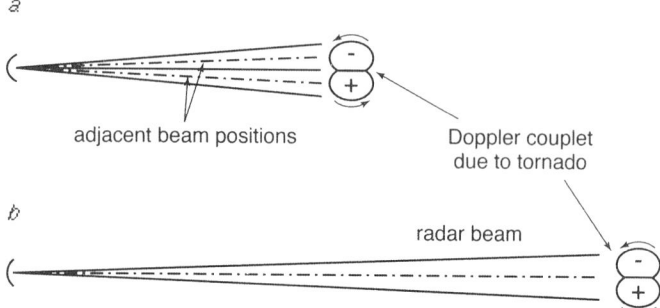

Figure 10.8 *Schematic representation of the compact Doppler couplet due to a tornado causing sharp change in the radial velocity sensed by adjacent beam positions (a). Far away from the radar (and/or for small tornadoes) the beam may cover almost the entire couplet whose lobes cannot be resolved by the beam, as shown in (b)*

radials at a given range location, together with high values of the Doppler spectral width (as explained above), are strong indicators of the presence of tornadoes or circulation that can develop into tornadoes.

Radar resolution volumes become wider in directions orthogonal to the radial with increase in distance from the radar (Section 6.2.1). Thus when tornadoes are observed at a significant distance from the radar, the resolution volume may cover the entire effective width of a tornado (Fig. 10.8*b*). Then the positive and negative lobes of the Doppler couplet cannot be resolved, and an important component of the tornado signature would be lost. In such cases, the Doppler spectrum of the tornado as a whole, as obtained from the radar resolution volumes (there may be more than one in the range or radial direction) that coincide with the tornado, provides the primary clue to its detection. Considering the Rankine vortex model of the tornado, the strong approaching velocities would produce a peak at a positive Doppler shift, and the receding velocities would cause a second peak at a negative Doppler shift. The result is a pronounced bimodal spectrum which can act as a signature for the recognition of tornadoes. The bimodal spectrum is symmetric about the zero-Doppler frequency if the tornado itself is rotationally symmetric and is symmetrically located with respect to the centre of the radar resolution volume; otherwise the spectrum is asymmetric. The centrifugal ejection of debris from the tornado also affects the spectral distribution. The bimodal tornado spectral signature predicted by the Rankine model has been repeatedly confirmed by the observation of actual tornado spectra (Zrnic' and Doviak, 1975; Zrnic' *et al.*, 1977).

One fact to remember in connection with the detection of tornadoes is that their reflectivities are often of a low order, perhaps only ~5 dBZ or less, essentially due to the dust and debris raised by the swirling winds. The spectral estimates, especially the second moment, may therefore have significant statistical uncertainty. Another fact to consider is that tornado shafts frequently display strong curvatures vertically, so that the locations of a given tornado as detected from radar scans at different heights may not coincide.

Because of the high destructive potential of tornadoes, prediction of their occurrence is of high value. Short-term prediction of tornado occurrence is possible by observing the circulation fields that act as precursors. Large tornadoes are generally preceded by circulation of a larger spatial scale starting at mid-tropospheric levels, i.e. at altitudes in the range of 6–8 km, and extending towards the ground. Such larger circulations can be readily mapped by radars, and detected through their characteristic Doppler couplets. Correlation with altitude can enhance the probability of success of predictions.

Once detected automatically, rotational phenomena such as tornadoes may also be tracked to record their motion history and extrapolate their position in the near future. The top frame of Colour Plate 13 shows the machine-derived tracks of some tornadoes.

10.5 Gust fronts

As discussed in Section 4.6, gust fronts are essentially low-altitude phenomena that are of serious concern from the point of view of aviation, especially during the crucial and vulnerable phases of low-altitude manoeuvres associated with landing and takeoff. These are also phenomena that have a fast progression, and may not always have strong visual clues indicating their presence. Detection and positional prediction of gust fronts is a strong requirement for terminal area weather surveillance systems.

In plan view, gust fronts would appear as linear features, and their distinguishing characteristic is a discontinuity (or rapid change) in horizontal wind velocity across the line. The velocity distribution on either side of the gust front is normally so as to produce a net horizontal convergence towards the linear feature. Since the primary information for sensor-based observation of gust fronts is related to wind velocity, the appropriate data field to use for its automatic detection is the Doppler velocity field of radar data.

Gust fronts may or may not be embedded in an immediate environment of precipitation. Hence the radar reflectivity at their location may vary greatly. Although their detection is primarily based on wind velocity data and not reflectivity, wide variation in reflectivity causes a corresponding variation in the signal-to-noise ratio of the echo signals, which affects the quality of the velocity estimates. Gust fronts occurring in the midst of rain would provide clean velocity data fields, while those occurring in clear air would yield somewhat noisy velocity data because of feeble signal return. Any program for automatic detection of gust fronts must be able to work with velocity data fields of such variable quality.

364 *Aviation weather surveillance systems*

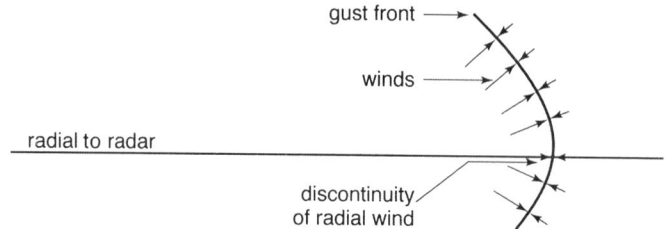

Figure 10.9 Schematic diagram showing discontinuity (or sharp gradient) in the radial wind as seen by a Doppler radar at the intersection of a radial with the gust front

Automatic gust front detection algorithms generally utilise the Doppler wind velocity data at low radar elevation angles, and examine the velocity variation along each radial of the radar scan (e.g. Uyeda and Zrnic', 1986). Any sharp or rapid velocity change exceeding a certain threshold value is noted, taking care to eliminate noise-like spikes either through a process of smoothing or by performing a run-length test. The location of the discontinuity on the radial is taken as a possible point on the linear feature representing the gust front (Fig. 10.9).

A more sophisticated algorithm may compute the 'true' winds using an appropriate method (see Section 6.6.5) before looking for discontinuities. Takahashi *et al.* (1991) use a Tangential Velocity Assumed Display (TVAD) technique to map the wind field, and use this field to locate frontal phenomena. The points of discontinuity corresponding to a gust front are shown in Fig. 10.10 in the plan view.

The next step in the gust front detection algorithm is to string together the points on successive radials which are suspected to be on the gust front. In an ideal situation, the points of Doppler velocity discontinuity on consecutive radials should arrange themselves into a clear linear feature representing the gust front. In practice, however, the organisation of the points is often not quite neat, as apparent from the example in Fig. 10.10. Some radials may not display any point of velocity discontinuity, in which case the curve representing the gust front will have a broken appearance, having segments separated by gaps (or 'washouts'). The causes of such washouts are discussed later in this section. It is also possible for some radials to show more velocity discontinuities than one. In such a case the apparent gust front curve will be branched or have multiple segments in certain sectors. Further, the locus of the discontinuity among the consecutive radials may itself show jumps or discontinuities. In general, the raw gust front representation would have a noisy and jagged appearance with jumps, breaks, branches and/or ambiguities. It requires considerable further processing, possibly with the use of artificial intelligence techniques, to construct a realistic representation of the gust front.

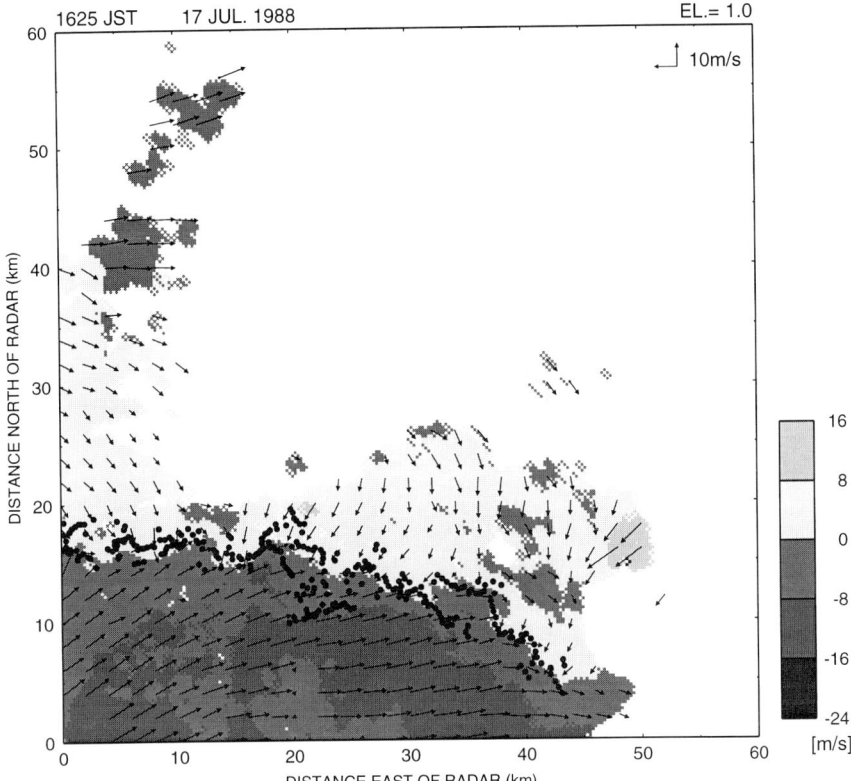

Figure 10.10 Points of velocity discontinuity (heavy dots irregularly arranged along the border between dark and light areas) representing a gust front, as obtained from single Doppler radar data. The arrows show the wind field obtained by the Tangential Velocity Assumed Display (TVAD) method, and the wind speeds are also contoured in grey shades. Notice the characteristic wind convergence along the gust front, as shown schematically in Figs. 10.9 and 4.5d. (Courtesy H. Uyeda, Hokkaido University, Sapporo, Japan)

A simple smoothing operation based on moving-averages or curve-fitting is usually inadequate to remove the aberrations present in the detected gust front. More potent methods of continuity enhancement and ambiguity resolution are necessary for obtaining a smooth and continuous arc that represents the gust front. Data association methods based on the nearest-neighbour principle are often useful for choosing the most likely one among multiple track segments over a given set of radials. Slope-matching techniques may be employed to eliminate multiple branchings of the gust front. Establishing continuity of line segments across significant gaps is a more complex problem, since the presence of large gaps makes it more difficult to associate data, resolve ambiguities, and match slopes.

The data quality is often such that all the problems associated with the reconstruction of the gust front cannot be overcome with processing techniques alone, and supplementary data are required to fill the gaps and resolve ambiguities. Such data can come from multilevel radar scans. Segments of the gust arc curve that are washed out in the data from one scan have a high probability of showing up on one or more other scans, and combining data from more scans than one is thus likely to give a much more complete and continuous picture of the gust front. Similarly, ambiguous or multiple branches can potentially be resolved by comparing data among multiple levels on the basis of coincidence or majority voting.

Certain factors must, however, be considered in the context of multilevel-scan data processing for automatic gust front detection. The first is that processing involving association of data from multiple levels is far more computation-intensive than single-scan processing. The time taken for the radar to execute multiple scans for data collection, together with the additional processing, causes an increase in the time taken for the gust front recognition process. The second factor arises from the geometry of gust front observation by radar, and is elaborated below.

As the gust front phenomenon is confined to the lower levels of the atmosphere, only the low-elevation scans of a normal radar scan cycle would intersect the phenomenon. The number of such scans that can provide data regarding gust fronts depends on the actual depth of the phenomenon and its distance from the radar installation. At close ranges, a few or even several of the lowest beam elevations may intersect the front, but as the range increases, the number of effective scan elevations for gust front observation may reduce to just one or two (at very long ranges, the gust front may altogether fall below the radar horizon). This is a distinct disadvantage because it is at long ranges that data quality is normally poor due to low signal-to-noise ratios (the problem being more serious when the gust front occurs in clear air), but the geometry of the problem is such that the freedom to combine data from multiple levels is curtailed at those ranges where it is most needed. Reliable gust front detection by radar thus requires observation at relatively close ranges. This is one of the major considerations that have led to the development of terminal area weather surveillance radars such as the TDWR (Section 7.5) even while longer-range radars of the class of the WSR-88D are available. An airport-located radar would be within a close range of gust fronts occurring within and in the vicinity of airport areas, where their effect is most hazardous for low-altitude flights.

There are other geometry-related factors of significant importance in the context of radar detection of gust fronts, particularly when machine recognition is involved. One major point to consider is *aspect sensitivity*. Since Doppler radars observe only the radial component of wind velocities, the apparent strength of the convergence or velocity discontinuity in a gust front, as sensed by a Doppler radar, depends on the orientation of the front relative to the direction to the radar. As shown schematically in Fig. 10.11, the full

Automatic detection and tracking of hazardous weather features 367

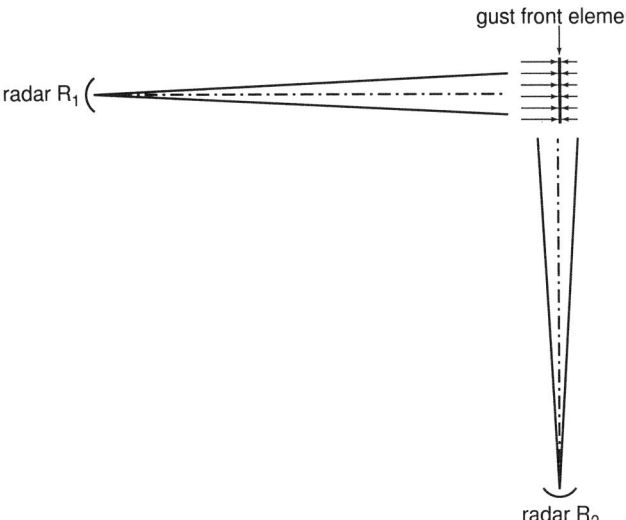

Figure 10.11   Schematic diagram illustrating aspect sensitivity in the detection of gust fronts. The radar R_1 looking at an element of the gust front normally will sense the full value of the convergence, while the radar R_2 along the front will not sense the winds which are normal to its radial

value of the convergence will be sensed by a radar when the local segment of the gust front is aligned perpendicular to the radial from the radar, while no convergence will be apparent to the radar when the front is along the radial. The assumption is made here that the convergence is normal to the gust front.

Aspect sensitivity has a significant bearing on the automatic detection of gust fronts. To avoid small velocity changes and noise effects from cluttering the gust front detection algorithm, the strength of the velocity jump at a candidate point is usually required to exceed a specified threshold before being considered as a part of the gust front. However, because of the cosine effect caused by aspect sensitivity, the apparent convergence along the parts of the gust front that are aligned nearly radially with respect to the radar would fall below the threshold, making the corresponding segments 'invisible' to the gust front detection algorithm.

Washouts of gust front segments are not due to aspect sensitivity alone. Many other factors can cause this phenomenon. The most obvious case is where the gust front is actually weak, i.e. has a velocity jump that falls below the threshold. However, it is possible that the velocity jump is significant, but the detection process is not proper because of weak echoes. This can happen when segments of the gust front have no accompanying precipitation, and must be detected in the clear-air mode. Distance from the radar accentuates this problem. The resulting weak echoes make the velocity estimates noisy, making it difficult to ascertain the point of velocity discontinuity. Yet another factor affecting gust front detection is related to data degradation due to the presence of ground clutter, which interferes strongly with the echo signals because of the proximity of the gust front to the ground. In summary, the causes of washouts of portions of gust fronts can be enumerated as: (i) weak convergence; (ii) weak reflectivity; (iii) ground clutter interference; and (iv) Doppler radar aspect sensitivity.

The gust front washout problem due to unfavourable aspects with respect to the Doppler radar can be remedied, at least in part, by looking for tangential (i.e. normal to the radial) discontinuities (or high gradients) in the radial velocity. It can, in principle, also be overcome by using data from other spectral moment fields. Gust fronts are often apparent in the reflectivity field, but almost invariably show up as a thin line in the turbulence (second moment of the Doppler spectrum) data field, since the shear and the resulting mixing process along the gust front causes significant localised turbulence. This observation can be combined with velocity field data to enhance the probability of gust front detection and reconstruction. However, the additional programming complexity and computational requirements are usually rather high, so that such data fusion has not been incorporated in operational real-time systems. Further, even if the front itself may be discernible in the turbulence data field, the associated wind parameters (strength of convergence, wind shift, and absolute wind speeds ahead of and behind the gust front) cannot be accurately estimated from single radar data for the parts of the gust front that are washed out.

As in the case of other types of hazardous weather features, one important use of automated detection capability is for tracking and future position extrapolation. Tracking of gust fronts, however, differs from that of phenomena such as thunderstorms or mesocyclones in the important aspect that the gust front is an extended linear feature while the others are localised phenomena for which a centroid can be defined. Thus, while the track of a thunderstorm can be shown as a line, the spatial progress of a gust front will appear as a family of shifting curves or arcs. Such prediction of gust front positions is extremely important. As mentioned in Section 7.5, information about wind shifts due to gust fronts should be available to controllers 20 min in advance of their occurrence over runways and in their close vicinity. Automated gust front tracking and position prediction helps in providing this information on a routine and sustained basis.

An example of a working algorithm for gust front detection is one developed for the Integrated Terminal Weather System (ITWS). The Initial Operational Capability Machine Intelligent Gust Front Algorithm developed by MIT Lincoln Laboratory (Troxel and Delanoy, 1995) utilises radar data from the TDWR and anemometer data from the LLWAS or ASOS (see Chapter 8), and knowledge-based signal processing techniques. The possibility exists of using data from additional sensors such as WSR-88D or other sources. Gust front signatures utilised for recognition include the reflectivity thin-line, velocity convergence and significant motion normal to the front.

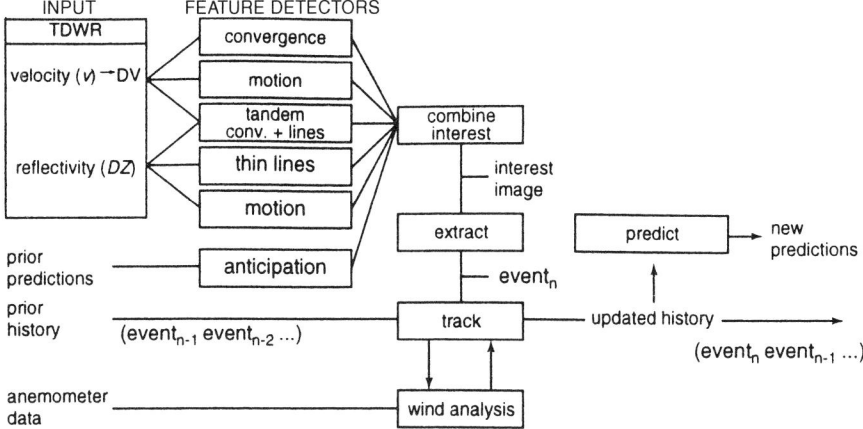

Figure 10.12 Block diagram of the Machine Intelligent Gust Front Algorithm for ITWS (from Troxel and Delanoy, 1995, reprinted with permission of MIT Lincoln Laboratory, Lexington, MA, USA)

Since these signatures occur in different data domains, techniques of data fusion are employed. In particular, each parameter in a pixel under consideration is converted into a normalised 'interest' signifying the probability of the pixel being a part of the gust front. These interests are then combined in rule-based ways, and pixels with high interest values are clustered to generate the gust front. The steps in the intelligent gust front detection process are shown in the block diagram of Fig. 10.12. In Fig. 10.13 are shown the individual interest fields, the combined interest image, and the features detected after thresholding.

10.6 Storm outflows and microbursts

The most serious forms of wind shear that endanger flight are associated with diverging wind fields at low altitudes, such as those caused by storm outflows and microbursts. The nature and hazard potential of such wind fields were discussed is Sections 4.5 and 4.7. The automatic detection of low-altitude divergence is therefore a strong requirement of modern aviation weather surveillance systems.

The signature of wind divergence is most clearly observed in the radial velocity data fields of Doppler radars. The streamlines of an ideal symmetric divergent wind field are shown in Fig. 10.14. When a Doppler radar observes such a wind field, the flow along the two streamlines aligned in the direction of the radial from the radar would produce the maximum Doppler shift, with the inward and outward streamlines yielding maxima of opposite polarities. The flow along the streamlines perpendicular to the radial from the radar

370 *Aviation weather surveillance systems*

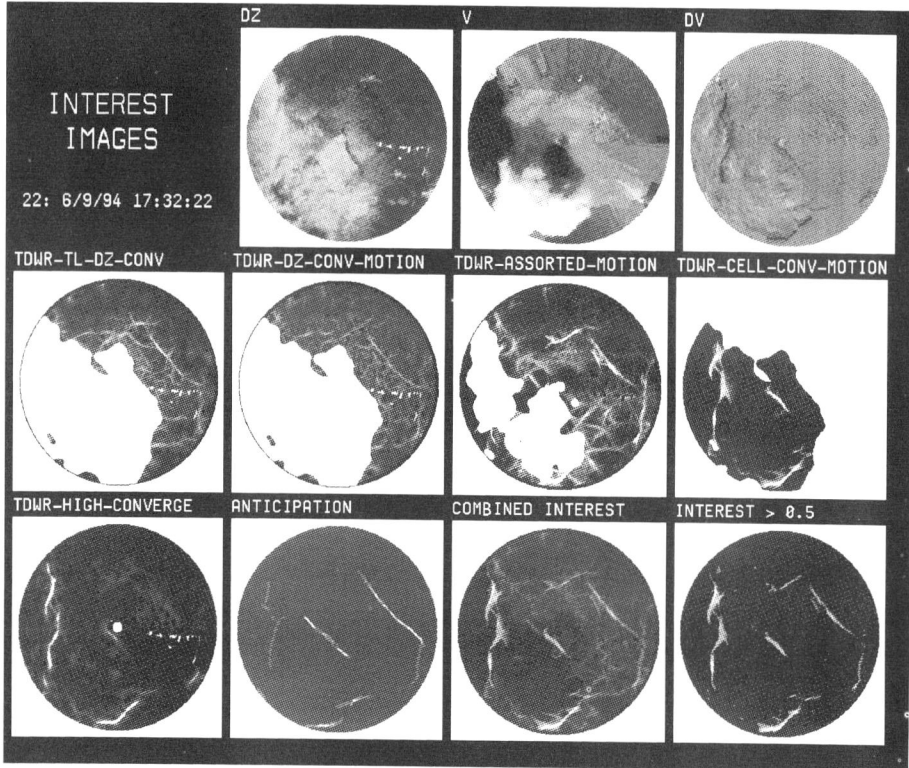

Figure 10.13 *Feature detection based on interest images. The top three fields are the input image data (DZ: reflectivity, V: velocity; DV: velocity change). The next six frames show the output interest images from individual feature detectors. The penultimate field is the combination of the various interest images, and the last frame shows the linear features detected by thresholding the combined interest at 0.5* (from Troxel and Delanoy, 1995, reprinted with permission of MIT Lincoln Laboratory, Lexington, MA, USA)

would produce no Doppler shift, and the Doppler velocity of the flow along the intermediate oblique streamlines would show a cosine dependence on their angles with respect to the radar radial. The result would be a Doppler couplet much like that due to a mesocyclone, but its orientation would be along the radar radial, unlike the case of the mesocyclone. Such a radially aligned Doppler couplet is the most characteristic signature of a divergent flow field.

Convergent wind fields can be recognised in a similar way through their radically oriented bilobed patterns, except that the polarities of the lobes would be opposite to those of divergent fields. Convergence is usually caused by the inflow of air feeding into thunderstorm updrafts. Such flow most often

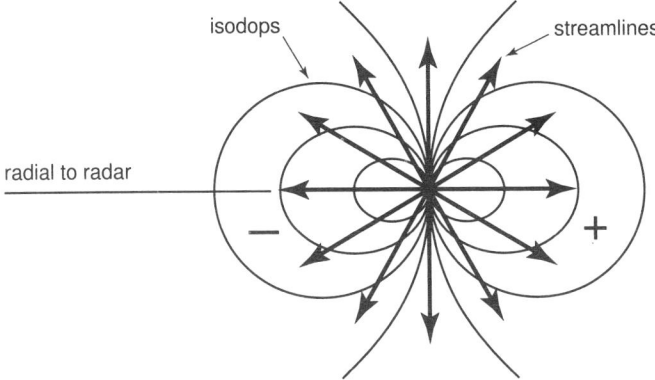

Figure 10.14 Streamlines and isodops of an ideally divergent wind field. The two Doppler lobes are of opposite sign, and the Doppler couplet is aligned along the radial to the radar

occurs without accompanying rainfall, and hence must be detected in the clear-air mode.

Microbursts can be detected in radar data fields through the Doppler signature of their low-altitude divergence (see Colour Plate 22). The principle of detecting microbursts is thus basically the same as that of any general storm outflow, and microbursts are distinguished essentially by their defining spatial scales. However, in practice certain special considerations arise in the observation and data collection for microburst detection. One important requirement for detecting microbursts is the ability to observe by radar down to very low altitudes, because the strongest horizontal winds due to microbursts generally occur in a shallow layer close to the ground (see Section 4.7.2). This means that the radar must be sufficiently close to the microburst so that radar horizon effects do not obscure the lowest altitudes from observation. It also calls for very good clutter rejection characteristics on the part of the radar, since the returns from low altitudes are more severely affected by ground clutter 'seen' through the main lobe of the antenna pattern. Another important requirement for microburst detection is frequent observation, at ground-grazing altitudes, of the space under surveillance so that the rapid growth and decay phases of the phenomenon (and hence those of its induced flow fields) are not undersampled or altogether missed.

As in the case of the detection of other phenomena discussed in the earlier sections of this chapter, the signatures of divergence (and convergence) can also depart considerably from the ideal. Two of the most common disturbing effects are asymmetry and the presence of vorticity or rotation along with divergence (or convergence). Asymmetry has the effect of making the lobes unequal and/or distorted, and vorticity rotates the alignment of the Doppler couplet from the radial direction with respect to the radar (see the related discussion in Section 10.4). A frequent cause of asymmetry is the obliquity of the shaft of downdraft air which causes the divergence. An oblique downdraft generally causes the strongest outflow in the direction of the shaft (or, more precisely, along the direction on the ground which makes the highest angle with respect to the shaft), and the weakest outflow in an opposite direction (see Fig. 4.9*b*). The contours of such an outflow field can be modelled as being ellipses in divergence recognition algorithms designed to take outflow asymmetry into account.

Once the Doppler couplet signifying the divergence field has been identified, the point of symmetry of the couplet can be assumed to be the centre of the divergence. In the case of asymmetric signatures, the centre is identified as the point with zero Doppler shift but maximum Doppler gradient. The locus of this point as determined over successive radar scans would constitute the track of the divergence field and hence that of its causative phenomenon such as the microburst. The size or extent of the divergent flow field is given by a circle (ellipse in the case of asymmetric outflows) centred at the centre of the divergence and enclosing both the lobes of the Doppler couplet up to a defined Doppler threshold.

10.7 Summary

In a scenario of extensive instrumentation for aviation weather surveillance, human involvement for routine interpretation of sensor data becomes expensive and prone to delays and errors. The timely generation of high-level data products that can be used by personnel without specialised meteorological training is facilitated by automatic detection of specific weather hazard features. For this reason, considerable research and development have been carried out for developing algorithms and software capable of detecting and recognising atmospheric features hazardous to aviation.

Recognition of atmospheric phenomena is based on the identification of characteristic features or signature(s) in the appropriate data field(s). For the simple storm cell, reflectivity is the pertinent data field, and closed contours with reflectivity increasing inward is an expected behaviour. In the case of mesocyclones and other rotational phenomena, the appropriate data is Doppler velocity, and the dominant signature consists of a Doppler doublet orthogonal to the line of sight. Divergence phenomena such as microbursts also produce Doppler doublets, but their orientation is along the radial. Another phenomenon considered is the gust front, which appears as a thin and weak line in the reflectivity field, a (generally curved) line of convergence in the velocity field, and displaying significant lateral motion. Processing the raw data to discern the respective signatures forms the core function of automatic detection algorithms. Algorithms based on a single radar scan (or other planar data arrays) and a single data field (e.g. reflectivity) are relatively simple to develop and faster to run, but may not be very robust and reliable. Multiscan and multiplane data are more difficult to process, but if handled properly, can yield robust detection and recognition of hazardous features. If multiparameter and/or multisensor data are utilised for detection, recognition and delineation, then the use of formal data fusion techniques may be necessary.

Feature tracking is a very useful follow-up of the detection process. It helps in extrapolating and providing short-term predictions of the positions of hazardous phenomena, and helps in planning flight and airport operations. Tracking is usually performed by following the motion of the centroid of the phenomenon of interest, or by a process of correlation between successive positions of a given phenomenon.

10.8 References

BURGESS, D.W. (1976): 'Single Doppler radar vortex recognition: Part I. Mesocyclone signatures'. Preprints of 17th Conference on Radar Meteorology, 1976, pp. 97–103

DASEY, T.J., DENNENO, A., and BOLDI, R. (1995): 'The integrated terminal weather system (ITWS) storm cell information algorithm'. Preprints of 6th Conference on Aviation Weather Systems, Dallas, TX, 15–20 January 1995 (American Meteorological Society, Boston), pp. 372–377

HANEDA, T., and UYEDA, H. (1997): 'A technique for detecting a vortex using gradient of Doppler velocity data'. Preprints of 28th Conference on Radar Meteorology, Austin, TX, 7–12 September 1997 (American Meteorological Society, Boston), pp. 345–346

TAKAHASHI, N., UYEDA, H., KIKUCHI, K., and OKAZAKI, M. (1991): 'A method to describe the fluctuation and discontinuity of horizontal wind fields by a single Doppler radar'. Preprints of 25th International Conference on Radar Meteorology, Paris, France, 24–28 June 1991 (American Meteorological Society, Boston), pp. 642–645

TROXEL, S.W., and DELANOY, R.L. (1995): 'Machine intelligent gust front detection for the Integrated Terminal Weather System (ITWS)'. Preprints of 6th Conference on Aviation Weather Systems, Dallas, TX, 15–20 January 1995 (American Meteorological Society, Boston), pp. 378–383

UYEDA, H., and ZRNIC', D.S. (1986): 'Automatic detection of gust fronts', *J. Atmos. Ocean. Technol.*, **3**, pp. 36–50

ZRNIC', D.S., and DOVIAK, R.J. (1975): 'Velocity spectra of vortices scanned with a pulse-Doppler radar', *J. Appl. Meteorol.*, **14**, pp. 1531–1539

ZRNIC', D.S., DOVIAK, R.J., and BURGESS, D.W. (1977): 'Probing tornadoes with a pulse Doppler radar', *Q. J. R. Soc.*, **103**, pp. 707–720

Chapter 11
Atmospheric turbulence and its detection by radar[1]

11.1 General

The role of turbulence as a frequent and serious atmospheric hazard to aviation was addressed in Section 3.3. It was pointed out in Chapter 4 that thunderstorms generate some of the strongest turbulence found naturally in the atmosphere. However, atmospheric turbulence is of widespread occurrence, and 'clear air turbulence' is a commonly experienced source of aviation hazard and discomfort. The basic capability of coherent weather radars to sense and measure turbulence through an estimate of the Doppler spectral width was discussed in Chapter 6. In view of the paramount importance of turbulence in the aviation context, this chapter is devoted to a closer and deeper study of some of the fundamental aspects of atmospheric turbulence, and its relationship with aviation on the one hand and radar-measured variables on the other.

Doppler radar is the only remote sensing instrument that can detect tracers of wind, both in the clear air outside precipitation zones and in low-visibility conditions such as those occurring inside heavy rainfall regions and cloud masses. Cloud and precipitation droplets hamper the operation of optical sensors such as lidars by almost completely extinguishing the optical radiation within a few metres of propagation distance. This unique 'all-weather' sensing capability makes the Doppler radar an instrument of choice to survey the wind fields traversed by aircraft. Of particular interest are the spatial and temporal variability of the wind. As mentioned before, wind shear and turbulence relate to two different aspects of this variability. After a short discussion on wind shear and turbulence in meteorological phenomena, this chapter briefly reviews the relation between the parameters of Doppler spectra and the wind and reflectivity fields in the field of view of the radar. This is then followed by an introduction to the statistical theory of turbulence to establish, on a quantitative basis, the relation between the Doppler spectrum width and the wind and turbulence fields. Because the eddy dissipation rate of turbulent energy is closely related to turbulence intensity, and because of the importance of thunderstorms in aviation safety and

---

[1] This chapter has been contributed by Dr. R.J. Doviak of the National Severe Storms Laboratory, Norman, OK, USA

efficiency, a short discussion is presented on the measurements and observations of eddy dissipation rate in thunderstorms. The chapter concludes with a discussion on turbulence effects on aircraft.

11.2 Wind shear and turbulence in meteorological events

Wind shear, the spatial change in wind on scales which can affect aircraft performance, is often associated with turbulence, and vice versa (see Section 3.3.2). To differentiate the two phenomena, wind shear and turbulence, we shall define wind shear as the spatial change in wind associated with specific meteorological events such as thunderstorms, tornadoes, etc., in which the wind field is ordered (i.e. it is either steady or changes slowly, and has a relatively simple dependence on spatial co-ordinates), whereas turbulence is a random and unpredictable distribution of wind velocities, with spatial and temporal scales shorter than those of the causative meteorological event.

Indeed, wind shear and turbulence are intertwined in meteorological events, with turbulence very often generated by the wind shear associated with these events. Wind shear, characterised by the presence of differential speeds between adjacent layers or masses of air, drives the formation of eddies of various sizes which constitute turbulence (Fig. 11.1). Thus the two (wind shear and turbulence) often exist side-by-side in meteorological phenomena, and if one examined the spatial spectrum of wind fluctuations in these phenomena, one would see a continuous distribution of wind variability, with scales or wavelengths typically ranging from the order of a kilometre or larger, corresponding to the ordered flow component of the meteorological events, to the order of centimetres and millimetres where the turbulent energy is eventually converted to heat by the viscosity of the air.

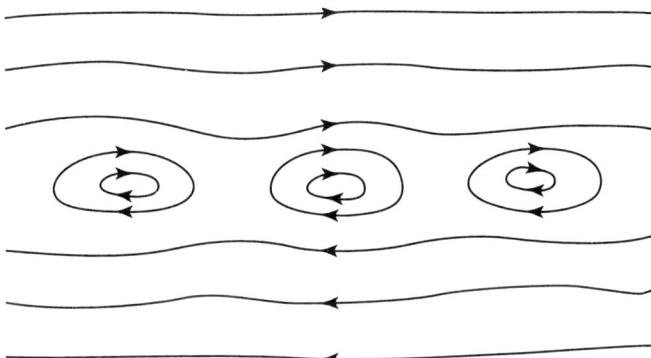

Figure 11.1 Schematic diagram showing the generation of eddies due to wind shear at the interface of layers of air moving relative to each other

The large-scale turbulence generated by meteorological events undergoes a continuous fracturing or scale-reduction process whereby the larger turbulent eddies generate those of smaller scales, which in turn generate even smaller scales until the energy of the eddies is dissipated by viscosity. The essence of this process was described by Lewis Richardson in his classic 1922 stanza (Stull, 1988):

> Big Whorls have little whorls,
> Which feed on their velocity;
> And little whorls have lesser whorls,
> And so on to viscosity.

Intense atmospheric turbulence is usually found in relatively large-scale transient meteorological events, such as thunderstorms, and also in events of smaller scale such as tornadoes and microbursts found within the thunderstorms. Relatively benign turbulence is found when the earth's surface is heated by solar radiation which causes the air to become hydrostatically unstable, which in turn causes fair weather meteorological phenomena such as *horizontally symmetric thermal plumes* (if the ambient wind is light), and *convective streets* or *rolls* (if the wind is strong); the latter are rendered visible by characteristic fair weather clouds that are aligned in relatively straight rows. Turbulence is also generated by the breaking of large-amplitude internal buoyancy waves. For example, Kelvin–Helmholtz (K-H) waves are generated in hydrostatically stable regions with strong vertical shear of horizontal wind. These sources of turbulence are briefly discussed below.

11.2.1 Thunderstorms

As mentioned in the earlier chapters, the most significant weather event which generates prodigious amounts of intense turbulence is the thunderstorm. The rapid ascent of the air shaft through a nearly static ambient air mass, and the side-by-side existence of ascending and descending shafts in mature storms (see Fig. 4.3), give rise to turbulence. The cauliflower structure of thunderstorm clouds is a manifestation of the presence of turbulence which mixes the updraft and outflow of the storm with the air of the clear environment. Fig. 11.2 shows such cauliflower structure in the anvil of clouds produced by tornadic thunderstorms over northern Oklahoma and southern Kansas in the USA. These storms had tops as high as 16.5 km. The superimposed light and dark contours in this Figure outline the area of thunderstorm rain, showing regions exceeding a moderate reflectivity factor of 45 dBZ (corresponding roughly to a rainfall rate $R > 20$ mm h^{-1}) and a lower reflectivity factor of 15 dBZ (corresponding to light rain, $R < 1$ mm h^{-1}), respectively. It is obvious from this Figure that thunderstorm turbulence (as shown by lobed billowing clouds) can reside far from the highest reflectivity region and is generally in a downstream direction relative to the environmental wind (which was from the WSW in this case).

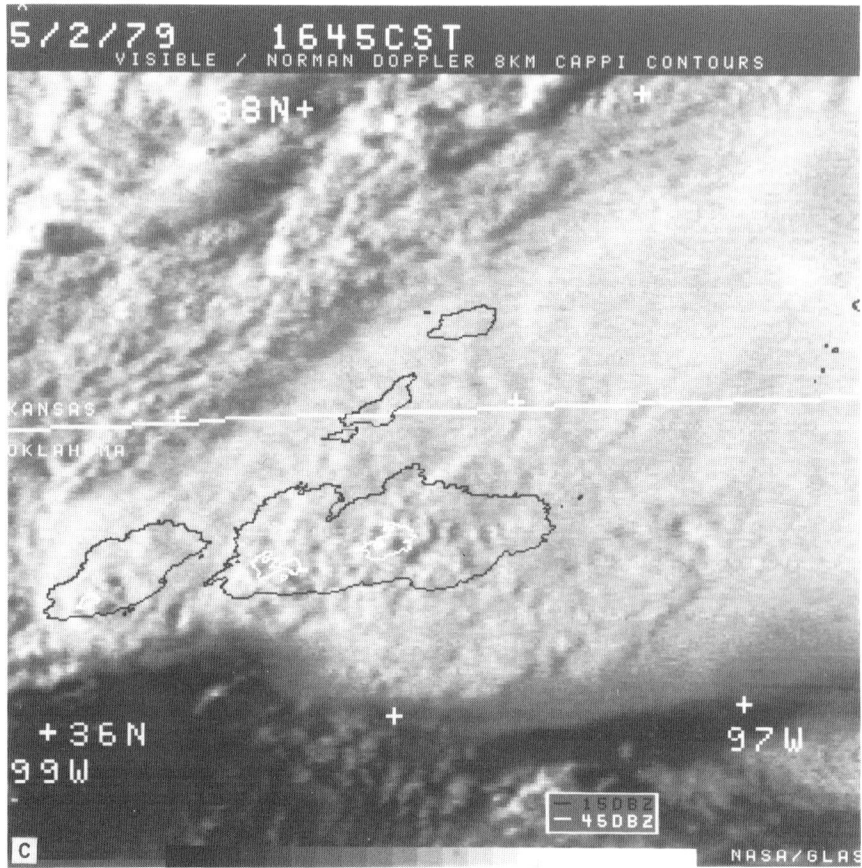

Figure 11.2 The anvil of clouds from storms on 2 May 1979 over Oklahoma and Kansas, USA. The Doppler radar reflectivity factor contours of 45 (white) and 15 (black) dBZ at the 8 km altitude are superimposed. (Courtesy G. Heymsfield, GLAS, NASA)

11.2.2 Thermal plumes

Relatively benign turbulence is commonly found in clear air convective plumes rising from the layer of air near the earth's surface heated by solar radiation. These plumes are responsible for the bumpiness that passengers often witness during the ascent from or descent on to airports when the aircraft is within a kilometre or two of the ground. Even though we consider plumes as meteorological events (although not necessarily significant ones) which generate turbulence on scales smaller than themselves, the plumes themselves are commonly referred to as the turbulent events. Figure 11.3 is a vertical cross-section through thermal plumes manifesting themselves in the reflectivity field observed with a Doppler radar. This figure shows that in this

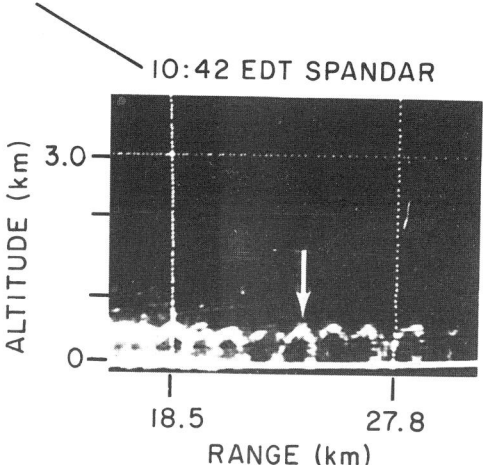

Figure 11.3 Vertical cross-section of the reflectivity structure produced by convective plumes at Wallops Island, Virginia, USA (from Rowland and Arnold, 1975)

summer afternoon in June when the solar radiation is intense convective plumes were rising to altitudes of ⩾500 m.

The boundaries, and especially the tops, of plumes have higher reflectivity because the refractive index irregularities, which are responsible for scattering the microwaves emitted by the radar, are more intense at the boundary between the plume's moist air and the relatively dry air into which the plume penetrates. This, in turn, is because the spatial moisture gradient is the highest at this boundary. Turbulent mixing along this boundary generates irregularities of many different spatial scales or wavelengths, but it is the scales of a half of the radar wavelength that principally scatter the microwaves; this phenomenon is called stochastic Bragg scatter (Doviak and Zrnic', 1993, Chapter 11).

11.2.3 K-H waves

Clear air turbulence, often called CAT in the aeronautical community, is usually generated by instabilities associated with strong vertical shear of the wind in zones of relatively strong static stability. The instability generates Kelvin–Helmholtz (K-H) waves if the gradient Richardson number, defined as

$$R_g \equiv \frac{g\dfrac{\partial \langle \theta \rangle}{\partial z}}{T_\ell \left| \dfrac{d\langle \mathbf{v}_h \rangle}{dz} \right|^2} \tag{11.1}$$

falls below 1/4. In this equation g is the acceleration due to gravity, $d\langle\theta\rangle/dz$ is the vertical (z) gradient of the expected ($\langle\bullet\rangle$ denotes ensemble or time average) potential temperature θ, T_ℓ is the temperature of the layer in which the wave evolves, and $\mathbf{v}_h$ is the horizontal wind vector.

Carefully controlled laboratory experiments with statically stable shear flow in aqueous solutions (Thorpe, 1969) and wind tunnels (Scotti and Corcos, 1969) have established convincingly that disturbances become dynamically unstable if R_g is reduced to 1/4, at which value K-H waves appear with a wavelength having a maximum growth rate. Comparison of typical values of the variables in the laboratory, ocean and atmosphere shows that the wavelength for maximum growth rate is $\sim 2\pi d\sqrt{2}$, where d is the thickness of the sheared layer (Stoeffler, 1972). Most interestingly, Thorpe's data suggest how fine-scale three-dimensional turbulence might be generated from large-scale two-dimensional waves. It is the fine-scale irregularities of refractive index generated by turbulence that are responsible for scattering of the radar waves. An example of a large-amplitude (~1.3 km) and long-wavelength (~3 km) K-H wave observed with radar is shown in Fig. 11.4. Turbulence is generated as the K-H waves roll up, producing a spiralling layer of fluid, and break. The transition to turbulence results from gravitational instability as the denser air becomes superimposed over the lighter air in the spiralling structure of the breaking wave. The spiralling structure generates patches of turbulence which are then elongated in the shear flow and coalesce, eventually forming a quasi-uniform layer of turbulence and refractive index irregularities.

Figure 11.4 also shows some relatively thin layers of reflectivity which are possibly refractive index irregularities associated with K-H waves of much smaller amplitudes and shorter wavelengths generated within thinner regions of sheared flow. These waves have not been resolved by the radar. A detailed view of such shallow-layer K-H waves are shown in Fig. 11.5. The wave in Fig. 11.5c occurs at 450 m above ground, and has small amplitude (~0.1 km) as well as short wavelength (estimated to be <0.6 km). These waves were observed as they advected over a radar having a fixed vertical beam. Because of the lower range (height) of observation, the horizontal resolution at the wave height in Fig. 11.5 is about three times better than that in Fig. 11.4.

11.3 Detection of turbulence with Doppler radar

As was discussed in Chapter 6, airborne and ground-based Doppler weather radars normally map fields of reflectivity, Doppler-derived radial velocity (i.e. the component of wind velocity either towards or away from the radar), and the width of the Doppler spectrum. These three Doppler spectral moments are related essentially to the intensity of precipitation, the wind, and the turbulence, respectively. The signal processor of the Doppler radar calculates these moments for each resolution volume along the beam. The location of

Atmospheric turbulence and its detection by radar 381

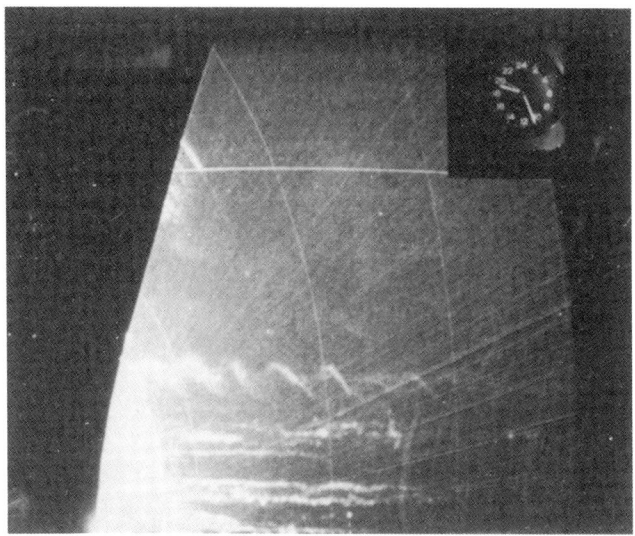

Figure 11.4 Range–height indicator (RHI) photograph at 120° azimuth taken at 14:26 Eastern Standard Time on 17 March 1969 with the 10.7 cm radar at Wallops Island, VA, USA. The height mark is at 12.2 km and the range marks are at 9.3 km intervals. More than seven separate horizontally stratified clear-air layers are visible below 6 km. Note the large-amplitude wave in the top layer. (Courtesy Jack Howard, NASA Wallops Island, VA, USA)

each resolution volume is determined by the sampling gates in the radar signal processor. The scatterers within each volume are weighted laterally by the beam pattern and radially (i.e. along the range) by the transmitted pulsewidth and the frequency response of the receiver. Many samples of the echo signal, usually several tens and sometimes over a hundred, are used to obtain reliable spectral estimates (see Section 6.6.3).

A typical Doppler spectrum for a resolution volume within a thunderstorm is plotted in Fig. 11.6. The spectral data indicated by crosses were obtained from a Fourier transform of 64 samples of a time series of complex video signals consisting of the in-phase and quadrature phase components, weighted with a von Hann window (Doviak and Zrnic', 1993). The unambiguous or Nyquist velocity $v_u = \lambda/4T_s$ is determined by the sample spacing T_s, equal to the radar pulse repetition time, and the radar wavelength λ. In this example $T_s = 922$ μs and $\lambda = 0.1052$ m, so $v_u = 28.5$ m s^{-1}; therefore the spacing of the spectral coefficients is 0.9 m s^{-1}. Both autocovariance and spectral analysis (with a threshold 15 dB below the spectral peak) were used to estimate the mean velocity and spectrum width (Doviak and Zrnic', 1993, Chapter 6). The mean Doppler velocity and spectrum width computed from the autocovariance are $\hat{v} = 14.9$ m s^{-1} and $\hat{\sigma}_v = 2.5$ m s^{-1}, where the ^ symbol

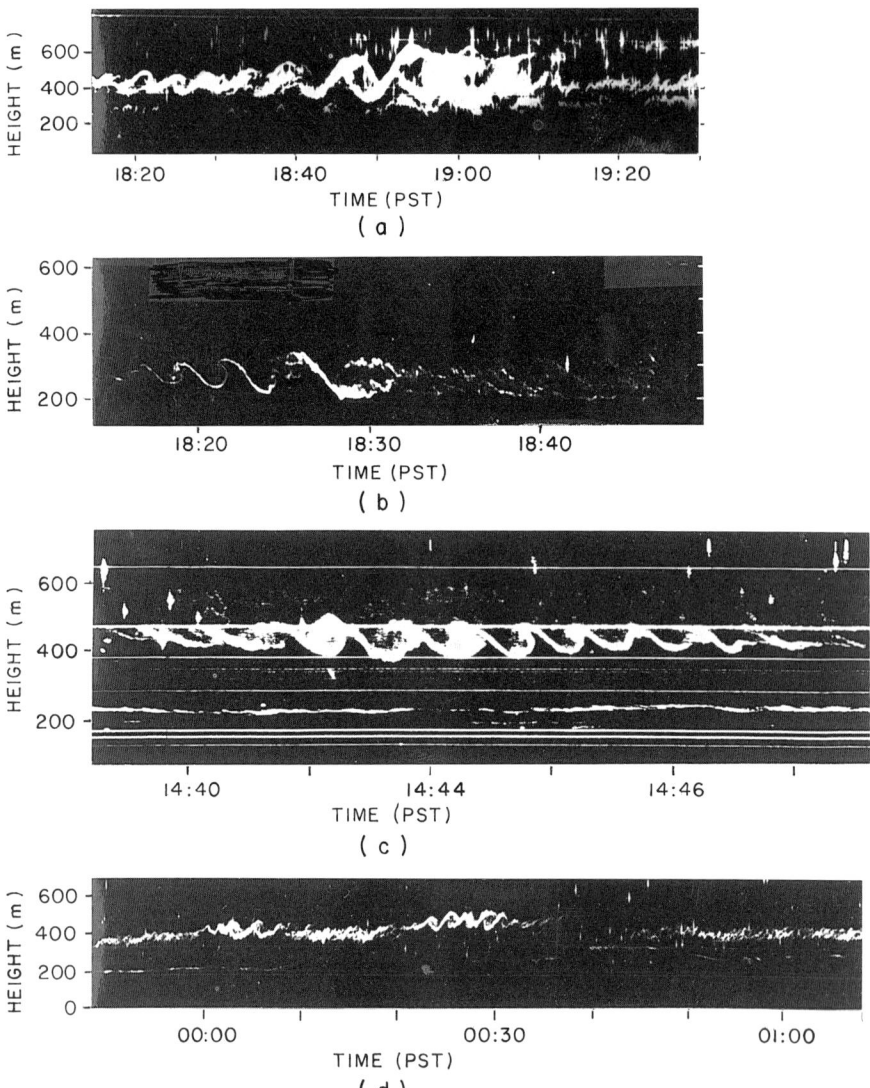

Figure 11.5 Wave perturbations and thin scattering layers observed by the FM-CW radar at San Diego. The remarkable resolution of this radar is evident from the vertical scale (a) 28 September 1971; (b) 6 August 1969; (c) 24 August 1970; (d) 14 July 1979 (from Gossard and Hooke, 1975)

designates an estimate. The Fourier method yields $\hat{v} = 15.4$ m s^{-1} and $\hat{\sigma}_v = 2$ m s^{-1}. These estimates were made from 64 complex echo signal samples obtained during a time period of 0.06 s. The uncertainty or standard deviation (SD) in the estimates was calculated using the theory given by

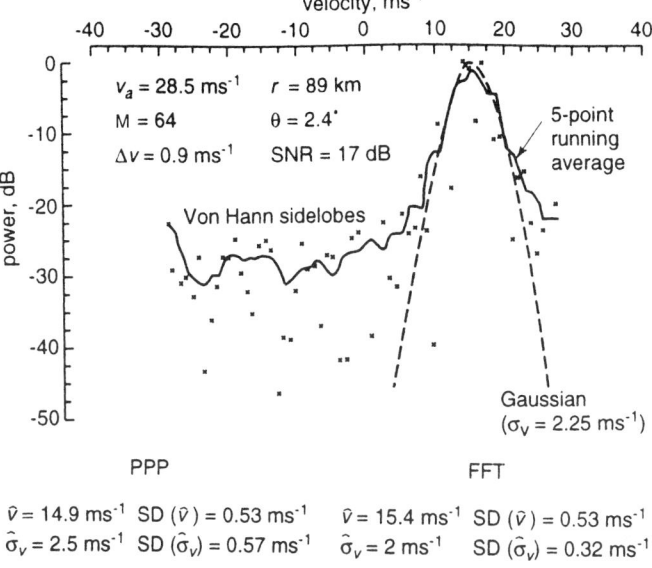

Figure 11.6 Doppler spectrum associated with a resolution volume in a storm (from Doviak and Zrnic', 1993)

Doviak and Zrnic' (op. cit.). The SDs shown in Fig. 11.6 are typical of modern Doppler weather radars, and the accuracy is adequate for most measurements in storms.

The radial velocity spectrum width σ_v, the square root of the second spectral moment about the mean velocity, is a function both of radar system parameters (antenna beamwidth, receiver bandwidth and transmitted pulsewidth) and of meteorological parameters that describe the distribution of hydrometeor density and velocity within the resolution volume. The parameter σ_v is usually associated with turbulence. This is because turbulence produces random radial motion of scatterers relative to the radar, which broadens the spectra. As discussed in Section 6.6.1, there are several other causes of spectral broadening. However, in most practical situations the dominant contribution to spectrum width comes from turbulence, with a component arising from the wind shear occurring within or across the resolution volume, which may be appreciable if the resolution volume is located in a high-shear zone. It is therefore a common practice to use the radial velocity spectrum width itself as an indicator or even a measure of turbulence, with appropriate corrections applied if significant inter-resolution-volume shear exists. We now proceed to analyse the way the physical process of turbulence is related to the radar-measured radial velocity spectrum width, and subsequently relate turbulence parameters with aviation hazard potential.

11.4 Statistical theory of turbulence

To relate radar measurements to turbulence, it is necessary to introduce elements of the statistical theory of turbulence. This theory establishes a quantitative relationship between the Doppler spectrum width and turbulence parameters including the spatial distribution of mean Doppler velocities.

The point-to-point changes in the actual wind velocities in a turbulent field are not easily determined or described. However, the properties of the wind field can be described in terms of statistical parameters which can be determined by turbulence theories.

A random variable that is a function of position $\mathbf{r}$ and time t defines a *random field*. If the random variable is a scalar quantity such as the refractive index $n(\mathbf{r}, t)$, the field is called a *scalar field*. When the random variable is a vector, it forms a *vector field*. Thus, the velocities in a turbulent flow form a random vector field such that at any point $\mathbf{r}$ (and time t) the velocity vector is a random variable $\mathbf{v}(\mathbf{r}, t)$.

A random field is said to be statistically *homogeneous* if its expected value is spatially uniform and if its correlation function between a pair of points $\mathbf{r}, \mathbf{r}'$ does not change when $\mathbf{r}$ and $\mathbf{r}'$ are both displaced by the same amount in the same direction. Thus, for a homogeneous random field in which the expected value (i.e. the component velocity associated with ordered flow) has been removed,

$$\langle v_i(\mathbf{r})\rangle \equiv 0 \qquad (11.2)$$

and the correlation function

$$R_{ij}(\mathbf{r}, \mathbf{r}') \equiv \langle v_i(\mathbf{r})v_j(\mathbf{r}')\rangle \qquad (11.3)$$

of two zero-mean velocity components v_i, v_j depends only on the distance $\boldsymbol{\rho} = \mathbf{r}' - \mathbf{r}$ between the two field points. The index $i = 1, 2$ or 3 identifies the velocity component along one of the three orthogonal directions. The components are in general time-dependent.

If the correlation does not depend on the direction of the *separation vector* $\boldsymbol{\rho}$, the field is said to be *isotropic*. The concept of isotropy is significantly different for vector fields than for scalar fields, as we now illustrate. Consider the correlation of the velocity at $\mathbf{r}$ with that at $\mathbf{r}'$, and suppose that the separation vector $\boldsymbol{\rho}$ is allowed to point in any direction from $\mathbf{r}$. Construct a coordinate system that is rigidly attached to $\boldsymbol{\rho}$ and has, without loss of generality, its axes parallel and perpendicular to $\boldsymbol{\rho}$ (Fig. 11.7). If the vector field is isotropic the correlations of the various components in the separation vector co-ordinates (e.g. $\langle v_\ell(\mathbf{r}')v_t(\mathbf{r})\rangle$, where v_ℓ, v_t are the longitudinal and transverse velocity components relative to $\boldsymbol{\rho}$) are independent of the orientation of $\boldsymbol{\rho}$. Although the velocity component correlations in the co-ordinate system tied to $\boldsymbol{\rho}$ (i.e. the rotating co-ordinate system) are independent of the direction of $\boldsymbol{\rho}$, it can be deduced that *the correlation of components in the*

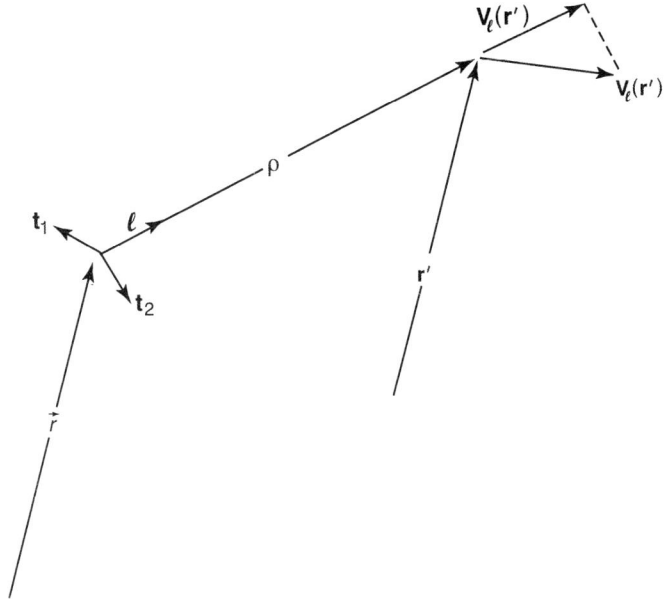

Figure 11.7 Co-ordinate system of orthogonal unit vectors ℓ, $\mathbf{t}_1$, $\mathbf{t}_2$ rigidly tied to the separation vector $\boldsymbol{\rho}$. The projection $V_\ell(\mathbf{r}')$ of the velocity vector $\mathbf{v}(\mathbf{r}')$ is onto the co-ordinate axis ℓ, which is parallel to $\boldsymbol{\rho}$

nonrotated frame (e.g. u, v, w, the horizontal and vertical wind components) do, in general, depend on the direction of $\boldsymbol{\rho}$. For example, for an isotropic field, the correlation $\langle u(\mathbf{r})u(\mathbf{r}')\rangle$ does depend on the direction of $\boldsymbol{\rho}$. Although a random vector field can be homogeneous but not isotropic, it can be shown that an isotropic field must be homogeneous.

For isotropic random vector fields, the correlation tensor is, in general,

$$R_{ij}(\mathbf{r}, \mathbf{r}+\boldsymbol{\rho}) = R_{ij}(\boldsymbol{\rho}) = P(\rho)\delta_{ij} + Q(\rho)\rho_i\rho_j/\rho^2 \qquad (11.4)$$

(Tatarskii, 1971), which does not depend on the location of $\mathbf{r}$, but does depend on the direction and magnitude of $\boldsymbol{\rho}$. In eqn. 11.4 i,j refer to the coordinates in the fixed frame, δ_{ij} is the Kronecker delta function ($\delta_{ij}=0$, $i\neq j$; $\delta_{ij}=1$, $i=j$), and ρ_i is the projection of $\boldsymbol{\rho}$ on the ith axis. $P(\rho)$ and $Q(\rho)$ can be any functions of the magnitude ρ of the separation vector $\boldsymbol{\rho}$. If $\boldsymbol{\rho}$ lies along the kth co-ordinate of the fixed frame, it can be deduced from eqn. 11.4 that

$$R_{ii} = R_{jj} = P(\rho) \qquad (11.5a)$$

which is then called the *transverse correlation function* $R_{tt}(\rho)$, and

$$R_{kk} = P(\rho) + Q(\rho) \qquad (11.5b)$$

is called the *longitudinal correlation* $R_{\ell\ell}$. Then eqn. 11.4 can be expressed as

$$R_{ij}(\boldsymbol{\rho}) = (1/\rho^2)(R_{\ell\ell} - R_{tt})\rho_i\rho_j + R_{tt}\delta_{ij} \qquad (11.6)$$

The Fourier transform of the correlation function of an isotropic velocity field to the wavenumber **K** space[2] gives the *spectral density tensor,*

$$\Phi_{ij}(\mathbf{K}) = \frac{1}{(2\pi)^3} \int R_{ij}(\mathbf{\rho}) \, e^{-j\mathbf{K}\cdot\mathbf{\rho}} \, dV_\rho \tag{11.7}$$

Its inverse retrieves the correlation function

$$R_{ij}(\mathbf{\rho}) = \int \Phi_{ij}(\mathbf{K}) \, e^{j\mathbf{K}\cdot\mathbf{\rho}} \, dV_K \tag{11.8}$$

where V_ρ and V_K are volumes in separation distance $\mathbf{\rho}$ space and wavenumber **K** space, respectively, and the integrals are over these volumes. The magnitude $K \equiv 2\pi/\Lambda$, where Λ is the Fourier wavelength or scale of the velocity perturbation. For an isotropic vector field, the spectral density tensor, like the correlation tensor, must be independent of rotation, so it can also be expressed in a form analogous to eqn. 11.4,

$$\Phi_{ij}(\mathbf{K}) = F(K)\delta_{ij} + G(K) K_i K_j / K^2 \tag{11.9}$$

or to eqn. 11.6,

$$\Phi_{ij}(\mathbf{K}) = (1/K^2)(\Phi_{\ell\ell} - \Phi_{tt}) K_i K_j + \Phi_{tt}\delta_{ij} \tag{11.10}$$

where $\Phi_{\ell\ell}(K)$ and $\Phi_{tt}(K)$ are called the longitudinal and transverse spectral densities. They are related to $R_{\ell\ell}(\mathbf{\rho})$ and $R_{tt}(\mathbf{\rho})$ through eqns. 11.6–11.10 and are not simply Fourier transform pairs of their corresponding correlation functions [e.g. $\Phi_{\ell\ell}(K) \neq F\{R_{\ell\ell}(\mathbf{\rho})\}$]. Panchev (1971, p. 102) gives the relations between $\Phi_{\ell\ell}(K)$, $\Phi_{tt}(K)$ and $R_{\ell\ell}(\mathbf{\rho})$, $R_{tt}(\mathbf{\rho})$ for isotropic vector fields. These relations show, for example, that $\Phi_{tt}(K)$ is a function of both $R_{\ell\ell}(\mathbf{\rho})$ and $R_{tt}(\mathbf{\rho})$.

If the flow is incompressible, the continuity equation puts an additional constraint on the random vector field, resulting in

$$R_{tt} = \frac{1}{2\rho} \frac{d}{d\rho}(\rho^2 R_{\ell\ell}) \tag{11.11}$$

The condition of incompressibility results in $\Phi_{\ell\ell}(K) \equiv 0$ (Panchev, 1971, p. 108), so that eqn. 11.10 reduces to

$$\Phi_{ij}(\mathbf{K}) = \left(\delta_{ij} - \frac{K_i K_j}{K^2}\right) \frac{E(K)}{4\pi K^2} \tag{11.12}$$

---

[2] There are two definitions of wavenumber **K**: one in which $K_i = (1/\Lambda_i)$ m^{-1}, where Λ_i is the wavelength in the *i*th direction, and the other in which $K_i = (2\pi/\Lambda_i)$ rad m^{-1}. Throughout this section we use the latter definition

where

$$E(K) = 4\pi K^2 \Phi_{ii}(K) \qquad (11.13)$$

is the one-dimensional spectral density that characterises isotropic incompressible turbulent flow. $E(K)\,dK$ represents the contribution to the total kinetic energy per unit mass (or a half of the total velocity variance) from waves with wavenumbers in the interval between K and $K+dK$ in the turbulent medium.

Measurements of three-dimensional spectra in the past were quite impractical because the available sensing instruments, such as those mounted on aircraft, could at best estimate the one-dimensional counterpart $S_{ij}(K_\ell)$, where K_ℓ is the wavenumber along the path ℓ of the aircraft, and $S_{ij}(K_\ell)$, the variance of velocities per unit wavenumber, is the Fourier transform of $R_{ij}(\ell)$ for ℓ displacements along the path. In contrast, radars can obtain a volume of velocity data in a very short time (i.e. of the order of a few minutes). Even though the three-dimensional spectrum of radial velocity variance can be obtained from radar data, it is considerably simpler to compute spectra along lines; such spectra are the variances of velocities per unit wavenumber, and serve as a measure of turbulence intensity or velocity variance. Moreover, one-dimensional spectra provide a convenient visual picture of the wave components along the line. By applying the formula $2\pi\delta(K) = \int e^{jK\rho}\,d\rho$ (which essentially states that a Dirac delta function is the inverse Fourier transform of a flat spectrum) to eqn. 11.7, the following relation can be derived:

$$S_{ij}(K_1) = \frac{1}{2\pi} \int_{-\infty}^{+\infty} R_{ij}(\rho_1, 0, 0)\, e^{-jK_1\rho_1}\, d\rho_1 = \int\int_{-\infty}^{+\infty} \Phi_{ij}(\mathbf{K})\, dK_2\, dK_3 \qquad (11.14)$$

where the separation vector $\boldsymbol{\rho}$ is now assumed to be directed along the co-ordinate axis for which $i=1$. Equation 11.14 can be verified by starting from the right-hand side, substituting for the integrand from eqn. 11.7 and performing the multiple integration operations over appropriate limits (Panchev, 1971).

For incompressible flow, the one-dimensional longitudinal and transverse spectra of isotropic vector fields are related to $E(K)$ through eqns. 11.12–11.14 as

$$S_\ell(K_1) = \frac{1}{2} \int_{K_1}^{\infty} \left(1 - \frac{K_1^2}{K^2}\right) \frac{E(K)}{K}\, dK \qquad (11.15)$$

and

$$S_t(K_1) = \frac{1}{4} \int_{K_1}^{\infty} \left(1 + \frac{K_1^2}{K^2}\right) \frac{E(K)}{K} dK \qquad (11.16)$$

where $S_\ell(K_1) = S_{11}(K_1)$ and $S_t(K_1) = S_{22}(K_1) = S_{33}(K_1)$. Note that any wavenumber K_i can replace K_1 in eqns. 11.15 and 11.16, because of the fact that S_ℓ and S_t are independent of direction if turbulence is isotropic.

11.4.1 Correlation and spectral functions in the inertial subrange

Here the functional forms of the correlation and the spectra for longitudinal and transverse components of turbulence are discussed. Kolmogorov in 1941 used dimensional analysis to derive the correlation function for an isotropic velocity field (Tatarskii, 1971). Briefly, he hypothesised a range of eddy sizes or scales in which there is no creation or dissipation of energy; energy only cascades (flows) from larger scales, where the source of turbulent energy resides, to smaller scales as eddies fragment. For example, large amplitude waves generate turbulence if they break, thus generating smaller-scale perturbations from the larger-scale waves as discussed in Section 11.23. Under equilibrium conditions this continuous flux of energy numerically equals the energy dissipation rate ε, which is the rate at which the turbulent energy is ultimately transferred to heat the air due to viscosity. Dimensional analysis then reveals that the longitudinal correlation function $R_{\ell\ell}(\rho)$ must have the form (Tatarskii, 1971, p. 54)

$$R_{\ell\ell}(\rho) = R(0)\left[1 - \left(\frac{\rho}{\rho_{ol}}\right)^{2/3}\right] \qquad (11.17)$$

if $\rho_i < \rho < \rho_{ol}$. In this inequality, ρ_i is the inner scale which corresponds to separations so small (typically millimetres to centimetres in the lower atmosphere) that viscous forces cannot be ignored, and ρ_{ol} is a separation corresponding to the outer scale or the largest scales (usually a few hundred metres in storms, much smaller in fair weather) for which the velocity perturbations are isotropic. The range of ρ satisfying the inequality defines the inertial subrange wherein turbulent energy is transferred without loss from large to small eddies (Tennekes and Lumley, 1972). $R(0)$ in eqn. 11.17 is the velocity variance of all the turbulence scales in the inertial range. Using eqn. 11.11 it can be shown that

$$R_{tt}(\rho) = R(0)\left[1 - \left(\frac{\rho}{\rho_{ol}}\right)^{2/3}\right] \qquad (11.18)$$

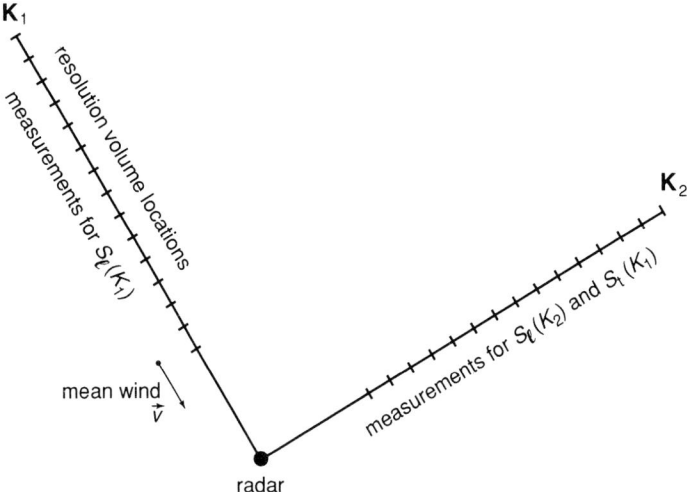

Figure 11.8 Geometry of data collection with a single Doppler radar to compute $S_\ell(K_1)$, $S_\ell(K_2)$ and $S_t(K_1)$

where, because of eqn. 11.11 it can be shown that

$$\rho_{ot} = \left(\frac{3}{4}\right)^{3/2} \rho_{ol} \qquad (11.19)$$

The correlation function of the longitudinal and transverse velocity components in planes parallel to the earth's surface sometimes exhibits two-thirds-power behaviour with respect to scales much larger than ρ_{oi} (Vinnichenko and Dutton, 1969; Doviak and Berger, 1980), where the subscript i is either ℓ or t, designating the outer scales of longitudinal or transverse correlation functions, respectively.

The velocity variance spectrum for the inertial subrange, where $K\rho_{oi} \gg 1$, is (Doviak and Zrnic', 1993)

$$S_i(K) = \frac{C_i^2 \Gamma(5/6) K^{-5/3}}{3 \times 2^{1/3} \Gamma(2/3) \sqrt{\pi}} \qquad (11.20)$$

where

$$C_i^2 = \frac{3 \times 2^{1/3} \Gamma(2/3)}{\Gamma(1/3)} R(0) \rho_{oi}^{-2/3} \qquad (11.21)$$

is a measure of the intensity of turbulence having scale sizes within the inertial subrange. Thus, the one-dimensional two-sided spectra of the velocity components parallel to and perpendicular to ρ have a $K^{-5/3}$ dependence.

To help visualise how spectra are derived from observations with radar, Fig. 11.8 is presented. The means (i.e. averages over each of the resolution

390 Aviation weather surveillance systems

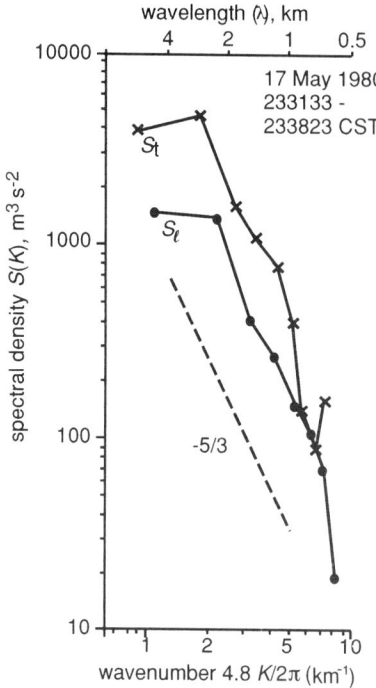

Figure 11.9 Spatial spectral densities. S_t is the transverse and S_ℓ is the longitudinal spectrum of vertical velocity variance. The dashed line represents the -5/3 law expected from isotropic turbulence in the inertial subrange

volumes) of radial velocities are obtained by pointing the radar beam along and perpendicular to the direction of the mean wind (this is typically the time average of the horizontal wind at a fixed height), so that spectral analysis of data along these two directions (K_1 is along the mean wind direction and K_2 is orthoginal to it) yields $S_\ell^f(K_1)$ and $S_\ell^f(K_2)$, where the superscript f signifies that the spectra are filtered by the radar's weighting function (see the heading below for detailed discussion). A third spectral component $S_t^f(K_1)$ can be computed from the time series of mean radial velocities at any one of the range locations along the beam when it is perpendicular to the mean wind. This computation requires the assumption that turbulent eddies are advected (i.e. translated) by the mean wind (the so-called *Taylor hypothesis*); it suffices to record the temporal change of the mean radial velocity, transform time to space, and perform spectral analysis.

Figure 11.9 shows sample spectra of atmospheric air motions measured by a Doppler radar operated with its beam fixed in a vertically upward-looking direction (i.e. corresponding to K_2). The longitudinal and transverse spectra, $S_\ell^f(K_2)$ and $S_t^f(K_1)$, respectively, were calculated using Doppler velocity measurements along a line perpendicular to the mean wind as depicted in

Fig. 11.8 (Brewster and Zrnic', 1986). It is seen from Fig. 11.9 that the smaller scales of turbulence follow the 5/3 law predicted by turbulence theory, as indicated by eqn. 11.20, and that the transverse spectrum is broader than the longitudinal one. This is in accordance with eqn. 11.19, which predicts a narrower correlation function along the transverse direction than the longitudinal, noting the inverse relationship between the widths of the spectrum and the correlation function.

11.4.2 Filtering by the radar's weighting function

The spectra of turbulence measured with a radar are filtered by the weighting functions associated with the radar's finite angular and range resolutions, and hence the variance of the mean radial velocities (i.e. the averages over each resolution volume) is attenuated. Velocity variance is a measure of the intensity of turbulence. If the range r to a given resolution volume is large compared to the resolution volume's dimensions, and the reflectivity is uniform, the finite angular and range resolutions of the radar give, for each resolution volume, a spatially averaged mean Doppler velocity (Doviak and Zrnic', 1993; Section 5.2)

$$\bar{v}(\mathbf{r}) \equiv \int v(\mathbf{r}_1) I_n(\mathbf{r}_1 - \mathbf{r}) \, dV_1 \qquad (11.22)$$

where $dV_1 = r_1^2 \, dr_1 \, d\phi_1 \, d\theta_1$ is an elemental volume, $\mathbf{r}_1$ is the vector distance from the radar to a general point (r_1, θ_1, ϕ_1), and $\mathbf{r}$ is the vector distance to the centre of the resolution volume. I_n in eqn. 11.22 is the normalised weighting function, defined as

$$I_n(\mathbf{r}_1 - \mathbf{r}) \equiv \frac{I(\mathbf{r}_1 - \mathbf{r})}{\int I(\mathbf{r}_1 - \mathbf{r}) \, dV_1} \qquad (11.23)$$

in which

$$I(\mathbf{r}_1 - \mathbf{r}) = \frac{C f^4(\theta_1 - \theta, \phi_1 - \phi) \, | W(r_1 - r) |^2}{L_a^2(\mathbf{r}_1) r_1^4} \qquad (11.24)$$

C is a radar constant, $f^4(\theta_1 - \theta, \phi_1 - \phi)$ is the angular weighting function of the antenna (i.e. the normalised two-way power gain of the radiation pattern), $W(r_1 - r)$ is the convolution of the receiver's impulse response and the transmitted waveform, and L_a^2 is the two-way atmospheric attenuation or loss factor of the radar signals due to scattering and absorption (see Section 6.3.1).

The spectrum $\Phi_v(\mathbf{K})$ of point *radial* velocities can be related to the spectral density tensor $\Phi_{ii}(\mathbf{K})$. If the size of the resolution volume is small compared to the range, the radial velocities can be considered parallel to K_1, the wavenumber along the beam axis, everywhere within the resolution volume. Then

$$\Phi_v(\mathbf{K}) = \Phi_{\ell\ell}(\mathbf{K}) \qquad (11.25)$$

Because eqn. 11.22 is a convolution product, the spatial spectrum $\bar{\Phi}_v(\mathbf{K})$ of averaged radial velocities equals

$$\bar{\Phi}_v(\mathbf{K}) = (2\pi)^6 \Phi_v(\mathbf{K}) \,|F(\mathbf{K})|^2 \qquad (11.26)$$

where $F(\mathbf{K})$ is the Fourier transform of the weighting function I_n. A three-dimensional Gaussian shape for $|F(\mathbf{K})|^2$, given by the expression

$$|F(\mathbf{K})|^2 = (2\pi)^{-6} e^{-(K_2^2 + K_3^2)r^2\sigma_\theta^2 - K_1^2\sigma_r^2} \qquad (11.27)$$

is a good approximation to most weighting functions. In eqn. 11.27 σ_θ^2 is the second central moment of the two-way antenna weighting function $f^4(\theta_1 - \theta, \phi_1 - \phi)$, and σ_r^2 is the second central moment of the range weighting function $|W(r_1 - r)|^2$ (Doviak and Zrnic', 1993).

The combination of eqns. 11.12, 11.15 or 11.16, and 11.26 produces the measured (i.e. filtered by the weighting function) spectra $S_\ell^f(K_1)$ and $S_t^f(K_1)$ of the longitudinal and transverse velocities as

$$S_\ell^f(K_1) = \frac{(2\pi)^6}{4} \int_{K_1}^{\infty} \left(1 - \frac{K_1^2}{K^2}\right) \frac{E(K)}{K} |F(K_1, K')|^2 \, dK \qquad (11.28)$$

and

$$S_t^f(K_1) = \frac{(2\pi)^6}{4} \int_{K_1}^{\infty} \left(1 - \frac{K_1^2}{K^2}\right) \frac{E(K)}{K} |F(K_1, K')|^2 \, dK \qquad (11.29)$$

where we have explicitly separated K_1 from $K' = (K^2 - K_1^2)^{1/2}$ in $F(K_1, K')$. Such dependence on two wavenumbers results from circularly symmetric antenna beams.

Using eqns. 11.15 or 11.16, the energy spectrum $E(K)$ of isotropic turbulence can be obtained from either of the two one-dimensional radar-measured spectra for scale sizes whose observed variance is not strongly attenuated by the weighting function. A comprehensive illustration of the effects of filtering on the longitudinal and transverse one-dimensional spectra of turbulence is given by Srivastava and Atlas (1974). They consider a Kolmogorov–Obukhov energy spectrum

$$E(K) = \begin{cases} K^{-5/3} & \text{for } K \geq K_0 \\ 0 & \text{for } K < K_0 \end{cases} \qquad (11.30)$$

with a cutoff at wavenumber K_0.

Then, from eqn. 11.15, the one-dimensional spectrum of point velocities is

$$S_\ell(K_1) = \begin{cases} \dfrac{9}{55}\pi K_1^{-5/3} & \text{for } K_1 \geq K_0 \\ 0.3\pi K_0^{-5/3}\left[1 - \dfrac{5}{11}\left(\dfrac{K_1}{K_0}\right)^2\right] & \text{for } K_1 < K_0 \end{cases} \quad (11.31)$$

It is interesting to note that, even though the three-dimensional spectrum $E(K)$ has no energy for $K < K_0$, the one-dimensional spectrum does have values for $K < K_0$. The energy at wavenumbers smaller than the cutoff (i.e. $K < K_0$) is contributed by the scales (K_1, K_2, K_3) for which $K^2 > K_0^2$, i.e. by wavenumbers which are larger than K_0 but whose projection along the K_1 axis (the antenna beam axis) is smaller than K_0.

Filtered spectra are obtained from eqn. 11.28, on substitution of eqn. 11.27, as

$$S_\ell^f(K_1) = \frac{1}{2}e^{-K_1^2(\sigma_r^2 - r^2\sigma_\theta^2)}\int_a^\infty \left(1 - \frac{K_1^2}{K^2}\right)\frac{E(K)}{K}e^{-K^2 r^2 \sigma_\theta^2}\,dK \quad (11.32)$$

where

$$a = \begin{cases} K_1 & \text{for } K_1 \geq K_0 \\ K_0 & \text{for } K_1 < K_0 \end{cases}$$

An interesting fact to observe here is that the filtered one-dimensional spectrum is attenuated in a direction along which there may have been no filtering, as apparent from eqn. 11.32 by noting that there is degradation in the K_1 direction even when $\sigma_r = 0$. This is due to filtering in the orthogonal (i.e. K_2, K_3) directions and the superposition (integration) of the resulting attenuated three-dimensional spectral density. The effect of filtering on radar-measured turbulent spectra has been illustrated graphically by Srivastava and Atlas (1974).

An illustration of the one-dimensional spatial spectrum of the earth's boundary layer winds (i.e. winds close to the ground) is shown in Fig. 11.10. The spectra are calculated from wind fields synthesised from Doppler measurements using two radars spaced 40 km apart. The angle between the radar beams at the centre of the area of analysis is 65°. In this example, taken from Reinking et al. (1981), the spatial spectra are multiplied by the wavenumber K_y of the y wind component (i.e. the component of air motion transverse to the mean wind direction) across convective rolls which were observed on a particular day. Plotted in the Figure are the radar-derived

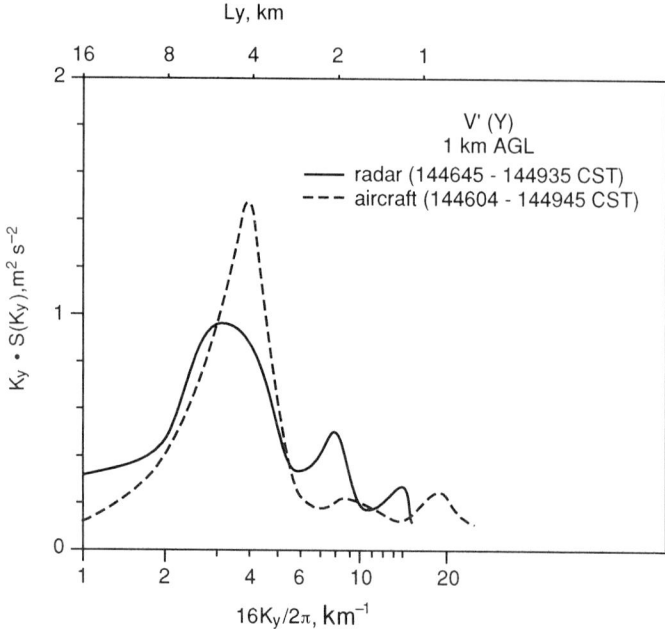

Figure 11.10 Example of time-coincident spectra $S_\ell(K_y)$ (multiplied by K_y) of the crossroll wind component v' at 1 km AGL from dual-Doppler radar and gust probe measurements. The spectrum deduced by radar is an average of 16 one-dimensional spectra within the radar synthesis region at the time the aircraft was flying a path across the rolls (from Reinking et al., 1981)

spectrum and the spectrum obtained from wind fluctuation measurements using gust probes mounted on an aircraft whose flight path was in the y direction. Both the spectra exhibit peaks at wavelengths of ~4 km, which corresponds to the spacing of the convective rolls. The agreement between radar and aircraft measurements is reasonable, considering the basic difference between the two measurement processes. The gust probe on the aircraft, with a sampling volume of only a few centimetres on a side, essentially makes point measurements along a line, whereas the radar data are averaged over a few hundred metres, the order of the dimensions of the resolution volume, along the corresponding line. This averaging explains why the peak of the radar-derived spectrum is rather blunt, while that of the gust-probe-derived spectrum is sharp.

Reinking *et al.* (1981) also provide plots of the crossroll wind variance $\sigma_{v'}^2$ computed from radar and aircraft measurements. Here again, the wind field variance as measured by the gust probe was usually higher than the values computed from radar observations, because of the filtering effects of the much larger resolution volume of the radar.

Although the turbulence spectra measured by radar are not exactly the same as the 'true' spectra, as obtained from *in situ* point-measured wind velocities with instruments such as airborne gust probes, radar measurements can still serve as reliable indicators, or even measures, of the turbulence parameters. This is because the volume-averaging or filtering process, which is the main factor responsible for the difference between the two, is a known process amenable to modelling. The relationship between the two methods of wind variance measurement is discussed below.

11.4.3 Variance of point and average velocities

To define the relationship between the variance of velocities at a point and the Doppler spectrum width measured by radar, let the variance of the velocity v at a point be denoted by σ_p^2. This is obtained from the ensemble average as

$$\sigma_p^2 = \langle v^2 \rangle - \langle v \rangle^2 \qquad (11.33)$$

The Doppler spectrum width σ_t due to turbulence is given by Doviak and Zrnic' (1993, eqn. 5.51), under the assumption that no steady wind is present (i.e. ignoring the spectrum broadening due to wind shear of the ordered flow (Section 11.2)), as

$$\sigma_t^2 = \overline{v^2} - (\bar{v})^2 \qquad (11.34)$$

where the overbar denotes a spatial average of velocities weighted by $I_n(\mathbf{r}, \mathbf{r}_1)\eta(\mathbf{r}_1)$, where $\eta(\mathbf{r}_1)$ is the reflectivity. The variance of the mean Doppler velocity $\bar{v}$ is, by definition,

$$\sigma_{\bar{v}}^2 \equiv \langle (\bar{v})^2 \rangle - \langle \bar{v} \rangle^2 \qquad (11.35)$$

Note that $\sigma_{\bar{v}}^2$ does not include the variance associated with the statistical uncertainty of the estimate of $\bar{v}$. Assuming turbulence to be locally homogeneous and the weighting function symmetrical, and recognising that ensemble and spatial averaging operations are commutative, eqn. 11.35 can be rewritten as

$$\sigma_{\bar{v}}^2 = \langle (\bar{v})^2 \rangle - \langle v \rangle^2 \qquad (11.36)$$

Finally, the ensemble average of eqn. 11.34, added to eqn. 11.36, after commuting ensemble and spatial averages, produces

$$\sigma_p^2 = \langle \sigma_t^2 \rangle + \sigma_{\bar{v}}^2 \qquad (11.37)$$

which indicates that the variance of the velocity at a point in a wind field is equal to the sum of the ensemble average of σ_t^2 and the variance of the turbulent velocities weighted by $I_n(\mathbf{r}, \mathbf{r}_1)$.

This very general result requires turbulence to be locally homogeneous, although not necessarily isotropic. In addition to being proportional to the turbulent kinetic energy, the two variances σ_t^2 and $\sigma_{\bar{v}}^2$ have relative magnitudes that depend on the weighting function $I_n(\mathbf{r}, \mathbf{r}_1)\eta(\mathbf{r}_1)$. These two variances describe how the kinetic energy is partitioned between subresolution-volume scales and scales larger than the resolution volume V_6.

If V_6 is small (i.e. the radar has high resolving power), $\langle \sigma_t^2 \rangle$ would be relatively small compared with $\sigma_{\bar{v}}^2$. Then the measure of turbulence dangerous to aircraft would more naturally be deduced from the spatial variability of $\bar{u}$. On the other hand, if V_6 is sufficiently large, $\langle \sigma_t^2 \rangle$ would be larger than $\sigma_{\bar{v}}^2$ and thus turbulence would best be gauged from measurements of σ_t. Weather radars have resolution that typically falls into the second category; hence turbulence is best inferred from measured spectrum width σ_v, although other contributions to σ_v might need to be accounted for in order to deduce σ_t (see eqn. 6.25). In thunderstorms, however, σ_v is principally due to σ_t as is shown in Section 11.7.

11.5 Doppler spectrum width and eddy dissipation rate

The eddy dissipation rate ε is a measure of the rate of energy transfer from larger to smaller scales of eddies, and eventual viscous dissipation as heat. It is an important parameter describing turbulence intensity. The various levels of turbulence, from the point of view of effects on aircraft, are often gauged by the value of ε as in Table 11.1 (Trout and Panofsky, 1969). The Doppler spectrum width σ_t due to turbulence is determined from measurements of σ_v, using eqn. 6.25. Usually, the only other term that may be significant, and even comparable to σ_t, is the spectrum width σ_s due to the shear of ordered flow. Equation 11.26 states that the spatial spectrum of the point Doppler velocity is filtered by the weighting function. This, together with eqn. 11.37, will be used to obtain a relationship between the Doppler spectrum width σ_t due to turbulence and the eddy dissipation rate ε under the assumption that the outer scale of turbulence is larger than the resolution volume.

First, we express the variances of point and averaged velocities in terms of their spectra:

$$\sigma_p^2 = \int \Phi_v(\mathbf{K}) \, dV_K \qquad (11.38)$$

Table 11.1 *The eddy dissipation rate as an indicator of turbulence level for aviation*

Value of eddy dissipation rate ε	Level of turbulence
$\varepsilon \geqslant 6.75 \times 10^{-2}$ m² s⁻³	Severe turbulence
$6.75 \times 10^{-2} \geqslant \varepsilon \geqslant 8.5 \times 10^{-3}$ m² s⁻³	Moderate turbulence
$8.5 \times 10^{-3} \geqslant \varepsilon \geqslant 3.0 \times 10^{-3}$ m² s⁻³	Light turbulence

$$\sigma_{\bar{v}}^2 = \int \Phi_{\bar{v}}(\mathbf{K}) \, dV_K \qquad (11.39)$$

Now, eqn. 11.37 shows that $\langle \sigma_t^2 \rangle$ is the difference between eqns. 11.38 and 11.39. After this difference is taken and $\Phi_{\bar{v}}(\mathbf{K})$ substituted from eqn. 11.26, we obtain the following formula, which connects σ_t to $\Phi_v(\mathbf{K})$:

$$\langle \sigma_t^2 \rangle = \int [1 - (2\pi)^6 |F(\mathbf{K})|^2] \Phi_v(\mathbf{K}) \, dV_k \qquad (11.40)$$

It is assumed that the reflectivity within the resolution volume is uniform (a good assumption for modern high-performance weather radars with narrow antenna beams and fine range resolution) and that $|F(\mathbf{K})|^2$ is given by eqn. 11.27.

To treat eqn. 11.40 further, we consider turbulence in an incompressible fluid, so that eqns. 11.12 and 11.13 are valid. Furthermore, we assume that the turbulent energy principally resides within the *inertial subrange*. Under these conditions the spectrum function $E(K)$ becomes

$$E(K) = A\varepsilon^{2/3} K^{-5/3} \qquad (11.41)$$

where A is a universal dimensionless constant between 1.53 and 1.68 (Gossard and Strauch, 1983, p. 262), and ε is the turbulent energy dissipation rate, normalised to unit mass.

To tie eqn. 11.41 to $\langle \sigma_t^2 \rangle$, it must be assumed that all $\langle \sigma_t^2 \rangle$ contributions come from velocity scales within the inertial subrange. This is quite true provided that contributions from turbulence scales larger than the radar resolution volume, which are not part of the inertial subrange, and the shear associated with ordered flow are removed (Istok and Doviak, 1986). Sinclair (1974) has observed the upper wavelength of the inertial subrange in a severe storm to vary from 150 to ~2000 m. It appeared to Sinclair that this variability is related to the storm intensity and the measurement altitude, i.e. the upper limit of the inertial subrange is largest in the upper half of the storm, where vertical velocities are usually largest and the turbulence is most intense. Such variability of the upper limit is not uncommon. Other investigators (Rhyne and Steiner, 1964; MacCready, 1962, 1964; Reiter and Burns, 1966; Reiter, 1970) have either shown or suggested that the upper limit of the inertial subrange may vary from 300 to 800 m, depending on the location of measurement. Figure 11.9 suggests that the outer scale of isotropic turbulence in a thunderstorm is larger than 1 or 2 km.

If the outer scale of the inertial subrange is much larger than the dimensions of the resolution volume, eqns. 11.12 and 11.41 can be substituted into eqn. 11.40 and the integration performed to obtain an expression giving ε as a function of $\langle \sigma_t^2 \rangle$ and the parameters describing $|F(\mathbf{K})|^2$ (Labitt, 1981; Gossard and Strauch, 1983). For a range resolution equal to or smaller than the radar beamwidth (i.e. $\sigma_r \leqslant r\sigma_\theta$), this expression is

$$\langle \sigma_t^2 \rangle = A\Gamma(2/3)(\varepsilon r \sigma_\theta)^{2/3} F\left(-\frac{1}{3}, \frac{1}{2}; \frac{5}{2}; 1 - \frac{\sigma_r^2}{r^2 \sigma_\theta^2}\right) \quad (11.42)$$

where $\Gamma(x)$ is the Gamma function of argument x, and F is the hypergeometric function, which is bounded between 0.918 and 1.

If the beamwidth of the radar is smaller than its range resolution (i.e. $r\sigma_\theta \leq \sigma_r$), then $\langle \sigma_t^2 \rangle$ has the same form as in eqn. 11.42, but with $r\sigma_\theta$ and σ_r interchanged. The hypergeometric function is then bounded by (Labitt, 1981)

$$\frac{27}{55} < F\left(-\frac{1}{3}, 2; \frac{5}{2}; 1 - \frac{r^2 \sigma_\theta^2}{\sigma_r^2}\right) < 1 \quad (11.43)$$

11.6 Eddy dissipation rates in thunderstorms

The turbulent kinetic energy dissipation rate is a useful quantity for characterising the intensity of turbulence within the inertial subrange. Taking a clue from the fact that the dissipation rate ε of the turbulent kinetic energy can be estimated from the Doppler spectrum width due to turbulence if and only if the largest scale of the inertial subrange is larger than the largest dimension of the radar resolution volume, Brewster and Zrnic' (1986) used a vertically pointing Doppler radar with a resolution finer than 100 m. They compared ε computed from spatial spectra with that from σ_t. The energy dissipation rates computed from spatial spectra of velocity fields agreed very well with the ε values computed from the Doppler spectral width.

However, often shear due to ordered flow and/or that due to the larger-scale turbulence (scales outside the inertial subrange and larger than the size of the radar resolution volume) are significant contributors to the overall spectrum width. These components need to be estimated from the spatial variation of the mean radial velocity $\bar{v}$. Then, to arrive at the component due to turbulence, one must subtract the contribution due to shear from the overall spectrum width (σ_v, eqn. 6.25). An estimate of the shear across the resolution volume can be obtained by least-squares fitting a plane surface that relates the mean radial velocity to the azimuth and elevation position surrounding the resolution volume. Once the shears are obtained, the spectrum width due to them can be readily computed (Doviak and Zrnic', 1993, eqn. 5.74).

Istok and Doviak (1986) have applied the procedure described here to estimate σ_t and calculated ε from eqn. 11.42 for data from a severe storm. One example from their analysis is shown in Fig. 11.11. This Figure presents the cumulative probability of ε in regions of a tornadic storm where the reflectivity factor is >30 dBZ, which corresponds to rainfall rates heavier than the order of 3 mm h^{-1}. This particular cutoff was chosen to avoid spurious spectrum widths generated by effects of sidelobes in regions of large reflectivity gradients. The storm region considered includes most if not all significant events, e.g. mesocyclone, updrafts, etc. As can be seen from this Figure, over 70% of the storm volume contains severe turbulence, as characterised by the threshold $\varepsilon > 0.0675$ m^2 s^{-3} (see Table 11.1).

The field of ε in the storm at a height of 9.3 km is presented in Fig. 11.12, along with the horizontal wind field (arrows) and reflectivity contours at that height. In comparing the fields of ε at lower heights, it is found that ε increases with height, and the areas of large ε increase as well (Istok and Doviak, 1986). This is consistent with previous turbulence studies (e.g. Donaldson and Wexler, 1969; Sinclair, 1974; Frisch and Strauch, 1976). The core of maximum reflectivity (inside the 50 dBZ contour) was found to be located north-north-east of a 45 m s^{-1} updraft located where there is strong divergence of the horizontal flow (Fig. 11.12). The area of largest ε is located in the vicinity of the updraft; this appears to be a common feature of thunderstorms (e.g. Lee, 1977).

To summarise, there exists an association between areas of large velocity gradients and large kinetic energy dissipation rates ε. Areas where the velocity gradients are largest either coincide with, or are upstream from, the zones of largest ε. Conversely, the values of ε are smaller where the velocity gradients are weaker.

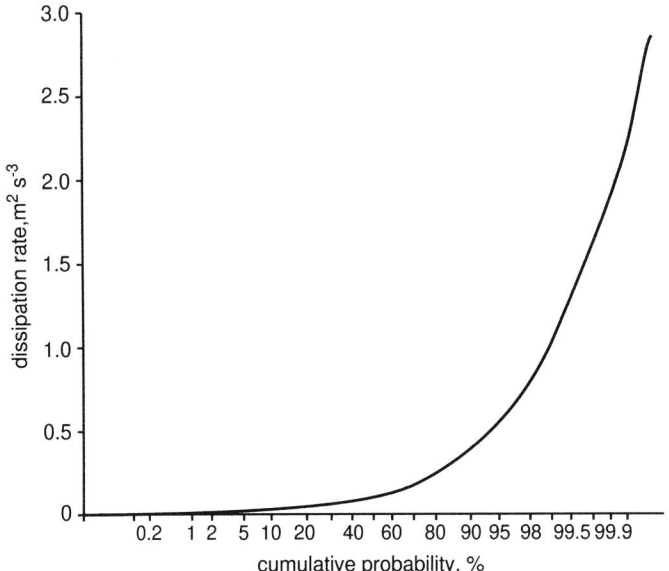

Figure 11.11 Cumulative probability of the eddy dissipation rate ε for a thunderstorm on 8 June 1974 at 1420 h Central Standard Time

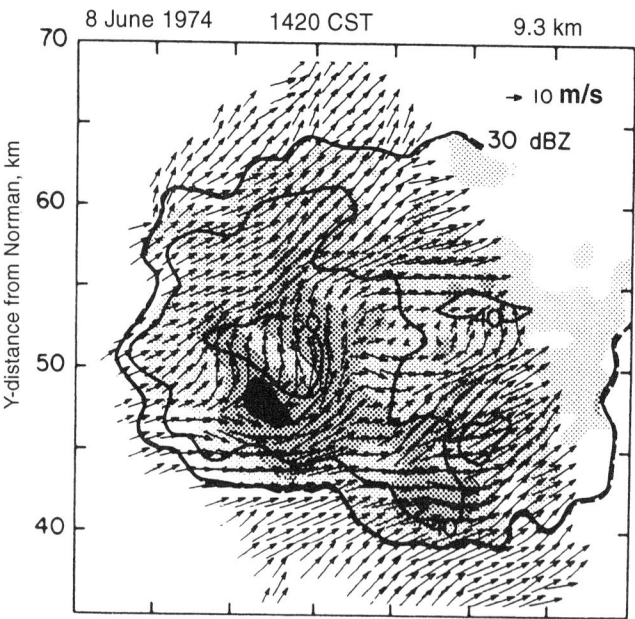

Figure 11.12 *Horizontal wind vectors (arrows, scale in the upper right corner), reflectivity factor contours (in dBZ, heavy solid lines), and eddy dissipation rates ε (in $m^2\ s^{-3}$, shaded areas) at an altitude of 9.3 km for a thunderstorm over Oklahoma, USA on 8 June 1974. The stippling indicates the following ranges of eddy dissipation rates: no stippling: ε < 0.1 $m^2\ s^{-3}$; light: 0.1 < ε < 0.3 $m^2\ s^{-3}$; medium: 0.3 < ε < 0.7 $m\ s^{-3}$; and dark: ε > 0.7 $m^2\ s^{-3}$* (from Istok and Doviak, 1986)

11.7 Avoiding turbulence

According to Lee (1977), there is a strong connection between the overall spectrum width σ_v and aircraft penetration measurements of turbulence. His data show that when aircraft-derived gust velocities exceeded 6 m s^{-1}, corresponding to moderate or severe turbulence, the radar-derived spectrum width exceeded 5 m s^{-1} in every case for aircraft within 1 km of the radar resolution volume. Although he did not remove the contribution of shear to the spectrum width, the cumulative probability of the total spectrum width and the width due to shear and turbulence (Fig. 11.13) suggests that, in thunderstorms, turbulence is the principal contributor to spectrum width. Spectrum widths >5 m s^{-1} exist in ~50% of a severe thunderstorm volume, suggesting that half the storm volume contains moderate or severe turbulence. Thus, it is prudent for aviators to avoid severe thunderstorms by a large margin.

Based on turbulence measurements by aircraft penetrating thunderstorms, and reflectivity measurements with non-Doppler weather radars, the Federal

Aviation Administration (FAA, 1978) of the USA has recommended that aircraft remain more than 20 km from the edge of the storm, usually defined as the 20 dBZ reflectivity contour. More recent measurements of turbulence with Doppler radar supports this recommendation. Unfortunately, these criteria are often ignored, sometimes with devastating consequences.

The failure to heed such advice may be due to the unacceptably high false alarm rates. If so, then it becomes imperative that more precise criteria be found to pinpoint hazardous regions without compromising flight safety. There is no doubt that the safest path is one that avoids storms by a wide margin, but often this is impractical, particularly around airports congested with aircraft and storms. Present guidelines for flights around thunderstorms claim so much useful and safe air space that pilots may be tempted to ignore these advisories in the interests of timely arrival at their destination. Doppler radars should help increase the credibility of the warnings without losing valuable air space, time, and margin of safety.

Not all scales of turbulence and/or wind shear are hazardous to aircraft. Furthermore, the level of threat of a meteorological event is conditioned by the response characteristics of aircraft being threatened. For example, a Boeing 727 aircraft has enhanced response to wind perturbation at frequencies near its phugoid frequency of 3×10^{-2} Hz, which corresponds to wind changes in a period of ~30 s along the aircraft's flight path. Thus, considering takeoff and landing speeds, it can be deduced that wind perturbations or eddies of ~3 km wavelength can have a more deleterious

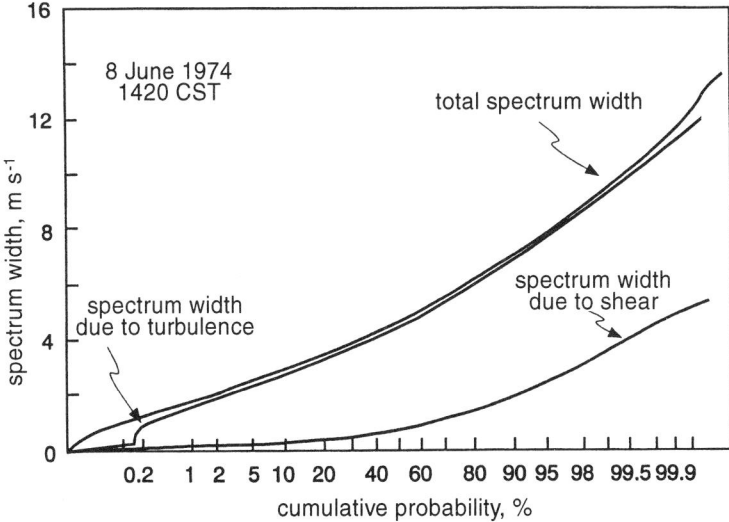

Figure 11.13   *Cumulative probability of the total spectrum width and the width due to linear radial velocity shear and turbulence (8 June 1974 at 1420 h Central Standard Time)*

effect on the performance of this aircraft than other wavelengths of similar amplitude

Large wind shear by itself is not necessarily very dangerous to aircraft except at takeoff or landing, because at higher altitudes there is ample space and time for the aircraft to recover from any loss of altitude. However, large shear, especially vertical shear of horizontal wind, can result in large-amplitude K-H waves, producing severe turbulence which is dangerous to aircraft. Hence pilots should avoid regions harbouring such shear. As stated earlier, the phenomenon producing turbulence might be more hazardous to aircraft than the turbulence itself. For example, the K-H wave shown in Fig. 11.4 has a wavelength of ~3.3 km, which is close to the phugoid wavelength of large aircraft, and could excite this mode of instability. Thus, this wave train itself could be more dangerous to aircraft than the turbulence it generates.

Turbulent eddies have a lifetime which increases roughly proportionally to their wavelength Λ_0, and decreases as their characteristic velocity (i.e. the RMS velocity of turbulence on scales $<\Lambda_0$) increases. Assuming a characteristic velocity of 10 m s^{-1} for a 3 km scale eddy, a lifetime of ~5 min is to be expected. On the other hand, if wind perturbations are associated with ordered flow such as microbursts, much longer lifetimes are found even though the size of a meteorological event is small. For example, typical horizontal dimensions of microbursts are 1 to 3 km and their lifetimes range from 5 to 15 min, but the period of severe wind shear within the microburst lasts from 2 to 4 min with an average velocity difference of 25 m s^{-1} across the microbursts' divergent flow (McCarthy and Serafin, 1984).

Usually aviation hazards are due to small-scale weather events embedded in phenomena of larger scales that are more easily detected and monitored. For example, although the US network of large-area weather surveillance radars (WSR-88Ds) and/or the more specialised airport terminal Doppler weather radars (TDWRs) will not resolve concentrated hazards such as tornadoes at all operating ranges, they will much more readily resolve the mesocyclones from which tornadoes frequently evolve. Furthermore, turbulence and short-lived intense drafts are found in waves and convection initiated along the leading edges of thunderstorm outflows, which are much larger scale events easily 'visible' with the Doppler radar if the outflows are not below the radar horizon. Detection of the large-scale events that spawn smaller-scale hazards provides longer lead times for warning of potential aviation hazards.

As pilots can take evasive action to avoid or minimise storm hazards, timely and accurate warnings should increase flight safety. However, because pilots have many tasks to perform during the critical stages of landing and takeoff, it is unrealistic to expect them to monitor weather data continuously in order to thread their way among weather hazards in the airport environment. Furthermore, since it takes time for a radar to survey a volume of space (at least a minute or two for a one-degree pencil beam radar) and additional time to arrive at a decision concerning the existence of threat and to communicate it, it seems unlikely to expect a response from pilots to hazards lasting only a few minutes or less. Even worse, if the radar is committed to scanning a large volume of space to provide surveillance of entire thunderstorms from near the ground to their tops, several minutes could pass before the radar's beam cycles back to probe any specific location. Thus, even if the Doppler radar detects these transient hazards, it may be more prudent to use the information on larger-scale, more persistent phenomena (e.g. gust fronts, mesocyclones, thunderstorm cells, etc.) to identify regions of *potential* hazard.

Thus warnings of potentially hazardous wind shear and turbulence based on observations of events of larger scale should lead to greater safety even if some of the warnings may not be associated with verified aviation hazards. For example, wind shear associated with microbursts had caused, or been a contributory cause of, crashes of commercial aircraft at a rate of about once every two years. After weather radar research and flight simulations confirmed these hazardous events as the likely principal cause of the crashes, and pilot training in flight simulators showed the imprudence of attempting to fly through such events, there has been at most one fatal crash due to microbursts in the past 11 years (1985–1996). This improvement in flight safety was attained even before the modern Doppler weather radars to detect and warn of the presence of microbursts were deployed at airports. Thus, simply avoiding dangerous thunderstorms, which are easily detected and tracked, and areas of potential hazards (e.g. regions of localised heavy precipitation, wind shear etc.) can lead to improved safety if pilots are also trained in recognising the dangers.

11.8 Summary

Since atmospheric turbulence is among the most prevalent causes of interference to aircraft flight, this chapter is devoted to understanding in some detail the qualitative and quantitative aspects of the phenomenon and its detection by radar. Turbulence essentially consists of random motions of air, and may be visualised as a combination of three-dimensional eddies of a continuum of sizes. The turbulent energy is continuously handed down from the larger scales to succeeding smaller ones, until it is lost as heat through dissipative processes. This dissipation rate is an indicator of the turbulence level, and may be used for classifying the intensity of turbulence according to its hazard potential for aviation.

As turbulence is a random process, it is best described in statistical terms. Some of the basic equations of turbulence have been discussed in this chapter. When a radar is used to detect atmospheric turbulence, as is the trend with modern aviation weather surveillance systems, the measured parameters of turbulence (in particular the echo signal and its spectral distribution) depend on its basic nature as well as that relative to characteristics of the sensing radar. Some of these relationships have been highlighted in the chapter. Radar observation of turbulence is also likely to be influenced by the presence of wind shear, the effects of which may be separated to a fair extent by appropriate data processing.

The detection, using Doppler radars, of atmospheric turbulence within air spaces used for aviation would greatly help in the optimal avoidance of this widespread phenomenon. Thus, using radar data, turbulence zones can be skirted by aircraft with a safe separation, but not by so large a distance that large volumes of the air space would be out of bounds. The latter is the case with rule-of-thumb methods of turbulence avoidance. Radar delineation of turbulence helps in recognising not only turbulence itself but also many other phenomena which often have characteristic turbulence signatures.

11.9 References

BREWSTER, K.A., and ZRNIC', D.S. (1986): 'Comparison of eddy dissipation rates from spatial spectra of Doppler velocities and Doppler spectrum widths', *J. Atmos. Ocean. Technol.*, **3**, pp. 440–452

DONALDSON, R.J., and WEXLER, R. (1969): 'Flight hazards in thunderstorms determined by Doppler velocity variance', *J. Appl. Meteorol.*, **8**, pp. 128–133

DOVIAK, R.J., and BERGER, M.J. (1980): 'Turbulence and waves in the optically clear planetary boundary layer resolved by dual-Doppler radars', *Radio Sci.*, **15**, pp. 297–317

DOVIAK, R.J., and ZRNIC', D.S. (1993): 'Doppler radar and weather observations' (Academic Press)

FEDERAL AVIATION ADMINISTRATION (1978): 'Thunderstorms'. Advisory Circular 00-24A, FAA, Department of Transportation, Washington, DC, 10pp

FRISCH, A.S., and STRAUCH, R.G. (1976): 'Doppler radar measurements of turbulent kinetic energy dissipation rates in a northeastern Colorado convective storm', *J. Appl. Meteorol.*, **15**, pp. 1012–1017

GOSSARD, E.E., and HOOK, W.H. (1975): 'Waves in the atmosphere: Atmospheric infrasound and gravity waves – Their generation and propagation' (Elsevier)

GOSSARD, E.E., and. STRAUCH, R.G (1983): 'Radar observations of clear air and clouds' (Elsevier)

ISTOK, M.J., and DOVIAK, R.J. (1986): 'Analysis of the relation between Doppler spectral width and thunderstorm turbulence', *J. Atmos. Sci.*, **43**, pp. 2199–2214

LABITT, M. (1981): 'Coordinated radar and aircraft observations of turbulence'. Project Report ATC 108. Massachusetts Institute of Technology, Lincoln Laboratory, Cambridge, MA

LEE, J.T. (1977): 'Application of Doppler radar to thunderstorm measurements which affect aircraft'. Final Report FAA-RD-77-145, FAA Systems Research and Development Service, Washington, DC

MacCREADY, P.B., JR. (1962): 'Turbulence measurements by sailplane', *J. Geophys. Res.*, **67**, pp. 1041–1050

MacCREADY, P.B., JR. (1964): 'Standardization of gustiness values from aircraft', *J. Appl. Meteorol.*, **3**, pp. 439–449

McCARTHY, J., and SERAFIN, R. (1984): 'The microburst: hazard to aviation', *Weatherwise*, **37**, pp. 120–127

PANCHEV, S. (1971): 'Random functions and turbulence' (Pergamon Press, Oxford)

REINKING, R.F., DOVIAK, R.J., and GILMER R.O. (1981): 'Clear-air roll vortices and turbulent motions as detected with an airborne gust probe and dual-Doppler radar', *J. Appl. Meteorol.*, **20**, pp. 678–685

REITER, E.R. (1970): 'Recent advances in the study of clear-air turbulence (CAT)', *Rev. Meteorol. Aeronaut.*, **30**, pp. 10–13

REITER, E.R., and BURNS, A. (1966): 'The structure of clear-air turbulence derived from "TOPCAT" aircraft measurements', *J. Atmos. Sci.*, **23**, pp. 206–212

RHYNE, R.H., and STEINER, R. (1964): 'Power spectral measurements of atmospheric turbulence in severe storms and cumulus clouds'. NASA Technical Note NASA TD D-2469, pp. 1–48
ROWLAND, J.R., and ARNOLD, A. (1975): 'Vertical velocity structure and the geometry of clear air convective elements'. Preprints of 16th Radar Meteorology Conference, pp. 296–303
SCOTTI, R.S., and CORCOS, G.M. (1969): 'Measurements on the growth of small disturbances in a stratified shear layer', *Radio Sci.*, **4**, pp. 1309–1313
SINCLAIR, P.C. (1974): 'Severe storm turbulent energy structure'. Preprints of 6th Conference on Aerospace and Aeronautical Meteorology, El Paso, TX (American Meteorological Society)
SRIVASTAVA, R.C., and ATLAS, D. (1974): 'Effect of finite radar pulse volume on turbulence measurements', *J. Appl. Meteorol.*, **13,** pp. 472–480
STOEFFLER, R.C. (1972): 'Additional research on instabilities in atmospheric flow'. NASA Report CR-1985, April 1972
STULL, R.B. (1988): 'An introduction to boundary layer meteorology' (Kluwer)
TATARSKII, V.I. (1971): 'The effects of the turbulent atmosphere on wave propagation' (Translated from the Russian edition by Israel Program for Scientific Translations Ltd. IPST Cat. No. 5319), (Available from the US Department of Commerce, UDC 551.510, ISBN 07065 0680 4, National Technical Information Service, Springfield, VA)
TENNEKES, H., and LUMLEY, J.L. (1972): 'A first course in turbulence' (MIT Press)
THORPE, S.A. (1969): 'Experiments on the stability of stratified shear flows', *Radio Sci.*, **12**, pp. 1327–1331
TROUT, D., and PANOFSKY, H.A. (1969): 'Energy dissipation near the tropopause', *Tellus*, **21**, pp. 355–358
VINNICHENKO, N.K., and DUTTON, J.A. (1969): 'Empirical studies of atmospheric structure and spectra in the free atmosphere', *Radio Sci.*, **4**, pp. 1115–1126

Chapter 12
Lightning and aviation[1]

12.1 General

Atmospheric electricity manifests itself most visibly and violently as lightning. As aircraft fly through air spaces with actual or potential lightning activity, a good understanding of the interaction between lightning and aircraft is imperative. The statistical fact that lightning has not caused too many fatal accidents has often led to the assumption that lightning may not be a serious hazard for aviation. However, lightning can and does cause significant functional impairment of modern aircraft. The increasing use of digital devices in navigational and control systems of contemporary aircraft is likely to make the aircraft more susceptible to direct and indirect lightning effects.

Several systematic experimental studies conducted in recent years have provided quantitative information about the impact of lightning on aircraft and served to dispel some preconceived notions regarding the susceptibility of aircraft to lightning strikes. Such knowledge is likely to be of great value in designing aircraft and their flight paths, and in mitigating the effects of lightning on aircraft.

12.2 Lightning, electric fields and atmospherics

Thunderstorms, also called electrical storms, are produced by cumulonimbus clouds and are always accompanied by lightning, usually with strong wind gusts, heavy rain, and sometimes hail. A unique property of thunderstorms is their electrical activity which results from space charge generation and separation. Such activity occurs during the strong convective updraft stage of the cumulonimbus (see Section 4.3), in the early phase of the storm, and during the strong downdraft-producing stage, characterised by precipitation, in the dissipating phase of the storm. The electrical structure of a thunderstorm changes with its evolution.

A simplified electrostatic model of thunderstorm space charges in the early convective stage is an electrical dipole formed by positive charges in the upper part of the cloud and negative charges in the lower part. With the downdraft forming during the mature phase of the storm, the region of

---

[1] This chapter has been contributed by Dr. Vladislav Mazur of the National Severe Storms Laboratory, Norman, OK, USA

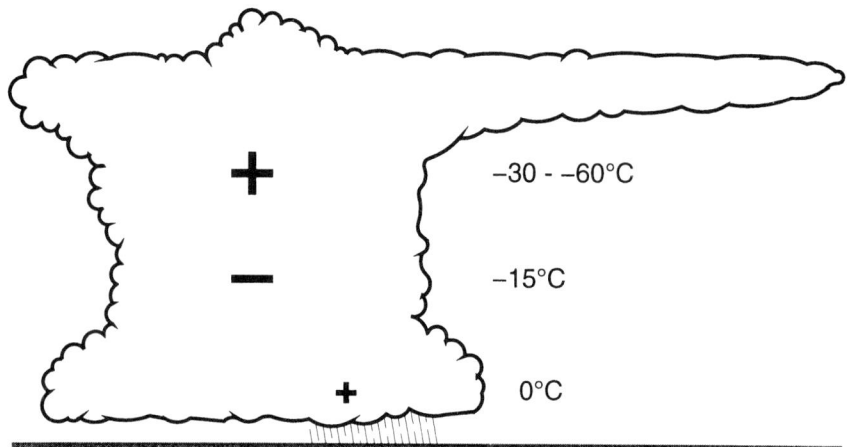

Figure 12.1 Tripole electrostatic model of a thunderstorm at the mature stage. Temperatures correspond to the altitude levels of the centres of different space charge regions

negative space charge descends to the vicinity of the $-15°C$ isotherm and remains there. The centre of the positively charged region, depending on the degree of the convective instability of the atmosphere, may be located between the isotherms at $-30°$ and $-60°C$, and the small positive space-charge region associated with precipitation appears near the $0°C$ isotherm (Fig. 12.1). Thus the mature phase of the storm has a tripole structure of space charges.

The charge structure determines the electric field distribution within the storm and hence the location of regions of lightning initiation and the type of lightning discharges. In fair weather, the atmospheric electric field near the earth's surface is typically of the order of 100 V/m, and is directed vertically in such a sense as to drive positive charges downwards to the earth. The fair-weather electric field decreases with altitude, falling, for example, to only ~ 5 V/m at an altitude of ~ 10 km. Near and under thunderstorms, the surface electric field varies widely in magnitude (in the scale of several kilovolts per metre) and direction, depending on the distance from the storm and the stage of storm evolution. Within localised regions of thunderstorms, the electric field can reach values as high as 400 kV/m.

Lightning starts in cloud regions with electric field high enough to produce electrical breakdown under the local environmental conditions and to support the development of a flash. Lightning is a self-propagating and electrodeless atmospheric discharge that transfers (through the induction process) the electrical energy of an electrified cloud into the induced charges and current in the conducting lightning channel (Mazur and Ruhnke, 1993). Natural lightning starts as a bidirectional and bipolar leader process, although unidirectional and unipolar leader development may occur at

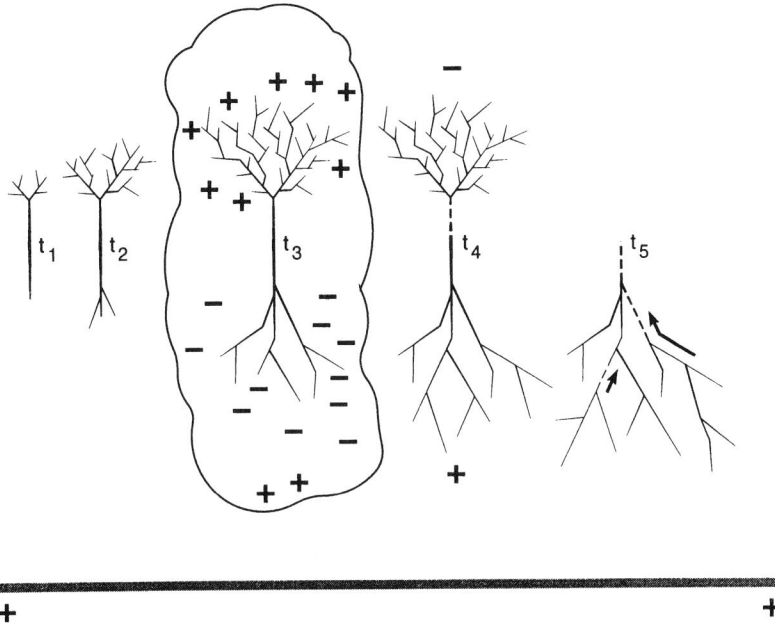

Figure 12.2 Conceptual sketch of processes in intracloud lightning flashes. Initiation of a bipolar and bidirectional leader occurs at time t_1. During the period t_2 to t_3, there is branching and progression of the bidirectional leader; the upper part is negative and the lower part positive. At time t_4, current cutoff in a trunk channel connecting the positive and negative parts of the lightning 'tree' takes place. Time t_5 marks the progression of the positive leader, with intermittent occurrence of negative recoil streamers (shown with arrows)

certain subsequent stages of the process. Lightning is composed initially of positive and negative leaders like a double-ended tree with one end made of negative leaders and the other of positive ones. The majority of lightning flashes occur within the cloud; they are called intracloud flashes (Fig. 12.2). Only when a leader reaches the ground, as is the case in cloud-to-ground flashes, does the ground potential wave (return stroke) affect the lightning process. Most of the lightning energy is dissipated as heat and light, with a small portion being transformed into sonic energy (thunder) and electromagnetic radiation.

The lightning channel is a branched plasma channel of length comparable with the dimensions of the thundercloud. The channel's electric potential can reach several tens of megavolts. Because of the electrodeless nature of the lightning discharge, however, the total charge in the channel, before it touches the ground, is zero, while the charges of the positive and negative leaders are equal and nonzero, and may reach a few coulombs. In cross-section, the channel has a current-carrying core with a diameter of several

410 *Aviation weather surveillance systems*

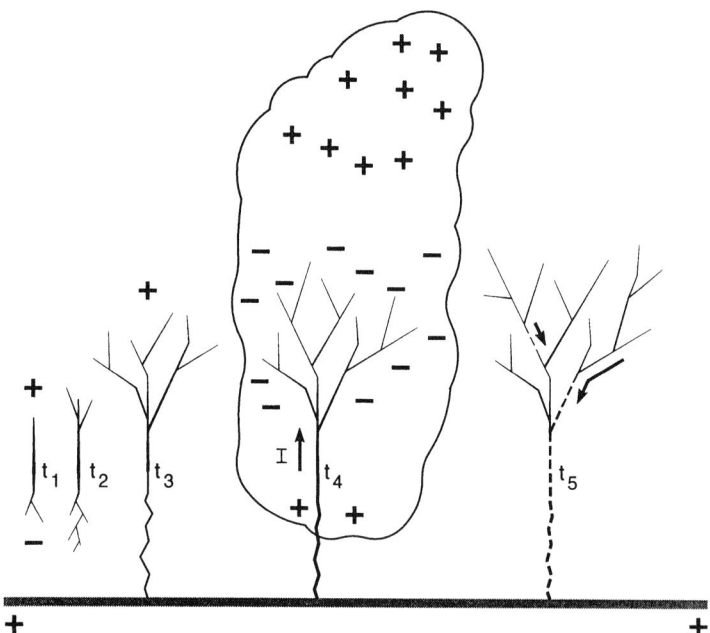

Figure 12.3 Conceptual sketch of processes in negative cloud-to-ground flashes. Initiation of a bipolar and bidirectional leader occurs at t_1; the upper part of the leader is positively charged, and the lower part negatively charged. At time t_2, the bidirectional leader progresses towards the ground. Ground contact of the negative leader occurs at t_3 after which the ground potential wave (return stroke) propagates upwards along the channel (t_4). The return stroke current is marked I. Cutoff of the continuing current in the channel to the ground occurs at t_5. Arrows mark negative dart leaders descending to the ground to start the next cycle (leader/return stroke) in the multistroke process

centimetres, surrounded by a corona envelope. The diameter of the corona envelope, which is in the scale of metres, is determined by the magnitude and spatial distribution of the radial electric field around the channel.

After a descending leader in the cloud-to-ground flash makes contact with ground, the ground potential wave, called a return stroke, propagates upwards, along the existing ionised leader channel, with a speed as high as 2×10^8 m/s. The tremendous potential gradient between the leader channel and the ground produces a return stroke current of tens of kiloamperes. Both the current and the propagation speed of the return stroke decrease with height. The entire return stroke process is completed within a few tens of microseconds.

In negative cloud-to-ground flashes (Fig. 12.3), the return stroke deposits positive charge of several coulombs on the preceding negative leader channel, thus charging the earth negatively. The opposite happens in the case of positive cloud-to-ground flashes, with the return stroke depositing negative charge of several tens of coulombs on the preceding positive leader channel, and charging the earth positively. Widespread global cloud-to-ground lightning activity, which consists predominantly of negative flashes, is the principal source of the negative charge of the earth that maintains the potential difference of several hundred thousand volts between the earth and the ionosphere.

Multiple return strokes are common in negative cloud-to-ground flashes, but positive cloud-to-ground flashes have only one return stroke. Bringing the zero ground potential to the upper end of the lightning tree by the return stroke explosively intensifies the breakdown process there. This new leader development produces a continuing current flow of hundreds of amperes into the channel (and thus to the ground) immediately following the return stroke. The period of continuing current may last from 1 ms to hundreds of milliseconds.

Lightning progression is in essence a continuing electrical breakdown consisting of a series of transient phenomena with varying timescales (Fig. 12.8 referred to in Section 12.3 shows the sequence of events in relation to an aircraft lightning strike). Therefore, lightning produces electromagnetic radiation in a wide frequency band ranging from the order of kilohertz to hundreds of megahertz. This radiation is called 'atmospherics' or, in short, 'spherics'. In the frequency range from 10 kHz to 100 MHz the E-field amplitude spectral density of the spherics in V/m per Hz is well approximated by a $1/f$ variation with respect to the frequency f (Volland, 1982, p. 268, Fig. 6). At frequencies > 100 MHz the curve drops, showing a variation closer to $1/f^2$. The power spectral density associated with the spherics would vary as the square of this amplitude spectral relationship. The most powerful radiated signal lies in the VLF band, and is caused by the return strokes of cloud-to-ground flashes.

12.3 Lightning–aircraft interaction

Fully metallic skin of the aircraft of previous generations made an almost perfect Faraday cage that shielded all internal systems from direct and indirect lightning effects. The increasing use of composite materials for primary aircraft structure, and of digital avionics for flight and engine control and systems management in modern aircraft makes them more susceptible to lightning-related hazards and damage.

Statistically, commercial aircraft experience approximately one direct lightning strike every 3000 flight hours (Plummer et al., 1985). The consequences, however, are generally benign. The damage is usually confined to burn marks on the skin and the trailing edges. The minimal amount of damage experienced by many aircraft is attributed to the widespread use of aluminium, an excellent electrical conductor, for the skins and the primary structure, and the use of mechanical and hydraulic control systems which are relatively immune to the adverse effects of lightning. However, even in aircraft utilising these traditional and proven design techniques, lightning-induced catastrophes are known to have occurred (Corbin, 1983).

The risks from lightning strikes to aircraft can be managed by in-flight lightning avoidance and also by vehicle hardening. The aircraft can be operated so as to avoid thunderstorms and related conditions most conducive to lightning–aircraft interaction. However, the conditions under which lighting strikes aircraft are not always predictable or readily identifiable. For instance, the 'nonthunderstorm' lightning strikes are not easily avoided because they are difficult to anticipate.

Three major research programmes for studying lightning–aircraft interaction, namely (i) NASA Storm Hazards Program (1980–1986), (ii) USAF-FAA Lightning Characterization Program (1984–1985, 1987), and (iii) French Transall Programme (1984, 1988), focused on in-flight measurements of lightning strike parameters and also produced significant new data about environmental conditions conducive to lightning strikes to aircraft in summer thunderstorms. Thunderstorms were the obvious choice to conduct such measurements because of the belief that lightning hazards occur where lightning activity naturally exists.

The experimental programmes, however, produced results quite contrary to this belief. The important thing learnt about lightning–aircraft interaction was that all lightning strikes in storms at flight altitudes > 6 km (high altitudes), and ~ 90% of strikes at < 6 km (low altitudes) are triggered by the aircraft itself (Mazur *et al.*, 1984; Mazur *et al.*, 1986; Reazer *et al.*, 1987). The main factor contributing to lightning strikes to aircraft is not the presence of natural lightning activity, but of an ambient electric field sufficient to initiate a discharge on an aircraft of a given size, configuration and speed. It was shown that only ~ 10% of strikes at low flight altitudes are actually natural lightning flashes intercepted by the aircraft (Mazur *et al.*, 1986; Reazer *et al.*, 1987).

Figure 12.4 presents experimental results to show that natural lightning activity is not the primary cause of lightning strikes to aircraft. The bar graph on the left, based on FAA data (Plummer *et al.*, 1985), indicates that the largest fraction (61%) of observed lightning strikes on aircraft did not have any natural lightning activity either before or after the strike. NASA data (Fisher *et al.*, 1986), shown on the right, also indicate that at all flight altitudes (shown split into two parts, above and below 6 km) the natural lightning activity is negatively correlated with aircraft lightning strikes, i.e. the probability of direct strikes to aircraft decreases with increase in natural lightning flash rate.

The concept of aircraft-triggered lightning was introduced by Clifford and Kasemir (1982), who suggested that the lightning strike is initiated by bipolar coronas and high-voltage streamers resulting from enhanced electric fields at the extremities of the aircraft. The first instrumental proof of triggered strikes

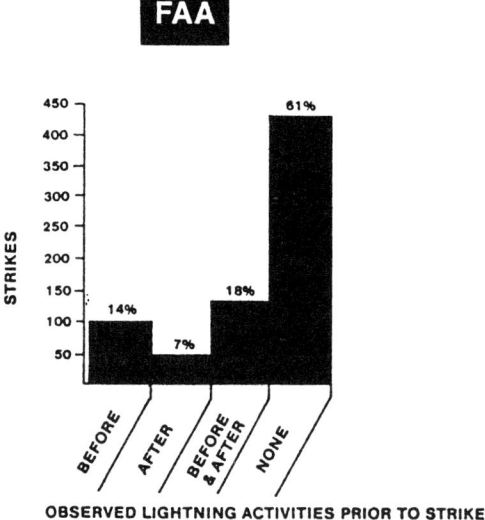

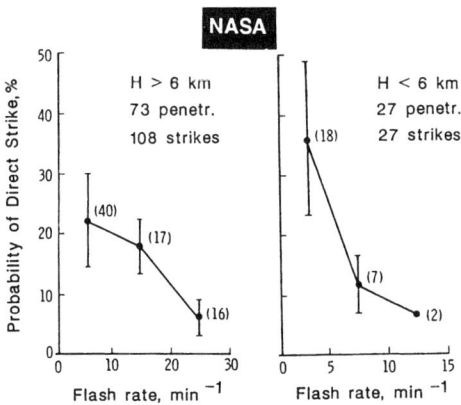

Figure 12.4 Experimental results showing the relationship between aircraft lightning strikes and the ambient natural lightning activity. Note that the two show a negative correlation. The lightning events in the NASA study were monitored and counted by using radar. The probability of a strike to an aircraft is defined as the ratio of the number of aircraft strikes to the total number of flashes passing through the radar resolution volume containing the aircraft

was obtained from the analysis of UHF-band radar echoes of aircraft during thunderstorm penetrations (Mazur *et al.*, 1984). Moreau *et al.* (1992), from airborne electric field and current measurements and high-speed video observations, provided additional evidence to view the lightning–aircraft interaction as a bidirectional leader process, and conclusive evidence of interception of aircraft by natural lightning at low altitudes.

To trigger a lightning discharge, a conductive aircraft should be in a strong electric field either inside or outside an electrified cloud. A conductor placed in the electric field is polarised through the induction mechanism so that the induced electric charges produce an electric field to compensate the ambient field. The local electric field on the tips of the aircraft's extremities, where the charges are maximum, can reach the breakdown values to start a self-propagating leader. The breakdown electric field to start a positive leader is several times smaller than that required to start a negative leader. This is a reason why lightning strikes on aircraft usually start with a positive leader. As an electrodeless discharge, the leader takes energy from the ambient electric field while maintaining conservation of charge within the aircraft–lightning system during lightning development. Therefore, with extension of the positive leader attached to one extremity, the negative charge on the opposite aircraft extremity grows, bringing the field there to the breakdown point to start the negative leader. This bidirectional leader development on the aircraft is the initiation stage of a lightning discharge that can be visualised as a double-ended tree with leaders developing at each end (Mazur, 1989).

After initiation on the aircraft, the triggered lightning develops inside the cloud in a manner similar to natural lightning discharge, and may become either an intracloud or cloud-to-ground flash, depending on the environmental electrical conditions and flight altitude. During such development, the aircraft remains part of the lightning channel for most of the flash duration (up to hundreds of milliseconds) or until the aircraft flies out of the channel. With the aircraft moving, the forward point of lightning attachment progresses along the fuselage towards the tail. The aircraft actually 'drags' the lightning channel while being attached to it at two points.

If a leader on the negative end of the 'lightning tree' reaches the ground, the return stroke(s) from the ground can reach the aircraft by traversing the ionised leader channel. This makes the aircraft involved in a single- or multistroke cloud-to-ground flash. About 25% of lightning strikes at altitudes of 6 km and below develop into cloud-to-ground flashes (Mazur and Fisher, 1990). The amplitudes of return strokes reaching the aircraft are much smaller than those measured at the ground level. They are comparable to, or less than, the amplitudes of both the recoil streamers intercepting the aircraft during intracloud development of the strike and the negative leader pulses during the initiation process (Mazur and Fisher, 1990).

A sequence of photographs vividly depicting the stages in the evolution of a lightning strike on an F-106B aircraft, taken in 1986 during the NASA Storm Hazards Program, may be found in a paper by Mazur *et al.* (1990). A few telling frames from this sequence are reproduced in Figs. 12.5–12.7. The Figures show some of the important subevents during the lifecycle of a single event of lightning discharge through the aeroplane, with the airframe actually forming a part of the charge flow path.

Lightning and aviation 415

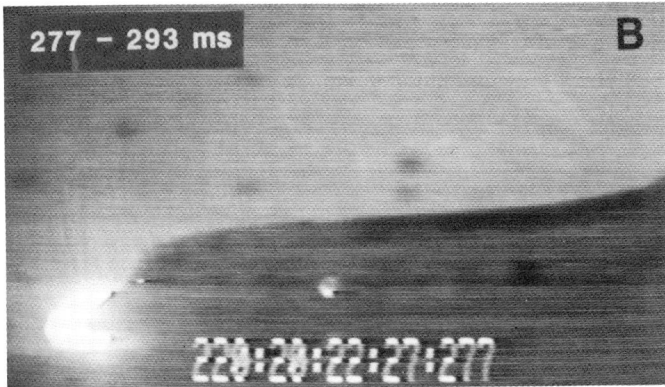

Figure 12.5   *Views of the forward part of the NASA F-106B aircraft during lightning strike, taken from a video camera mounted on the left wing tip: (A) luminosity around the nose tip due to the approaching positive leader; (B) burst of luminosity in the same area, attributable to the occurrence of negative stepped ladder prior to attachment to the nose; (C) luminosity due to continuous current flow on the attached channel. The time interval of each video frame is shown in the top left-hand corner*

416 *Aviation weather surveillance systems*

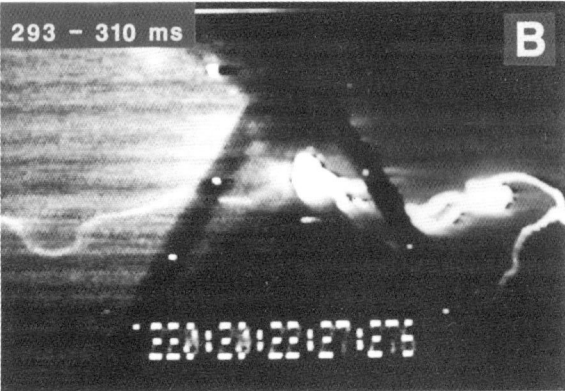

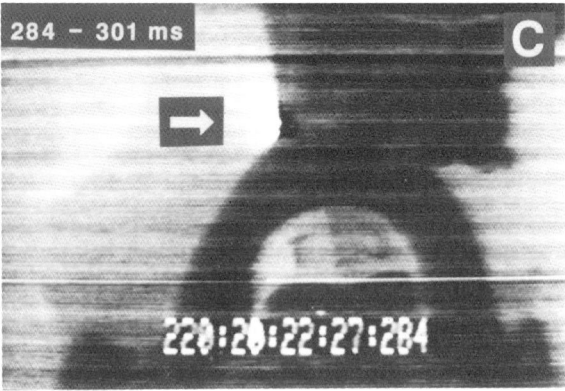

Figure 12.6   Views of lightning strike to NASA F-106B aircraft: (A) and (B) consecutive frames, taken from the forward-looking cockpit-mounted video camera, of the lightning channel attached to the nose after interception of the positive leader by the aircraft; (C) attachment of the channel to the vertical stabiliser, seen from the rear-looking cockpit-mounted video camera

Lightning and aviation 417

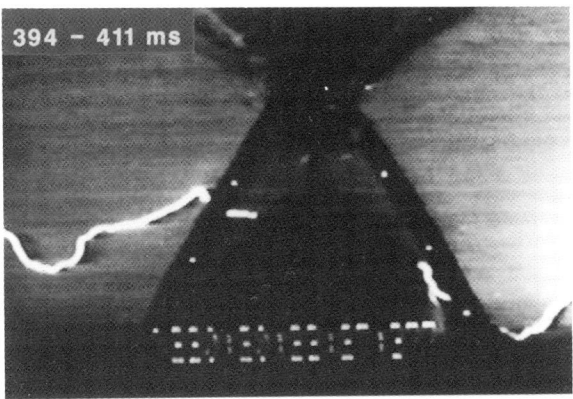

Figure 12.7 Burst of luminosity from the dart leader and a sequential return stroke during the lightning strike to NASA F-106B aircraft. The aircraft attachment points of the stroke are the vertical stabiliser and the right wing tip. Top: View from the forward-looking cockpit-mounted camera. Bottom: View from the rear-looking cockpit-mounted camera

Mazur and Moreau (1992), from the analysis of airborne electromagnetic records of lightning strikes to the FAA CV-580 and the French C-160 instrumented aeroplanes, interpreted the processes occurring during the intracloud propagation of a lightning flash triggered or intercepted by the aeroplane. Fig. 12.8 shows a sample of the electric current and field measured on the CV-580 aircraft as a result of a lightning strike. These processes (recoil streamer, dart leader/return stroke sequences, and the secondary initiations of new discharges), with their high current pulse amplitudes, may present a greater lightning threat to aircraft than those during strike initiation.

Indeed, Fig. 12.8 provides an example of another important finding that aircraft–lightning interaction produces sufficient current and/or voltage in the aircraft skin and/or structure to pose a threat to the electrical/electronic

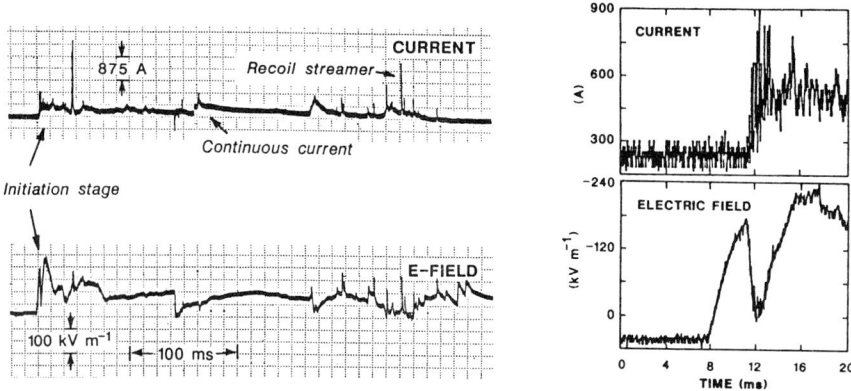

Figure 12.8   *A sample of the electric current and field measured on the CV-580 aircraft as a result of a lightning strike, showing the signatures of the sequence of events during the flash. Note that (i) the phenomenon lasted over 400 ms, (ii) a series of current pulses are noticeable during the initiation stage, (iii) a continuous current flows during most of the discharge, and (iv) recoil streamers appear during the later stages of the discharge. The boxed graphs at the right show the current and electric field variation during the initiation stage in expanded form*

systems of the aircraft. The Figure shows very strong current and field spikes > 2500 A and 250 kV per metre, respectively.

12.4 Weather conditions and lightning strikes to aircraft

For avoidance of lightning strikes, we need to know where in electrified clouds lightning strikes to aircraft may take place, and with what probability. This information is useful to pilots, aviation meteorologists, and possibly to air traffic controllers.

The earlier beliefs that lightning strikes to aircraft somehow relate to the presence of turbulence were dismissed. Most lightning strikes in summer storms occur in light rain and light turbulence conditions (Mazur *et al.*, 1986). The statistics of aircraft lightning strikes plotted in Figs. 12.9 and 12.10 lend credence to these general observations. From data obtained with the NASA F-106B instrumented airplane, it follows that the probability of triggering lightning decreases with increasing rate of natural lightning, as already shown in Fig. 12.4. A similar conclusion was made in rocket-triggered lightning experiments: the probability of triggering lightning is very low when the rate of natural lightning is high, and *vice versa*. This seemingly paradoxical phenomenon may be understood if we consider a natural triggering mechanism in storms and compare it with an artificial triggering by aircraft.

Lightning and aviation 419

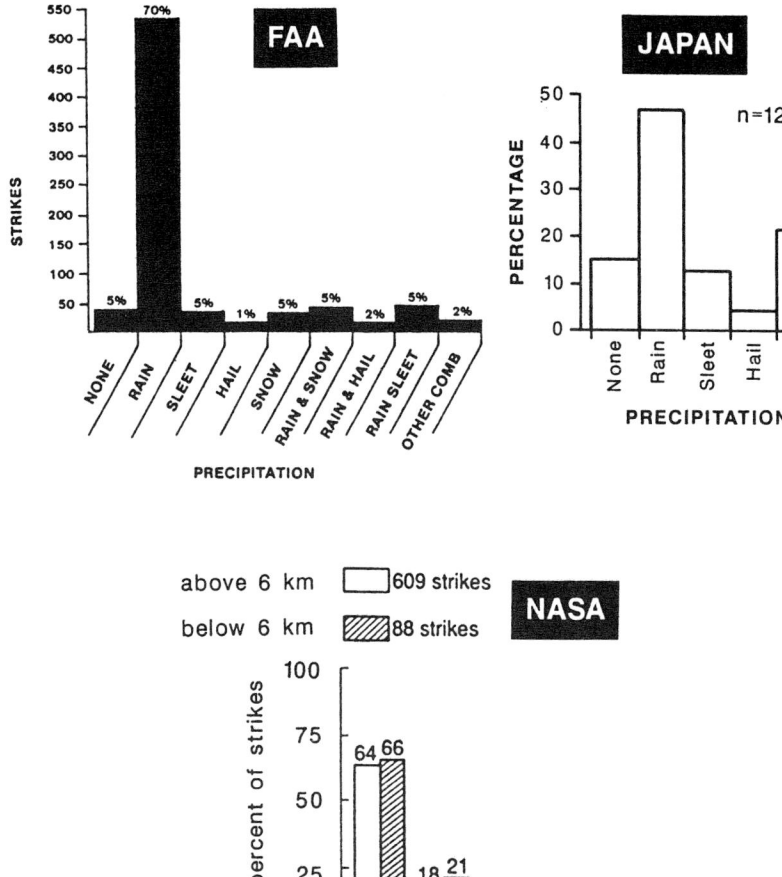

Figure 12.9 A correlation of the number of aircraft lightning strikes with the ambient precipitation conditions, carried out in three different studies, shows that the largest number of strikes is associated with rain, followed by snow. However, the intensity of precipitation during the strikes is light or negligible at all altitudes. The precipitation intensities have been estimated visually

When the natural triggering mechanism is active, it works like a control electrode for a gas discharge in a thyratron (an analogue of a lightning flash), starting it each time when an electric field in the anode-cathode gap (an analogue of an ambient electric field inside the cloud) is ready to support propagation of this discharge. In such an analogy, an additional source of

420 *Aviation weather surveillance systems*

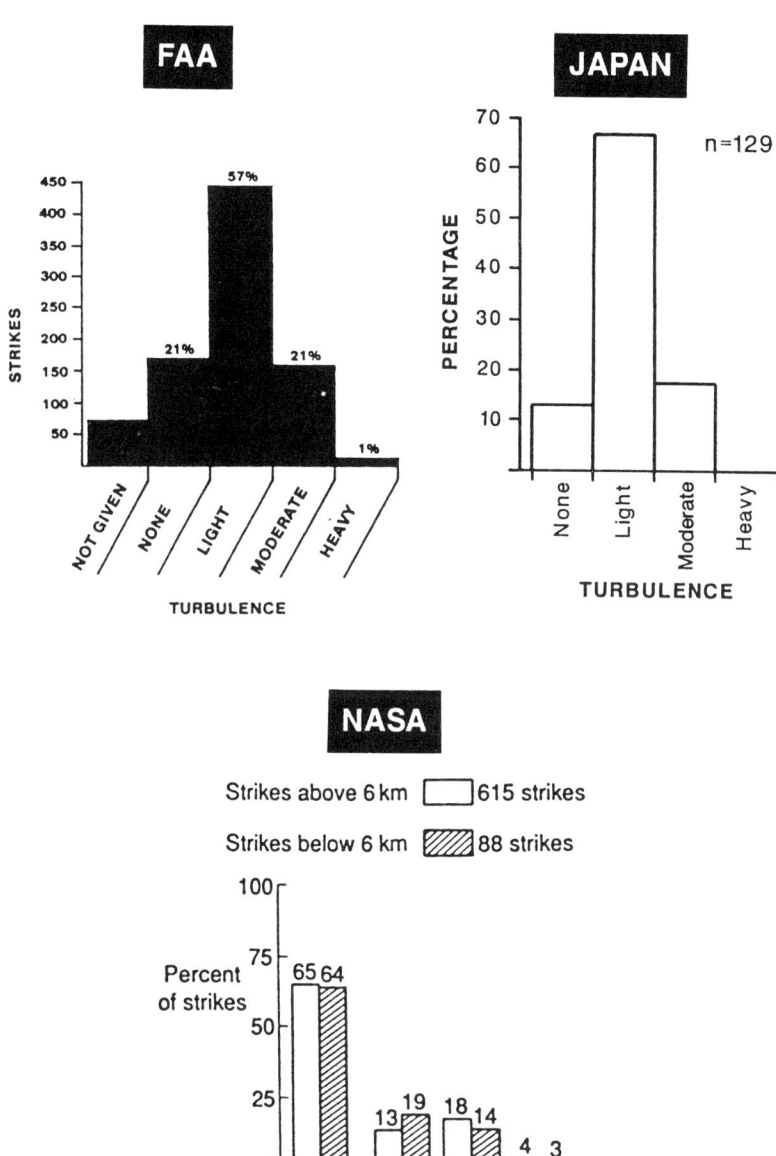

Figure 12.10   *A correlation of the number of aircraft lightning strikes with the ambient turbulence levels, carried out in three different studies, shows that the largest number of strikes is associated with light or negligible turbulence at all altitudes. The turbulence intensities have been classified according to the pilot's perception*

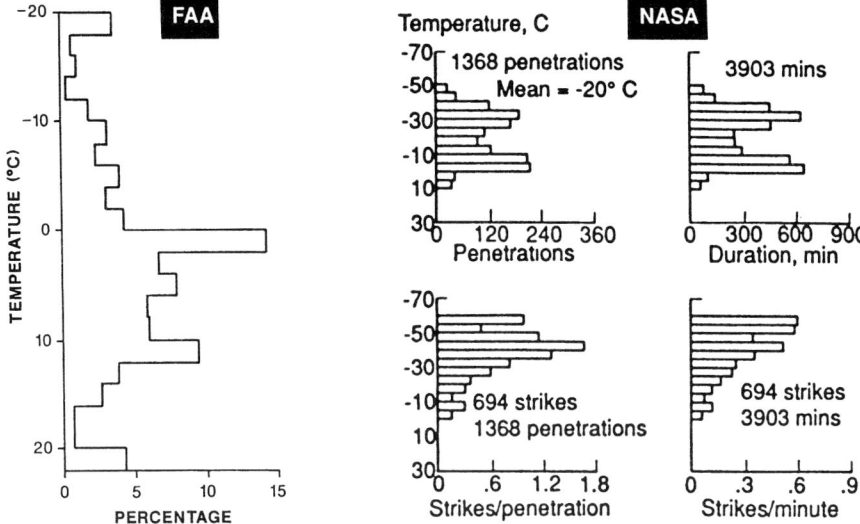

Figure 12.11 Analysis of aircraft lightning strikes as a function of the ambient temperature, based on two different studies

triggering (aircraft or rocket with a trailing wire) does not produce lightning initiation, being out of phase with the large-scale cloud electric field.

In line with this explanation are observations showing that the decaying stage of the storm is the most dangerous for penetrating aircraft; the probability of aircraft-triggered lightning as well as chances to succeed with a rocket-triggered lightning are the highest at this time. Although the cloud is still electrified during the decaying stage, the natural triggering mechanism there is very weak, evidence of which is a lack of natural lightning activity. Thus an aircraft or a rocket with a trailing wire becomes a likely source of lightning initiation due to its conductivity and size.

The experience of the NASA Langley Research Center's Storm Hazards Program and the USAF-FAA Lightning Characterization Program indicates that the longest flight time needed to obtain a lightning strike to aircraft is at altitudes near and below the cloud base. These altitudes are most frequently penetrated by aircraft in the terminal areas during landing and takeoff. The question as to whether these altitudes may be considered the safest for aircraft from the point of view of lightning hazards remains to be answered.

The dependence of aircraft lightning strikes on certain environmental factors is shown in Figs. 12.11–12.13. In Fig. 12.11, the analysis of aircraft lightning strikes as a function of the ambient temperature, performed in two different studies, shows marked differences in the findings. In the FAA study, presented over the limited temperature range between -20 and $+20°C$, the interval from -10 to $+10°C$ was found to be the most conducive to aircraft lightning strikes, with the subrange 0–10°C accounting for the majority of

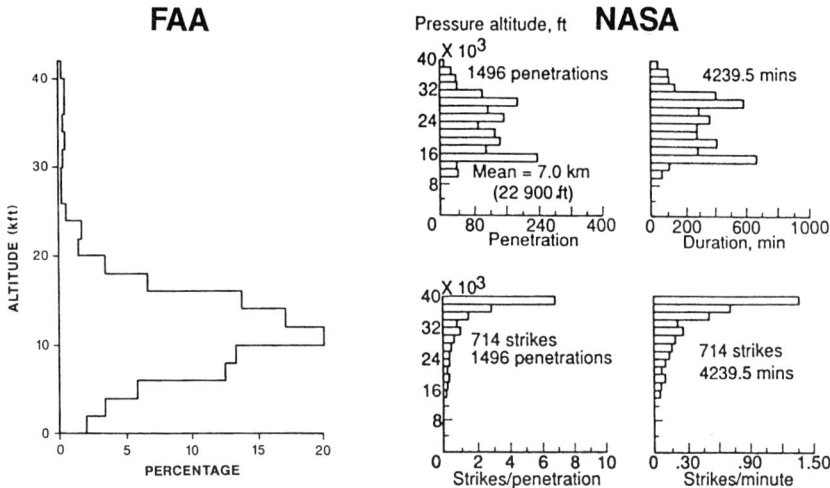

Figure 12.12 Analysis of aircraft lightning strikes as a function of the flight altitude, based on two different studies

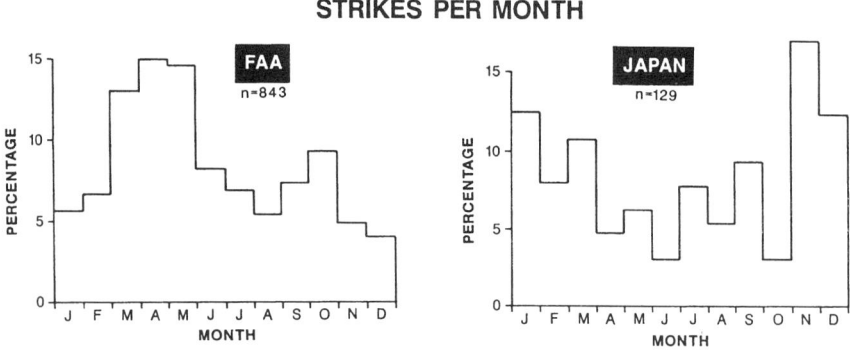

Figure 12.13 Variation of aircraft lightning strike frequency around the year in USA and Japan

strikes within this interval. The NASA data, on the other hand, show the strike rates (with respect to both time and number of flights) to be the highest in a colder temperature range of -30 to $-60°C$. The strike rates decrease rapidly with increasing ambient temperature, becoming negligible in the 0–10°C range. However, it must be pointed out that the FAA data are not normalised with respect to the duration of storm penetration (the part of the flight intended to study lightning hits), and are based on flights in diverse weather situations. In contrast, the NASA data were taken during summer-type thunderstorms only, over a wider temperature range (from −60 to +10°C), and are normalised with respect to the duration of penetration. This data set may therefore be considered to be the more complete and reliable of the two.

The altitude-dependence of aircraft lightning strike rates, presented in Fig. 12.12, also shows a similar difference between the two studies. The profile based on FAA data has a pronounced bulge in the altitude slab from 0 to 20 000 ft, peaking at about 10 000 ft altitude. Above 20 000 feet, the FAA data show negligible lightning strikes to aircraft. On the other hand, NASA data show the highest strike rates (within the height interval of 10 000 to 40 000 ft over which flight experiments were conducted) to occur at 40 000 ft altitude. This difference in the findings between the two agencies is similar to their difference in the case of temperature-dependence of lightning strikes, as was shown in Fig. 12.11. Such behaviour is only to be expected, since the ambient temperature in the earth's atmosphere is a monotonic function of the altitude in the height interval under consideration.

Finally, the variability of the aircraft lightning rate within the annual cycle is shown in Fig. 12.13 at two well separated locations on the earth. While the US analysis shows the strike frequency peak to be during the March–May period, the peak for Japan is found to occur during the winter months. In comparing these results, however, one must remember the difference between the conditions under which the respective data were collected. The most important difference is that the US data represent an integrated average over the different climatological conditions across a large country, while the Japanese data (All Nippon Airways, 1984) pertain to a relatively small area which has a less diverse climate.

12.5 Detection and surveillance of lightning phenomena

When an aircraft enters an electrically active zone of a weather system, the pilot has little control over the process of lightning strikes to his aircraft. The avoidance of lightning-prone zones of the air space is therefore the best way of minimising lightning encounters by aircraft. This establishes the need for a lightning surveillance and warning system for aviation. The coverage of such a system may be of national or regional scales, or at least of the scales of terminal areas, so that zones of concentrated aviation activity may be served by warning about lightning hazards.

A lightning warning system may be of a general type, indicating the potential for lightning generation by a weather system, based on meteorological forecast and/or radar observations. Such information, if made generally available to pilots, can be used to avoid or bypass thunderstorms. However, more accurate delineation of lightning-hazardous zones may be possible by sensing the actual occurrence of lightning flashes.

Lightning phenomena offer themselves for detection by both passive and active sensing methods. Since lightning flashes emit strong electromagnetic radiation, they are readily sensed by merely receiving the radiated signals through proper antennas. By appropriate processing of the received signals it is possible to estimate the position of lightning radiation sources. Lightning may also be detected by long-wavelength radars using the scattering property of the plasma channels formed during the process (Williams *et al.*, 1990). However, passive methods of lightning detection and location are simpler and less expensive, and hence are more commonly employed.

Ground lightning detection networks that provide data about the locations of cloud-to-ground flashes (i.e. the points of their attachment to the ground) have recently been implemented in the USA and Europe. These data complement meteorological radar observations in identifying lightning-producing weather systems for the purpose of providing general warning to aviation.

As mentioned in Section 12.2, intracloud lightning flashes represent the majority of lightning activity. In addition, these flashes usually precede cloud-to-ground flashes by as much as 20 min and cover significant regions of the storm. For these reasons, mapping of intracloud lightning is important. Mapping techniques for intracloud flashes have been used by the lightning research community for more than two decades. One example of such a technique is the time-of-arrival (TOA) technique (Proctor, 1971), which uses the difference in the time of arrival of the electromagnetic signal from a lightning flash at different antennas to calculate the location of each flash. A TOA system is operational at the NASA Kennedy Space Center in the USA (Lennon and Maier, 1991). Another method of lightning location is based on the interferometric technique, which utilises the difference in the phase of the radio-frequency signal from lightning, as received by different antennas, to calculate the locations of the radiation sources (Richard *et al.*, 1986; Rhodes *et al.*, 1994). An interferometric system for two-dimensional mapping of intracloud lightning flashes is commercially available (Richard *et al.*, 1988) and is already in use in France, Belgium, the Netherlands, at the European Space Agency Centre in French Guyana, and in Japan.

Although any lightning mapping system is useful in identifying an existing thunderstorm, none of these detection systems is helpful in identifying the potential threat of triggering lightning flashes by aircraft penetrating either a non-lightning-producing cloud or even a decaying storm. Decaying thunderstorms should be regarded as being as dangerous as, or perhaps more dangerous than, active thunderstorms for penetration by aircraft. Thus what would be most useful for preventing lightning strikes to aircraft is a warning about the presence of a high electric field in the air space being navigated.

The concept of measuring the ambient electric field using *in situ* sensors is realistic, but such information would be of little use in enhancing aviation safety against lightning strikes. This is so because, on the one hand, the aircraft would have to be in the vicinity of the high-field zone for the *in situ* instruments to sense the field, and, on the other, the high speed of the aircraft would make it difficult for the pilot to avoid the zone without endangering the aircraft. Thus the only useful solution is to be able to sense strong electric fields ahead from a distance using remote sensors. Fortunately, this idea appears to be feasible. There is enough evidence already that remote detection of regions of high electric field can be achieved with the use of polarimetric radars (Zrnic' *et al.*, 1984; Metcalf, 1995; Krehbiel *et al.*, 1996). However, much more work remains to be done in this area to make such detection robust and reliable enough for aviation use.

12.6 Lightning threats to aircraft: what else do we need to know?

Although there is a strong indication that the physics of lightning initiation in winter thunderstorms, stratiform and mixed-phase clouds would be the same as in thunderstorms, there are no scientific data on the characteristics of electrical discharges to aircraft under these conditions. Therefore, presently, we project our knowledge of lightning strikes to aircraft in summer thunderstorms on those experienced in other environmental conditions. This assumption is not justified yet, and therefore may be misleading or possibly erroneous. It is expected that, owing to the different strength, spatial extent and distribution of electric fields inside winter storms and nonthunderstorm clouds, the strikes there may differ from those in summer storms. For the same reasons, the cloud regions with the highest probability of lightning strikes to aircraft may be quite different from those in summer storms.

While storm penetrations with instrumented aircraft have provided scientific data on environmental conditions conducive to lightning hazards, the data on lightning hazards to aircraft in winter storms, marginally electrified stratiform, and mixed-phase clouds that do not produce natural lightning are totally lacking. What is known from pilots' surveys is that the majority of reported strikes to civil aircraft and space vehicles in the US occurred in marginally electrified and mixed-phase clouds (Plummer *et al.*, 1985), and in Japan the majority of reported strikes occurred in winter storms (Goto and Narita, 1986). Since the problem of lightning hazards to aircraft in most common weather conditions of marginally electrified clouds remains unsolved, the Atlas Centaur accident of 1987 at the NASA Kennedy Space Center (Christian *et al.*, 1989) should be a constant reminder to us that this could happen again.

12.7 Summary

Lightning strikes to aircraft, which are atmospheric electrical discharge phenomena involving aircraft, occur frequently. However, their harmful effects are confined to local skin damages and failure of individual subsystems; overall aircraft fatality is rare with the current all-metal aircraft, the structure of which offers natural shielding against strong electrical fields.

However, lightning hazards to aircraft can become more pronounced with the increasing use of nonmetallic composites in aircraft construction, and sensitive electronic devices in aircraft navigational and control systems.

A few co-ordinated research programmes have been conducted to study the aircraft lightning phenomenon physically and statistically. The most important finding of these studies, contrary to popular belief, is that it is not naturally occurring lightning (i.e. lightning that would have occurred even if the aircraft were not present in the vicinity) that usually 'strikes' aircraft passing through them, but that most often the aircraft itself actually triggers the process of lightning, and becomes a part of the electric discharge path. Indeed, in thunderstorms, aircraft lightning strikes have a negative correlation with natural lightning activity.

Statistically, lightning strikes to aircraft have been found to have a correlation with certain environmental parameters. A majority of strikes are associated with rain or snow, but the precipitation intensity is negligible or light during the strikes. Aircraft lightning strikes have a similar dependence on turbulence, with a majority of strikes being associated with negligible or light turbulence. In terms of flight altitude, aircraft have the most susceptibility to lightning strikes at ~40 000 ft, and ambient temperatures between -30 and $-60°C$ appear to be most conducive to such strikes. The highest rates of aircraft lightning strikes in the USA are found to occur during the early spring months, while in Japan the peak rates occur in winter.

Although much knowledge regarding lightning–aircraft interaction has been generated by careful experiments, the phenomenon of electrical hazards to aircraft in marginally electrified clouds and weather conditions other than summer thunderstorms is a complex one, and still remains to be understood.

12.8 References

ALL NIPPON AIRWAYS (1984): *All Nippon Airways Crews News*, No. 85, pp. 9–14

CHRISTIAN, H.J., MAZUR, V., FISHER, B.D., RUHNKE, L.H., CROUCH, K., and PERALA, R.A. (1989): 'The Atlas/Centaur lightning strike incident', *J. Geophys. Res.*, **94**, pp. 13169–13177

CLIFFORD, D.W., and KASEMIR, H.W. (1982): 'Triggered lightning', *IEEE Trans. Electromagn. Compat.*, **EMC-24**, pp. 112–122

CORBIN, J.C. (1983): 'Lightning interaction with USAF aircraft'. Proceedings of the International Aerospace and Ground Conference on Lightning and Static Electricity, Ft. Worth, TX, June 1983, DOT/FAA/CT-83/25, pp. 66-1–66-6

FISHER, B.D., BROWN, P.W., and PLUMMER, J.A. (1986): 'Summary of NASA storms hazards lightning research, 1980–1985'. Proceedings of the International Aerospace and Ground Conference on Lightning and Static Electricity, Dayton, OH, 24–26 June 1986, pp. 4-1–4-15

GOTO, Y., and NARITA, K. (1986): 'Lightning interaction with aircraft and winter lightning in Japan', *Res. Lett. Atmos. Electr.*, **6**, pp. 27–34

KREHBIEL, P., CHEN, T., MCCLARY, S., RISON, W., GRAY, G., and BROOK, M. (1996): 'The use of dual polarization radar observations for remotely sensing storm electrification', *Meteorol. Atmos. Phys.*, **59**, pp. 65–82

LENNON, C.L., and MAIER, L. (1991): 'Lightning mapping system'. NASA Conference Publication 3106, pp. 89-1–89-10

MAZUR, V. (1989): 'A physical model for lightning initiation on aircraft in thunderstorms', *J. Geophys. Res.*, **94**, pp. 3326–3340

MAZUR, V., and FISHER, B.D. (1990): 'Cloud-to-ground strikes to the NASA F-106 airplane', *J. Aircr.*, **27**, pp. 466–468

MAZUR, V., and MOREAU, J.-P. (1992): 'Aircraft-triggered lightning: Processes following strike initiation that affect aircraft', *J. Aircr.*, **29**, pp. 575–580

MAZUR, V., and RUHNKE, L.H. (1993): 'Common physical processes in natural and artificially triggered Lightning', *J. Geophys. Res.*, **98**, pp. 12913–12930

MAZUR, V., FISHER, B.D., and BROWN, P.W. (1990): 'Multistroke cloud-to-ground strike to the NASA F-106B airplane', *J. Geophys. Res.*, **95**, pp. 5471–5484

MAZUR, V., FISHER, B.D., and GERLACH, J.C. (1984): 'Lightning strikes to an airplane in a thunderstorm', *J. Aircr.*, **21**, pp. 607–611

MAZUR, V., FISHER, B.D., and GERLACH, J.C. (1986): 'Lightning strikes to a NASA airplane penetrating thunderstorms at low altitudes', *J. Aircraft*, **23**, pp. 499–505

METCALF, J. (1995): 'Radar observations of changing orientations of hydrometeors in thunderstorms', *J. Appl. Meteorol.*, **34**, pp. 757–772

MOREAU, J.P., ALLIOT, J.C., and MAZUR, V. (1992): 'Aircraft lightning initiation and interception from *in situ* electric measurements and fast video observations', *J. Geophys. Res.*, **97**, pp. 15903–15912

PLUMMER, J.A., RASCH, N.O., and GLYNN, M.S. (1985): 'Recent data from airline lightning strike reporting project', *J. Aircr.*, **22**, pp. 429–433

PROCTOR, D.E. (1971): 'A hyperbolic system for obtaining VHF radio pictures of lightning', *J. Geophys. Res.*, **76**, pp. 1478–1489

REAZER, J.S., SERRANO, A.V., WALKO, L.C., and BURKET, H.D. (1987): 'Analysis of correlated electromagnetic fields and current pulses during airborne lightning attachments', *Electromagn.*, **7**, pp. 509–539

RHODES, C.T., SHAO, X.M., KREHBIEL, P.R., THOMAS, R.J., and HAYENGA, C.O. (1994): 'Observations of lightning phenomena using radio interferometry', *J. Geophys. Res.*, **99**, pp. 13,059–13,082

RICHARD, P., and SOULAGE, A. (1988): 'The SAFIR lightning monitoring and warning system, first operational results and applications', Proceedings of 8th International Conference on Atmospheric Electricity, Uppsala, Sweden, pp. 687–692

RICHARD, P., DELANNOY, A., LABAUNE, G., and LAROCHE, P. (1986): 'Results of spatial and temporal characterization of the VHF-UHF radiation of lightning', *J. Geophys. Res.*, **91**, pp. 1248–1260

VOLLAND, H. (1982): 'Handbook of atmospherics' (CRC Press, Boca Raton, FL)

WILLIAMS, E.R., MAZUR, V., and GEOTIS, S. (1990): 'Lightning investigation with radar', in D. Atlas (Ed.): 'Radar in meteorology' (American Meteorological Society, Boston) pp. 143–150

ZRNIC', D.S., DOVIAK, R.J., and MAHAPATRA, P.R. (1984): 'The effect of charge and electric field on the shape of rain drops', *Radio Sci.*, **19**, pp. 75–80.

Chapter 13
Polarisation diversity radars[1]

13.1 General

It is clear from the discussion hitherto in this book that the primary sensor of the modern aviation weather surveillance system is the Doppler weather radar in its different forms. It not only estimates the various reflectivity-derived parameters such as the intensities of rainfall and hailfall, but also detects and provides qualitative and quantitative pictures of wind velocity and turbulence fields in the atmosphere. Modern aviation weather surveillance systems, by virtue of their ability to measure these parameters, and automatically derive higher-level weather products and perform their interpretation from the aviation management point of view, represent a quantum jump over the state of the art existing till recently. Yet there is scope for further improvement, and some courses of action for effecting these are visible and being pursued already. One of the major directions of possible improvement is the development and deployment of *polarisation diversity radars*.

13.2 Description

Conventional weather radars, as well as the modern Doppler weather radars being deployed for advanced aviation weather surveillance, have the common characteristic that they employ a uniform polarisation for all signals radiated by them. The reception of the reflected signal is also most often at a common polarisation, usually the same as that of the transmitted signal. The difference in the polarisation characteristics of different scatterers is sometimes used to detect them more optimally, as for example in the case of the Dopplerised weather channel of the ASR-9 radar discussed in Section 7.6. However, even here the radar switches between two different polarisation modes only on a macroscopic timescale, i.e. at any given time and in any given operating mode the polarisation of the transmitted pulses is uniform.

In polarisation diversity radars (Bringi and Hendry, 1990), in contrast, the polarisation is made to vary from pulse to pulse. Although there are endless possibilities for the way that the polarisation can be varied between pulses, both in transmission and reception, a common strategy that is relatively

---
[1] This chapter has been contributed by Dr. Dusan S. Zrnic' of the National Severe Storms Laboratory, Norman, OK, USA

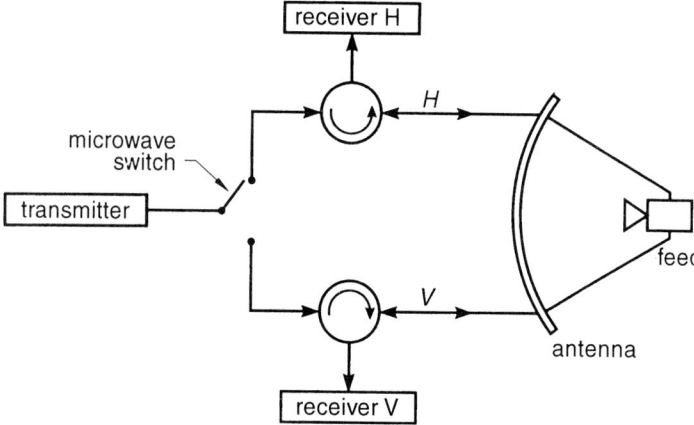

Figure 13.1 A simple schematic block diagram of a polarisation diversity radar

simple to operate and yet can provide much of the essential polarisation information consists of radiating the alternate pulses with orthogonal linear polarisations. The two orthogonal polarisations are normally horizontal and vertical. The return from each pulse is processed through two receivers, one for the horizontally polarised component of the signal and the other for the vertically polarised component. A schematic block diagram of a polarisation diversity radar is given in Fig. 13.1. Because radars with polarisation diversity perform their measurements based on the polarised components of signals, they are also called *polarimetric radars*.

There are four different reception–transmission combinations in the commonly used polarisation diversity scheme described above: horizontal–horizontal (hh), horizontal–vertical (hv), vertical–horizontal (vh) and vertical–vertical (vv). Because of the additional information available in these four polarised components, polarisation diversity radars have the capacity to provide more accurate characterisation of the phenomena they observe than radars that employ only a fixed polarisation in transmission and reception. In particular, these polarised components of the received signal, and other parameters derived from them, can be used to make more accurate and robust estimates of local and instantaneous rainfall rates compared with single-polarisation radars. This is of significance to aviation.

Another use of polarisation diversity radars which is of even greater importance for aviation is in the sensing, estimation and delineation of hail at flight altitudes. Single-polarisation radars can also help in inferring the presence of hail, but they essentially sense the reflectivity of the scatterers for this purpose. As discussed before, high reflectivities indicating hail can also arise from heavy rain. Polarisation diversity radars can provide additional information not only to distinguish hail from rain with greater confidence, but even to estimate the fraction of the two different phases of water in a population of mixed-phase scatterers such as melting hail.

13.3 Basic definitions

Intrinsically, radar measurements involve weighted averaging of parameters within the resolution volume. In a polarisation diversity radar employing horizontal and vertical polarisations the following statistically well defined second-order moments are the main source of information regarding the physical characteristics of the scatterers and/or the medium of propagation (Doviak and Zrnic', 1993):

(i) reflectivity factor at horizontal polarisation

$$Z_h = \frac{4\lambda^4}{\pi^4 |K_w|^2} \langle |s_{hh}|^2 \rangle \tag{13.1}$$

(ii) reflectivity factor at vertical polarisation

$$Z_v = \frac{4\lambda^4}{\pi^4 |K_w|^2} \langle |s_{vv}|^2 \rangle \tag{13.2}$$

(iii) linear depolarisation ratio

$$\text{LDR}_{hv} = 10 \log \frac{\langle |s_{hv}|^2 \rangle}{\langle |s_{vv}|^2 \rangle} \tag{13.3}$$

(iv) correlation coefficient at zero lag

$$\rho_{hv}(0) = \frac{\langle s_{vv} s_{hh}^* \rangle}{\langle |s_{hh}|^2 \rangle^{1/2} \langle |s_{vv}|^2 \rangle^{1/2}} \tag{13.4}$$

In these definitions s refers to the complex scattering coefficient of a hydrometeor particle, with the subscripts respectively denoting the polarisation of the scattered and incident electric field components being considered, and K_w depends on the complex dielectric constant of water. The $\langle \bullet \rangle$ operator denotes ensemble averaging over the scatterers, normally within a resolution volume of the radar.

Suitable combinations of the four moments given by eqns. 13.1–13.4 can be used to gain further insight about hydrometeor characteristics. One important parameter is the *differential reflectivity*, given as

$$Z_{DR} = 10 \log \frac{Z_h}{Z_v}$$

$$= 10 \log \frac{\langle |s_{hh}|^2 \rangle}{\langle |s_{vv}|^2 \rangle} \tag{13.5}$$

which in practice is computed as a difference, in the dBZ scale, between the reflectivity factors. Physically, Z_{DR} is a measure of the reflectivity-weighted oblateness (Jameson, 1983) of the scatterers. Raindrops assume a somewhat flattened shape due to the mutually opposing gravity and drag forces distorting the spherical shape induced by the surface tension. As the distorting forces are essentially in the vertical direction, the oblateness is such that the major axis is predominantly horizontal, and the minor axis vertical. The ensemble of raindrops therefore presents different radar cross-sections to horizontally and vertically polarised radar waves. Further, the oblateness of raindrops is a function of their size, with larger drops being more oblate because gravity and drag forces are stronger for such drops vis-à-vis the surface tension, which is size-independent. The value of Z_{DR} thus provides a measure of the median drop size in a population of raindrops. It is a particularly convenient measure of this parameter because it is independent of the number concentration of hydrometeors in the population, unlike the reflectivity factor Z_h or Z_v alone, which is measured by radars without polarisation diversity.

Another useful composite parameter is the reflectivity difference,

$$Z_{DP} = 10 \log(Z_h - Z_v) \qquad (13.6)$$

where the reflectivity factors are in $mm^6 \, m^{-3}$. Golestani et al. (1989) mention a unique relation between Z_{DP} and Z_h in pure rain, which can be used to infer the presence of such rain. Deviations from this relation are attributed to other types of hydrometeors, and can be used to obtain the fraction of Z_h contributed by ice. Reinking et al., (1997) have utilised depolarisation ratios in a dual polarisation radar to identify drizzle that may potentially cause aircraft icing.

13.4 Propagation effects

Just as the shape and other parameters of hydrometeors affect their scattering characteristics, they also affect the propagation of electromagnetic waves. Thus inferences about hydrometeor properties can be drawn by measuring the changes undergone by electromagnetic radiation while propagating through a medium containing the hydrometeor population.

As mentioned in the last subsection, falling raindrops display a degree of oblateness depending on their size. As the drops generally have a larger horizontal dimension than vertical, the horizontal electric field associated with radar waves is attenuated more than the vertical component of the field during propagation through rain. For the same reason the phase shifts of the horizontal and vertical fields along the propagation path are different, with the horizontally polarised fields experiencing larger phase shifts in rain. The differential phase shift between these two signals therefore contains important information regarding the nature of the weather phenomena present in the propagation medium.

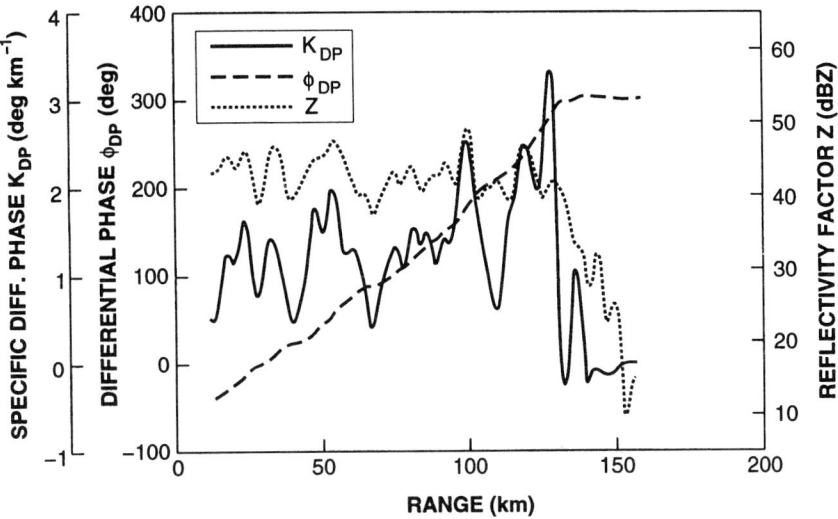

Figure 13.2 *Reflectivity factor, differential phase and specific differential phase along a radial in the storm of 9 June 1993* (after Ryzhkov and Zrnic', 1995a)

The two-way *differential phase* ϕ_{DP} corresponding to a particular resolution volume is given as

$$\phi_{DP} = \phi_{hh} - \phi_{vv} \tag{13.7}$$

where ϕ_{hh} is the two-way phase shift of the radar signal from that resolution volume with horizontally polarised transmission and reception, and ϕ_{vv} is the two-way phase shift with vertically polarised transmission and reception. The cumulative phase shift ϕ_{DP} corresponding to a radar resolution volume is influenced by the phenomena present all along the radial path to the resolution volume. To gain insight into the local behaviour of weather phenomena, the range rate of the phase shift is the pertinent measurement. This quantity is called the *specific differential phase*, and is defined as

$$K_{DP} = \frac{d\phi_{DP}}{2dr} \tag{13.8}$$

In practice this may be deduced from measurements of the ϕ_{DP} and using a difference form of the definition in eqn. 13.8.

An example of the cumulative phase shift along a radial is shown in Fig. 13.2 which was obtained with the polarimetric radar of the US National Severe Storms Laboratory. The radar system, located at Cimarron in central Oklahoma, is described by Zahrai and Zrnic' (1993). The Figure shows that a phase difference of 340° between horizontal and vertical polarisations can phase difference of 340° between horizontal and vertical polarisations can accrue over a 120 km path if the reflectivity is >40 dBZ. Note the general agreement between the features of the reflectivity factor and the specific differential phase. This corroborates the understanding that zones with heavier rain rates cause more rapid phase changes between the polarised components. However, the peaks of the specific differential phase curve rise along the range axis, while those of the reflectivity factor curve are found to remain rather flat. This difference is primarily because of the attenuation of the radar signals along the range in heavy rain. Differences in the details of the two curves may also partly be due to the variability of drop size distribution in rain at different locations (Section 6.3.2). The effect of attenuation on the reliability of reflectivity values was discussed earlier in Section 7.5. In general both attenuation and drop size distribution variability have a much stronger effect on reflectivity factor than specific differential phase. When accurately measured, the latter parameter may therefore provide a more robust estimate of the local rainfall rates.

13.5 Rainfall measurement

Conventional radar estimation of rainfall rate R is based on a measurement of the reflectivity factor Z. However, the relationship between these two variables is not unique, as Z is proportional to the 6th power of drop size (diameter), while R depends on the 3.67th power of the size (see Section 6.3.2). To explain further, it is possible to have the same rain rate by having either a small number of large raindrops or a larger number of smaller raindrops, but the value of the reflectivity factor will be different for the two types of rain. Natural rainfall involves a range of particle sizes. In such a case Z is proportional to the 6th moment of the drop size distribution while R depends on the 3.67th moment. Since even the simplest drop size distribution requires a minimum of two parameters, rain rate determination from Z alone would leave at least one parameter undetermined. Without a knowledge of the exact size distribution law, R cannot be definitely inferred from Z.

One direct way of overcoming this problem is to relate R to Z empirically. Particular $R(Z)$ relations are representative of average conditions for specific rainfall types. Thus there are numerous different relations to cater for different rainfall situations. Battan (1973) lists 69 different $R(Z)$ relations which yield a spread in the rainfall rates greater than an order of magnitude for certain Z values. In addition to uncertainties in the drop size distribution, errors in the radar calibration affect the empirical determination of $R(Z)$ relations and their application for rain rate determination in specific situations. Radar polarimetry offers great promise to alleviate this problem by providing a robust means of rain rate estimation.

Several polarimetric techniques have been proposed for estimating rain rate (Ryzhkov and Zrnic', 1995a,b). The earliest, suggested by Seliga and

Bringi (1976), used the reflectivity factor and the differential reflectivity to eliminate the two parameters of the exponential drop size distribution. For this rainfall estimator to be precise, the radar must be well calibrated and the nature of the drop size distribution should not deviate significantly from that of typical rain, i.e. it should have a basically exponential form. Contamination by hail can degrade the estimates unless it is accounted for.

Sachidananda and Zrnic' (1987) suggest the following single-parameter relationship between the specific differential phase K_{DP} and rain rate R that is valid for Rayleigh scattering:

$$R = 5.1(K_{DP}\lambda)^{0.866} \text{ mm h}^{-1} \tag{13.9}$$

in which K_{DP} is measured in deg km^{-1} and λ is the radar wavelength in cm. Equation 13.9 has been tested for a wavelength of 10 cm, but its validity for shorter wavelengths (e.g. 5 and 3 cm) needs further scrutiny (Tan et al., 1991).

Rain rate estimates based on K_{DP}, although independent of radar calibration, are accurate only at relatively high intensities of rain (>18 mm h^{-1}) if range resolution is to be kept better than ~ 4 km. To preserve accuracy at low rain rates, data averaging over 10 km in range is required (Ryzhkov and Zrnic', 1996a), causing a high degree of blurring of the weather picture (see resolution requirements in Fig. 5.3).

As a general rule, parameter estimation based on frequency and phase measurements is preferred to estimation based on amplitude measurements because of the relative immunity of the former to additive noise and system nonlinearities. A common example is the superior clarity of the sounds conveyed by the commercial FM radio stations as compared to the crackling reception from AM radios. Because specific differential phase is obtained from phase measurements, it retains several advantages over power measurements (Zrnic' and Ryzhkov, 1996): (i) it is independent of receiver and transmitter calibrations; (ii) it is not affected by microwave power attenuation due to rain; (iii) it is immune from beam blockage effects; (iv) it is little biased by the filtering due to ground clutter cancellers; (v) it is insensitive to variations in drop size distribution; (vi) it is little biased by the presence of hail; (vii) it can be used to detect anomalous propagation.

The ability of polarimetric radars to perform rain rate estimates essentially free from attenuation effects is depicted in Fig. 13.3, which pertains to a squall line in a particularly advantageous position. At the time of observation the squall line was passing through the polarimetric radar at Cimarron, Oklahoma, as mentioned above. Thus, along the direction of the squall line, this radar saw a long stretch of rainfall, resulting in heavy attenuation of its signal. An operational WSR-88D radar located in the area was outside the squall line at the time. This radar observed the squall line from the side. Its signal did not have to penetrate deep into the rain area and hence did not experience much attenuation. With this background, the contours presented in Fig. 13.3 are revealing. The polarimetric estimates of rainfall $R(K_{DP})$ agree

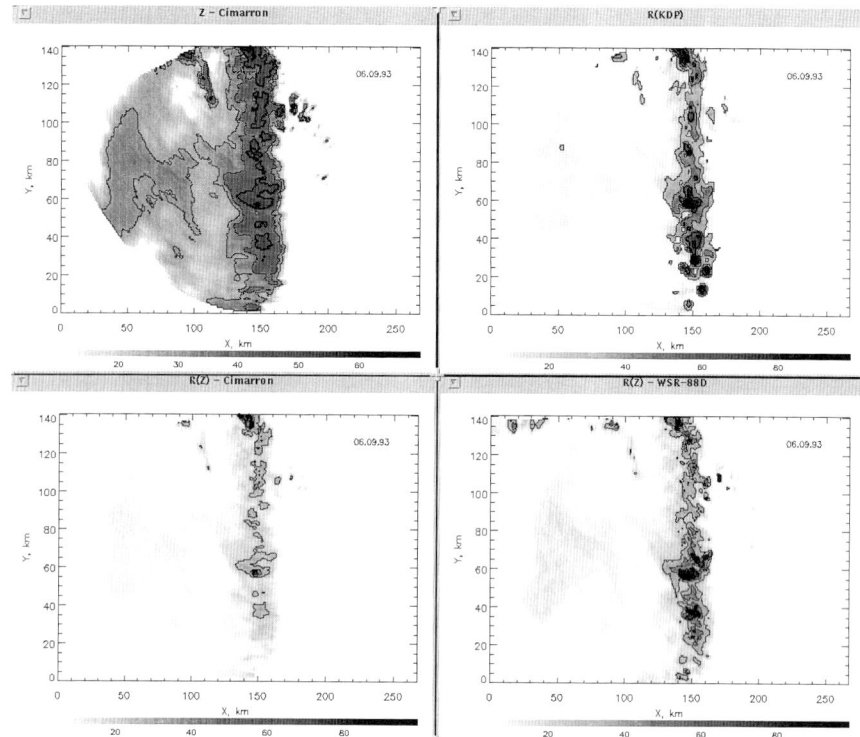

Figure 13.3   Comparison of rainfall intensities associated with a squall line as derived from three different sources. Left: The reflectivity field obtained from the polarimetric Doppler radar at Cimarron, OK, USA (top), and the rainfall rate estimated from this reflectivity (bottom). Right: Rainfall rate estimated from the specific differential phase K_{DP} measured by Cimarron radar (top), and the same rate derived from the reflectivity measured by a WSR-88D radar observing the squall line from the side (bottom). Note that the K_{DP}-based estimate from the Cimarron radar is much closer to the WSR-88D estimate (assumed to be the true estimate) than is the reflectivity-based estimate from the same radar

very well with the reflectivity-based estimate from the WSR-88D, which is not significantly affected by attenuation. The $R(Z)$ estimate from the polarimetric radar, in contrast, is different from both of these, and vastly underestimates the rain rate. This shows that the specific differential phase K_{DP} is a more reliable indicator of rain rates than the conventionally used reflectivity factor Z in the presence of attenuation.

The relative insensitivity of rain rates derived from K_{DP} to variations in drop size distribution may be explained by the fact that K_{DP} is a forward scattering parameter unlike Z, which depends on the backscattering by hydrometeors. Thus, for Rayleigh scattering, K_{DP} is proportional to the 4.24th moment for the larger drops in a typical drop size distribution, and to about the 5th moment for smaller drops (Ryzhkov and Zrnic', 1996a). The 3.67th moment of the distribution, on which the rain rate depends, is much closer to these moments than to the 6th moment, which is derived from reflectivity measurements. Thus, also from the point of view of robustness against variability of drop size distributions, the polarimetric parameter K_{DP} is more reliable for use in rain rate estimation than the conventional reflectivity parameter Z.

It was mentioned in Section 6.3.2 that a commonly used drop size distribution for rain rate estimation from radar reflectivity data is the Marshall–Palmer distribution. Colour Plate 20 shows the contours of rain accumulation over an area in central Oklahoma for a case in which the assumption of the Marshall–Palmer distribution results in a significant underestimation of the rainfall (Ryzhkov and Zrnic', 1996b). The actual total rainfall values shown in the top left quarter of the Figure are derived from rain gauge measurements by spatially interpolating the readings from 42 rain gauges distributed over the area. The reflectivity-based rain accumulation estimates $R(Z)$ obtained from two radars covering the area are shown in the right half of Colour Plate 20. These two estimates are in good agreement between themselves, but are much lower than the actual rainfall values. In contrast, the polarimetric estimates $R(K_{DP})$ obtained from the Cimarron radar, shown in the bottom left quarter of Colour Plate 20, are in much better general agreement with the actual rainfall amounts. The difference in the fine structure of the actual rainfall and $R(K_{DP})$ fields is due to the much finer spatial resolution of the radar relative to the 42-point rain-gauge network. Thus the polarimetric rainfall picture of the area, like the other two radar pictures, shows the streaks corresponding to the tracks of the thunderstorm cells while providing accurate quantitative estimates of the rainfall accumulation.

The above discussion makes it clear that the use of polarisation diversity can significantly enhance the capability of ground-based weather surveillance radars to accurately estimate the quantity, intensity and nature of precipitation. This is of great advantage from the aviation point of view. Another application where polarisation diversity holds great potential relates to airborne severe weather detection. This is explained next.

Airliners, transport aeroplanes and the larger and more sophisticated among the executive aircraft are usually fitted with radars that provide a picture of rainfall areas in a sector ahead of the aircraft. These radars typically operate in the X-band (wavelength in the region of 3 cm) because of space and weight constraints. As discussed before, at such high frequencies radar energy is rapidly attenuated by rain. Airborne weather radars are therefore severely limited in their ability to look into thick layers of rain and provide acceptable quantitative estimates of rainfall intensities. In heavy rain the penetration depth is low, giving the pilot a false impression of only a small depth of intense rain ahead. The rain beyond this depth appears to be lighter than its true intensity as attenuation causes less power to be returned to the receiver than is expected from the local hydrometeor density. Such a misleading picture of the rain field can tempt the pilot to fly through the apparently thin rain band, which actually gets the aircraft deeper into areas of heavy rain. The use of polarimetric radars, and particularly the specific differential phase measured by such radars, would provide accurate rain intensity pictures ahead of aircraft up to distances from which a detectable signal is returned.

13.6 Hail detection

Heavy rain and hail have overlapping ranges of reflectivity values, making it difficult to distinguish between the two. The detection of hail can be made more definite by using one or more differential polarimetric measurement(s) in addition to reflectivity. Leitao and Watson (1984) used linearly polarised radar data to establish a boundary in the Z_h-Z_{DR} space that separates rain from hail. Bringi et al. (1984) presented convincing evidence of hail detection using the (Z_h, Z_{DR}) pair measurement. Analysis of simulated radar data derived from disdrometer[2] measurements led Aydin et al. (1986) to propose a *hail detection signal* H_{DR}, which is the distance of the observed value of Z_h from a hail–rain boundary in the Z_h-Z_{DR} space, i.e.

$$H_{DR} = Z_h - F(Z_{DR}) \text{ dB} \tag{13.10}$$

where

$$F(Z_{DR}) = \begin{cases} 60 & Z_{DR} > 1.74 \\ 19 Z_{DR} + 27 & 0 < Z_{DR} \leq 1.74 \\ 27 & Z_{DR} \leq 0 \end{cases} \tag{13.11}$$

In an analogous manner, Balakrishnan and Zrnic' (1990a) proposed a boundary in the Z_h-K_{DP} space. The empirical expression for this boundary that distinguishes pure rain from mixed-phase precipitation and hail is given as

$$Z_h = 8 \log(2 K_{DP}) + 49 \tag{13.12}$$

The lower the measured value of K_{DP} from that given by the above equation, the higher the probability that the precipitation contains hail.

In addition to the discrimination between rain and frozen precipitation, the two parameters K_{DP} and Z_h may also provide quantitative estimates of liquid water and ice in the mixture. This is because mixed-phase precipitation is anisotropic with respect to radar wave propagation, causing a differential phase shift between the vertical and horizontal polarisations, and K_{DP} depends mainly on rain in a mixture of rain and hail. Balakrishnan and Zrnic' (1990a) and Aydin et al. (1995) have shown that robust measurements of rain in a mixture of rain and hail are possible.

---

[2] An instrument for measuring the sizes of raindrops

These studies *prima facie* have established that radars with polarisation diversity are aptly suited for the detection of, and discrimination between, the ice (hail) and liquid (rain) phases of precipitation. However, considerable further studies and developmental activities are required before these systems and the related algorithms can be used operationally for aviation applications.

13.7 Automatic classification and quantification of precipitation

A further generalisation of the problem of hail detection involves the automatic delineation of the volumes of space which are populated with different types or mixtures of hydrometeors, the identification of the type(s) of hydrometeors within each volume, and the quantification of each species of hydrometeor within the volume. The hydrometeor species include rain, hail (further classified according to size and wet/dry state), graupel (or 'soft hail') and snow.

Intuition suggests that precise measurement of precipitation parameters should involve at least two steps: first a correct classification of the type of precipitation at a location of interest needs to be made, and then appropriate relations are to be applied to each class to estimate the amounts of the constituent hydrometeor species. The applicable relations are usually semi-empirical. The present practice of spatial delineation using radar reflectivity alone treats this problem in a rather trivial manner, wherein the choice is between a few relations, and the operator decides if the precipitation is frozen or liquid. A far more sophisticated regime of precipitation characterisation is possible with polarimetric radars.

Put simply, classification requires the partition of the measurand space into regions of corresponding hydrometeor types. At present approaches based on decision trees (tables), established relations between variables, and fuzzy classifiers are being developed.

Straka and Zrnic' (1993) have presented an algorithm based on an extended table (Doviak and Zrnic', 1993) that relates polarimetric variables to hydrometeor types. The table is obtained from measurements, scattering simulations and literature search. The algorithm uses weights assigned to the various multiparameter variables to classify hydrometeors. The choice of weights is founded on previous measurements, physical reasoning, modelling and other inputs. Typically, the weights are nonzero for more than one hydrometeor type because several species may produce the same values of some polarimetric variables. The hydrometeor type chosen by the algorithm is the one with the highest mean score (average of weights corresponding to each polarimetric variable), termed the *confidence factor*. In this sense the algorithm is a fuzzy logic classifier (Mendel, 1995), though the algorithm cannot 'improve itself by learning' as neural networks do.

440 Aviation weather surveillance systems

Figure 13.4 Vertical cross-section of a supercell storm that occurred on 13 June 1983 in Colorado. The classification of precipitation types was done without human intervention. The zone marked 'graupel' includes graupel and small hail less than 1 cm in diameter. Large hail is between 2 and 4 cm, and very large hail refers to sizes > 4 cm (Courtesy J. Straka)

The hydrometeor identification scheme consists of two steps. First a set of thresholds are used to determine proper weights which are applied to map the polarimetric variables into candidate subsets of hydrometeor populations. The weighted thresholds divide the multidimensional parameter space into regions which correspond to particular hydrometeor types. The values of the thresholds have been obtained from previous measurements (e.g. Balakrishnan and Zrnic', 1990a,b) and modelling studies (e.g. Aydin and Zhao, 1990; Balakrishnan and Zrnic', 1990a,b).

In the second step of the identification process, pairs of variables are examined to further refine the classification. Typically these are (Z_h, Z_{DR}) and (Z_h, K_{DP}), as described in the preceding subsection. After classification, the precipitation rates (or amounts) are estimated. This estimation requires the evaluation of functional relations between the radar variables and the hydrometeor size distributions.

The classification procedure was first tested on a supercell storm that occurred in Colorado on 13 June 1984 (Conway and Zrnic', 1993). The available data consisted of the reflectivity factors using 10 cm and 3 cm radars, a linear depolarisation ratio at 3 cm, and differential reflectivity at 10 cm. The result of the classification process is shown in Fig. 13.4, in which the coherent spatial structure of precipitation is clearly visible. Furthermore, the details of precipitation distribution are physically reasonable from meteorological considerations. However, in this experiment, there is no further independent corroboration of the hydrometeor distribution via *in situ* observations other than hail confirmation on the ground. In a subsequent experiment in 1995, *in situ* data have been obtained using storm-penetrating aircraft for comparison with the algorithm, but these have not been analysed.

Automatically interpreted polarimetric measurements offer unprecedented detail of bulk hydrometeor properties to be observed. Generally the locations of the principal types (rain and hail) are in agreement with conceptual models. Conceptual models are derived from many remote and *in situ* observations, and are generic in nature, whereas the automated algorithm provides specific qualitative and quantitative information about individual storms. For example, details from the algorithm reveal the locations of mixed-phase precipitation, as well as the median rain and hail sizes.

13.8 Status and prospects for aviation use

To summarise the capabilities of polarisation diversity radars, they permit estimation of more than one independent variable that is related to the bulk hydrometeor properties such as the solid or liquid phase, and the mean values and spread of the distribution of size, shape and spatial orientation. However, the additional measurements using polarisation diversity radars are not sufficient to determine uniquely all the properties of precipitation. This is because the polarimetric variables are not independent among themselves, and their number is small compared with the number of hydrometeor types. At least as many independent measurands as the number of scatterer types are needed to solve the inverse scattering problem. This is not feasible, and other methods are sought.

The problem of determination of the types and amounts of precipitation is not completely solved, although a strong case can be made for application of polarimetric radars to measurements of rain, hail and their mixtures. Particularly attractive from the aviation point of view are prospects of overcoming attenuation effects on rainfall estimated by short-wavelength radars such as airborne weather radars. Even more intriguing is the possibility of mapping precipitation fields and following the evolution of hydrometeors in fast-developing storms.

Radar polarimetry is gaining broad acceptance in the meteorological community, albeit rather slowly given that its hardware technology has been established and many of its capabilities have been proven. The reason for this, apart from the moderate additional hardware complexity and cost, might be that the fields of polarimetric variables are nonintuitive; further formal processing is required before these fields can be transformed into meteorologically useful pictures. Algorithms for such processing and inference are just beginning to be developed. When they mature to a level that they can deliver robust and reliable weather products at a speed commensurate with aviation needs, polarisation diversity radars should significantly enhance the quality of weather information available for the management of aviation.

13.9 Summary

Just as the Doppler weather radar adds more dimensions of information to the conventional reflectivity-only radars by providing wind and wind-related data, the addition of polarimetric capability to coherent radars further extends their information-generating capacity. By incorporating the capability of radiating and receiving electromagnetic signals with different types of polarisation, it is possible to observe certain atmospheric hazards more accurately and robustly than with an ordinary Doppler weather radar which utilises only one fixed polarisation for all its operations.

Different combinations of the measured amplitudes and phases of the received signals can yield a wide variety of polarimetric parameters for a given radar resolution volume. However, some of these are more important than others, and are more frequently used. Prominent among these are the differential reflectivity, the linear depolarisation ratio, and the specific differential phase. The first of these parameters is a measure of the oblateness (and hence the size) of raindrops, the second is influenced by the average orientation of the drops with respect to the vertical, and the third one serves as a robust indicator of the local rainfall rate, unaffected by the attenuation of the radar signals in the intervening medium. The last-mentioned advantage is particularly important in the aviation context because conventional radars, and even some Doppler radars without polarisation diversity, especially those operating at C- or X-band frequencies, can grossly underestimate the rainfall intensity at locations deep inside heavy rainfall areas (as seen from the radar), and can thus misguide pilots to venture into those locations.

In addition to being insensitive to attenuation effects, polarimetric observations can also improve rain intensity estimation in the presence of significant uncertainties in drop size distribution and radar calibration errors. Another major advantage of polarimetric radars for potential use in aviation weather surveillance is the ability to distinguish hail from rain more reliably, and even automatically classify different types of precipitation and quantify the proportion of the constituent phases in mixed-phase precipitation.

Polarimetric radars have not yet been put to aviation use operationally. Further research into polarimetric signatures of aviation-significant weather phenomena and development of software for reliable use of the additional capabilities of such radars are necessary before they can be accepted for routine aviation use. However, the considerable research in this field conducted hitherto has already shown the device to be potentially capable of improving aviation weather surveillance to a level beyond what is possible with the current Doppler radars.

13.10 References

AYDIN, K., and ZHAO, Y. (1990): 'A computational study of polarimetric radar observables in hail', *IEEE Trans. Geosci. Remote Sens.*, **28**, pp. 412–422

AYDIN, K., BRINGI, V.N., and LIU, L. (1995): 'Rain-rate estimation in the presence of hail using S-band specific differential phase and other radar parameters', *J. Appl. Meteorol.*, **34**, pp. 404–410

AYDIN, K., SELIGA, T.A., and BALAJI, V. (1986): 'Remote sensing of hail with a dual linear polarization radar', *J. Clim. Appl. Meteorol.* **25**, pp. 1475–1484

BALAKRISHNAN, N., and ZRNIC', D.S. (1990a): 'Estimation of rain and hail rates in mixed phase precipitation', *J. Atmos. Sci.*, **47**, pp. 565–583

BALAKRISHNAN, N., and ZRNIC', D.S. (1990b): 'Use of polarization to characterize precipitation and discriminate large hail': *J. Atmos. Sci.*, **47**, pp. 1525–1540

BATTAN, L.J. (1973): 'Radar observations of the atmosphere' (University of Chicago Press, Chicago, IL), 324 pp

BRINGI, V. N., and HENDRY, A. (1990): 'Technology of polarization diversity radars for meteorology', *in* D. Atlas (Ed.): 'Radar in meteorology', American Meteorological Society, Boston, MA, Chap. 19a, pp. 153–190

BRINGI, V.N., SELIGA, T.A., and AYDIN, K. (1984): 'Hail detection using a differential reflectivity radar', *Science*, **225**, pp. 1145–1147

CONWAY, J.W., and ZRNIC', D.S. (1993): 'A study of embryo production and hail growth using dual-Doppler and multiparameter radars', *Mon. Weather Rev.*, **121**, pp. 2511–2528

DOVIAK, R.J., and ZRNIC', D.S. (1993): 'Doppler radar and weather observations' (Academic Press, San Diego, CA), 562 pp

GOLESTANI, Y., CHANDRASEKAR, V., and BRINGI, V.N. (1989): 'Intercomparison of multiparameter radar measurements'. Preprints of 24th International Conference on Radar Meteorology, Tallahassee, FL, American Meteorological Society, pp. 309–314

JAMESON, A.R. (1983): 'Microphysical interpretation of multi-parameter radar measurements in rain, Part I: Interpretation of polarization measurements and estimation of raindrop shapes', *J. Atmos. Sci.*, **40**, pp. 1792–1802

LEITAO, M.J., and WATSON, P.A. (1984): 'Application of dual linearly polarized radar data to prediction of microwave path attenuation at 10–30 GHz', *Radio Sci.*, **19**, pp. 209–221

MENDEL, J.M. (1995): 'Fuzzy logic systems for engineering: A tutorial', *Proc. IEEE*, **83**, pp. 345–377

REINKING, R.F., MATROSOV, S.Y., MARTNER, B.E., and KROPFLI, R.A. (1997): 'Dual polarization radar to identify drizzle, with application to aircraft icing avoidance', *J. Aircr.* **34**, pp. 778–784

RYZHKOV, A., and ZRNIC', D. (1995a): 'Precipitation and attenuation measurements at a 10-cm wavelength', *J. Appl. Meteorol.*, **34**, pp. 2121–2134

RYZHKOV, A., and ZRNIC', D. (1995b): 'Comparison of dual-polarization radar estimators of rain', *J. Atmos. Ocean. Technol.*, **12**, pp. 249–256

RYZHKOV, A., and ZRNIC', D. (1996a): 'Assessment of rainfall measurement that uses specific differential phase', *J. Appl. Meteorol.*, **35**, pp. 2080–2090

RYZHKOV, A., and ZRNIC', D. (1996b): 'Rain in shallow and deep convection measured with a polarimetric radar', *J. Atmos. Sci.*, **53**, pp. 2990–2995

SACHIDANANDA, M., and ZRNIC', D.S. (1987); Rain rate estimates from differential polarization measurements', *J. Atmos. Ocean. Technol.*, **4**, pp. 588–598

SELIGA, T.A., and BRINGI, V.N. (1976): 'Potential use of radar differential reflectivity measurements at orthogonal polarizations for measuring precipitation', *J. Appl. Meteorol.*, **15**, pp. 69–76

STRAKA, J., and ZRNIC', D.S. (1993): 'An algorithm to deduce hydrometeor types and amounts using multiparameter radar data'. Proceedings of 26th Conference on Radar Meteorology, Norman, OK (American Meteorological Society, Boston), pp. 513–515

TAN, J., HOLT, A.R., HENDRY, A., and BEBBINGTON, D.H.O. (1991): 'Extracting rainfall rates from X-band CDR radar data by using differential propagation phase shift', *J. Atmos. Ocean. Technol.*, **8**, pp. 790–801

ZAHRAI, A., and ZRNIC', D.S. (1993): 'The 10-cm wavelength polarimetric weather radar at NOAA's National Severe Storms Laboratory', *J. Atmos. Ocean. Technol.*, **10**, pp. 649–662

ZRNIC', D.S., and RYZHKOV, A. (1996): 'Advantages of rain measurements using specific differential phase', *J. Atmos. Ocean. Technol.*, **13**, pp. 454–464

Index

75 MHz marker beacon 25

aborted landing 12
absorption cross-section 159
accelerometer 22, 310, 316
active stabilisation 20
adjoint method (of wind retrieval) 207, 208
advanced traffic management system (ATMS) 343
aerodrome control *see* control, aerodrome
aerodynamic roughness 59
aerofoil 15, 17, 20, 61, 63, 64
aerosols 161, 163, 164, 183
agility 122, 129
airborne collision avoidance system (ACAS) 28
aircraft separation 3, 14, 27–30, 35, 40, 74, 130, 340
  procedural 29
  radar 29
aircraft situation displays (ASDs) 343
air-data system 23
air pocket 39
airport surface detection radar (ASDR) 30
airport surveillance radar (ASR) 30, 32
  ASR-7 30
  ASR-8 30, 281
  ASR-9 30, 281–284
    with weather channel 7, 30, 281–283, 289, 339, 341, 429
airport visibility (AV) 325, 329, 331
air route surveillance radar (ARSR) 30, 274, 275
  ARSR-2 30
  ARSR-3 30
  ARSR-4 30

air space
  controlled 28–30, 72, 73, 123
  special rule 28
  uncontrolled 28
  upper 28
air traffic control (ATC) 3, 4, 6, 7, 12, 26–35, 41, 73, 96, 130, 134, 137, 181, 245, 274, 275, 279, 281, 282, 326, 328, 339–341, 347
  procedural 29
  radar beacon system (ATCRBS) 31, 34
air traffic management 4, 245, 340
air traffic service 6, 27–29, 32, 35, 122, 125
aliasing interval 255, 259, 264
altimeter
  barometric 23
  radio 25
ambiguity 7, 73, 194, 247, 252–254, 256–258, 260–262, 264, 268, 271, 278, 365
  Doppler 256, 257, 260, 264, 278
  range 247, 252–54, 256, 257, 261, 267, 271, 277
  range-Doppler 256, 257, 278
  resolution 247, 258, 260, 261, 272, 364
  velocity 252, 256, 261, 264, 285
amplifier
  klystron 191, 283
  logarithmic 191
  low-noise 192
analogue-to-digital (A/D) converter 192, 283
anemometer 62, 134, 137, 140, 290, 306, 307, 330, 368
  centre-field 307, 308
  cup type 306
  vane type 122, 306

angle of attack 15–18, 46, 52, 63, 64, 98, 102, 310
angular resolution 145, 163, 201
antenna
  dual-beam 282
  fan-shaped beams 25, 26, 31, 191, 274, 281–283
  gain 147, 169, 245, 391
  scan direction 152
  scan rate 148, 189, 191, 195, 201, 245, 264, 271, 272, 283, 324
anvil 83
approach control *see* control, approach
arcus (cloud) 100
area control *see* control, area
ARINC Communication and Retrieval System (ACARS) 341, 342
atmospheric eddies 183
atmospheric electricity 37, 69, 407
atmospheric profilers *see* wind profilers
atmospherics 95, 97, 407, 411
attenuation coefficient 159–161, 163
automated radar terminal system (ARTS) 32
automated surface observing system (ASOS) 318, 321, 329, 339, 368
automated weather observation system (AWOS) 7, 317, 318, 320, 329, 331
automatic classification (of precipitation) 439
automatic detection
  based on interest images 368, 370
  gust fronts 363
  mesocyclones 356
  microbursts 369
  storm outflows 369
  thunderstorms 353
  tornadoes 360
  weather features 352
automatic direction finder (ADF) 24
automatic gain control (AGC) 191
autopilot 21, 22, 46
aviation corridors 3
Aviation Gridded Forecast System (AGFS) 7, 342, 343, 347
aviation impact variables (AIVs) 342, 345
aviation terminal forecast 48
Aviation Weather Products Generator (AWPG) 7, 238–240, 343, 346
  National (NAWPG) 347, 348
  products 348
  Regional (RAWPG) 347, 348

AWOS data acquisition system (ADAS) 318
axis
  fore-and-aft 19, 20
  lateral 17, 19
  pitch 19
  roll 19, 20
  yaw 19

backscattering cross-section 155, 166
batch mode (of signal transmission) 257
beam filling
  incomplete 171
  nonuniform 171
  partial 171
bore 103, 104
  undular 104, 217
Bragg
  effect 303
  scatter 379
  scatterer 158
  scattering 157
buffeting 66

carburettor 63
chop 53, 54, 56
chord length 20
clear air 157 (definition) 6
  detection 140
  mode 367, 371
  observation 171, 247, 301, 315
  phenomena 157, 191, 266
    detection 269
  radar 295
  reflectivity (or backscattering) 157, 158, 295, 297, 315
  returns (or echoes) 157, 158, 258, 291, 330
  scan modes (of WSR-88D) 272
  turbulence (CAT) 56, 113, 315, 375
  wind shear 315
cloud base 68, 69
clutter 143
  canceller 249, 435
  ground 183, 187, 188, 194, 201, 259, 267, 269, 272, 282, 309, 315, 368, 435
  rejection 201, 247, 257, 261, 276
  weather 143
coherent radar *see* radar, coherent
cold front 224
collection efficiency 65

collision avoidance 3, 27
collision-coalescence 173
comfort 2, 6, 38, 42
communication 1–3, 5, 6, 39
  simplex 33
  aeronautical 32
  broadcast mode 33
  selective mode 33
conditionally stable atmospheric layer 82
confidence factor 439
control
  aerodrome 28
  approach 28
  area 28
  forces 20
  moments 20
  surface 64
  zones 28
convergence (definition) 87
  linear 87
  radial 87
convolution 172
corona 310
correlation 388
  coefficient at zero lag 431
  longitudinal 385
  transverse 385
crosswind 40, 43, 48
cumulonimbus convection 82
cumulus clouds 81

data
  basic 194, 221
  display 134
  high level 6, 74, 133
  integration 7, 339
  interpretation 8, 134
  low level 133
  processing
    intelligent 74
    multisensor 133
    refreshment rate 6, 122
    single-sensor 133
    speed 6
  products
    meteorological 5
      computation 194
      primary 185
      smoothing 153
  rate 33, 34, 268
    reduction 54, 153
  WSR-88D 250
dead-reckoning 23

defruiter 31
density current 101
derived gust velocity 52
differential phase 433
  specific 235, 433
dihedral 20
dipole (in thunderstorms) 407
direct user access terminal (DUAT) services 348
discrete phase coding 263
distance measuring equipment (DME) 24, 26
divergence (definition) 87
  generalised 87
  linear 87
  radial 87, 88
Doppler
  doublet (or couplet) 227, 230, 357, 362, 371
  frequency 182
  navigator *see* navigator, Doppler
  radar 6-8, 51
  shift 28, 182
  spectral moment estimation 191, 195, 263
    pulse-pair method 197, 199
    spectral method 198, 200
  spectrum width 187 (definition), 199 (determination)
downburst 91
downdraft 43
drag 18, 59–61
  coefficient 18, 66
  induced 18
  lift-dependent 18
  pressure 18
  wave 18
drop size
  distribution 173, 174
    Marshall-Palmer 174, 437
  statistics 8
droplet size distribution (in clouds) 64
dry line 224
dual-Doppler-radar mapping 205, 206
Dutch roll 21
dynamic modes 43
  rigid body 51

economy 6, 40, 42
eddy dissipation rate 8, 316, 396–400
efficiency 1, 6, 11, 40, 42
  aerodynamic 60

448 *Aviation weather surveillance systems*

electroluminescence 2
engine
  extinction 59
  turbofan 19
  turbojet 19
enhanced traffic management system (ETMS) 348
en route
  flight advisory service (EFAS) 28
  operations 34
extinction coefficient 159, 329

false alarm 251, 279
fast Fourier transform (FFT) 261
F-factor 49
  equivalent 312
Federal Aviation Regulation (FAR) 318
field
  random 384
    isotropic 384
    statistically homogeneous 384
  scalar 384
  vector 384
flameout 59, 61
flap 17
flare 12
flight data processing (FDP) system 32
flight delays 39, 41
flight mechanics 15
flight plan 28
flight rules
  instrumented (IFR) 27, 29, 41
  visual (VFR) 27, 29, 59
flight service station 28
flow
  streamlined 16, 61
  turbulent 17
fly-by-wire 18
fore/aft scanning technique (FAST) 207
forward scatterometer 243, 244, 329, 330
free flight 135
frontal shear 113
freezing altitude (or height) 61

geographic situation display 236
geostationary operational
    environmental satellites (GOES) 321
  GOES-7 323
  GOES-8 323
  GOES-9 321, 325
  GOES I-M 323, 324

glideslope angle 12
global positioning system (GPS) 24
graphic weather display system (GWDS) 348
gravity current 101
gust 43, 56
  alleviation factor 52
gust front 8, 98, 223
  asymmetric 100
  symmetric 100
  washout 367
gyroscope 23, 316

hail
  damage due to 62
  detection 438
    signal 438
  size 94
hailstone 61
hard landing 48
headwind 21, 46–48
heavier-than-air machines 2, 15
height of neutral buoyancy 82
height of neutral density 82
hydrometeorological phenomena 37, 57
hydrometeors 58, 60, 292
hydroplaning 59

ice accretion 63, 64
icing 61, 63–67
inertial navigator *see* navigator, inertial
inertial subrange (of turbulence) 388
inference generation 134
infinite-impulse-response filters 247
insects (airborne) 184
instrument flight rules (IFR) *see* flight rules, instrument
instrument landing system (ILS) 25, 26
  glideslope 25
  localiser 25
integrated terminal weather system (ITWS) 7, 338, 355
interim terminal Doppler weather radar (ITWR)
intermediate frequency (IF) 192
International Civil Aviation Organisation (ICAO) 27
International Maritime Satellite Organisation (INMARSAT) 34
isodops 357
irrigation 5

Index 449

jets
  low-level 111, 112
  nocturnal 111
jet stream 57, 113, 114
joint airport weather studies (JAWS) 106

Kelvin-Helmholtz instability 218
klystron amplifier, *see* amplifier
Kolmogorov-Obukhov energy spectrum 392

landing guidance 60
lapse rate 82
  saturated adiabatic 82
laser radar 163
lateral guidance 25, 26
lateral wind 48
lidar 163
lift 15, 17, 18, 51, 59–61, 63
  coefficient 16, 17, 52
lift-to-drag (L/D) ratio 60
lightning
  avoidance 412
  channel 409
  cloud-to-ground flashes 410
    negative 410
    positive 411
  detection 423
  in marginally electrified stratiform clouds 425
  in mixed-phase clouds 425
  in thunderstorms 407
  in winter storms 425
  intracloud flashes 96, 409
  leader 96, 409
  return strokes 411
  strike to aircraft 415–417
  'tree' 414
Lightning Characterization Program 412
linear depolarisation ratio 431
lines of position 24
liquid water content 64
Local Analysis and Prediction System (LAPS) 342
  for terminal areas (T-LAPS) 342
LORAN-C 24
loss factor 169
  atmospheric 169
  finite bandwidth 170
  system 169
low level wind shear alert system (LLWAS) 7, 305–309
    enhanced 307–309
    with network expansion (LLWAS-NE) 308

macroburst 105
magnetic compass 23
Marshall-Palmer distribution *see* drop size
mean radial velocity 188 (definition), 197 (determination)
median volume diameter (MVD) 66
memory matrix 196
mesocyclone 8, 227, 228
mesoscale analysis and prediction system (MAPS) 342
meteorologist's weather processor (MWP) 348
microburst 8, 48
  asymmetry 111
    shape 111
    strength 111
  dry 106, 107
  outflows 109
  risk image 344
  wet 106, 107
  wind index (WINDEX) 241
microwave landing system (MLS) 25, 26
mode S (beacon system) 31, 33, 344
  see also transponder
mountain wave 43
moving-Doppler-radar technique 207
multiple-time-around echoes 253

navigation 2, 3, 4, 6
  hyperbolic 24
navigational aids (navaids) 60
navigator
  Doppler 23
  inertial 23
NEXRAD 246
nondirectional beacon (NDB) 24
Northern Illinois Meteorological Research on Downbursts (NIMROD) 106
nowcasting 42
Nyquist frequency 147
Nyquist velocity 255

observation ceiling 128
OMEGA 24
optical horizon 126
oscillations 47, 52, 56 *see also* dynamic mode

outdoor sports 5

parameter estimation 7
parcel theory of convection 82
passenger discomfort 39, 40, 48
pattern recognition 8
pencil beam 145
periodogram 198
phugoid mode 21, 103
pilot fatigue 20
piston-engined aircraft 63
pitch plane 20
pitching moment 19
  coefficient 19
pitot-static tube (also pitot tube) 16, 22
pixel 152
plan position indicator (PPI) 31, 128
plasma channel 410
point target 6, 143
polarimetric radar *see* radar
position fixing 23, 24
preamplifier 192
precipitation
  intensity 7
precision approach radar (PAR) 30
predict 356
  gust fronts 363
  thunderstorms 356
  tornadoes 363
primary sensor 6, 138
principal user processor (PUP) 246
probability of detection 251, 279
profilers *see* wind profilers
Project Rough Rider 90
propeller 19
pseudo-dual-Doppler method (of wind mapping) 207, 209
pulse-pair processing *see* Doppler
pulse repetition frequency (PRF) 146, 248
pulse repetition interval (PRI) 146
pulse repetition time (PRT) 146
  staggered 262
pulse-to-pulse phase coherence 191

radar
  calibration 435
  coherent 6
  cross-section 154
  data acquisition (RDA) unit 246
  data acquisition status control (RDASC) 257
  data processor (RDP) 32
  equation *see* radar, range equation
  horizon 126
  polarimetric 8, 430
  polarisation diversity 429
  primary 30
  product generator (RPG) 246
  pulsed 144
  range equation 165
  secondary 30
    surveillance (SSR) 30
      monopulse 31
  separation 29
  signal
    in-phase (I) 190, 192
    quadrature (Q) 190, 192
  signatures 6
radio acoustic sounding systems (RASS) 7, 302–305, 325
  augmentation for sensing icing 304
radio-interference 247
rain
  accumulation 234, 235
  attenuation 8, 159
  cell 93
  orographic 175
  rate 59, 93, 174, 222
  stratiform 175
  thunderstorm 175
random phase transmission 263
range folding 232, 252 *see also* range ambiguity
range-height indicator (RHI) 203, 221, 226
range overlays 257
Rankine vortex 357, 358, 360
rate of climb 66
Rayleigh scatterers 154
Rayleigh scattering 435
real-time weather processor (RWP) 347
receiver losses
  atmospheric 169
  filtering 170
  lens effect 170
  system 169
reflectivity 155 (definition)
  difference 432
  differential 431
  determination 195
  factor 155
    equivalent 156
    horizontal polarisation 431
    vertical polarisation 431
  gradient 191

Index 451

regional atmospheric modelling system (RAMS) 342
resolution volume (definition) 149
  symmetrising 153
Reynolds number 63
roll 102
  solitary 102
roll cloud 100
roll mode (of oscillation) 21, 48
runway
  icing 40
  visual range (RVR) 27, 326, 327, 329–332

satellite
  GOES *see* geostationary operational environmental satellites
  imagery 4
    infrared band 321
    multispectral 321
    visible band 321
  observation 320
  radiometric 320
scan cycle 270, 273
scanning strategies 270
  stacked-beam 271
schedule-keeping 6, 39–42
search and rescue services 29
second trip echo 253
sidelobe effects
  in ATC 31
  in weather radar 193
signal processing 186, 260
signal-to-noise ratio (SNR) 148
sink rate 89
siting (of radar) 7, 266
skin friction 18
slope of the lift curve 52
snow
  dry 60, 185
  wet 60
snowstorms 60
solitons *see* waves, solitary
space missions 14
space shuttle 60
space-time grid 7
specific differential phase
  see differential phase
specific extinction 159
spectral decomposition 260
spectral density tensor 386
spectral functions 388
spectral method (of moment estimation) *see* Doppler

speed
  air 21, 47, 52
  ground 21
  inertial 22
  stall *see* stall
  wind *see* wind
spherics *see* atmospherics
squall line 436
stability 61
  dynamic 21
  static 20
stable local oscillator (STALO) 192
stall 17, 47, 59, 63
  angle-of-attack 17
  margin 66
  speed 12, 17
state-of-the-atmosphere variables (SAVs) 342
storm cell identification and tracking (SCIT) algorithm 229
storm cells 8
Storm Hazards Program 71, 412
storm top 83
stratosphere 81
supercell 84, 215
sweep (of wings) 20
synchronous detector 192

tactical air navigation system (TACAN) 24
tailwind 21, 43, 46, 4
tangential velocity assumed display (TVAD) 207, 361
temperature profile 7
terminal air traffic control automation (TATCA) 341
terminal area surveillance system (TASS) 285
terminal Doppler weather radar (TDWR) 7, 275–280
  location 280
terminal radar control (TRACON) 12, 279
terminal speed (or velocity) 58, 173
threat alert and collision avoidance system (TCAS) 28
three-body effect 283
thrust 18, 19, 49, 50, 52, 59, 61
thunderstorm
  complex 85
  cumulus stage 83
  dissipating stage 83
  mature stage 83
time-domain computation (of spectral

time-domain computation (of spectral moments) 197
time–height plot 56, 299–304
time-of-arrival (TOA) technology 424
tornado 8, 92
 ground track 228
 vortex signature (TVS) 225, 361
tracers (of atmospheric motion) 183, 206
tracking 355
 gust fronts 368
 thunderstorms 355
 tornadoes 363
traffic density 14
Transall Programme (of lightning study) 412
transmissometer 328
 detector 328
 projector 328
transmitted power
 average 146
 peak 146
transoceanic flying 3
transponder 24, 28, 30–34, 290, 344
 mode S 31, 33, 34, 344, 348
tropopause 81
troposphere 81
turbulence
 airborne measurement 316
 clear-air *see* clear air turbulence
 continuous 53
 extreme 56, 90
 in K-H waves 379
 in thermal plumes 378
 in thunderstorms 89, 90, 377
 intermittent 53
 light 53, 58, 90, 396
 moderate 53, 58, 90, 396
 occasional 53
 severe 54, 58, 90, 396
turbulent flow
 isotropic incompressible 387
turbulent structure parameter of refractive index 158

unambiguous range 147, 253
unambiguous velocity 147, 255
 interval 255
undershoot 48
updraft 83
user-friendliness 6

vector wind 204, 207
 retrieval 206
velocity aliasing 252

velocity azimuth display (VAD) 203, 219, 250, 291
velocity dealiasing 258
velocity discontinuities (or jumps) 8, 258
very-high frequency omnirange (VOR) 24
visual flight rules *see* flight rules
voice switching system 33

waves
 buoyancy 102, 321
 gravity 102, 321
 solitary 102
wavenumber 386
weather
 feature
  detection 8
  hazard estimation 8
  recognition 8
 products 7, 73, 279
 radar *see* radar and Doppler
 surveillance
  long-range 7
weathercock mechanism 21
 anemometer combination 306
wind
 field
  divergent 8
 gradients 40
 shear (definition) 44, 45
  aiborne detection 310
  forward 45
  hazard index *see* F-factor
  horizontal 45, 46, 108, 109, 165
  vertical 45, 49
 shift 22
 speed 7
 vector *see* vector wind
wind profilers 7, 56, 58
 405 MHz 296
 915 MHz 297
 boundary layer 196
 mesospheric-stratospheric-tropospheric (MST) 298
 radar 291
 stratospheric-tropospheric (ST) 297
 tripod configuration 293
 tropospheric 295
  lower 295
wing loading 18, 52
Winter Icing and Storms Project (WISP) 66
Wright brothers 2, 3

WSR-57 178–179
WSR-88D 7, 246–278, 285, 339, 341, 342, 347, 361, 366, 368, 402, 435

zone of blindness 267
Zooplankton 184
Z-R relationship 175–177